# TERRA MAXIMA

## THE ATLAS OF SUPERLATIVES

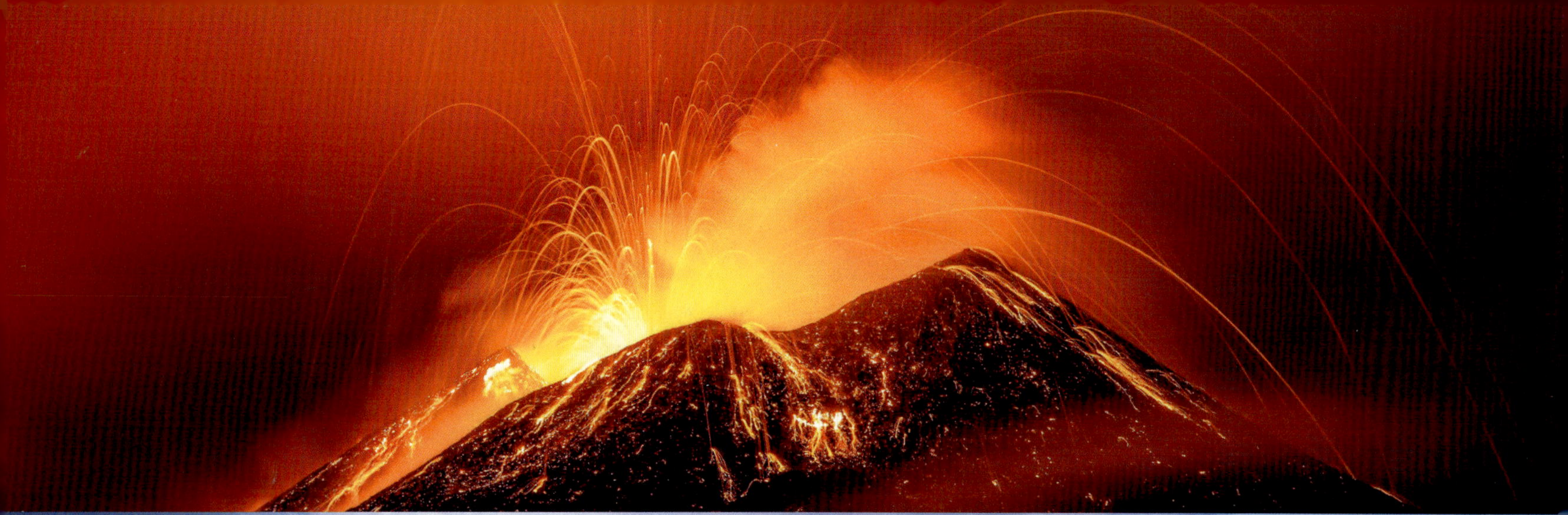

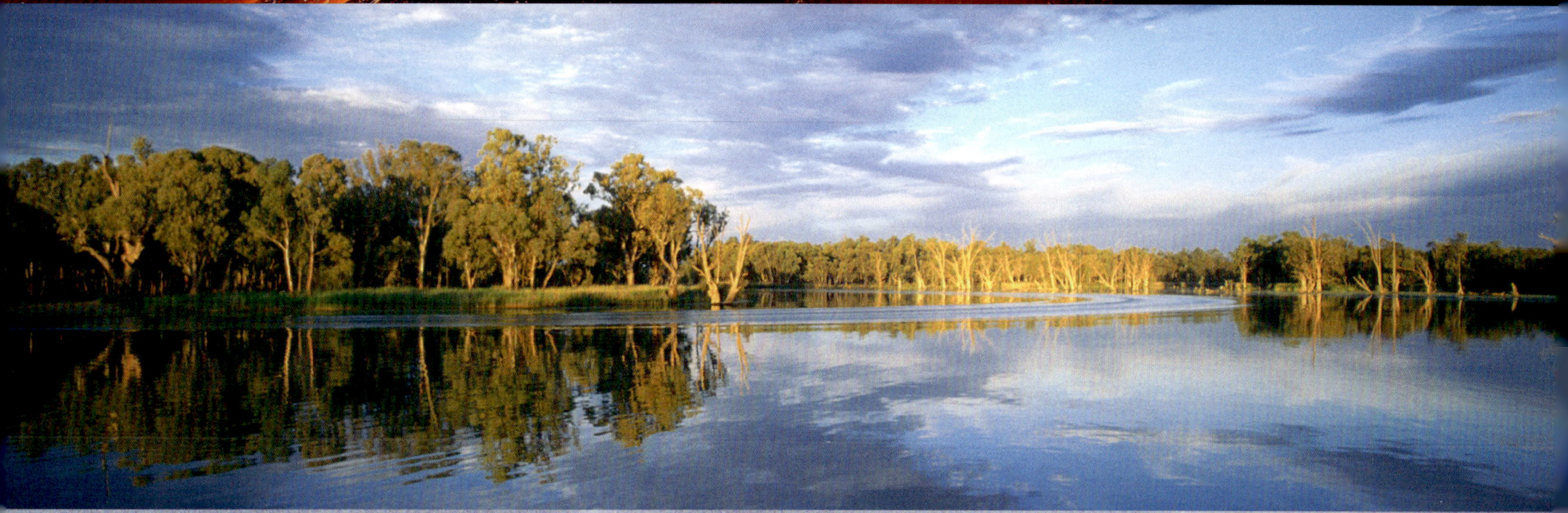

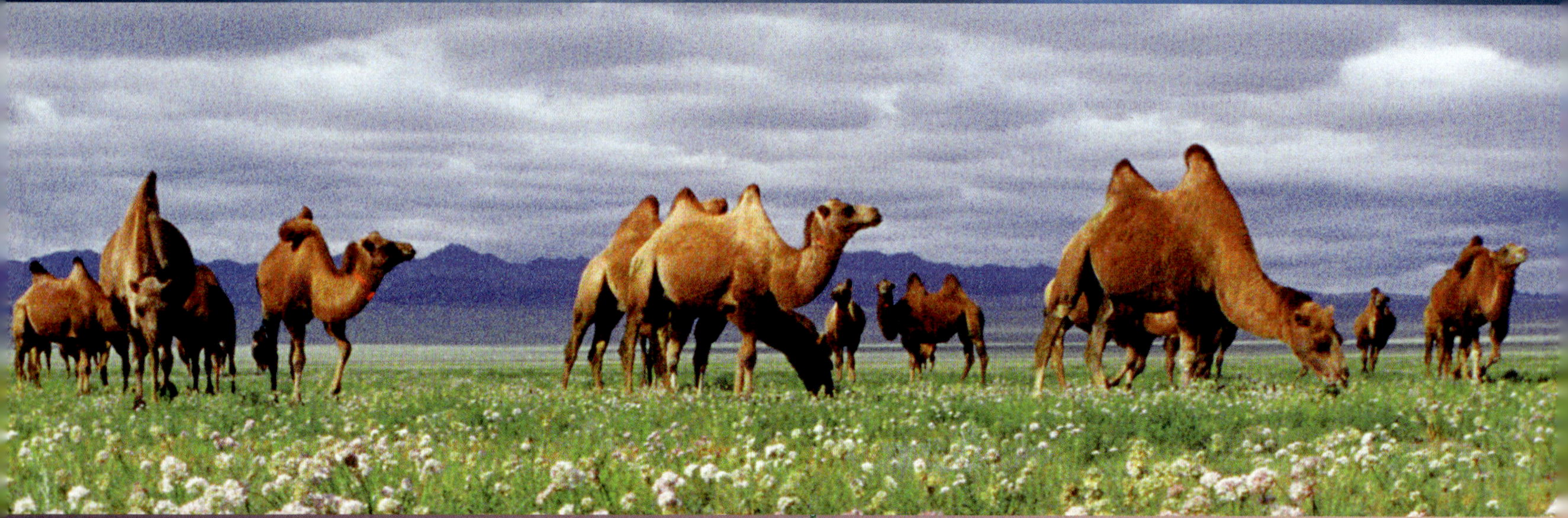

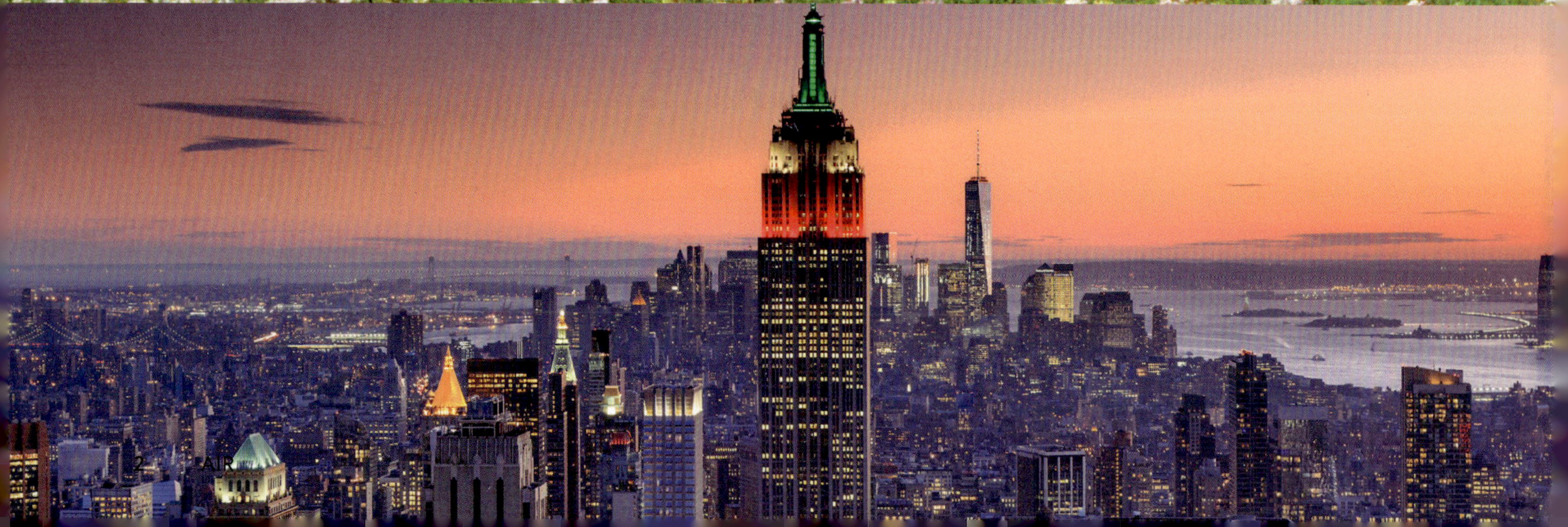

# INTRODUCTION

Our sense of wonder

Our world is full of wonders. And we want to marvel at them! Because the superlatives of our planet are not to be taken for granted – they are not mere add-ons to our existence. They are unique, breathtaking, incomparable. Yet – what would actually happen if we were to make some comparisons? Would these detract from the staggering records set by nature and humankind? Would they detract from the fantastic achievements? Not at all. On the contrary, if we occasionally make comparisons between natural and technological phenomena, then we acknowledge the marvellous assets and masterpieces surrounding us every day, which we also encounter on our travels.

The ATLAS OF SUPERLATIVES takes us on an exciting journey of discovery to the largest oceans, to the bottom of the deepest seas with their inhabitants and visitors, then up to the highest peaks and glacial heights and to volcanic craters spewing out lava. We will follow the tracks of the greatest and fastest land animals and also witness the power and sophistication in even the smallest creatures. We will allow ourselves to gaze upwards with yearning, and follow the spectacular aerial acrobatics of birds as well as planes. Sky-scraping monuments to contemporary architecture and magnificent constructions of modern engineering can be just as mesmerizing as the achievements of earth's smallest architects – the corals.

We have divided the themes into the four elements: fire, water, earth and air. According to the theory of the four elements from the days of alchemy, 'everything consists of specific proportions of the four elements'. Scientific progress has led to the periodic system – a list of all the chemical elements, arranged according to their nuclear charges. For this book, however, the original, philosophical and artistic principles will apply. In accordance with these, we have succeeded in rearranging the unbelievably diverse spectrum of natural and human phenomena from a surprising and entertaining perspective.

When volcanoes spew flames, then clearly fire is at work; when whales plunge into the sea, then they are enjoying the element of water. But what is an island? It consists of earth, yet is surrounded by water. Or the desert, where the hottest temperatures rage during the day, but it is bitterly cold at night. Let us give ourselves the freedom to allow the overall framework to unfold.Our main aim here is to view our multi-faceted world in wonder. We hope you enjoy it!

*Pictures from top: Mount Etna erupts like a firework display in 2002.*
*The slow-moving Murray River, with its tree-lined banks, is the longest river on the Australian mainland.*
*A herd of camels makes its way across the steppes of the Gobi Desert with its colourful display of flowers.*
*The Manhattan skyline with the Empire State Building in the centre and the One World Trade Center behind it, to the right.*

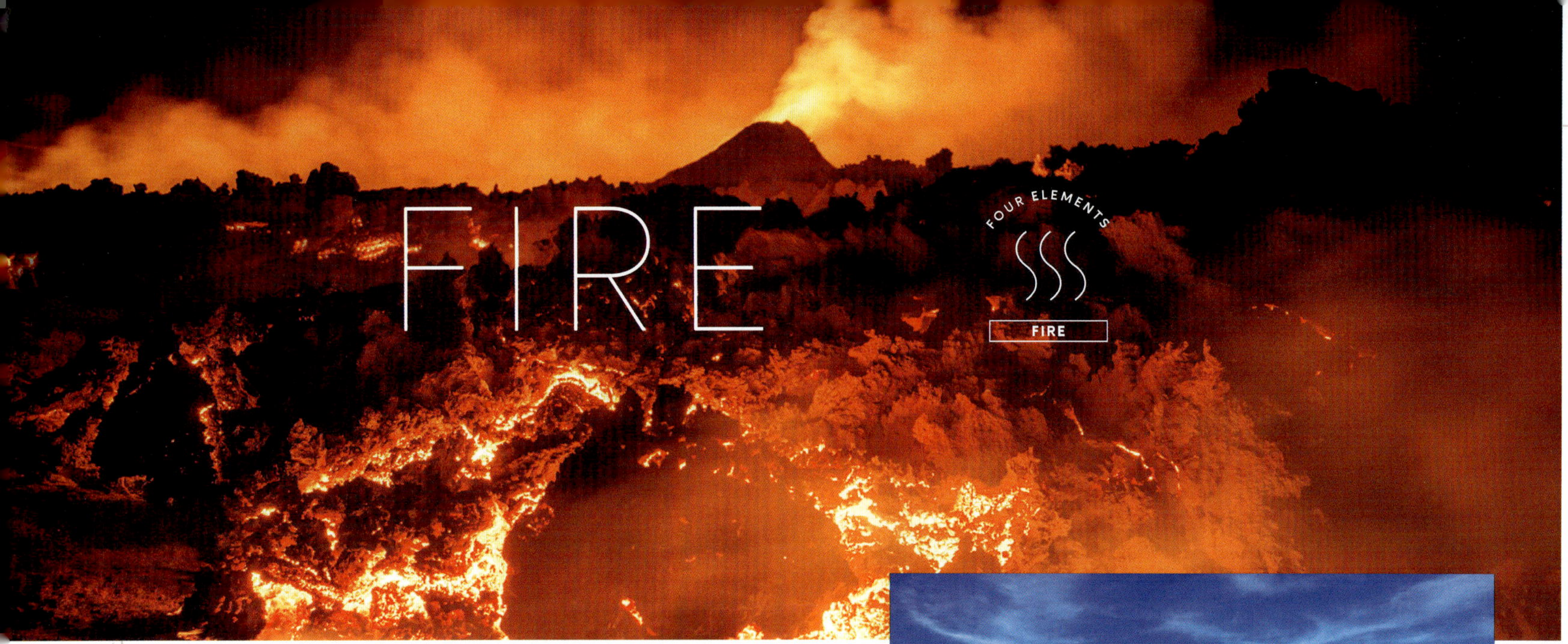


## VOLCANOES 10



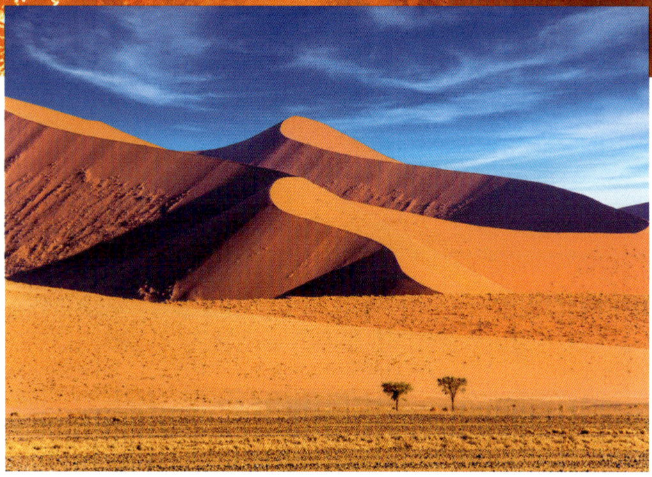

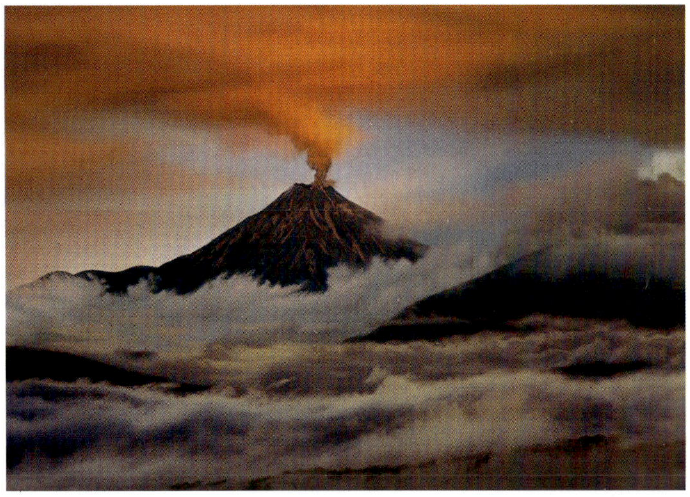

## FIRE RITUALS 44



## DESERTS 52

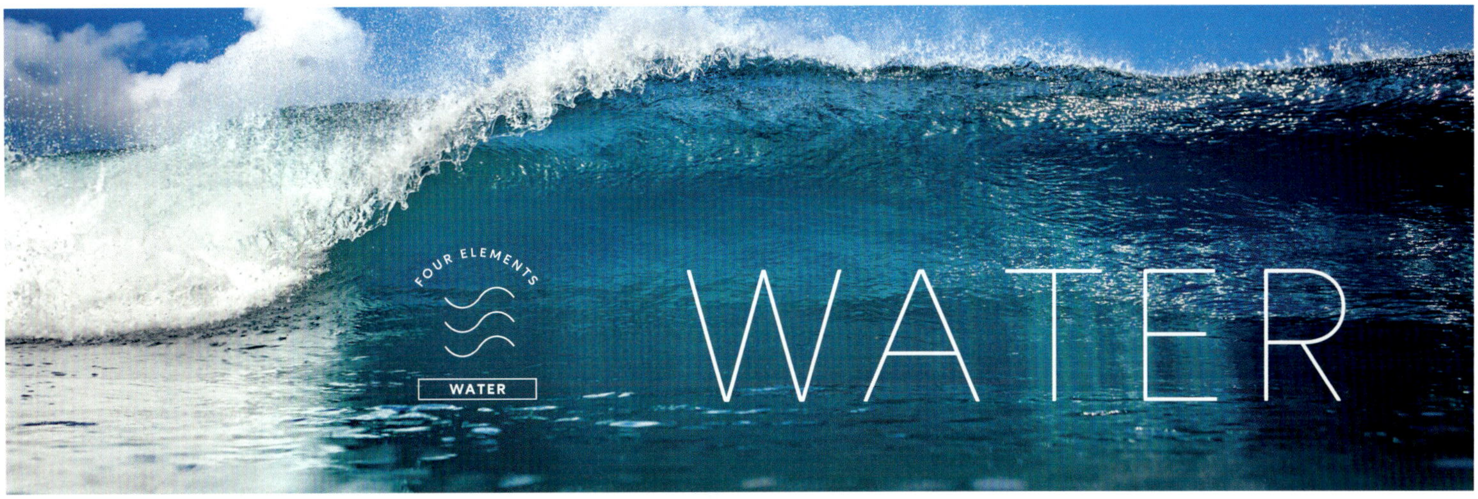
FOUR ELEMENTS
WATER
WATER



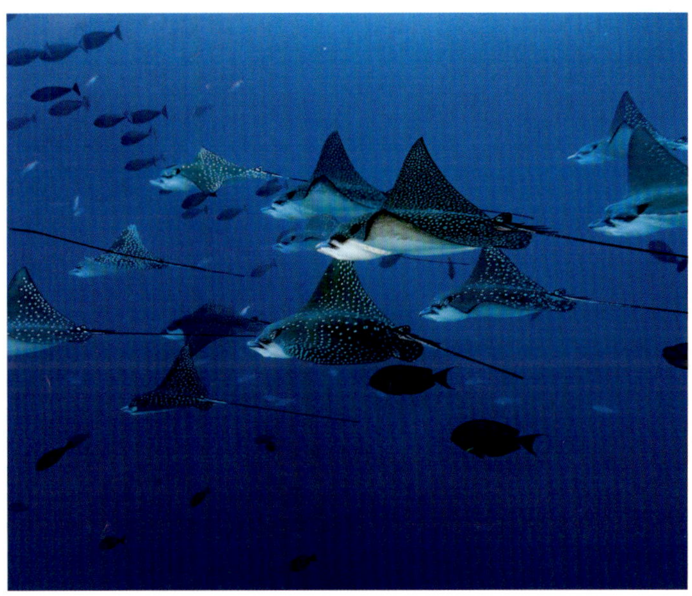

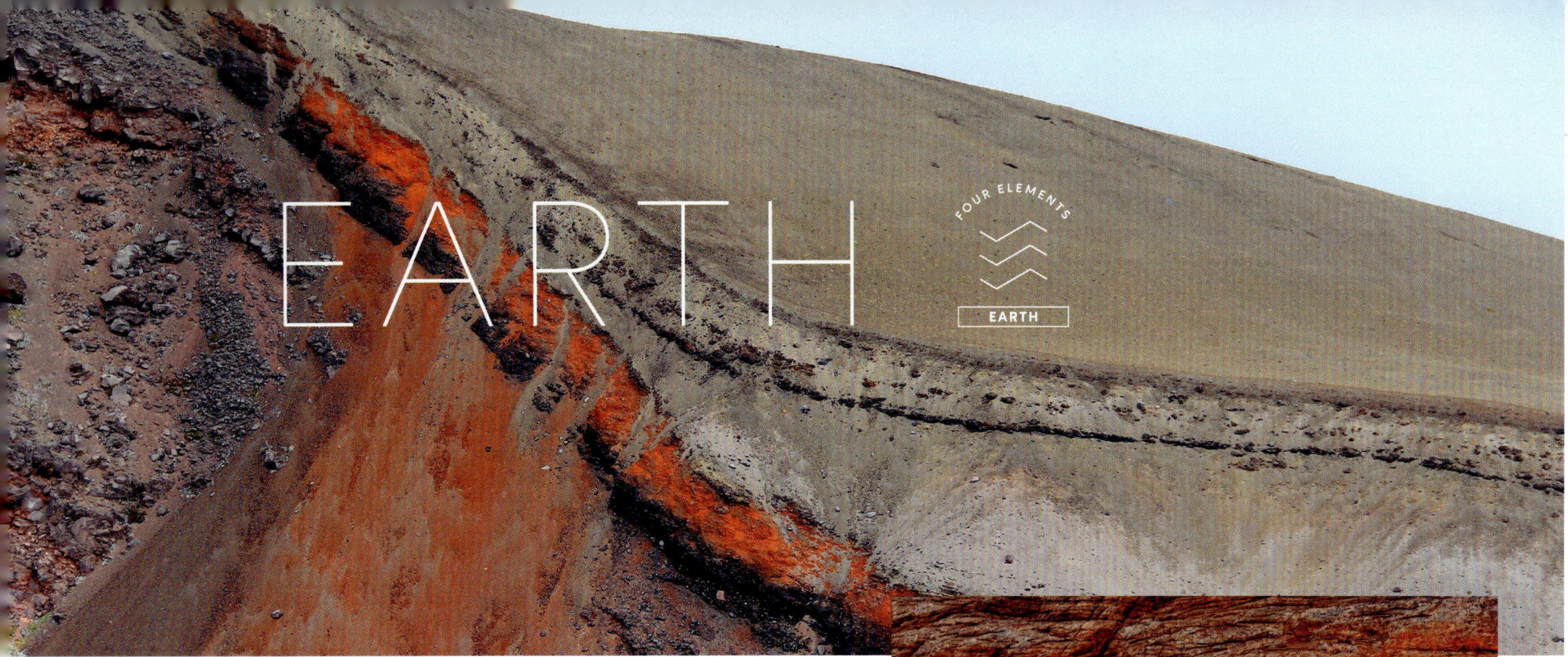


# EARTH



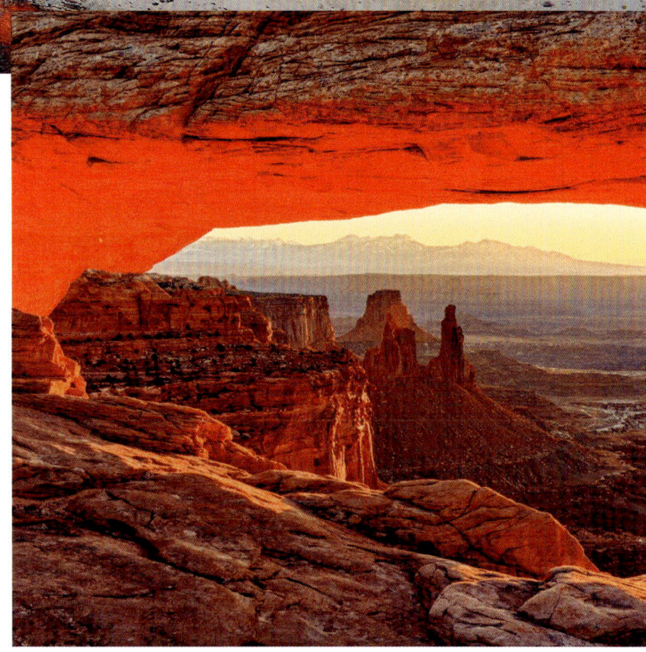

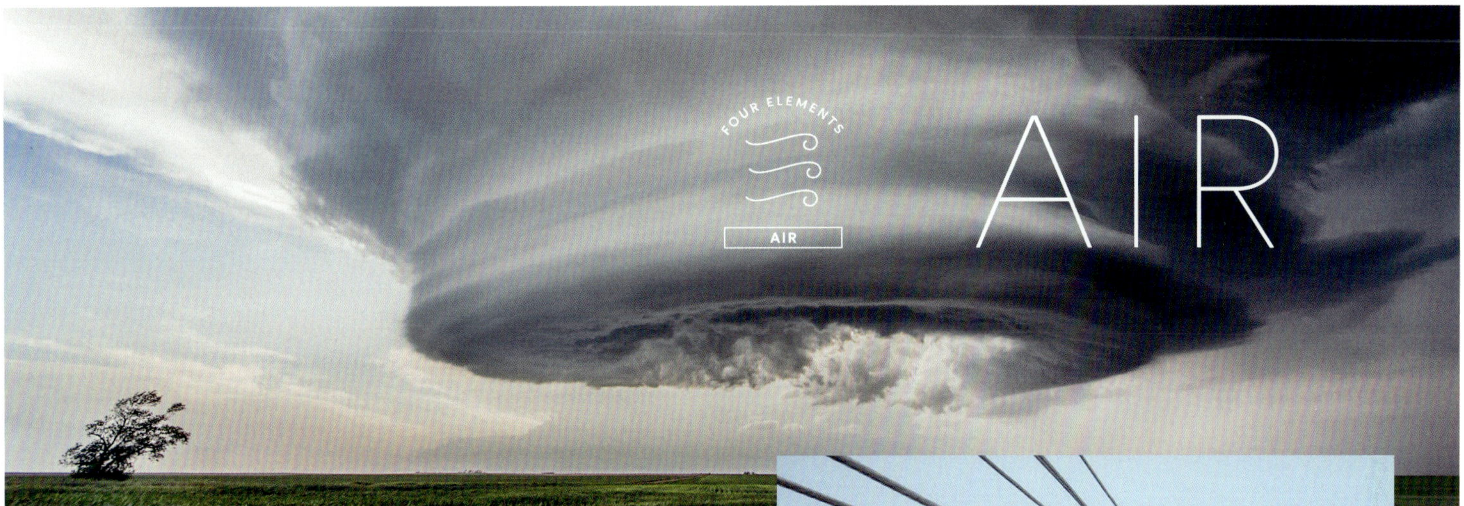

FOUR ELEMENTS
AIR
AIR




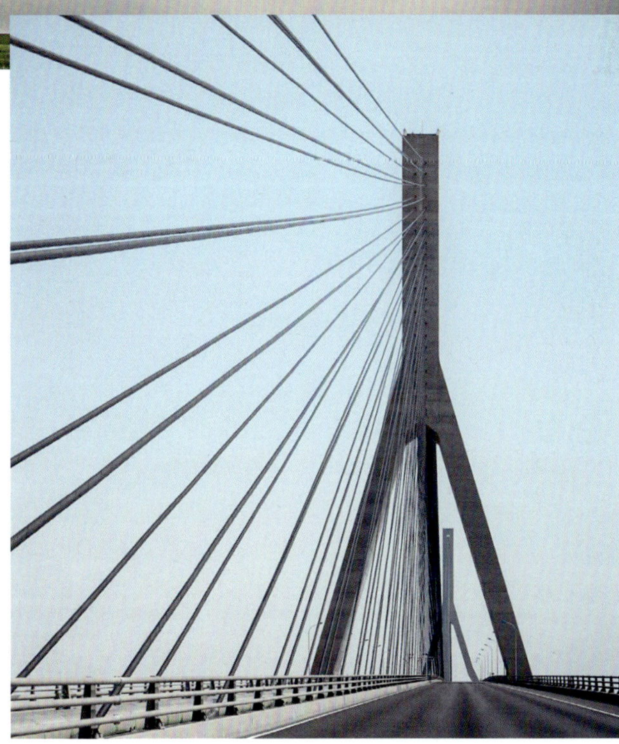

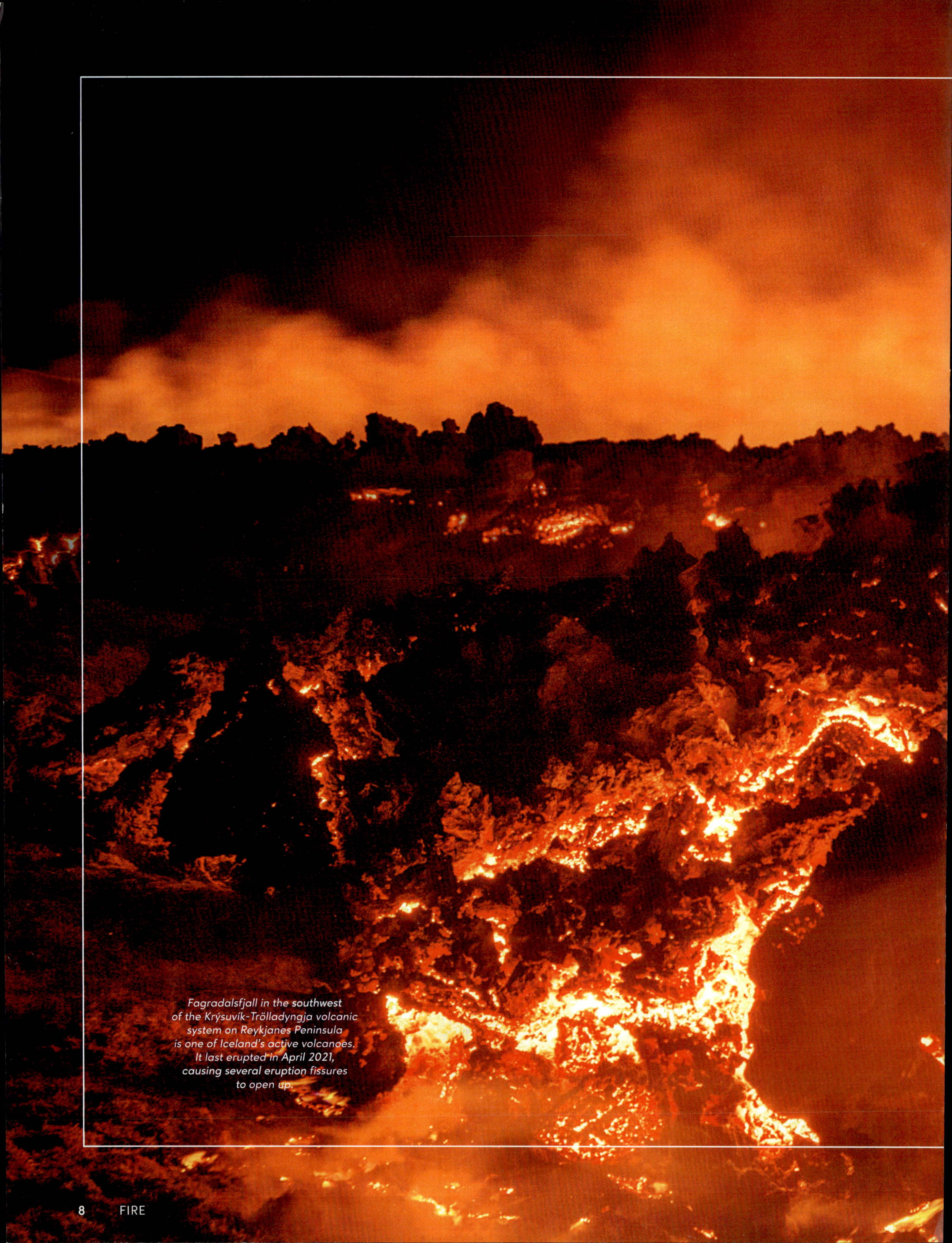

*Fagradalsfjall in the southwest of the Krýsuvík-Trölladyngja volcanic system on Reykjanes Peninsula is one of Iceland's active volcanoes. It last erupted in April 2021, causing several eruption fissures to open up.*

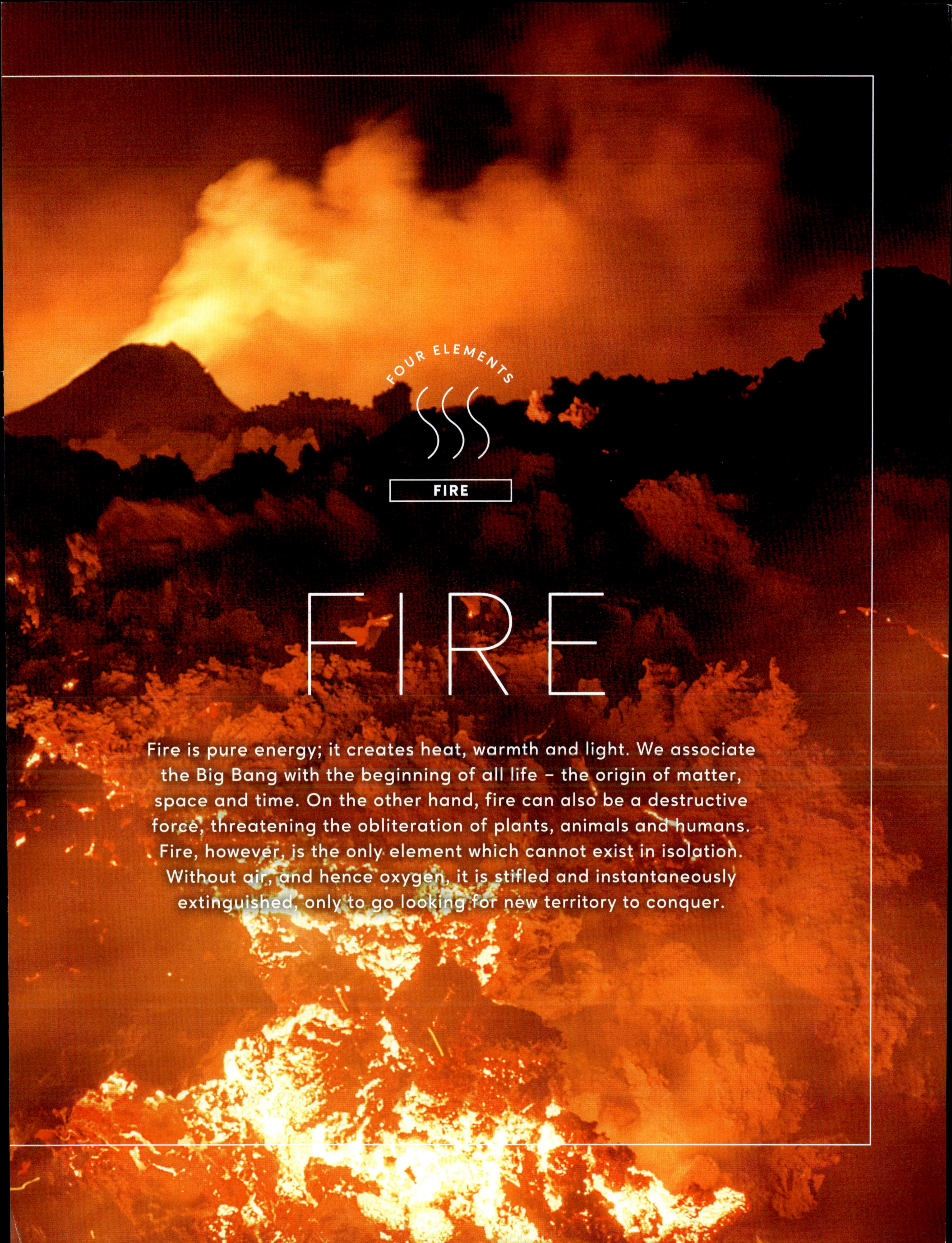

FOUR ELEMENTS

FIRE

# FIRE

Fire is pure energy; it creates heat, warmth and light. We associate the Big Bang with the beginning of all life – the origin of matter, space and time. On the other hand, fire can also be a destructive force, threatening the obliteration of plants, animals and humans. Fire, however, is the only element which cannot exist in isolation. Without air, and hence oxygen, it is stifled and instantaneously extinguished, only to go looking for new territory to conquer.

FOUR ELEMENTS

FIRE

# VOLCANOES

A large part of the earth's crust comes from molten rock, which rises up to the surface from the earth's interior, cools and then solidifies to form volcanic rock

Volcanic activity is almost exclusively confined to the margins of the tectonic plates. Of the 500 to 600 active volcanoes in the world, 80 per cent lie on convergent plates and 15 per cent on divergent plates. Around 5 per cent are 'hot spot volcanoes'. These volcanoes are fixed on an exceptionally hot area of the earth's mantle and always erupt in the same place, as the tectonic plate pushes away over them. Some of the most famous volcanoes of the world are of this type, such as those on the 2,400-km (1,491-mile) long Hawaii archipelago, for instance, which originated as fixed hot spot volcanoes below the Pacific plate, which was moving.

The fire-spewing smokestacks can be classified according to the type of magma supply: there are fissure volcanoes such as Mount Etna on Sicily, which consistently erupt in different places (fissures) and have a slow-moving downward lava flow. Central volcanoes (e.g. Mount Vesuvius near Naples) behave in a completely different way. In these, a pipe-like opening leads into the magma chamber. When the volcano erupts, the lava mass is hurled out with explosive force, causing the formation of calderas (cauldron-shaped indentations).

Volcanoes can also be classified according to their shape. There are composite or stratovolcanoes, which are conical (e.g. Mount Fuji in Japan); shield volcanoes (e.g. Mount Loa on Hawaii); plateau volcanoes, where the less viscous lava pushes up to the surface through fissures; cinder cones and ash cones, which are noticeably smaller and generally have a regular, conical shape (e.g. Mount Stromboli in Italy); maars, which are caused by an explosion of lava and are funnel- or basin-shaped (e.g. in the Volcanic Eifel); and calderas, which have cauldron- or bowl-shaped indentations (e.g. the Caldera de Taburiente on La Palma).

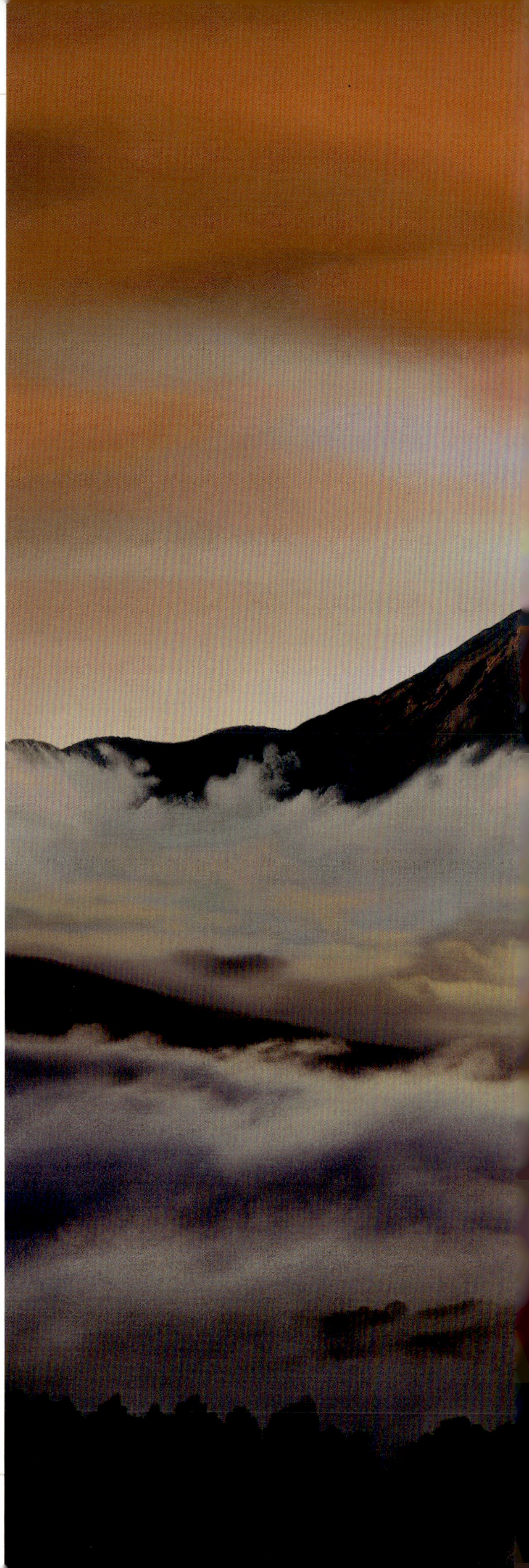

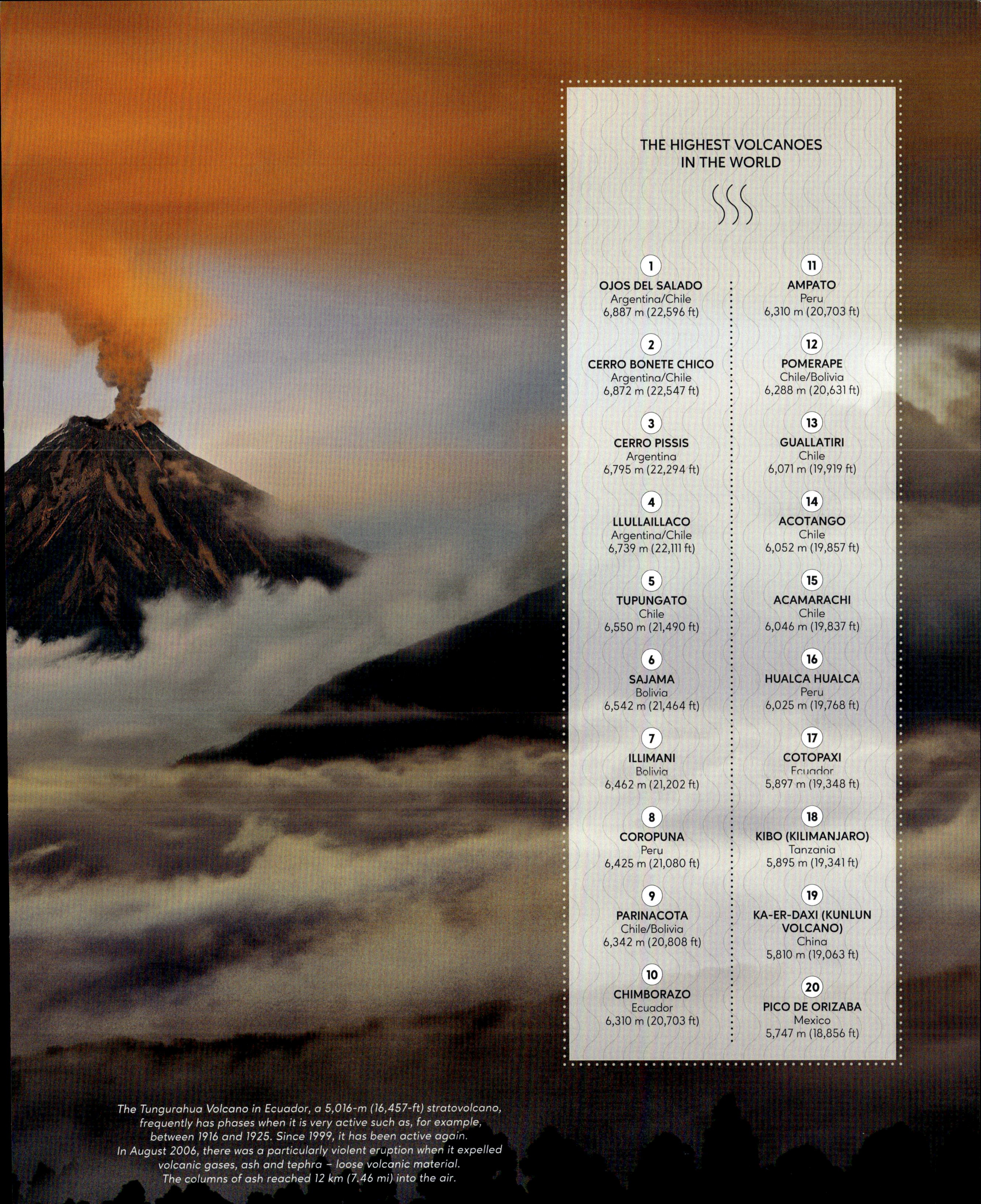

## THE HIGHEST VOLCANOES IN THE WORLD

1. **OJOS DEL SALADO**
Argentina/Chile
6,887 m (22,596 ft)

2. **CERRO BONETE CHICO**
Argentina/Chile
6,872 m (22,547 ft)

3. **CERRO PISSIS**
Argentina
6,795 m (22,294 ft)

4. **LLULLAILLACO**
Argentina/Chile
6,739 m (22,111 ft)

5. **TUPUNGATO**
Chile
6,550 m (21,490 ft)

6. **SAJAMA**
Bolivia
6,542 m (21,464 ft)

7. **ILLIMANI**
Bolivia
6,462 m (21,202 ft)

8. **COROPUNA**
Peru
6,425 m (21,080 ft)

9. **PARINACOTA**
Chile/Bolivia
6,342 m (20,808 ft)

10. **CHIMBORAZO**
Ecuador
6,310 m (20,703 ft)

11. **AMPATO**
Peru
6,310 m (20,703 ft)

12. **POMERAPE**
Chile/Bolivia
6,288 m (20,631 ft)

13. **GUALLATIRI**
Chile
6,071 m (19,919 ft)

14. **ACOTANGO**
Chile
6,052 m (19,857 ft)

15. **ACAMARACHI**
Chile
6,046 m (19,837 ft)

16. **HUALCA HUALCA**
Peru
6,025 m (19,768 ft)

17. **COTOPAXI**
Ecuador
5,897 m (19,348 ft)

18. **KIBO (KILIMANJARO)**
Tanzania
5,895 m (19,341 ft)

19. **KA-ER-DAXI (KUNLUN VOLCANO)**
China
5,810 m (19,063 ft)

20. **PICO DE ORIZABA**
Mexico
5,747 m (18,856 ft)

*The Tungurahua Volcano in Ecuador, a 5,016-m (16,457-ft) stratovolcano, frequently has phases when it is very active such as, for example, between 1916 and 1925. Since 1999, it has been active again. In August 2006, there was a particularly violent eruption when it expelled volcanic gases, ash and tephra – loose volcanic material. The columns of ash reached 12 km (7.46 mi) into the air.*

FOUR ELEMENTS

FIRE

# OJOS DEL SALADO

The highest volcano
on earth

Ojos del Salado lies in the Atacama desert on the border between Argentina and Chile and is the second highest peak in the Americas and the highest volcano in the world. Some of the measurements taken in the 20th century reported it as even higher than Aconcagua, making it the highest mountain in the Americas. This, however, has not been confirmed by more recent and precise GPS-based measurements. In 1994, one set of measurements claimed that Monte Pissis was 6,882 m (22,580 ft) high – this placed Ojos del Salado as the third highest peak in the Americas. But this was also shown to be incorrect. It is now generally agreed that Ojos del Salado, at 6,887 m (22,596 ft), is roughly 100 m (328 ft) higher than Cerro Pissis. Ojos del Salado is assumed to be extinct, as there have been no signs of explosive eruptions. There have, however, been repeated reports of gas emissions from the volcano. This type of activity was seen in 1937 and 1956 as well as on 14 November 1994 when steam and sulfur fumes issued from the crater. In 2007, a Chilean managed to ascend Ojos del Salado to a height of 6,688 m (21,943 ft) in a modified ATV, thereby achieving a world record; it was the first time anyone had ascended that high without the aid of an aircraft.

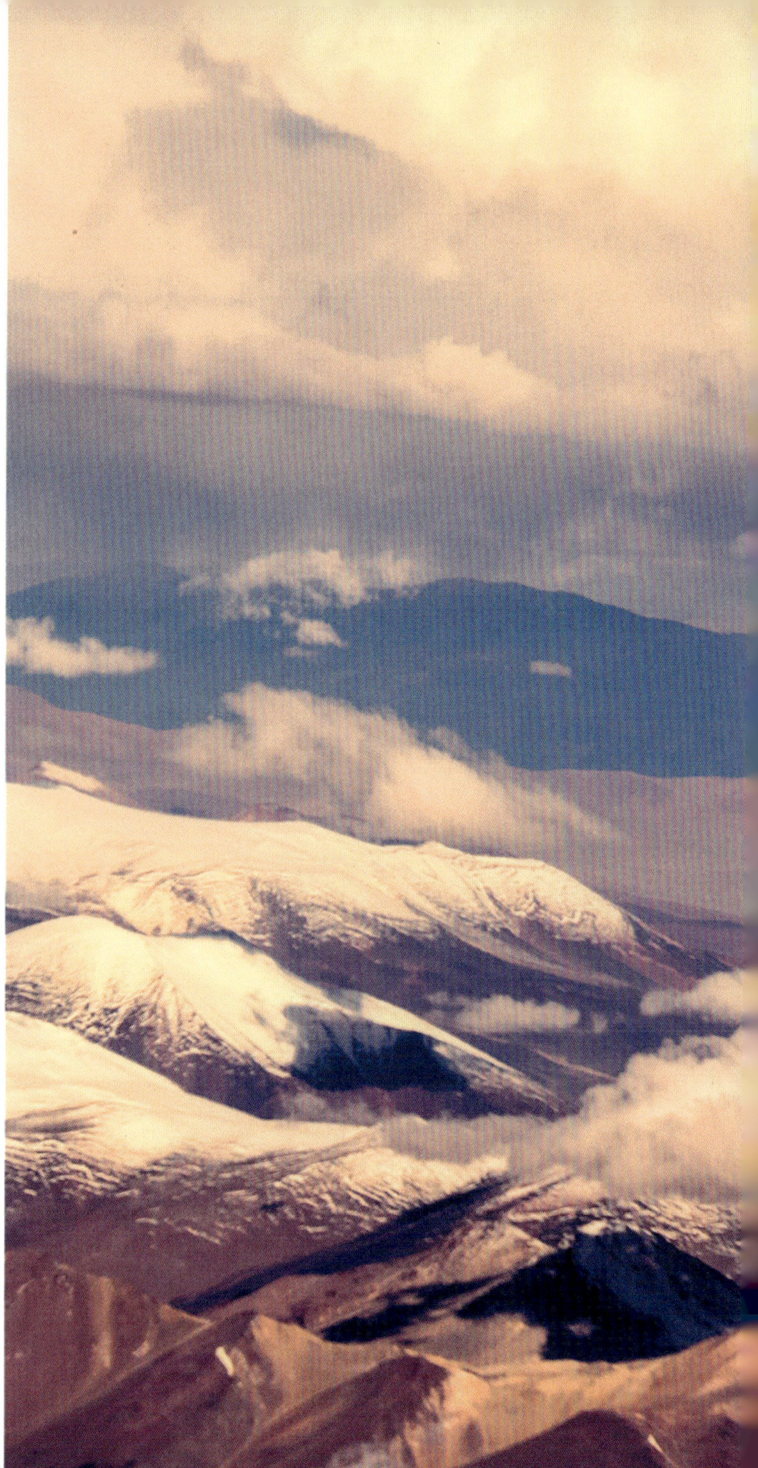

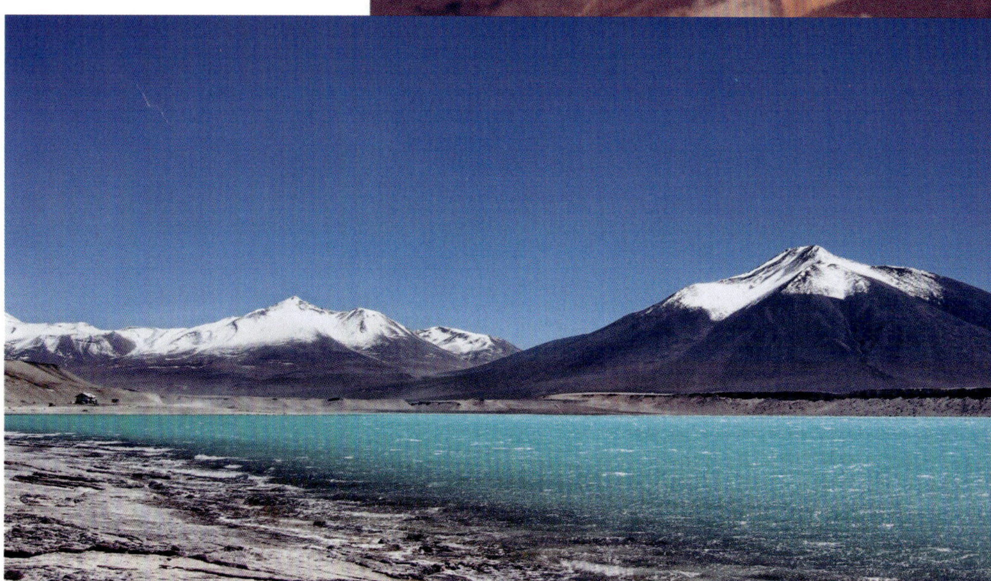

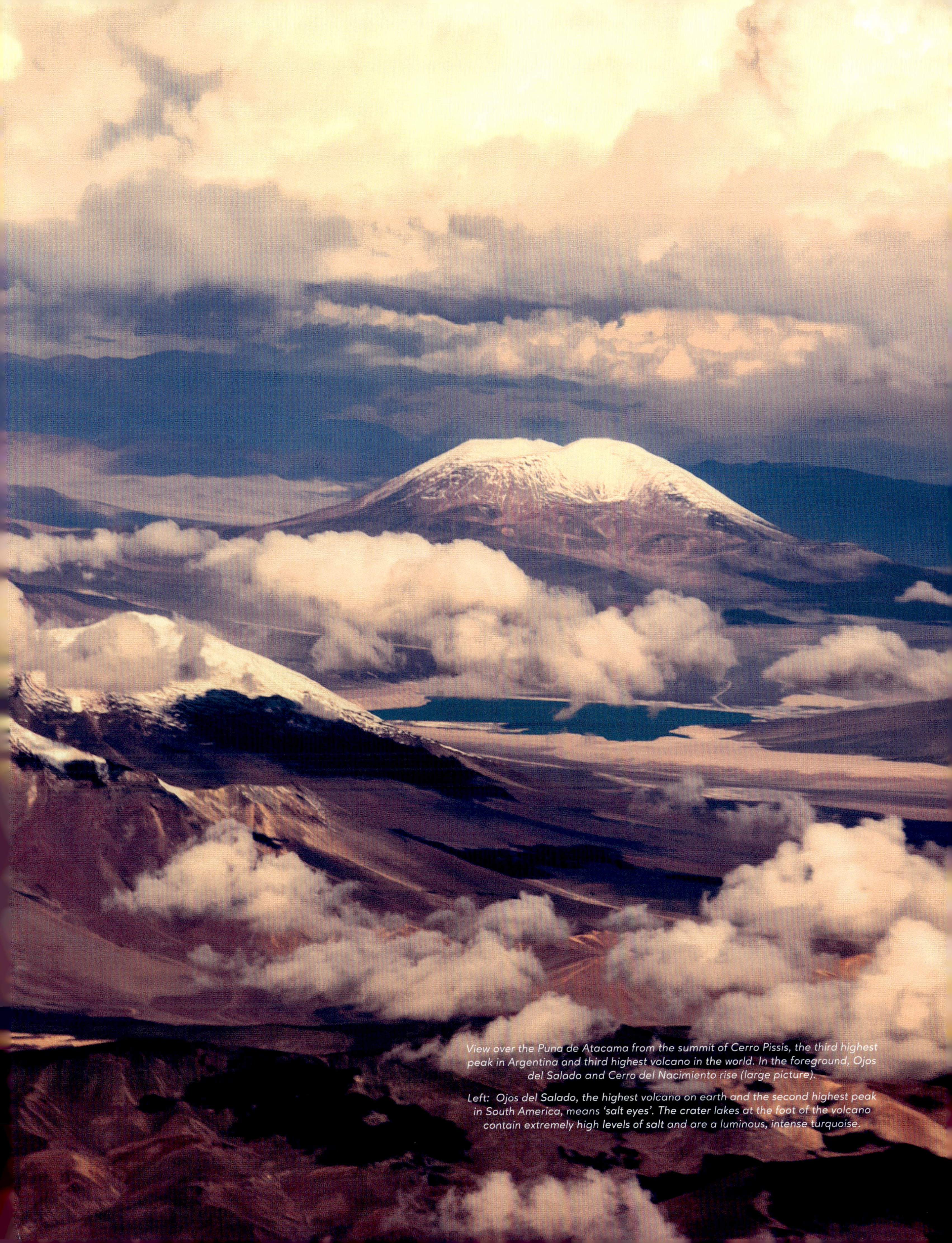

View over the Puna de Atacama from the summit of Cerro Pissis, the third highest peak in Argentina and third highest volcano in the world. In the foreground, Ojos del Salado and Cerro del Nacimiento rise (large picture).

Left: Ojos del Salado, the highest volcano on earth and the second highest peak in South America, means 'salt eyes'. The crater lakes at the foot of the volcano contain extremely high levels of salt and are a luminous, intense turquoise.

# VOLCANIC ERUPTIONS

The most powerful volcanic eruptions in history

## The largest volcanic eruptions, BC:

1 **La Garita Caldera**
approx. 27 million years ago – Colorado, USA

2 **Bruneau-Jarbidge**
approx. 11 million years ago – Idaho, USA

3 **Yellowstone**
approx. 2.2 million years ago – Wyoming, USA

4 **Yellowstone**
approx. 1.2 million years ago – Wyoming, USA

5 **Valle Grande**
approx. 1.1 million years ago – New Mexico, USA

6 **Long Valley Caldera**
760,000 years ago – California, USA

7 **Yellowstone**
640,000 years ago – Wyoming, USA

8 **Aso Kysh (Mount Aso)**
300,000-80,000 years ago – Kyushu, Japan

9 **Lake Toba**
75,000 years ago – Sumatra, Indonesia

10 **Phlegraean Fields**
35,000 years ago – Campania, Italy

11 **Lake Taupo**
26,500-22,500 years ago – North Island, New Zealand

12 **Aira Kysh**
22,000 years ago – Kyushu, Japan

13 **Kurile Lake**
circa 5700 years ago – Kamchatka, Russia

14 **Mount Mazama**
circa 4900 years ago – Oregon, USA

15 **Kikai**
circa 4300 years ago – Ryukyu Islands, Japan

16 **Santorini**
circa 1630 years ago – Cyclades, Greece

## The largest volcanic eruptions, AD:

1 **Ambrym**
50 – Vanuatu

2 **Vesuvius**
79 – Campania, Italy

3 **Lake Taupo**
181 – North Island, New Zealand

4 **Eldgjá**
936 – Iceland

5 **Baitoushan/Paektusan**
1050 – China/Korea

6 **Etna**
1169 – Sicily, Italy

7 **Kuwae**
1453 – Vanuatu

8 **Vesuvius**
1631 – Campania, Italy

9 **Etna**
1669 – Sicily, Italy

10 **Lanzarote**
1730 - 1736 – Canary Islands, Atlantic Ocean

11 **Laki/Lakagígar**
1783-1784 – Iceland

12 **Tambora**
1815 – Sumatra, Indonesia

13 **Krakatoa**
1883 – Indonesia

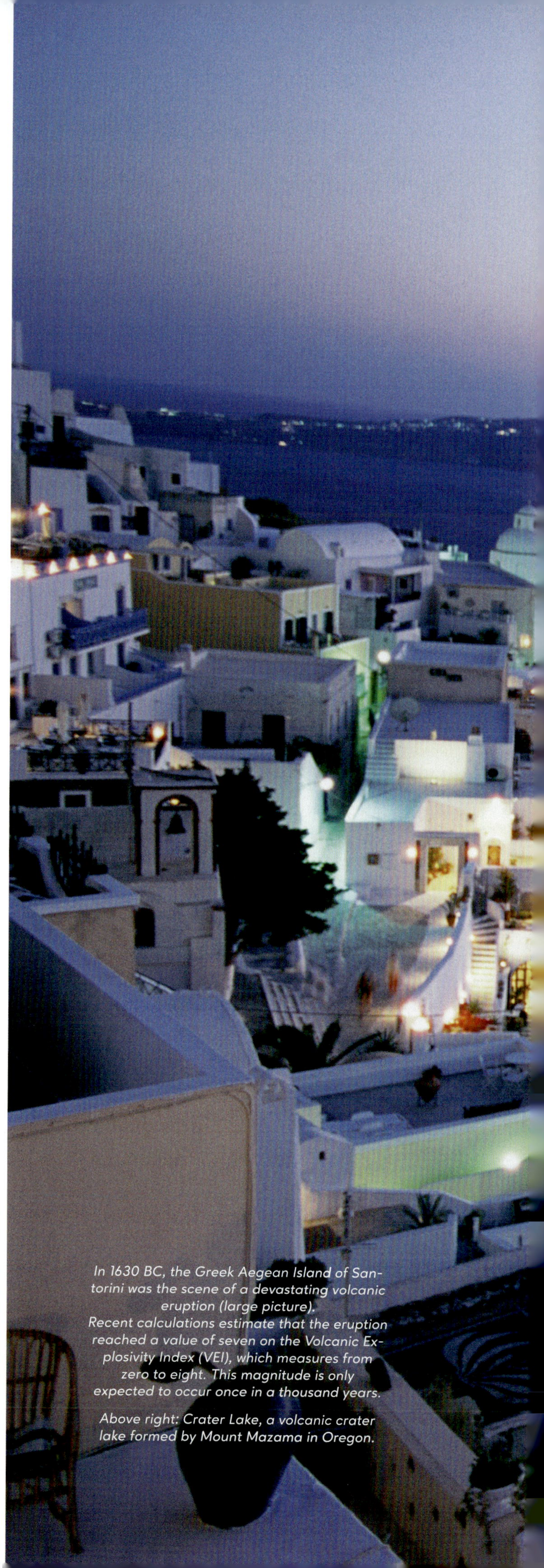

*In 1630 BC, the Greek Aegean Island of Santorini was the scene of a devastating volcanic eruption (large picture). Recent calculations estimate that the eruption reached a value of seven on the Volcanic Explosivity Index (VEI), which measures from zero to eight. This magnitude is only expected to occur once in a thousand years.*

*Above right: Crater Lake, a volcanic crater lake formed by Mount Mazama in Oregon.*

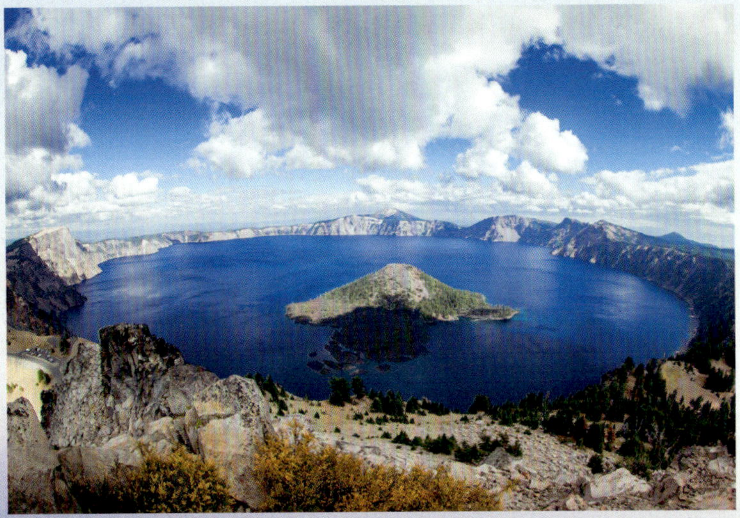

DID YOU KNOW?

### Vesuvius with museum

Mount Vesuvius is one of the most famous volcanoes in the world and is the only active volcano on the European mainland. It is well-known on account of the eruption in 79 AD, which destroyed the Roman cities of Pompeii, Herculaneum and Stabiae. Although it is one of the most dangerous volcanoes in the world, the city of Naples, with a population of over one million, lies only a few kilometres away. A new museum shows visitors how the volcano has changed throughout the centuries, when experts predict the next eruption and the impact the volcano has on the animal and plant life of the region. It also offers information on the Vesuvius National Park and the many questions relating to the famous volcano itself.

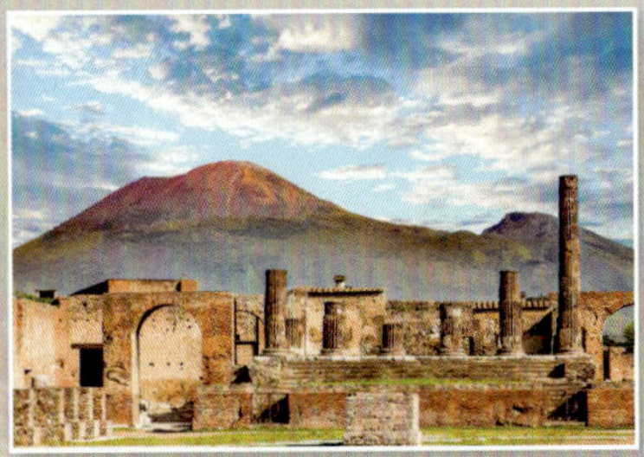

# VOLCANIC ERUPTIONS

The most violent volcanic eruptions in modern times

People have always settled near volcanoes. The inhabitants of towns and villages in the vicinity of volcanoes are aware of the threat – but, of course, they hope that the volcano on their doorstep will not erupt. The calm, however, is often deceptive. Even volcanoes that have lain dormant for several thousand years can erupt at any time. There are more than 550 active volcanoes in the world. Among these, the most well-known are Mount Vesuvius and Mount Etna in Italy, Pinatubo in the Philippines and Mount St Helens in the USA. During the last hundred years, eruptions from these volcanoes have destroyed landscapes and lives.

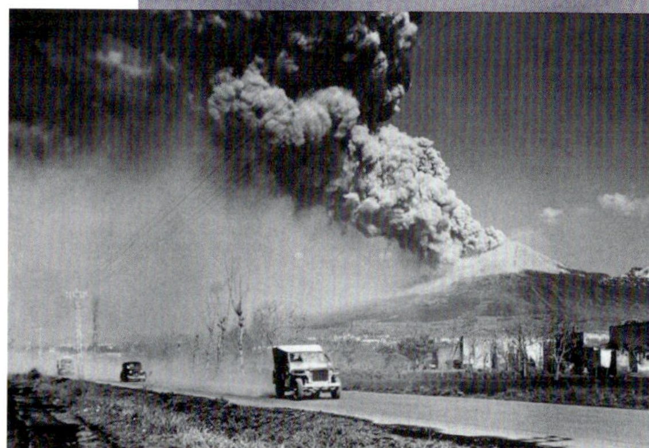

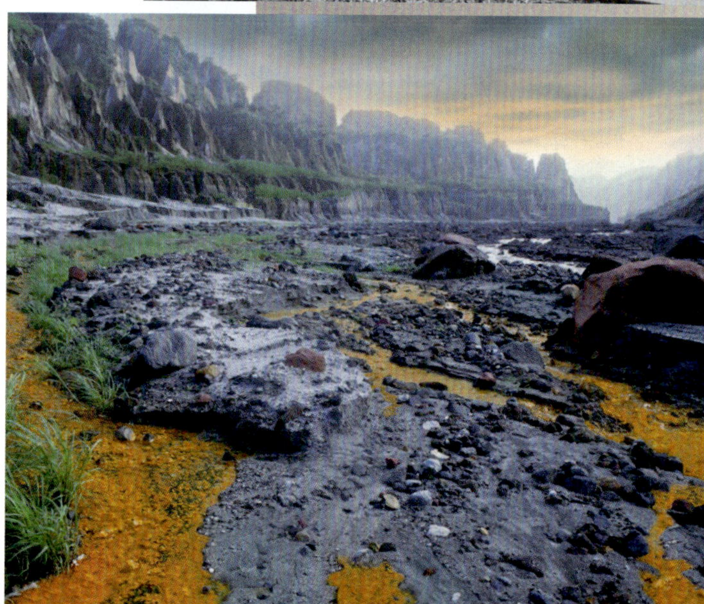

## THE LARGEST VOLCANIC ERUPTIONS IN MODERN TIMES:

1. **SANTA MARIA** 1902 – Guatemala
2. **MONT PELÉE** 1902 – Martinique
3. **NOVARUPTA** 1912 – Alaska Peninsula
4. **KELUD** 1919 – Java, Indonesia
5. **GUNUNG AGUNG** 1963 – Bali, Indonesia
6. **SURTSEY** 1963 – Iceland
7. **ELDFELL** 1973 – Iceland
8. **NYIRAGONGO** 1977 – Zaire
9. **MOUNT ST HELENS** 1980 – Oregon, USA
10. **NEVADO DEL RUIZ** 1985 – Colombia
11. **PINATUBO** 1991 – Philippines
12. **CERRO HUDSON** 1991 – Chile
13. **SOUFRIERE HILLS** 1995/1997 – Montserrat, Caribbean
14. **TUNGURAHUA** 2000 - 2006 – Ecuador

*Vesuvius, 1944 (above, top), Pinatubo, 1991 (above).*

*Gunung Agung on Bali, 1963 (right), Mount St. Helens, 1980 (far right).*

*Large picture: The Caribbean island of Montserrat lies on three volcanic zones: Silver Hills in the north, Centre Hills in the central area and Soufrière Hills in the south.*

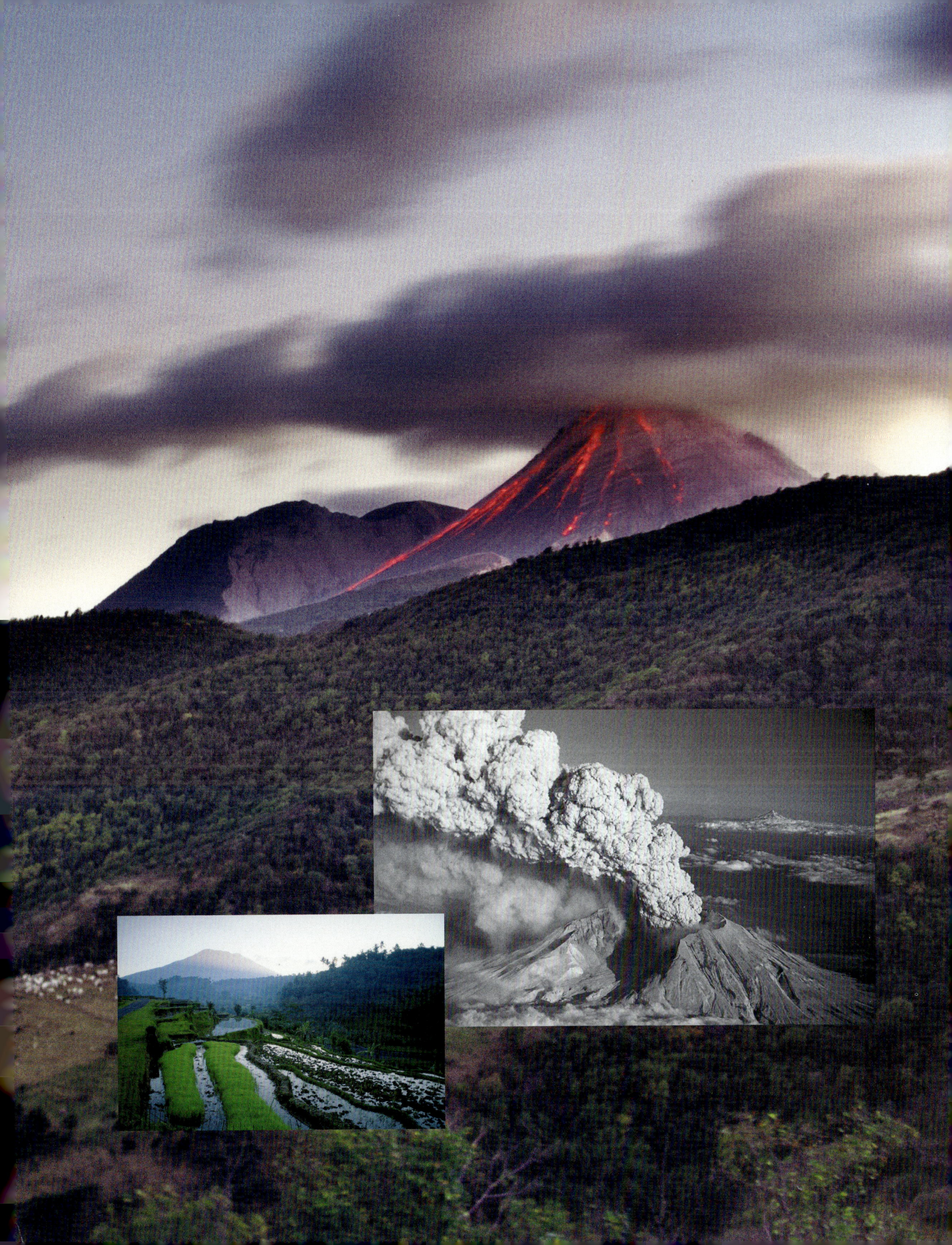

FOUR ELEMENTS

FIRE

# THE PACIFIC RING OF FIRE

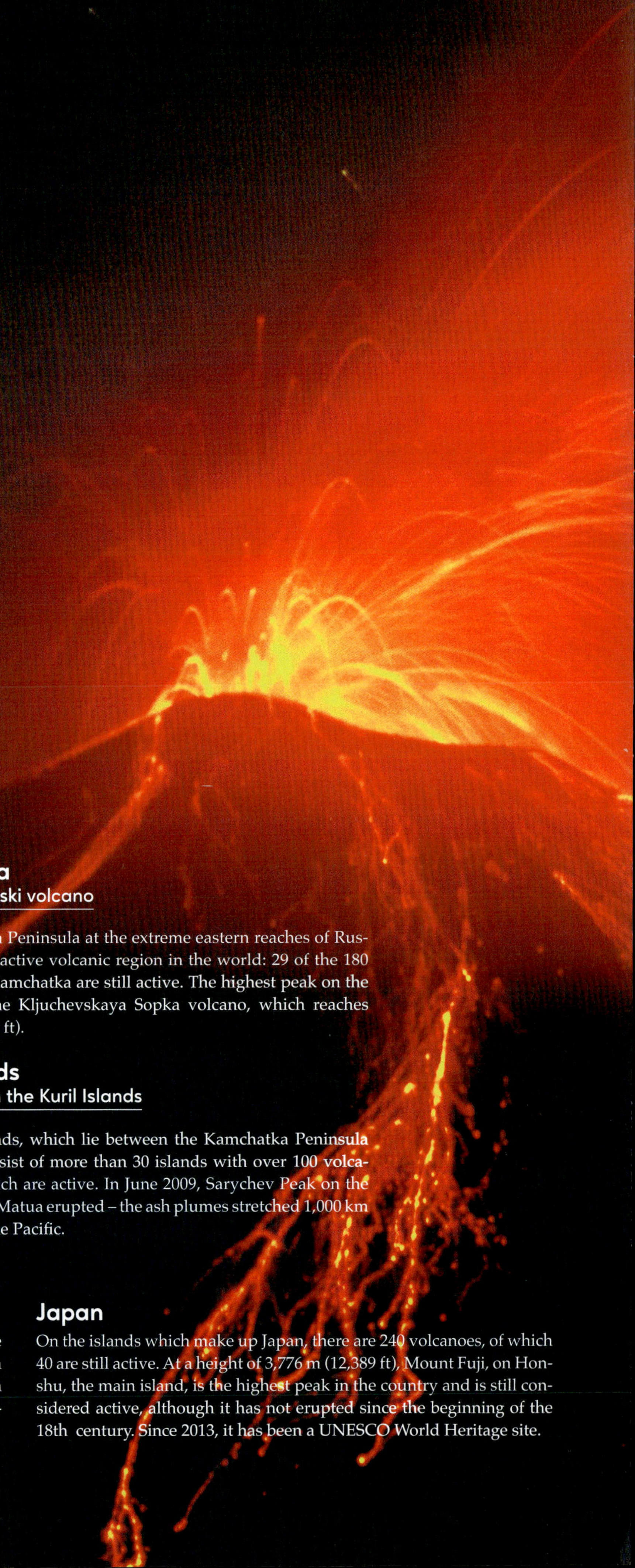

The largest volcanic system in the world

A 'ring of fire' comprising multiple series of volcanic islands and volcanoes extends for 32,500 km (20,195 mi) around the edge of the Pacific Ocean. It stretches from the Antarctic Peninsula via Tierra del Fuego to the series of volcanoes which form part of the cordilleras in South and Central America. The island arc, which starts with the volcanoes of the Aleutian Islands in Alaska, marks the northern edge of the Pacific plate. It then continues at the Kamchatka Peninsula in Siberia and the Kuril Islands. This is where the Pacific plate pushes under the Eurasian plate. The series then resumes with an island arc that extends from Japan to Taiwan, ending in the Mariana Islands in the south. It marks the edge of the Pacific plate where it borders the smaller Philippine plate, which itself pushes against three other plates, and thus forges the series of extremely active volcanoes in Indonesia and the Philippines at the plates' edges. It concludes with the volcanic chains along the Indo-Australian plate, which run across Melanesia and Fiji down to New Zealand and on to some of the sub-antarctic islands.

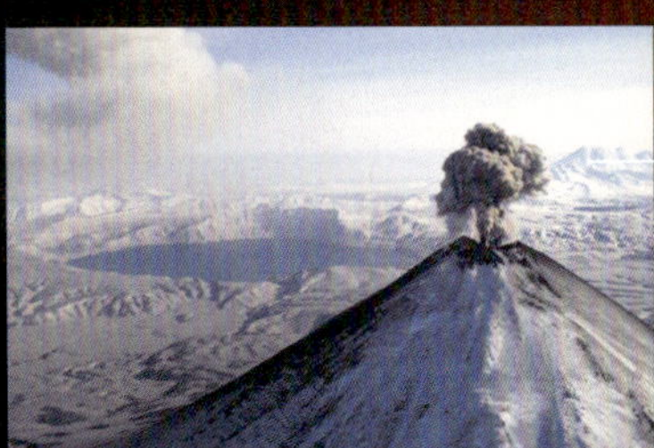

## Kamchatka

**Active Karymski volcano**

The Kamchatka Peninsula at the extreme eastern reaches of Russia is the most active volcanic region in the world: 29 of the 180 volcanoes on Kamchatka are still active. The highest peak on the peninsula is the Kljuchevskaya Sopka volcano, which reaches 4,750 m (15,585 ft).

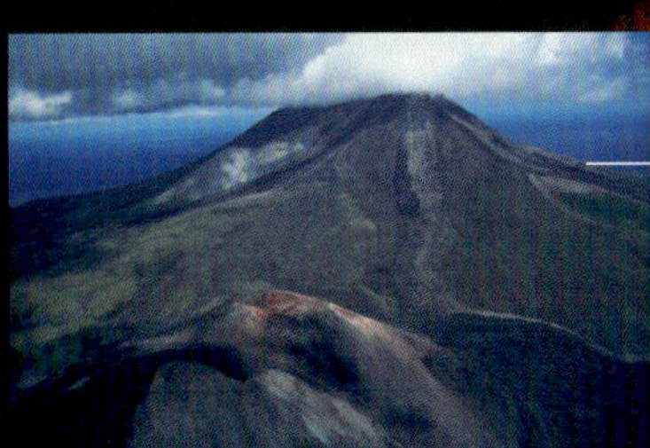

## Kuril Islands

**Volcanoes on the Kuril Islands**

The Kuril Islands, which lie between the Kamchatka Peninsula and Japan, consist of more than 30 islands with over 100 volcanoes, 39 of which are active. In June 2009, Sarychev Peak on the Kuril Island of Matua erupted – the ash plumes stretched 1,000 km (621 mi) into the Pacific.

## Philippines

The Philippines are volcanic peaks that rise up out of the ocean; the highest is Mount Apo on Mindanao at 2,954 m (9,692 ft). In 1991, Pinatubo on Luzon erupted, which was one of the major volcanic eruptions of the 20th century.

## Japan

On the islands which make up Japan, there are 240 volcanoes, of which 40 are still active. At a height of 3,776 m (12,389 ft), Mount Fuji, on Honshu, the main island, is the highest peak in the country and is still considered active, although it has not erupted since the beginning of the 18th century. Since 2013, it has been a UNESCO World Heritage site.

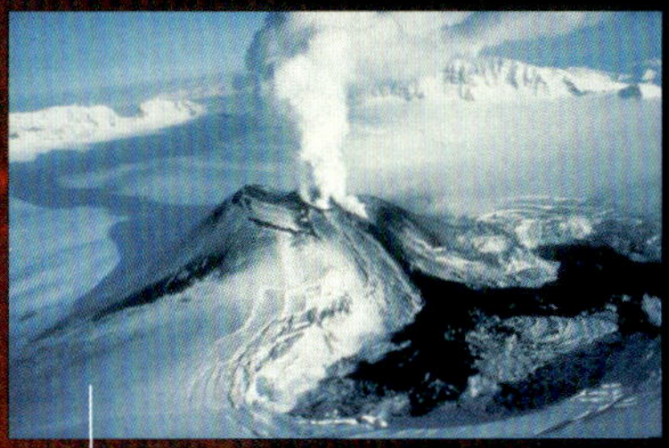

## Aleutian Islands / Alaska
### Eruption of Veniaminof, Alaska

The Aleutian archipelago, which lies between North America and Asia, has around 80 volcanoes, some of which are still active, such as Mount Cleveland on Chuginadak Island. Active volcanoes can also be found on the Alaskan mainland, from the peninsula to the Wrangell Mountains, among them Mount Katmai, which was the site of the region's most violent eruption in 1912.

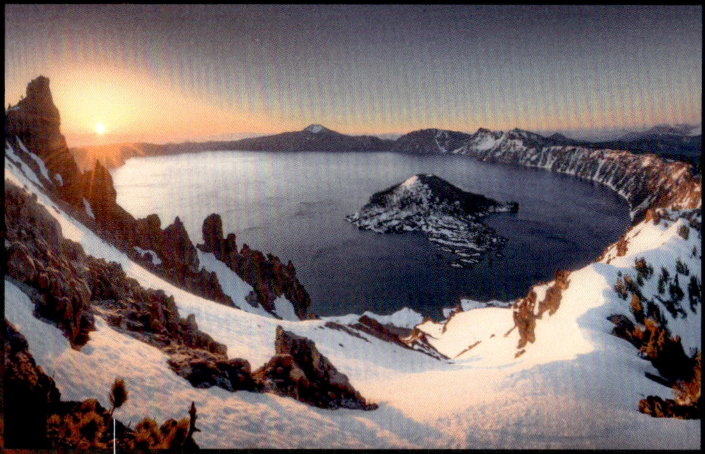

## Cascade Range
### Crater Lake National Park

The Cascade Range runs parallel to the west coast of North America, from British Columbia through the US states of Washington and Oregon and into northern California. The highest peak in the Cascade Range is Mount Rainier (4,395 m (14,420 ft)) near Seattle. The most famous volcano in the range, because it is also the most active, is Mount St Helens (2,549 m (8363 ft)), which lost its peak when it erupted most spectacularly in 1980.

## Central America

In Costa Rica, the volcanoes, some of which are active, are strung together as if on a chain: Turrialba, Irazu (at 3,432 m (11,260 ft), the highest volcano in the country), Barva, Poas and Arenal. The last of these, the Arenal, erupted in 1968 after a long period of dormancy, killing around 80 people. Since then, it has been permanently active. The Central American volcanic chain continues on northwards.

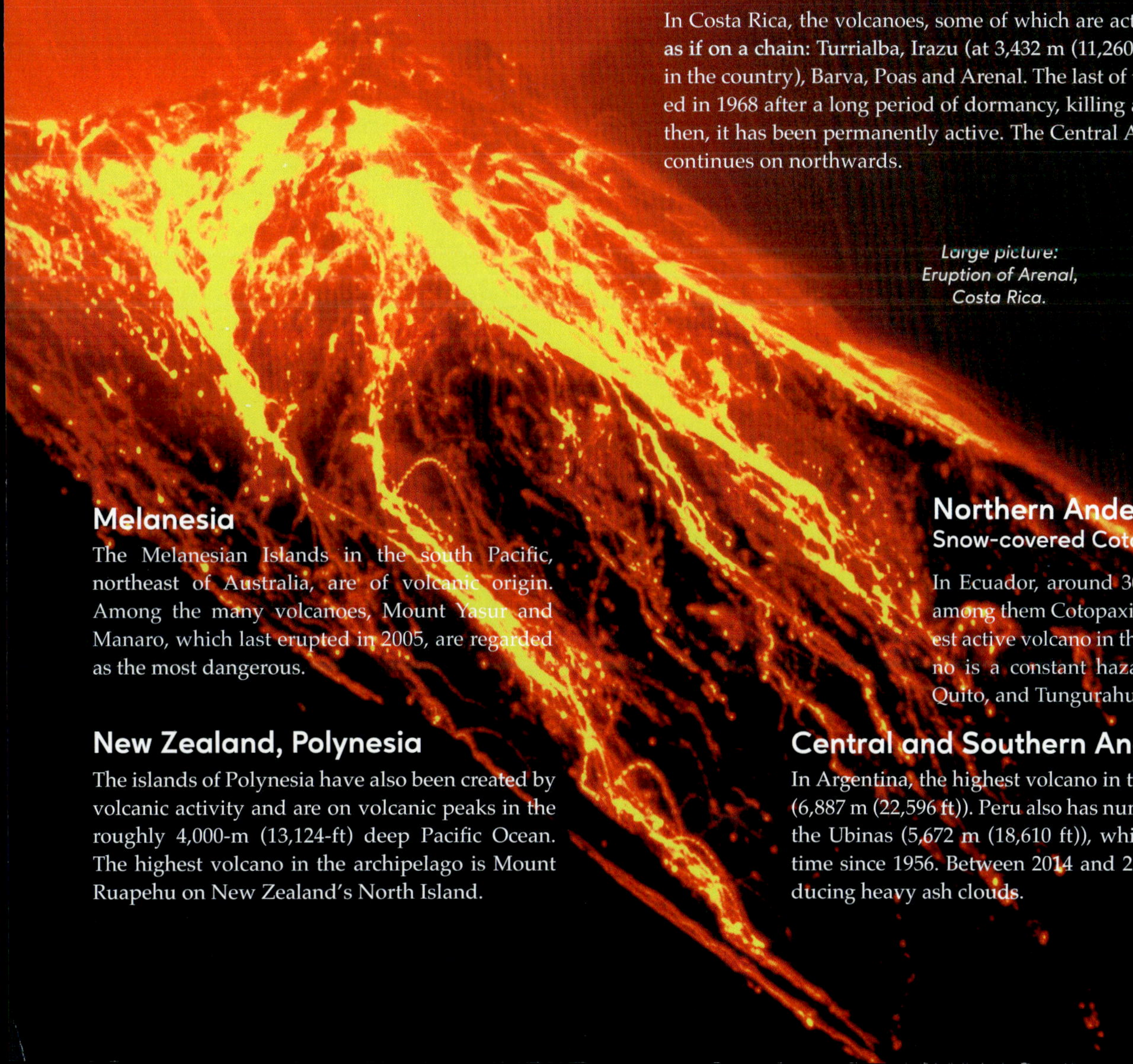

*Large picture: Eruption of Arenal, Costa Rica.*

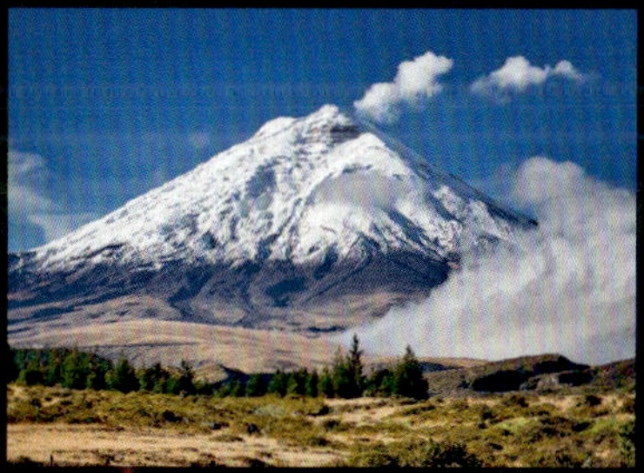

## Northern Andes
### Snow-covered Cotopaxi, Ecuador

In Ecuador, around 30 of the volcanoes are active, among them Cotopaxi (5,897 m (19,348 ft)), the highest active volcano in the world. The Pichincha Volcano is a constant hazard for the country's capital, Quito, and Tungurahua claimed many lives in 2006.

## Melanesia

The Melanesian Islands in the south Pacific, northeast of Australia, are of volcanic origin. Among the many volcanoes, Mount Yasur and Manaro, which last erupted in 2005, are regarded as the most dangerous.

## New Zealand, Polynesia

The islands of Polynesia have also been created by volcanic activity and are on volcanic peaks in the roughly 4,000-m (13,124-ft) deep Pacific Ocean. The highest volcano in the archipelago is Mount Ruapehu on New Zealand's North Island.

## Central and Southern Andes

In Argentina, the highest volcano in the world rises – Ojos del Salado (6,887 m (22,596 ft)). Peru also has numerous active volcanoes, such as the Ubinas (5,672 m (18,610 ft)), which erupted in 2006 for the first time since 1956. Between 2014 and 2015, it erupted once again, producing heavy ash clouds.

FOUR ELEMENTS

FIRE

# HAWAIIAN ISLANDS

The highest volcanic islands in the world

The Hawaiian Islands all originated from volcanic activity; the shield volcanoes in Hawaii are in fact the largest volcanoes in the world, if taking the sections under the ocean's surface into account, since the base of each extends to a depth of 5,400 m (17,717 ft). The youngest volcano, Loihi, is located under the ocean: its summit lies 969 m (3,179 ft) underwater. If it erupts with the same frequency as Mount Loa, then it will take tens of thousands of years to reach sea level.

## MAUNA KEA

**Big Island, 4,205 m (13,797 ft)**

Measured from its base, Mauna Kea has an overall height of more than 9,000 m (29,529 ft), making it the highest peak in the world. In contrast to the neighbouring Mauna Loa, Mauna Kea is not active at present and is classified as dormant.

## MAUNA LOA

**Big Island, 4,170 m (13,682 ft)**

The stratovolcano Mauna Loa, the second highest elevation in the Hawaii archipelago, developed on a hot spot and last erupted in 1984. At present, although it no longer expels lava, it is still classified as an active volcano.

## HUALALAI

**Big Island, 3,055 m (10,023 ft)**

Haulalai is now considered a dormant volcano, having last erupted on several occasions at the end of the 18th and the beginning of the 19th centuries. More recently, the Keahole-Kona airport was built on the lava flows created by those eruptions.

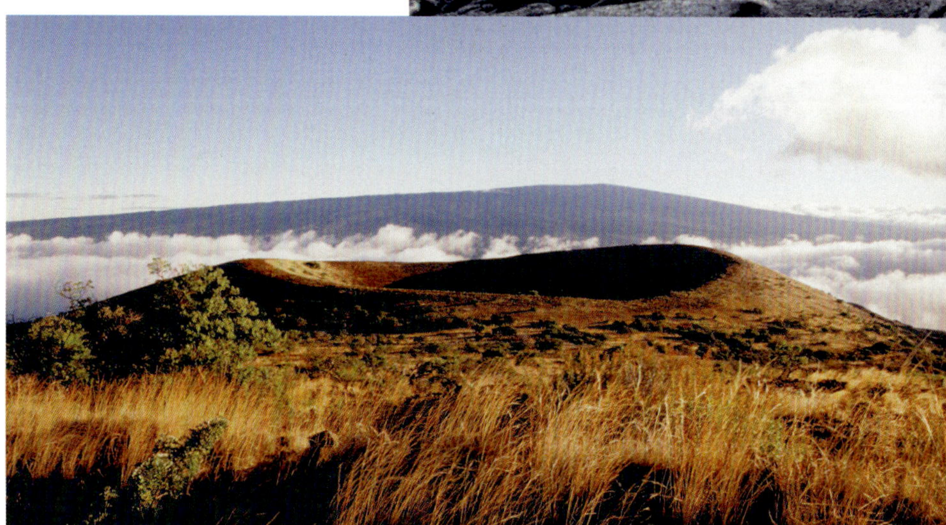

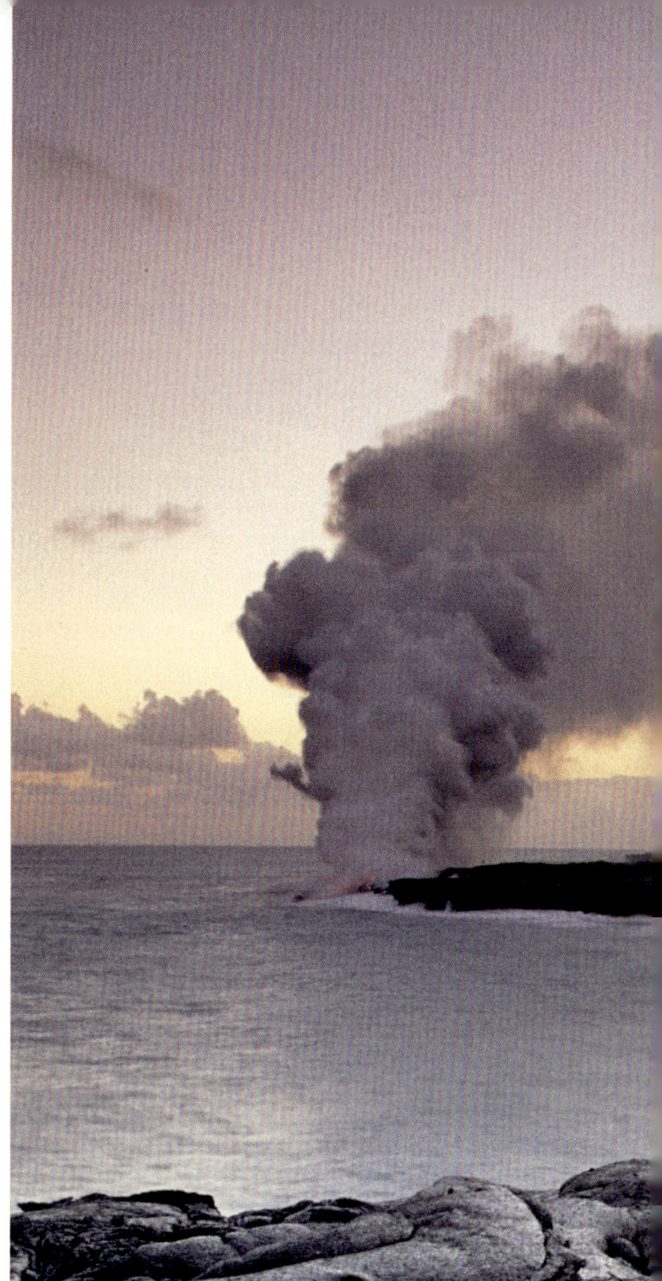

*Mauna Loa on Hawaii – one of the largest active volcanoes in the world (above)*

*Kilauea spews molten lava into the ocean (large picture). According to folklore, this happens when Pele, the goddess of volcanoes, becomes angry. Some of the glowing lava flows through lava tunnels.*

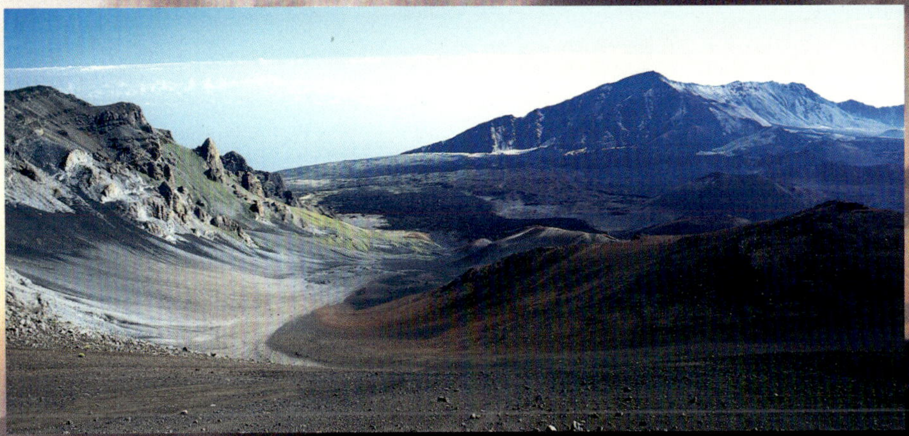

## HALEAKALA

**Maui Island, 3,055 m (10,023 ft)**

Haleakala, which occupies over two-thirds of Maui's total surface area, is the only volcano on the archipelago, other than Big Island, that has been active in the last 600 years (above). Its crater, with its 34-km (21-mile) long perimeter, is one of the largest in the world. Haleakala (meaning 'house of the sun') last erupted in 1790.

**TIP**

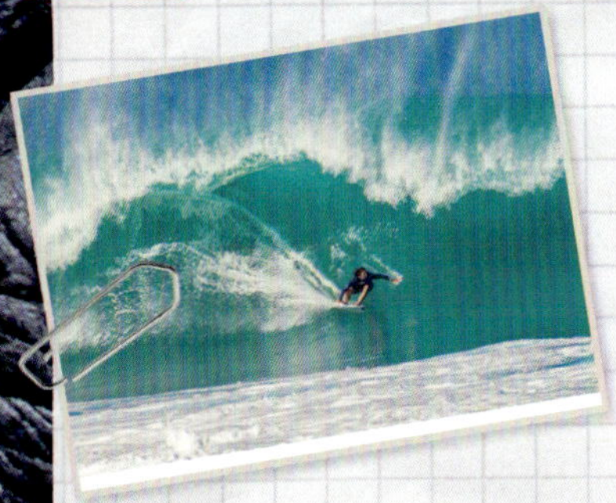

**BANZAI PIPELINE**
**SURF OR MARVEL**

For surfers, this sets the bar for everything else: the Banzai Pipeline is every surfer's dream. This is the gigantic wave which often features in commercials or in photos - a place that epitomizes a perfect surfing experience. The reef where the waves break is on Sunset Beach, off the Hawaiian island of Oahu. Low salt levels in the ocean and a strong current create the right conditions for many a surfer to realize their dream. True surfers look forward to winter when the waves are higher; the waters are shallower in the summer months. Generally, beginners should not risk going in the water here, but standing on the shore and watching is also quite an event. Every wave seems to offer a new photo opportunity.

FOUR ELEMENTS

FIRE

# EUROPE

Volcanism mainly occurs at Europe's margins. On Iceland, there are around 30 active volcanoes; the highest one on the European continent rises in the east – Mount Elbrus in the Caucasus

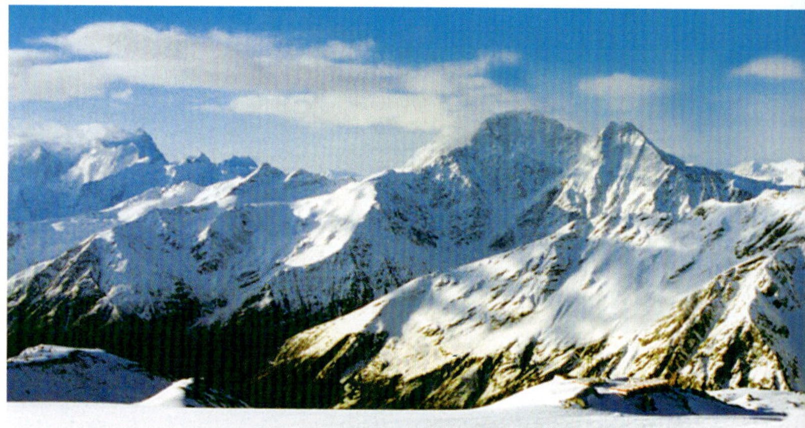

## KASBEK

**Georgia, 5,047 m (16,559 ft)**

The Caucasus Mountains – the isthmus between the Black Sea and the Caspian Sea – reach their highest elevation in the central Caucasus, between the Elbrus and Kasbek peaks. A total of 15 of the peaks here are over 4,800 m (15,749 ft). At 5,047 m (16,559 ft) high, Kasbek, which lies near the Georgian Military Road, is the third highest peak in Georgia and the eighth highest in the Caucasus. This stratovolcano, which is classed as extinct, was first climbed in 1868.

## ARAGATS

**Armenia, 4,095 m (13,436 ft)**

Mount Aragats, an isolated, extinct stratovolcano with four summits which form the edge of the volcanic crater, lies in the Aragatsotn province, 40 km (25 mi) northwest of the Armenian capital, Yerevan. It is the highest elevation in the Republic of Armenia. The peak, covered by glaciers, was formed during the Pliocene or Pleistocene periods. Massive lava flows once issued from the numerous fissures on the slopes of Mount Aragats when it erupted and flowed into the valley.

## KARBARGIN OTH

**Georgia, 3,650 m (11,976 ft)**

The volcanoes of the Kabargin Oth Group also form part of the Caucasus mountain range, which stretches between the Black Sea and the Caspian Sea for around 1,100 km (684 mi) on Russian, Georgian, Armenian, Azerbaijan and Turkish soil. The Kabargin Oth Group, which rises up in the Greater Caucasus in Georgian territory, very close to the border with Russia and southwest of the Kasbek volcano, is made up of a dozen cinder cones and lava domes.

## ELBRUS

**Russia, 5,642 m (18,511 ft)**

Mount Elbrus lies in the middle of the Caucasus, between the Caspian Sea and the Black Sea and is the highest peak in Russia and in Europe. The mount has two summits – the western summit is 5,642 m (18,511 ft) high while the eastern summit stands at 5,621 m (18,442 ft) – and is currently a dormant stratovolcano with slopes that are largely glaciated. It developed in the Holocene epoch in a vast caldera (Spanish for 'cauldron'). On the eastern summit, there is a volcanic crater measuring 250 m (820 ft) across. Mount Elbrus last erupted around 2,000 years ago; since then, it has been dormant. There are, however, some signs of volcanic activity, albeit faint ones, such as hot springs. If this activity were to increase, there is a danger that the massive ice cap could melt very rapidly and the meltwater and lahars (mudflows) could bury the surrounding area. Extensive glaciers and snow fields, which only partially thaw in spring, stretch over the slopes of the Elbrus.

## GEGHAM MOUNTAINS

**Armenia, 3,597 m (11,802 ft)**

The Gegham Mountains are a volcano field in the western part of central Armenia between Jerevan and Lake Sevan, with the highest peak reaching 3,597 m (11,802 ft). These volcanoes and lava flows originated in the Holocene epoch and cover an area of 65 km (40 mi) by 35 km (22 mi). Lava from the central and eastern volcanoes once flowed into Lake Sevan. The most recent activity of the Gegham volcanoes was approximately 2,000 years ago and created many more cinder cones and lava flows. Today, the Gegham volcanic chain is considered extinct.

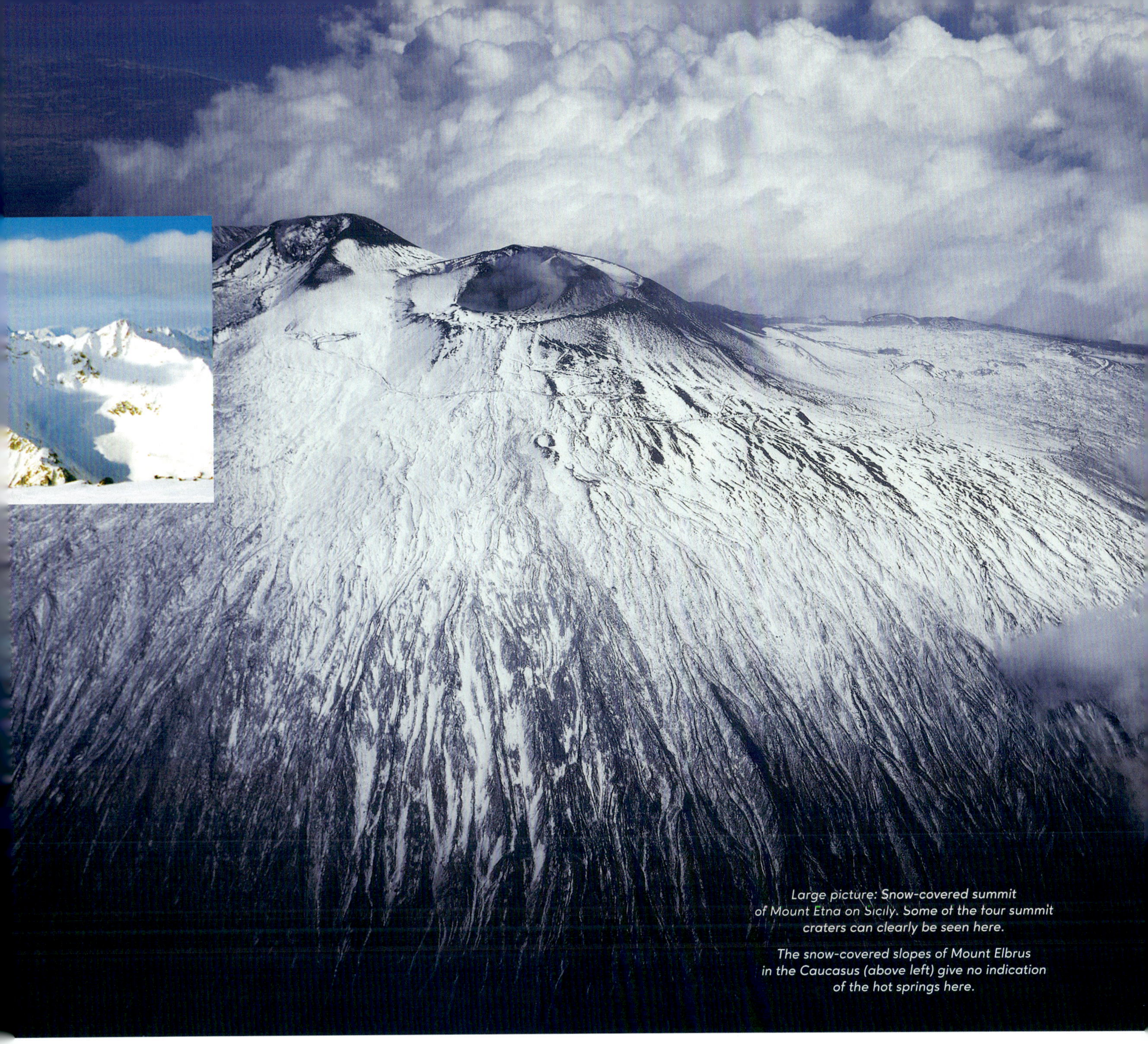

*Large picture: Snow-covered summit of Mount Etna on Sicily. Some of the four summit craters can clearly be seen here.*

*The snow-covered slopes of Mount Elbrus in the Caucasus (above left) give no indication of the hot springs here.*

## DAR-ALAGES

Armenia, 3,329 m (10,922 ft)

The Dar-Alages group, also known as the Daly-Tapa group, rises in southern Armenia, south of Lake Sevan in the western foothills of the Vardenis Ridge. It is a series of six cinder and lava cones dating from the Pleistocene to the Holocene periods. Long ago, the most recent being around 3,000 to 4,000 years ago, eruptions of the Vaiyots-Sar lava dome created massive lava flows which dammed the Arpa river and flowed up to 10 km (6 mi) to the west.

## ETNA

Italy, 3,323 m (10,903 ft)

Mount Etna, which developed 600,000 years ago in northeastern Sicily, is the highest active volcano in Europe. Its central cone is 300 m (1,641 ft) high; in addition, there are four summit craters as well as 300 secondary, or parasitic, craters. Etna experiences a major eruption virtually every year. Among the active volcanoes in the world, it takes first place with respect to the number of its eruptions that have historically been documented. The lava flows generally do not issue from the summit crater but from the flanks of its secondary craters.

FOUR ELEMENTS

FIRE

# ETNA

The most active volcano
in Europe

Mount Etna, on the Italian island of Sicily, lies on the western side of the major fault zone called the Messina Fault and, at 3,323 m (10,903 ft), is the highest active volcano on the European continent. It developed approximately 600,000 years ago at the point where a bay reached far inland. The exact height of the volcano at present is not known, as it often changes due to cinder cones and frequent eruptions. The Mount Etna massif occupies a vast area (approximately 1,250 km$^2$ (4,826 sq mi)); its perimeter measures around 250 km (155 mi).

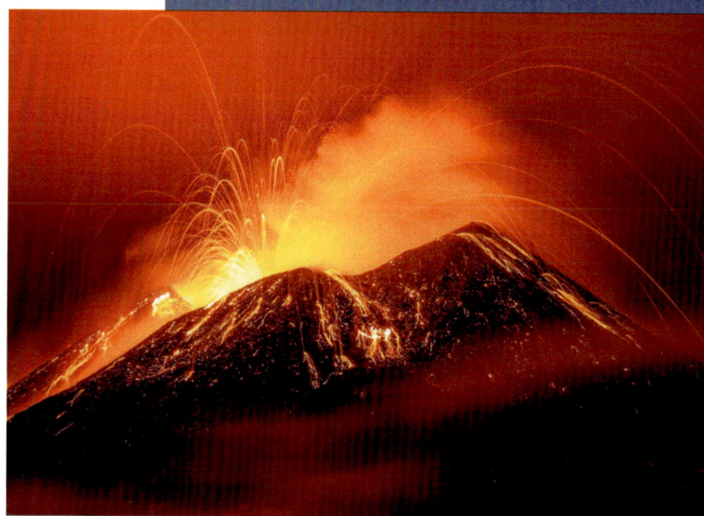

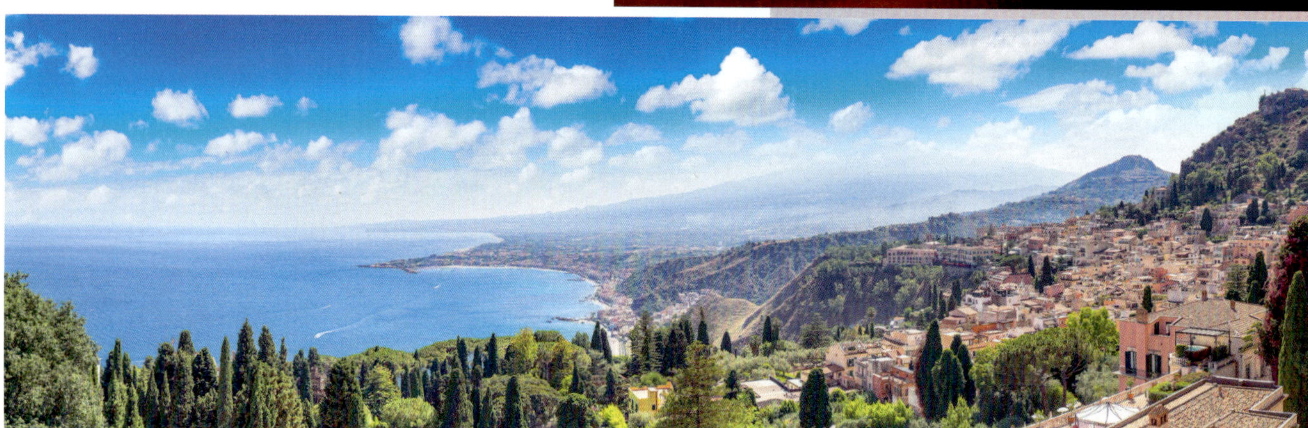

*Large picture: Dramatic pillar of smoke above Mount Etna.*

*Mount Etna erupts like a firework display in 2002 (above, top).*

*Taormina on the Ionian Sea. Mount Etna, 40 km (25 mi) away, is visible in the distance.*

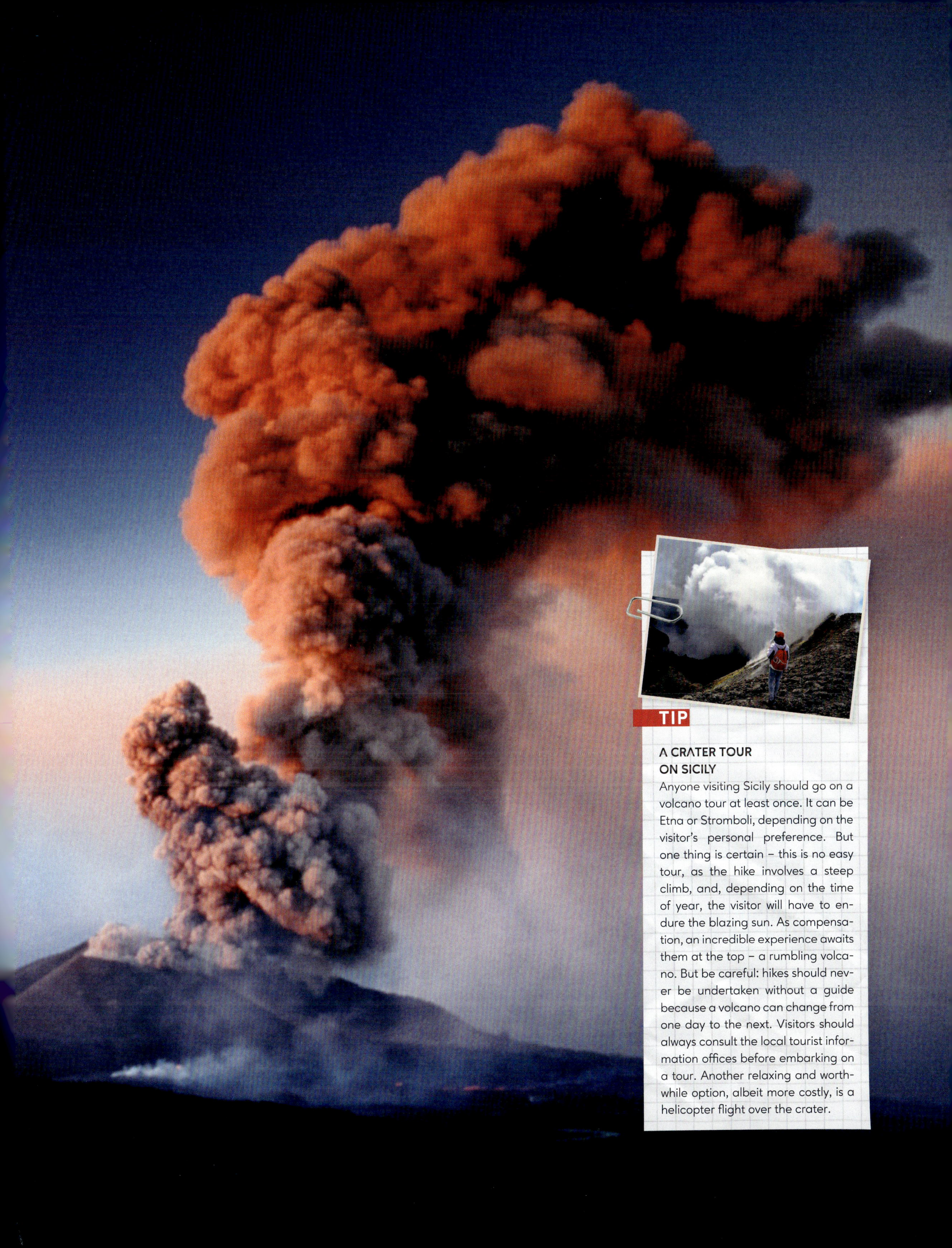

TIP

**A CRATER TOUR ON SICILY**

Anyone visiting Sicily should go on a volcano tour at least once. It can be Etna or Stromboli, depending on the visitor's personal preference. But one thing is certain – this is no easy tour, as the hike involves a steep climb, and, depending on the time of year, the visitor will have to endure the blazing sun. As compensation, an incredible experience awaits them at the top – a rumbling volcano. But be careful: hikes should never be undertaken without a guide because a volcano can change from one day to the next. Visitors should always consult the local tourist information offices before embarking on a tour. Another relaxing and worthwhile option, albeit more costly, is a helicopter flight over the crater.

FOUR ELEMENTS

FIRE

# ICELAND

## The largest volcanic island in the world

With a surface area of 103,000 km² (39,769 sq mi), Iceland is the second largest island in Europe and the largest volcanic island in the world. It lies in the north Atlantic, south of the Arctic circle. The island originated from volcanic activity, and approximately 30 of its 140 or more volcanoes, such as Hekla and Askja, are currently still active. Indications of ongoing volcanic activity are geysers and solfataras (areas where gases escape from the ground), which are particularly prevalent in the western half of the island. The geological features and volcanic activity make Iceland one of the most geologically interesting regions in the world.

### THE HIGHEST VOLCANOES IN ICELAND

1. **HVANNADALSHNÚKUR** 2,110 m (6,923 ft)
2. **BÁRÐARBUNGA** 2,000 m (6,562 ft)
3. **AUSTUR-KVERKFJÖLL** 1,920 m (6,300 ft)
4. **HOFSJÖKULL** 1,782 m (5,847 ft)
5. **ESJUFJÖLL** 1,760 m (5,775 ft)
6. **GRIMSVÖTN** 1,725 m (5,660 ft)
7. **EYJAFJÖLL** 1,666 m (5,466 ft)
8. **HERDUBREID** 1,682 m (5,519 ft)
9. **ASKJA** 1,516 m (4,974 ft)
10. **KATLA** 1,512 m (4,961 ft)
11. **HEKLA** 1,491 m (4,892 ft)

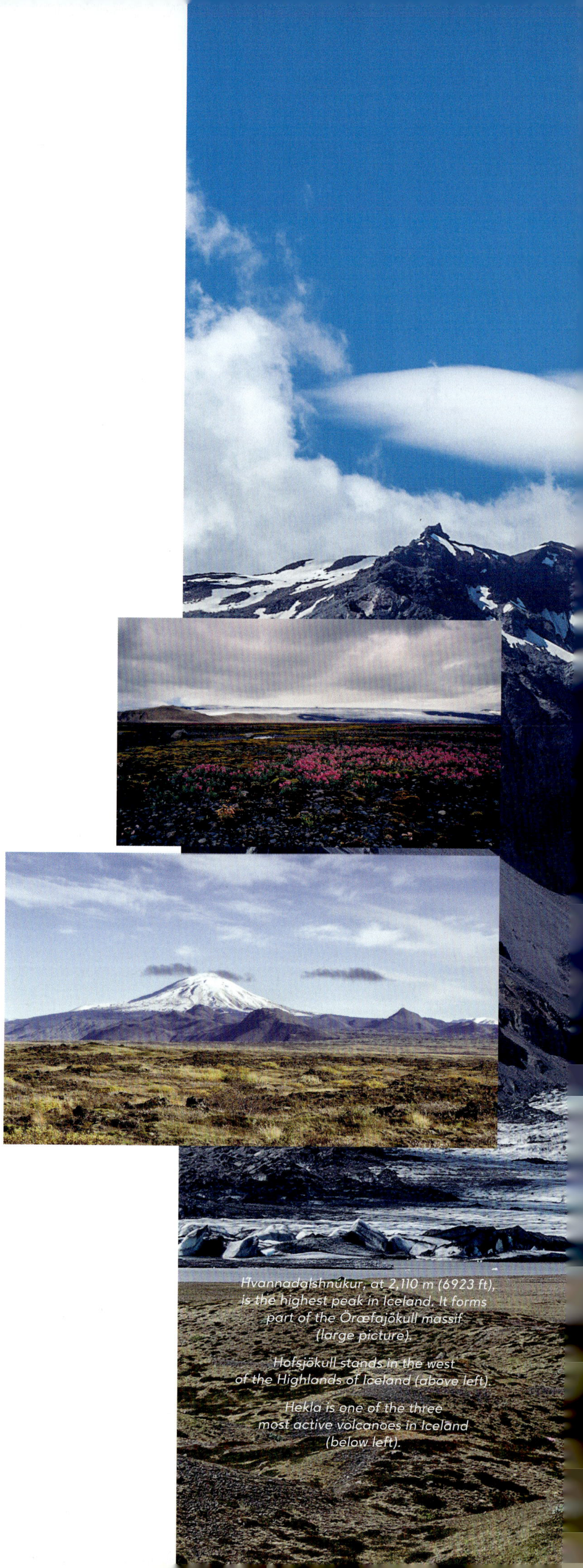

Hvannadalshnúkur, at 2,110 m (6923 ft), is the highest peak in Iceland. It forms part of the Öræfajökull massif (large picture).

Hofsjökull stands in the west of the Highlands of Iceland (above left).

Hekla is one of the three most active volcanoes in Iceland (below left).

## SURTSEY – THE YOUNGEST VOLCANIC ISLAND

Surtsey (right), a volcanic island with an area of 1.4 km² (0.5 sq mi), unexpectedly emerged on 14 November 1963 approximately 30 km (18.6 mi) off the Icelandic coast following an underwater volcanic eruption. The pillars of smoke from the eruption reached heights of 3 km (1.9 mi). The island is gradually being eroded by the wind, sea and rain. It has now already lost half of its original mass. The island is uninhabited and access is only permitted for scientific purposes.

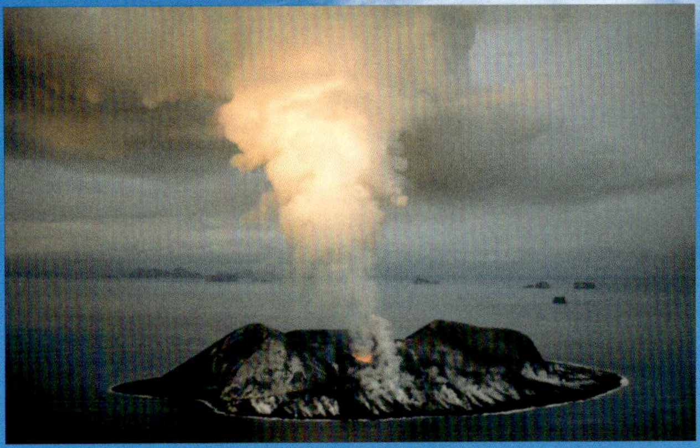

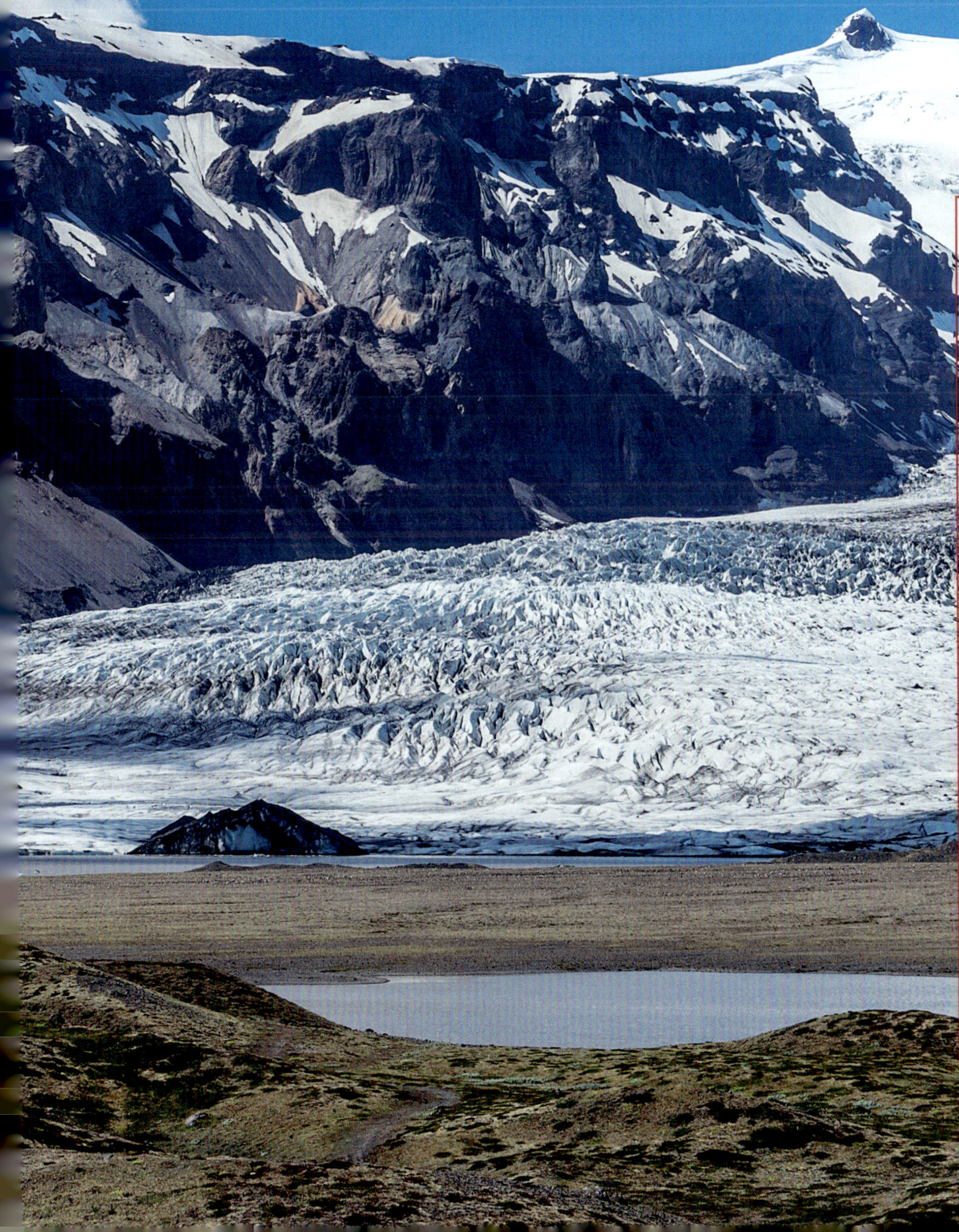

DID YOU KNOW?

### Heated pavement in winter

Underfloor heating is not considered a luxury on Iceland but part of everyday life. The turbulence underground, due to its massive volcanic energy, ensures that there is no shortage of hot water for the inhabitants, even during the harsh northern winters. This geothermal energy is one of the major assets of the island, and it is so abundant that it is used not just for heating homes and for producing electricity, but also in heating pipes that run under the pavement. The hot water from the thermal springs thus enables Reykjavik to have heated pavements, which means there is no need to clear the snow.

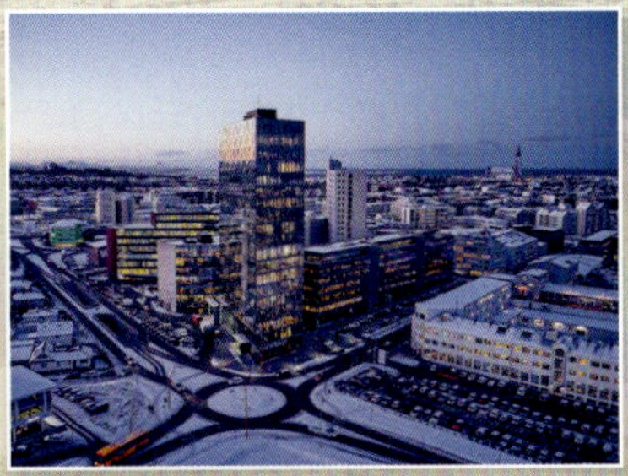

FOUR ELEMENTS

FIRE

# ASIA

Much of the landscape in Asia as well as the mountains and islands off the Pacific coast have been caused by plate tectonics

The Pacific and Philippine plates push against the Eurasian plate, causing increased volcanic activity. Tectonic instability is a feature of the Indonesian islands with their approximately seventy active volcanoes. And because the Japanese islands are in fact the elevated sections of underwater volcanic chains, they still have some active volcanoes in addition to several extinct ones.

## KA-ER-DAXI (KUNLUN)

**China 5,810 m (19,063 ft)**

The Kunlun Volcanic Group lies in the northwestern uplands of Tibet, within the Kunlun Mountains, which stretch for 3,000 km (1,864 mi) and reach heights of over 7,700 m (25,293 ft). In the west, Ashi Shan, also known as Ka-er-daxi or Vulkan, is 5,810 m (19,063 ft) high and the youngest and highest volcano in Asia. It consists of roughly sixty well-preserved pyroclastic cones (composed of rubble from different types of rocks expelled by the volcano), which are grouped around Lakes Aqqikkol and Wuluke. The main vent last erupted on 27 May 1951. The large collection of volcanic cones in this region is explained by the tectonic plate activity, as volcanoes only emerge at these 'weak points' in the earth's crust. The Indian sub-continent and the Eurasian plate collided in the region of the Kunlun Shan Range during the late Triassic period around 200 million years ago.

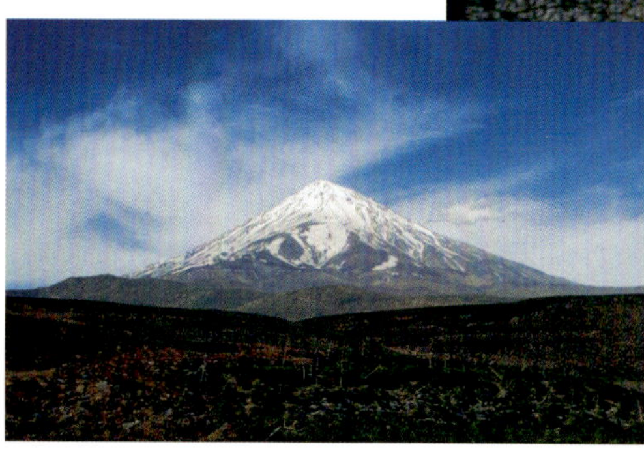

## DAMAVAND

**Iran, 5,670 m (18,603 ft)**

Mount Damavand ('the mountain from which smoke and ash arise' in Persian, small picture, above) is a dormant volcanic cone although there are outlets on its highest craters which regularly emit sulfurous fumes. At 5,670 m (18,603 ft) above sea level, it is the highest peak in the Elbrus mountain range as well as the highest elevation in Iran and the entire Near East. Alongside Mount Kilimanjaro, it is the highest freestanding peak in the world. The elevation difference – 4,700 m (15,421 ft) – from the foot of Mount Damavand to its summit is significantly greater than that of Mount Everest.

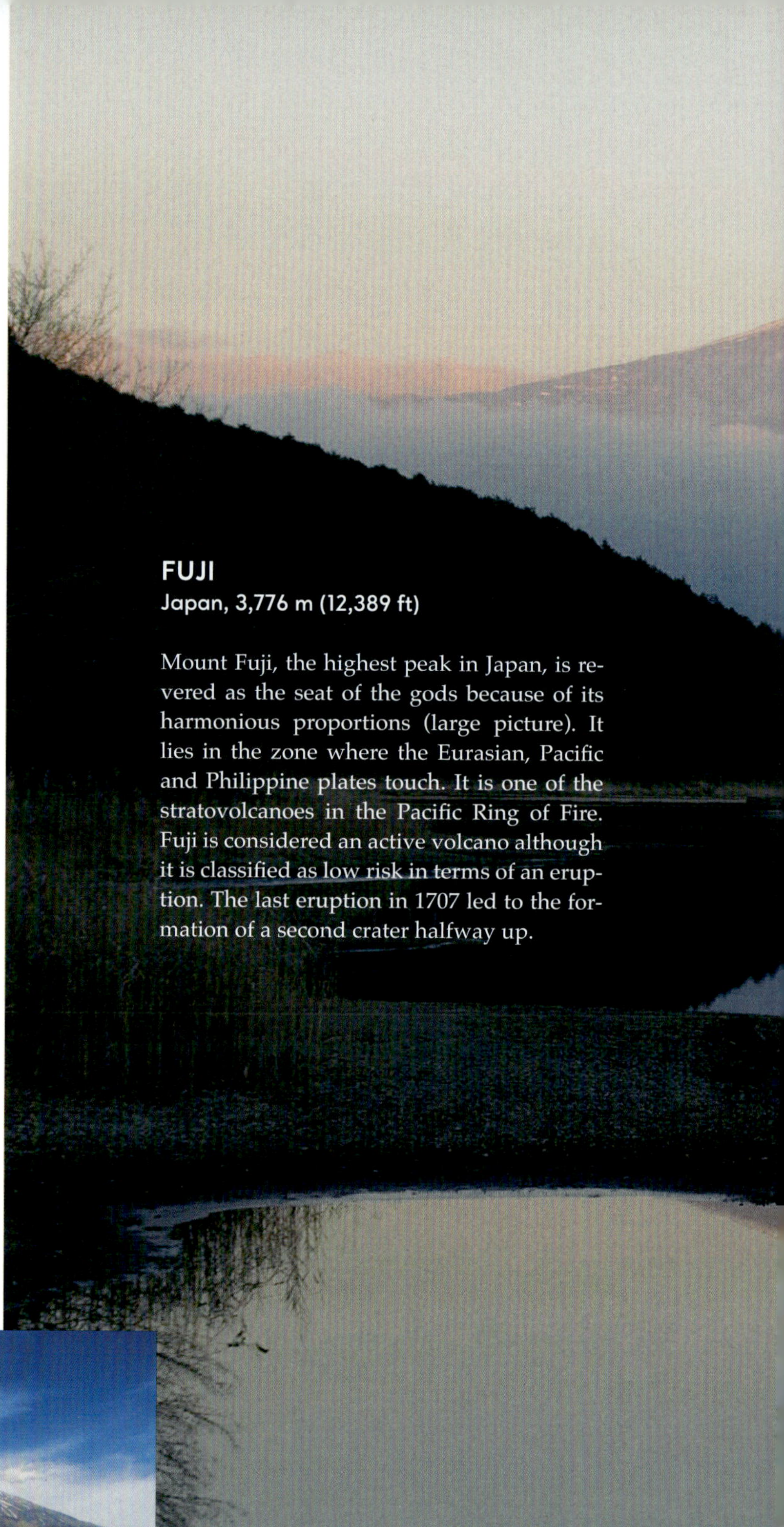

## FUJI

**Japan, 3,776 m (12,389 ft)**

Mount Fuji, the highest peak in Japan, is revered as the seat of the gods because of its harmonious proportions (large picture). It lies in the zone where the Eurasian, Pacific and Philippine plates touch. It is one of the stratovolcanoes in the Pacific Ring of Fire. Fuji is considered an active volcano although it is classified as low risk in terms of an eruption. The last eruption in 1707 led to the formation of a second crater halfway up.

## ARARAT

**Turkey, 5,165 m (16,946 ft)**

The Anatolian Plateau, close to the borders with Armenia and Iran, is dominated by the massive Mount Ararat, the highest elevation in Turkey. Contrary to its Kurdish name, *Çiyayê Agirî* (fire mountain), Ararat is now classed as a dormant volcano. It last erupted in 1840. Mount Ararat has only been accessible since 2001, albeit under certain conditions, as it is considered a sacred mountain where Noah landed his ark.

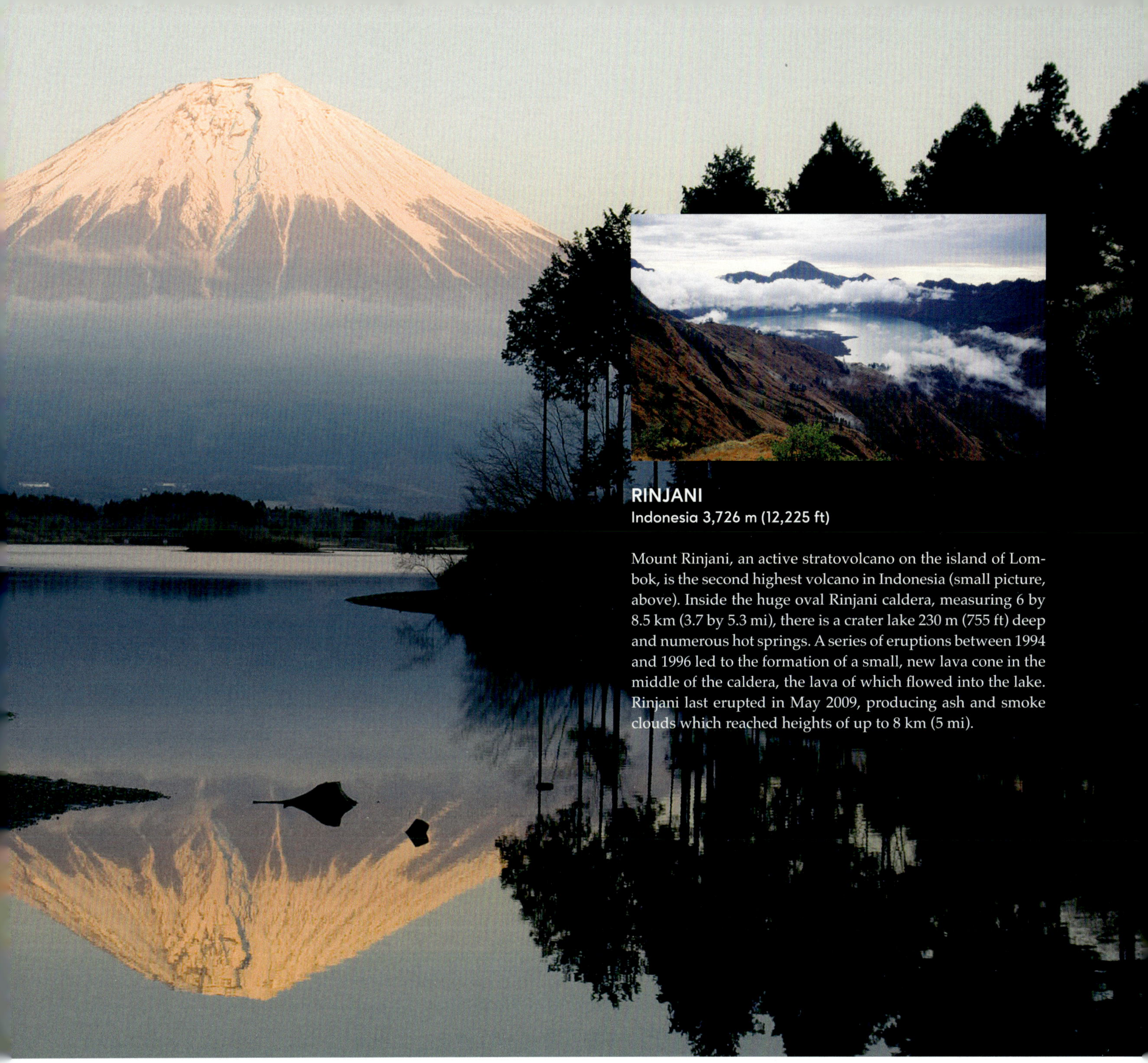

## RINJANI

Indonesia 3,726 m (12,225 ft)

Mount Rinjani, an active stratovolcano on the island of Lombok, is the second highest volcano in Indonesia (small picture, above). Inside the huge oval Rinjani caldera, measuring 6 by 8.5 km (3.7 by 5.3 mi), there is a crater lake 230 m (755 ft) deep and numerous hot springs. A series of eruptions between 1994 and 1996 led to the formation of a small, new lava cone in the middle of the caldera, the lava of which flowed into the lake. Rinjani last erupted in May 2009, producing ash and smoke clouds which reached heights of up to 8 km (5 mi).

## KLYUCHEVSKAYA SOPKA

Russia, 4,750 m (15,585 ft)

In the far east of Russia, the Klyuchevskaya Sopka volcano rises; at 4,750 m (15,585 ft), it is the highest peak in the Kamchatka Peninsula. The volcano consists of a truncated cone with a central crater as well as 70 secondary craters and cones on its slopes. A distinctive feature of this volcano, which has erupted more than 50 times since 1700, is the constant trail of smoke above its summit. At the foot of the volcano, an observation station has been built to monitor volcanic activity on the peninsula.

## KERINCI

Indonesia, 3,805 m (12,484 ft)

The volcanoes on the island of Sumatra form one of the largest volcanic complexes in Indonesia. Of these, Gunung Kerinci is, overall, the highest volcano of Indonesia at 3,805 m (12,484 ft). It towers about 3,000 m (9,843 ft) above the surrounding tea plantations of the Kayu Aro plateau. Kerinci has shown ongoing volcanic activity for many centuries, including both mild activity and sudden periodic eruptions. After a hiatus of several years, increased volcanic activity with eruptions and tremors as well as an increase in ash emissions last occurred in 1998 when the village of Palempok experienced multiple eruptions and ash fall and in 1999 when the entire surrounding area was covered in ash and sustained quakes (low, frequent tremors). The latter occurrence resulted in alterations in the summit's topography. Kerinci last erupted in 2004, and, at times, the pillar of smoke above its crater still reaches heights of up to 1 km (0.6 mi).

FOUR ELEMENTS

FIRE

# KAMCHATKA

## The largest concentration of volcanoes in the world

The Kamchatka Peninsula in Russia, 1,200 km (746 mi) long and up to 450 km (280 mi) wide, lies between the Bering Sea and the Sea of Okhotsk and is one of the most significant volcanic regions in the world. Geologically speaking, the peninsula is still relatively young: It was a mere two million years ago that the Pacific plate pushed under the edge of the Eurasian plate across a broad front. In the process, the peninsula was pushed upwards. The Western Range, with its marked subdivisions, runs lengthwise through Kamchatka and peaks in the Ichinskaya Sopka volcano (3,621 m (11,880 ft)). In the central section of the peninsula, the Eastern Range runs parallel to it, surrounded by 160 volcanoes. The highest of the 29 active volcanoes is Klyuchevskaya Sopka (4,750 m (15,585 ft)). On average, around six of the volcanoes on Kamchatka erupt every year. In 1996, the Kamchatka volcanoes were designated a UNESCO World Heritage Site.

### THE HIGHEST VOLCANO IN KAMCHATKA

1. **KLYUCHEVSKAYA SOPKA** 4,750 m (15,585 ft)
2. **KAMEN** 4,585 m (15,043 ft)
3. **USHKOVSKY** 3,943 m (12,937 ft)
4. **TOLBACHIK** 3,682 m (12,081 ft)
5. **KRONOTSKY** 3,528 m (11,575 ft)
6. **KORYAK** 3,456 m (11,339 ft)
7. **GAMCHEN** 2,576 m (8,452 ft)

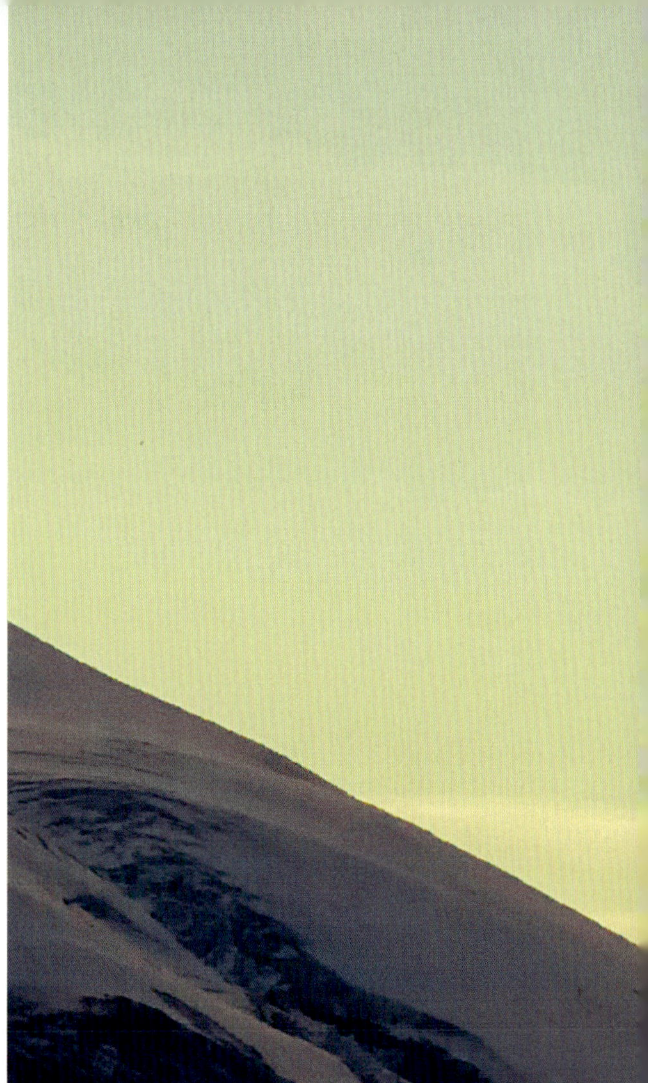

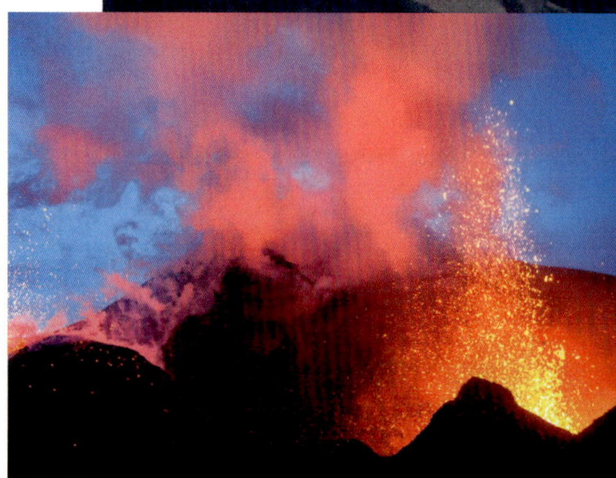

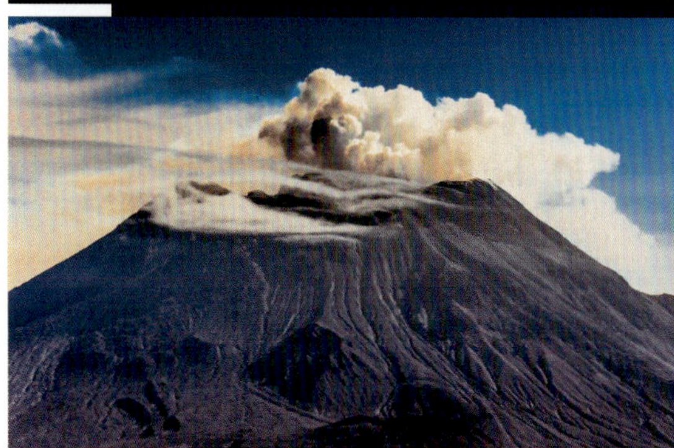

*Chains of several volcanoes are strung together on Kamchatka. Large picture: Klyuchevskaya Sopka, the highest volcano in Asia. Pictures from above: Tolbachik and Bezymianny, which erupted in 1956 and still smoulders today.*

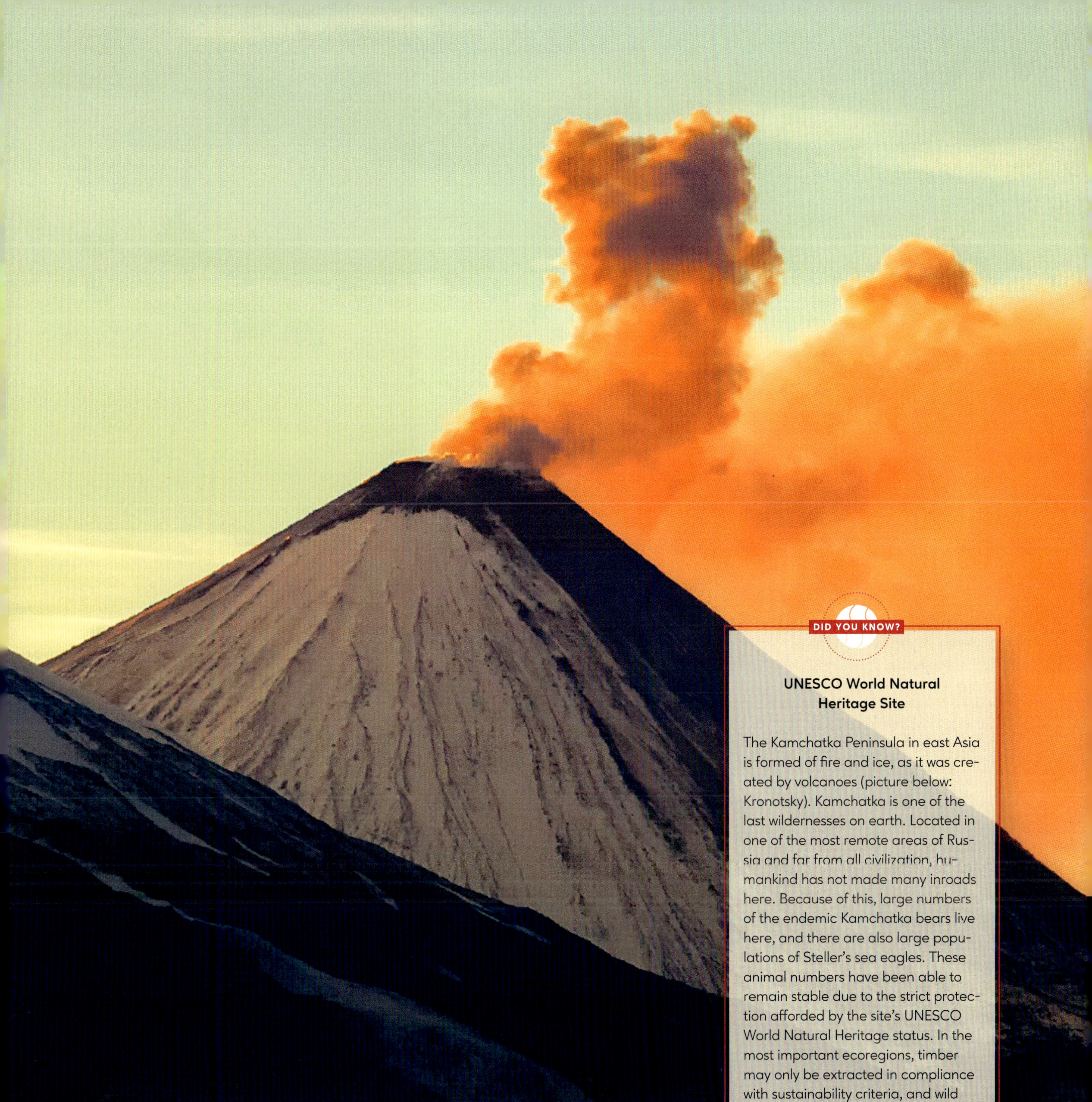

DID YOU KNOW?

**UNESCO World Natural Heritage Site**

The Kamchatka Peninsula in east Asia is formed of fire and ice, as it was created by volcanoes (picture below: Kronotsky). Kamchatka is one of the last wildernesses on earth. Located in one of the most remote areas of Russia and far from all civilization, humankind has not made many inroads here. Because of this, large numbers of the endemic Kamchatka bears live here, and there are also large populations of Steller's sea eagles. These animal numbers have been able to remain stable due to the strict protection afforded by the site's UNESCO World Natural Heritage status. In the most important ecoregions, timber may only be extracted in compliance with sustainability criteria, and wild salmon catches are also regulated.

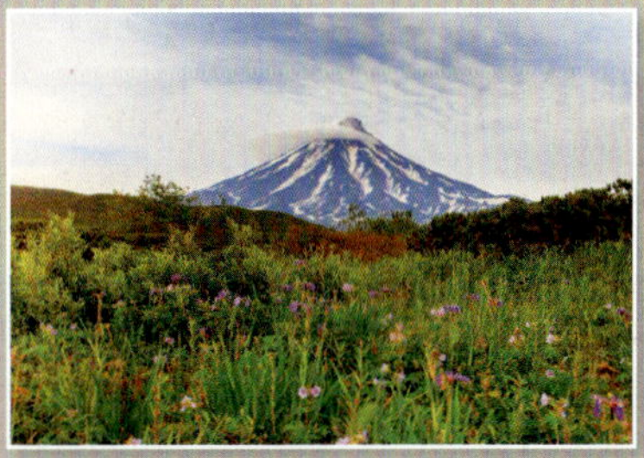

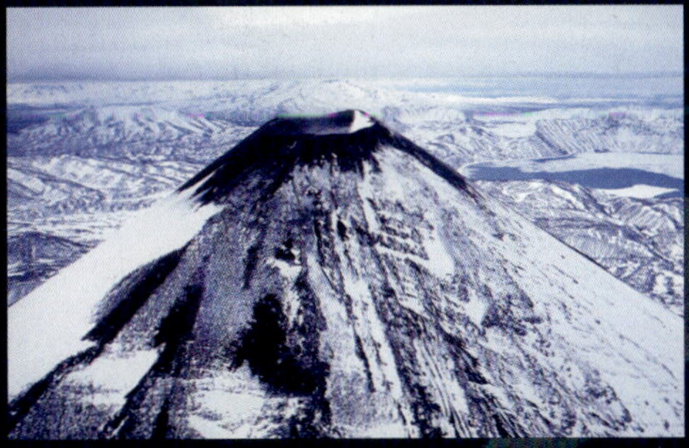

*The snow-covered Karymsky volcano.*

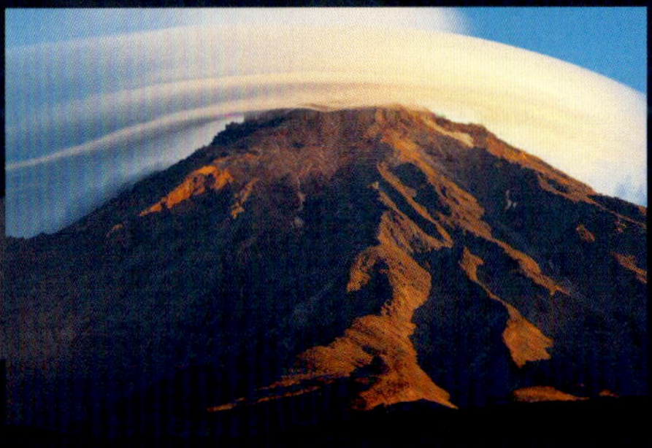

*The Koryak volcano, here without snow.*

FOUR ELEMENTS

FIRE

# AUSTRALIA/OCEANIA

Australia is a land that exhibits uniform natural formations with wide expanses and only slight altitude differences. The highest mountains and volcanoes in Oceania are on the islands, particularly New Zealand and Papua New Guinea. The highest volcano in Oceania rises in Papua New Guinea – Mount Giluwe at 4,368 m (14,331 ft) – followed by Mount Hagen at 3,778 m (12,396 ft). Papua New Guinea has several islands which contain volcanoes that are just as impressive – some are dormant, and some are active. New Zealand's North Island also has many volcanoes.

## NEW GUINEA

**Mount Giluwe, 4,368 m (14,331 ft)**
**Mount Hagen, 3,778 m (12,396 ft)**
**Doma Peaks, 3,568 m (11,707 ft)**
**Crater Mountain, 3,233 m (10,607 ft)**

Mount Giluwe, with its double summit, is the second highest peak in Papua New Guinea (after Mount Wilhelm, 4,509 m (14,794 ft)) and the highest volcano in the country and in the whole of Australia and Oceania. The second highest volcano, Mount Hagen, towers 35 km (22 mi) northeast of here. Both of these stratovolcanoes, whose summits are covered with glaciers, have experienced massive erosion over the course of thousands of years. New Guinea's smaller volcanoes – Doma Peaks and Crater Mountain – are also classed as dormant or extinct.

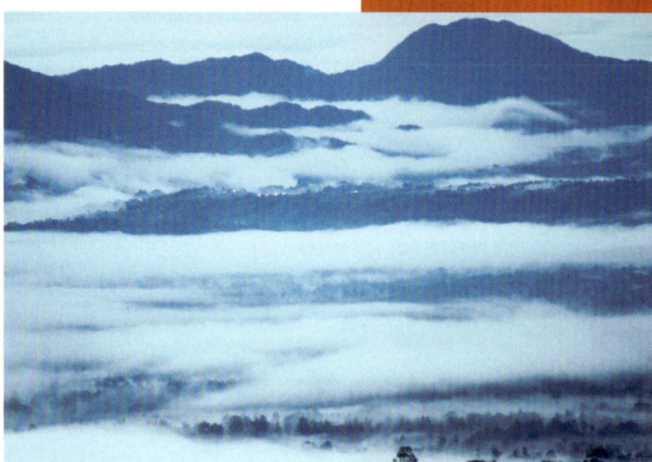

## NEW ZEALAND

**Mount Ruapehu, 2,797 m (9,177 ft)**
**Mount Taranaki, 2,518 m (8,262 ft)**
**Mount Ngauruhoe, 2,291 m (7,517 ft)**
**Mount Tongariro, 1,968 m (6,457 ft)**

Mount Taranaki, or Mount Egmont, rising on New Zealand's North Island, is a dormant volcano which last erupted in 1775. The isolated, snow-covered volcanic cone stands on an almost circular peninsula which measures 60 km (196,850 ft) across and was formed from its pyroclastic sediment. Three further volcano craters dominate the central region of the North Island and form the Tongariro uplands: Mount Ruapehu, the most active in recent times; Mount Ngauruhoe, the highest peak on the North Island, which last erupted in 1975; and Mount Tongariro, which is already extinct. The mountain region is largely covered in glaciers and cleft by numerous steep ravines. Bizarre cliff formations are also a feature of the landscape.

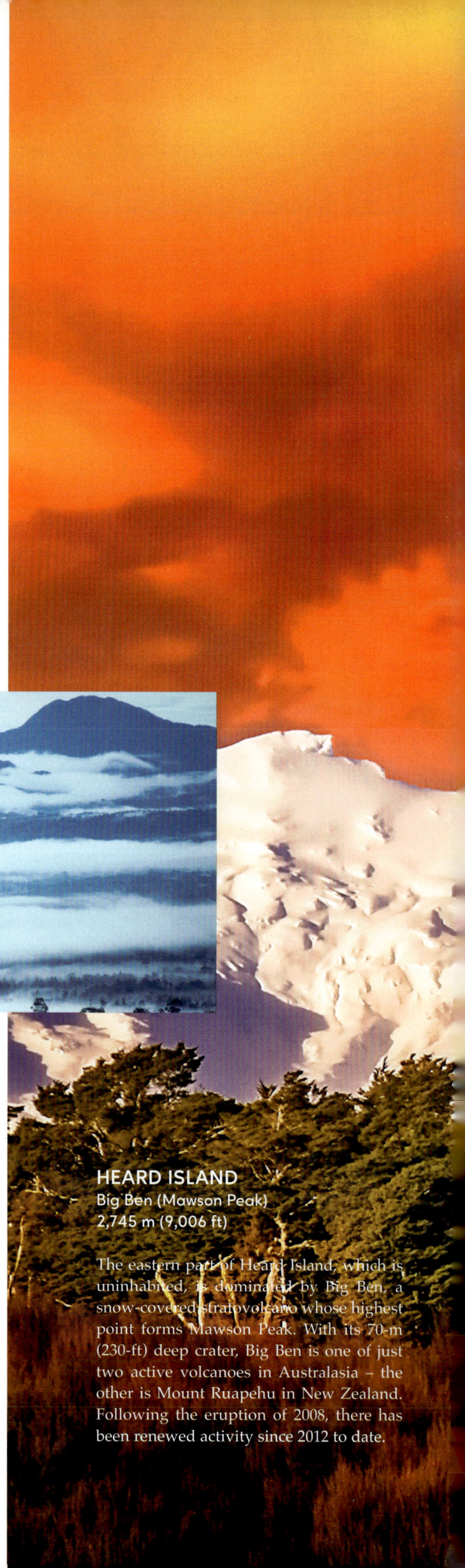

## HEARD ISLAND

**Big Ben (Mawson Peak)**
**2,745 m (9,006 ft)**

The eastern part of Heard Island, which is uninhabited, is dominated by Big Ben, a snow-covered stratovolcano whose highest point forms Mawson Peak. With its 70-m (230-ft) deep crater, Big Ben is one of just two active volcanoes in Australasia – the other is Mount Ruapehu in New Zealand. Following the eruption of 2008, there has been renewed activity since 2012 to date.

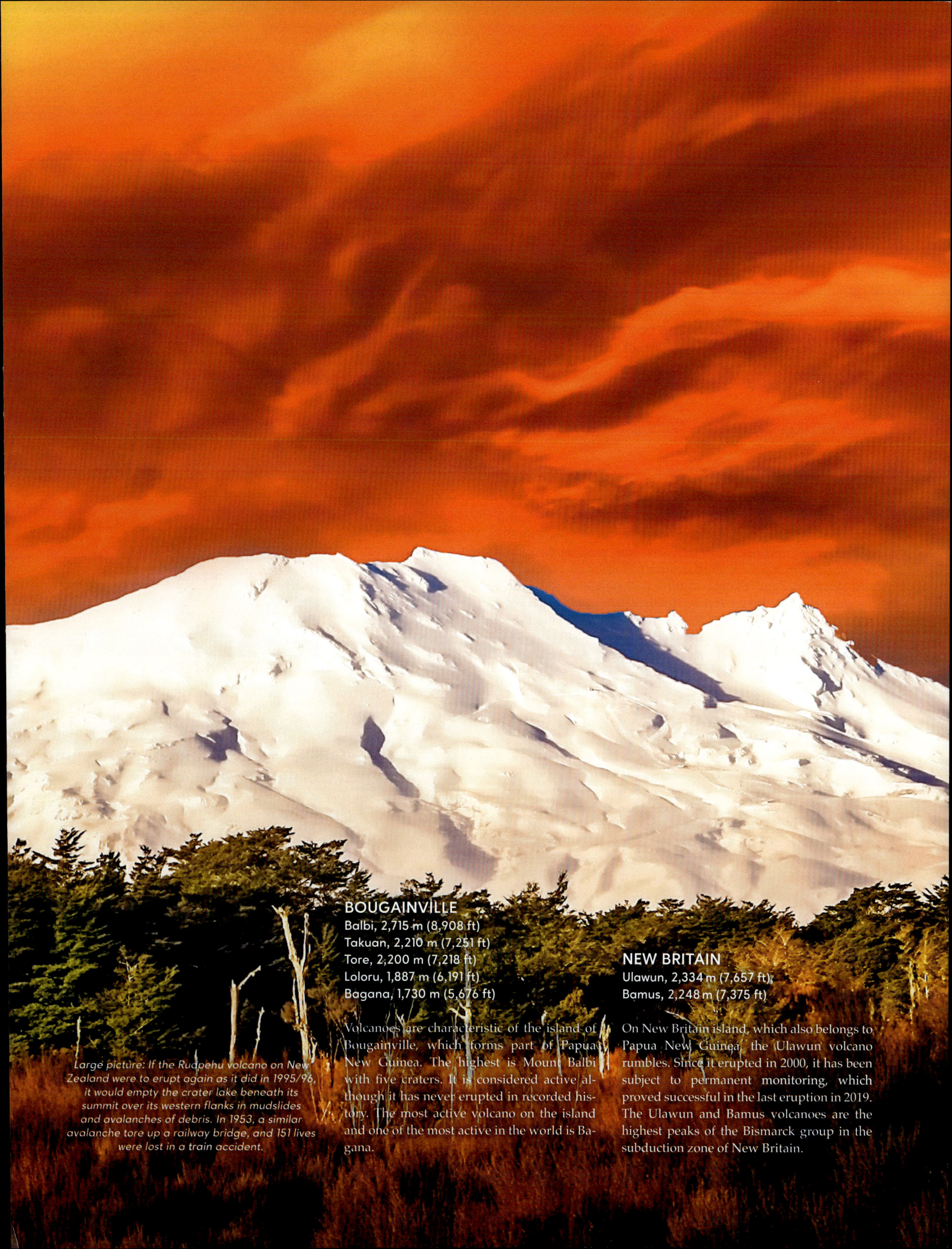

## BOUGAINVILLE

Balbi, 2,715 m (8,908 ft)
Takuan, 2,210 m (7,251 ft)
Tore, 2,200 m (7,218 ft)
Loloru, 1,887 m (6,191 ft)
Bagana, 1,730 m (5,676 ft)

Volcanoes are characteristic of the island of Bougainville, which forms part of Papua New Guinea. The highest is Mount Balbi with five craters. It is considered active although it has never erupted in recorded history. The most active volcano on the island and one of the most active in the world is Bagana.

## NEW BRITAIN

Ulawun, 2,334 m (7,657 ft)
Bamus, 2,248 m (7,375 ft)

On New Britain island, which also belongs to Papua New Guinea, the Ulawun volcano rumbles. Since it erupted in 2000, it has been subject to permanent monitoring, which proved successful in the last eruption in 2019. The Ulawun and Bamus volcanoes are the highest peaks of the Bismarck group in the subduction zone of New Britain.

*Large picture: If the Ruapehu volcano on New Zealand were to erupt again as it did in 1995/96, it would empty the crater lake beneath its summit over its western flanks in mudslides and avalanches of debris. In 1953, a similar avalanche tore up a railway bridge, and 151 lives were lost in a train accident.*

FOUR ELEMENTS

FIRE

# AFRICA

A still tectonically active region of the African continent is the East African Rift, with several volcanoes along its fracture zone

## KIBO (KILIMANJARO)

**Tanzania, 5,895 m (19,341 ft)**

Mount Kibo ('the bright one' in Swahili) in Tanzania is, at 5,895 m (19,341 ft) above sea level, the highest peak and thus also the highest volcano in Africa. It is one of the Seven Summits of the world – the highest peaks of the seven continents. The highest point on Kibo, called Uhuru ('freedom'), rises above the other peaks of the Kilimanjaro massif, which is the highest mountain range in the continent and runs along the edge of the eastern rift of the East African Rift. Kibo, lying approximately 340 km (211 mi) south of the equator, is the highest of the three volcanoes in the Kilimanjaro uplands. It is connected with the Mawensi (5,148 m (16,891 ft)), which lies around 10 km (6.21 mi) further west, via a saddle with an average height of 4,300 m (14,108 ft).

## BATIAN (MOUNT KENYA)

**Kenya, 5,199 m (17,058 ft)**

Batian, a dormant volcano lying around 140 km (87 mi) northeast of the Kenyan capital of Nairobi, stands in the centre of the country, approximately 15 km (9.32 mi) south of the equator. At 5,199 m (17,058 ft) high, it is the second highest peak on the African continent as well as the highest elevation of the Mount Kenya massif. In 1997, UNESCO also added this mountain range to its list of World Natural Heritage sites.

## MAWENSI (KILIMANJARO)

**Tanzania, 5,148 m (16,891 ft)**

After Mount Kibo, the second highest elevation in the Kilimanjaro massif in Tanzania is Mount Mawensi ('the dark one'), which lies approximately 340 km (211 mi) south of the equator. The highly weathered summit area of this extinct volcano is not glaciated, just covered in frost rubble, and has sheer, towering cliffs.

## MOUNT MERU

**Tanzania, 4,562 m (14,968 ft)**

In the north of Tanzania, the Meru volcano, the second highest peak in the country, rises. It lies in Arusha National Park, 65 km (40 mi) southwest of the Kilimanjaro massif. Meru was originally significantly higher, but it lost its peak during a prehistoric eruption. A smaller crater called an ash cone formed inside the large crater as a result of eruptions in about 1880. Meru has been dormant since 1910.

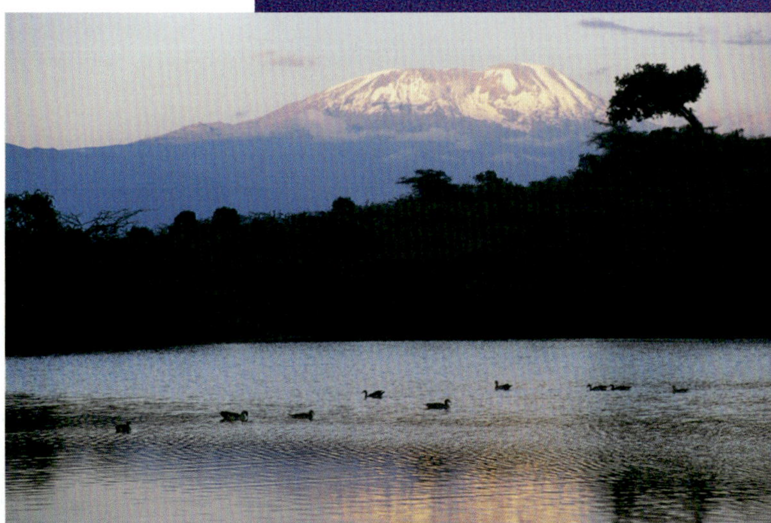

*Mount Meru with the Momela Lakes in the foreground (above). Its caldera is around 3.5 km (2.2 mi) in diameter.*

*Large picture: Mount Kenya was first scaled in 1899. Despite its ranking as the second highest summit in Africa, it has proved harder to conquer than the continent's highest peak.*

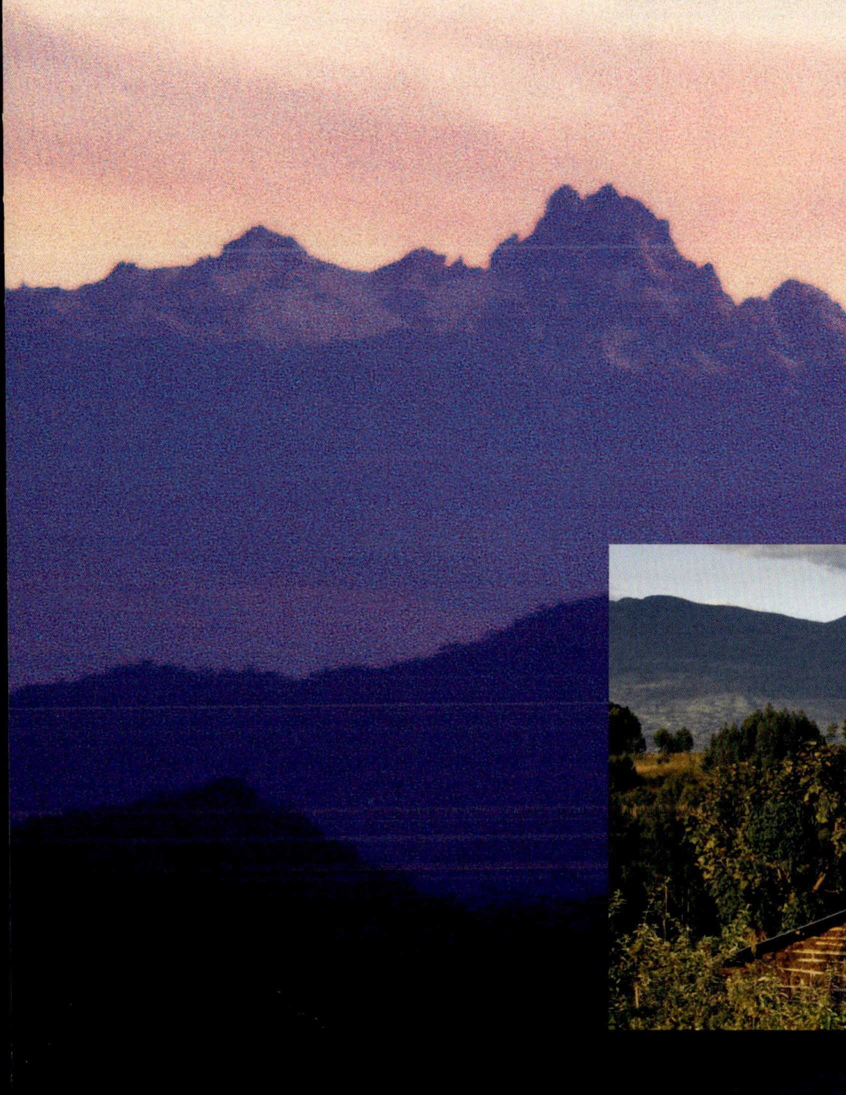

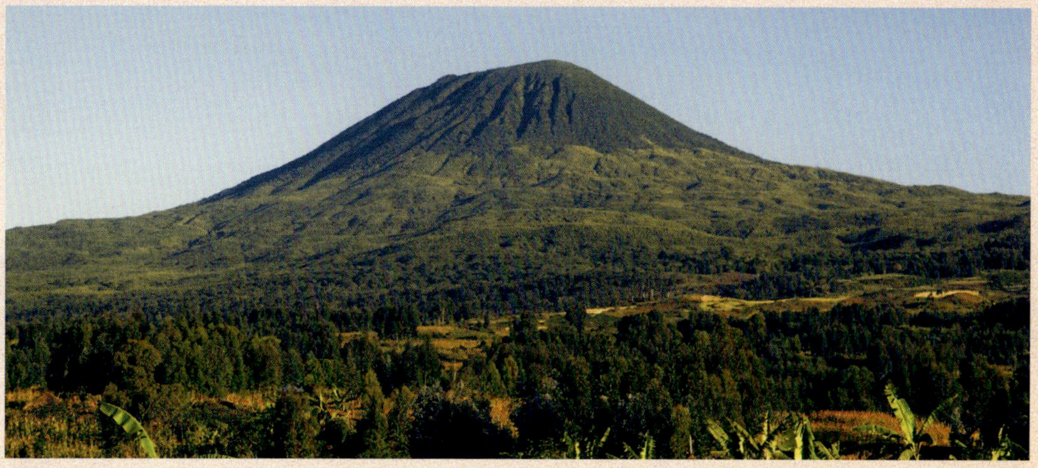

## MUHAVURA

**Rwanda, 4,127 m (13,540 ft)**

Mount Muhabura, also known as Muhavura, is an extinct stratovolcano at the eastern end of the Virunga mountain range on the border between Rwanda and Uganda. A small crater lake, only 40 m (131 ft) wide, lies at its summit. A col, or saddle, joins Muhavura to the neighbouring Mount Gahinga, which is also a dormant volcano. It is not known when Muhavura last erupted, but vast lava fields are evident.

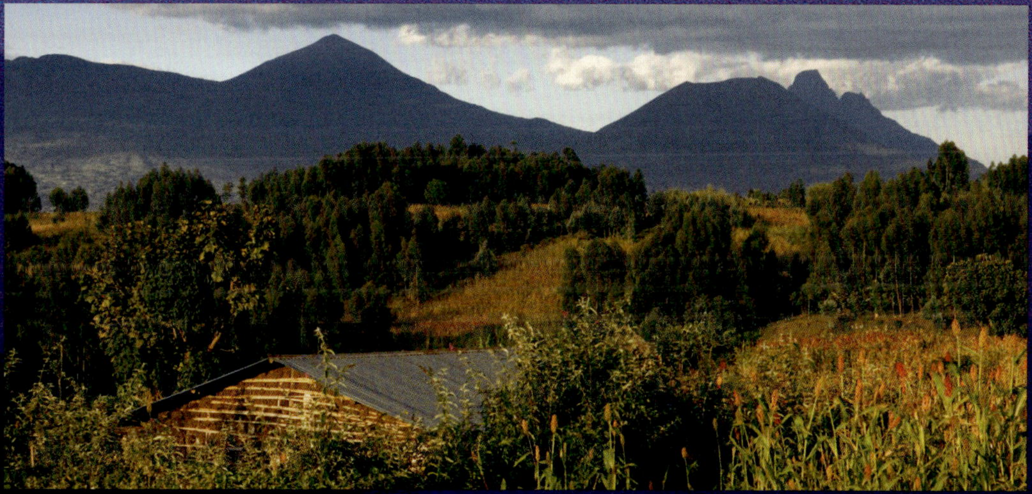

## MOUNT CAMEROON

**Cameroon, 4,095 m (13,435 ft)**

Mount Cameroon, also known as Fako, is the highest peak in western Africa. This active volcano in southwestern Cameroon lies on a 'hot spot' – a place on the earth's mantle where molten rock from the interior of the earth rises up to the surface and partially melts in the process. During a volcanic eruption, the magma emerges as lava. Mount Cameroon and the other active volcanoes in Cameroon lie on the remnants of one of these hot spots in the Atlantic.

## VISOKE

**Dem. Rep. of the Congo/Rwanda, 3,711 m (12,176 ft)**

Visoke, or Bisoke, is a symmetrical volcano and is the most northeasterly of a series of stratovolcanoes at the southern end of the Virunga range on the border between the Democratic Republic of the Congo (DRC) and Rwanda. When it erupted in 1957 two small cones formed on the northern flank of Visoke, approximately 11 km (6.8 mi) from the summit. The Visoke and Karasimbe volcanoes are connected by a saddle (picture above).

## MOUNT SABINYO

**Rwanda/Dem. Rep. of the Congo/Uganda, 3,634 m (11,923 ft)**

Mount Sabinyo sits in the tri-border region between the African states of Rwanda, Uganda and the DCR and is a long-extinct stratovolcano at the eastern end of the Virunga volcanic chain on the Central African Rift. A group of critically endangered mountain gorillas lives in the forests on the slopes of Mount Sabinyo – in the Volcanoes National Park in Rwanda, the Mgahinga Gorilla National Park in Uganda and the Virunga National Park in the DCR.

FOUR ELEMENTS

FIRE

# KILIMANJARO

The most powerful volcano in Africa

The volcanic massif of Kilimanjaro ('the mount of evil spirits' in Swahili) is considered the highest peak on the African continent. It consists of three volcanoes which have grown together: Shira (4,962 m (16,280 ft)), Mawensi (5,148 m (16,891 ft)) and Kibo. The latter, Kibo, reaches a height of 5,895 m (19,341 ft) at is summit, the Uhuru Peak. It is the only mountain with a crater with a diameter measuring 2 km (1.25 mi), in whose depths volcanic activity can be discerned. From its snow cap, sheer hanging glaciers reach down to heights of around 4,300 m (14,108 ft). Mount Kilimanjaro was declared a national park in 1971.

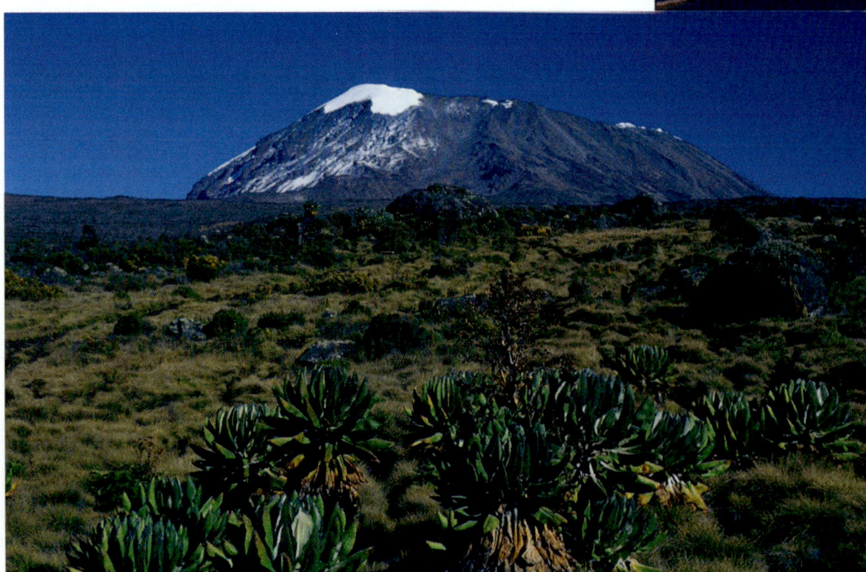

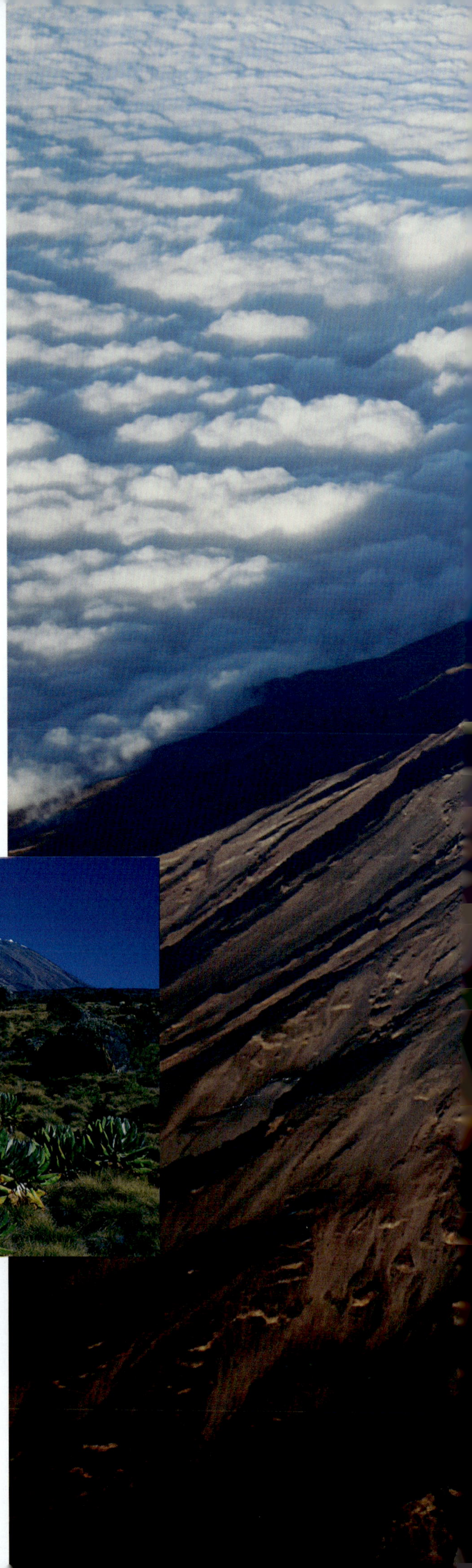

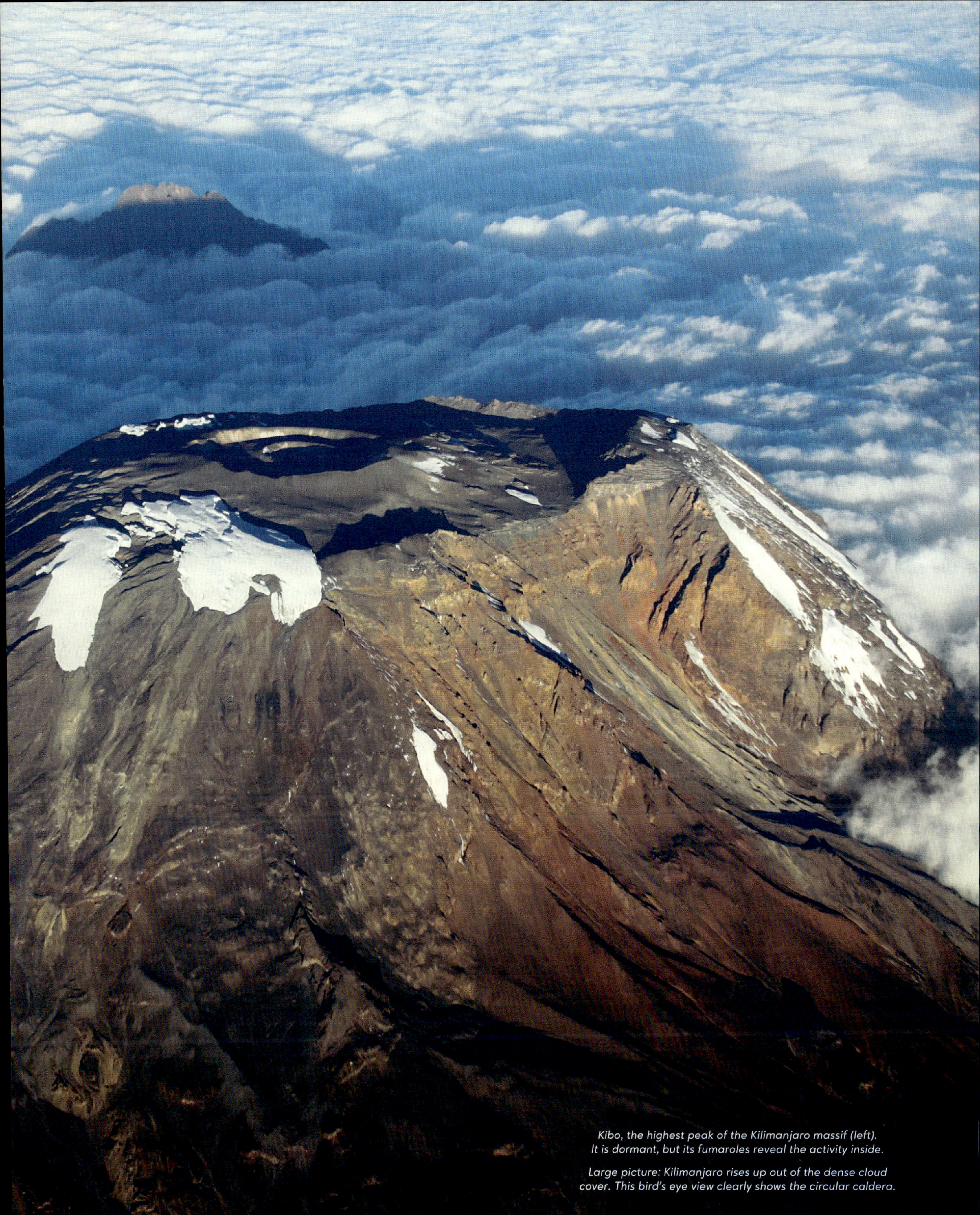

Kibo, the highest peak of the Kilimanjaro massif (left). It is dormant, but its fumaroles reveal the activity inside.

Large picture: Kilimanjaro rises up out of the dense cloud cover. This bird's eye view clearly shows the circular caldera.

# NORTH AND CENTRAL AMERICA

Western North America, which is part of the Pacific Ring of Fire, is repeatedly shaken by natural forces from the earth's interior. The highest peaks in Mexico lie in the Cordillera Volcánica.

*Like the entire St Elias mountain range, Mount Bona, a stratovolcano, is heavily glaciated – the ice fields run along the valleys for kilometres (below).*

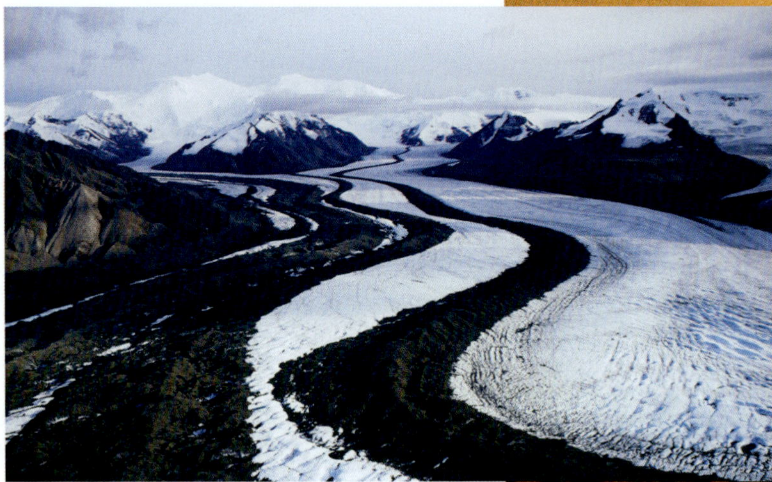

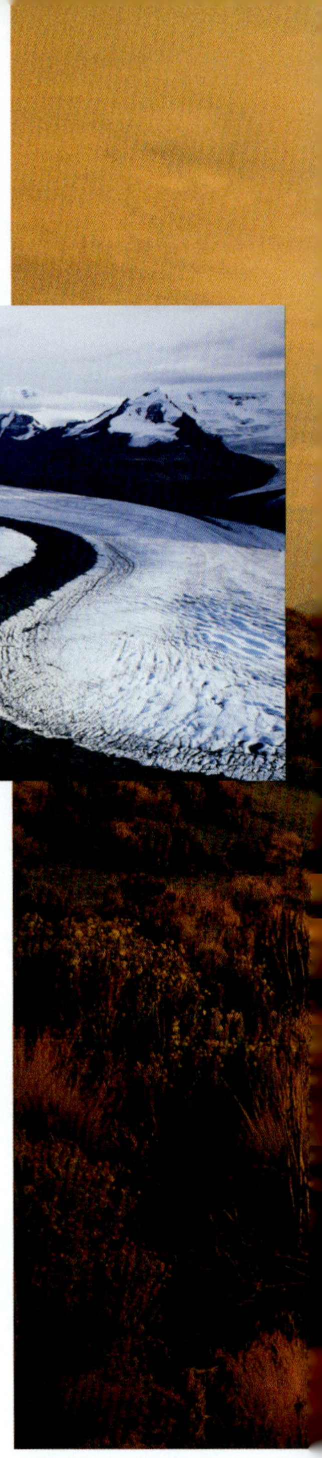

## MOUNT BONA

**USA/Alaska, 5,005 m (16,421 ft)**

The stratovolcano Mount Bona forms part of the St Elias mountain range in eastern Alaska. This mountain chain, the highest section of the North American Coast Mountains, stretches for more than 400 km (249 mi) from southeastern Alaska across the Canadian territory of Yukon as far as the extreme northwest of British Colombia. It forms the water divide between the Yukon River and the Pacific Ocean. Mount Bona, at 5,005 m (16,421 ft), is the fifth highest elevation in the United States and is also the highest volcano in North America after the three highest volcanoes in Mexico (Pico de Orizaba, Popocatépetl and Iztaccíhuatl). Like its neighbour, Mount Churchill, to which it is connected by a saddle, virtually the entire Mount Bona massif is heavily glaciated. The volcano acquired its name in 1897 when Luigi Amadeo von Savoyen discovered it on his first ascent of Mount St Elias and christened it after his ship, the Bona.

## MOUNT BLACKBURN

**USA/Alaska, 4,996 m (16,391 ft)**

At 4,996 m (16,391 ft), Mount Blackburn is the highest elevation in the Wrangell Mountains in Alaska and is the sixth highest peak and the second highest volcano in the USA. The stratovolcano, which is heavily eroded, is almost entirely covered in ice fields. It is named after the US Senator, Joseph Blackburn.

## MOUNT CHURCHILL

**USA/Alaska, 4,766 m (15,670 ft)**

Mount Churchill is best known for being the source of the White River Ash, a huge ash field measuring roughly 540,000 km² (208,495 sq mi) which was produced as a result of the largest volcanic eruption in North America during the last 2,000 years. The deposit in the northern section originated from an eruption about 1,900 years ago; the deposit in the eastern section originated about 1,250 years ago. Strong winds distributed the ash over a massive area. The ash layer is 60 cm (2 ft) deep in places.

## MOUNT RAINER

**USA/Washington, 4,395 m (14,420 ft)**

Mount Rainier is the highest elevation in the Cascade Range. The national park of the same name stretches out around this stratovolcano which lies southeast of Seattle in the US state of Washington. Above 1,800 m (5906 ft), Mount Rainier's slopes are covered with glaciers. The edges of the crater, however, remain free of ice due to the heat produced by the active volcano. The last documented eruption occurred between 1820 and 1854; according to eye-witness accounts, there were several more eruptions in the late 1800s.

## MOUNT WRANGELL

**USA/Alaska, 4,317 m (14,164 ft)**

Mount Wrangell, in the mountain range named after it, lies in southeastern Alaska, close to the border with the Canadian Yukon. Mount Wrangell is the only volcano in the Wrangell Volcanic Field with historically documented eruptions that have generally produced minimal smoke or ash. The peak has been emitting geothermal heat since the 1950s; this led to the most recent eruption in 2002.

## MOUNT ADAMS

**USA/Washington, 3,743 m (12,281 ft)**

Mount Adams is the second highest peak in the Pacific Northwest of the United States of America after Mount Rainier. Volcanic activity began around 940,000 years ago; the volcano last erupted roughly 1,000 years ago.

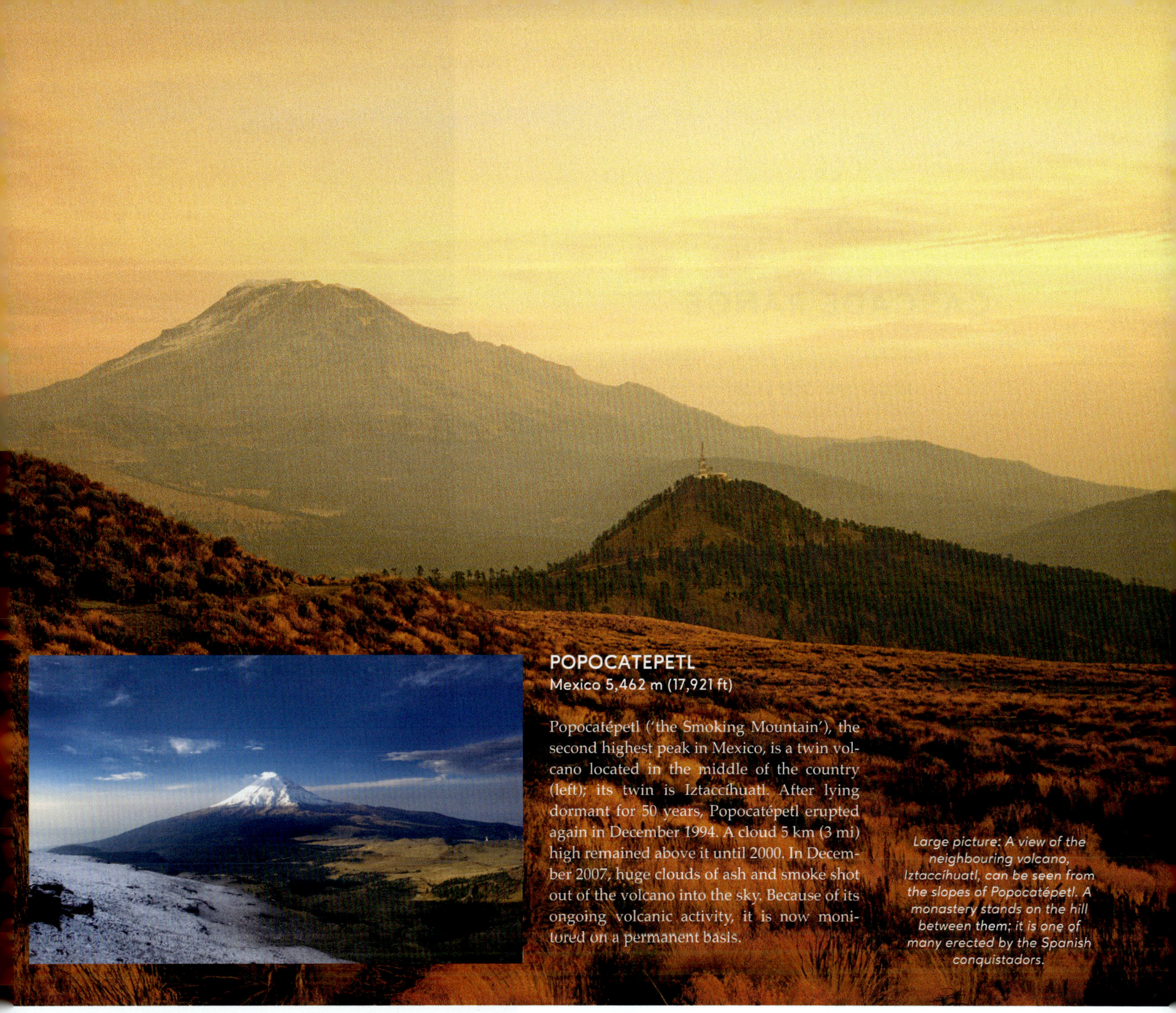

## POPOCATEPETL

**Mexico 5,462 m (17,921 ft)**

Popocatépetl ('the Smoking Mountain'), the second highest peak in Mexico, is a twin volcano located in the middle of the country (left); its twin is Iztaccíhuatl. After lying dormant for 50 years, Popocatépetl erupted again in December 1994. A cloud 5 km (3 mi) high remained above it until 2000. In December 2007, huge clouds of ash and smoke shot out of the volcano into the sky. Because of its ongoing volcanic activity, it is now monitored on a permanent basis.

*Large picture: A view of the neighbouring volcano, Iztaccíhuatl, can be seen from the slopes of Popocatépetl. A monastery stands on the hill between them; it is one of many erected by the Spanish conquistadors.*

## PICO DE ORIZABA

**Mexico 5,636 m (18,492 ft)**

Pico de Orizaba, or Citlaltépetl ('star mountain' in the Náhuatl language), is the highest peak and volcano in Mexico. The stratovolcano is part of the Cordillera Volcánica on the border between the Mexican states of Veracruz and Puebla. Pico de Orizaba is one of only three volcanoes in Mexico whose summits are constantly covered in snow and ice (the others are Popocatépetl and Iztaccíhuatl). Although it lies 110 km (68 mi) inland, it is visible from the Gulf of Mexico. A major eruption of Pico de Orizaba occurred in 1566; the last eruption was in 1846. After a period of dormancy since that time, there has recently been an increase in the number of fumaroles above the crater, which may point to a new phase of activity.

## IZTACCIHUATL

**Mexico 5,286 m (17,343 ft)**

The third highest peak in Mexico is Iztaccíhuatl ('White woman'), a volcano 70 km (43 mi) from Mexico City which can be seen from the city in clear weather. Mexicans associate its three summits with the head, breasts and feet of a sleeping woman, so they call the mountain the 'Mujer dormida'. In contrast to its neighbour, Popocatépetl, the Iztaccíhuatl volcano is extinct. Erosion has caused it to lose its characteristic volcano shape.

## COLIMA

**Mexico, 4,330 m (14,206 ft)**

Colima in fact consists of two peaks: the older inactive Nevado de Colima, and the younger Volcán de Colima, which is believed to be Mexico's youngest volcano. More than 40 eruptions have been recorded since 1576. The most serious eruption was in 1913, but, in 1991 and 2005, eruptions which produced vast amounts of ash also occurred. The UN considers Colima to be one of the most dangerous volcanos in the world.

FOUR ELEMENTS

FIRE

# CASCADE RANGE

The largest volcanic chain
in North America

The Cascade Range extends a total of 1,100 km (683 mi) from British Colombia in Canada in the north, across the US states of Washington and Oregon to California in the south. Its numerous peaks, all recorded as over 3,000 m (9,843 ft) high, are extinct volcanoes, in part covered by glaciers. Even now, volcanic activity can still be seen. The highest summits are Mount Rainier, Mount Shasta and Mount Adams.

## THE HIGHEST VOLCANOES IN THE CASCADE RANGE

1. **MOUNT RAINIER** 4,395 m (14,420 ft)
2. **MOUNT SHASTA** 4,317 m (14,164 ft)
3. **MOUNT ADAMS** 3,743 m (12,281 ft)
4. **MOUNT HOOD** 3,425 m (11,237 ft)
5. **MOUNT BAKER** 1,760 m (5,775 ft)
6. **GLACIER PEAK** 3,213 m (10,541 ft)
7. **MOUNT JEFFERSON** 3,199 m (10,495 ft)
8. **LASSEN PEAK** 3,189 m (10,463 ft)
9. **MOUNT ST HELENS** 2,549 m (8,363 ft)

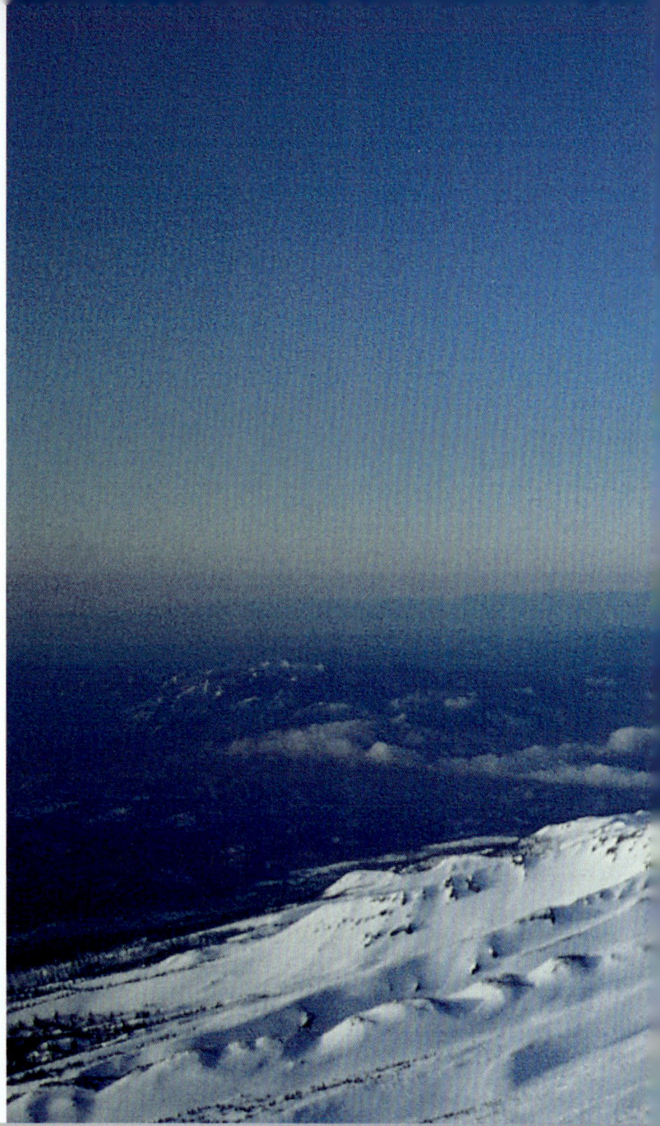

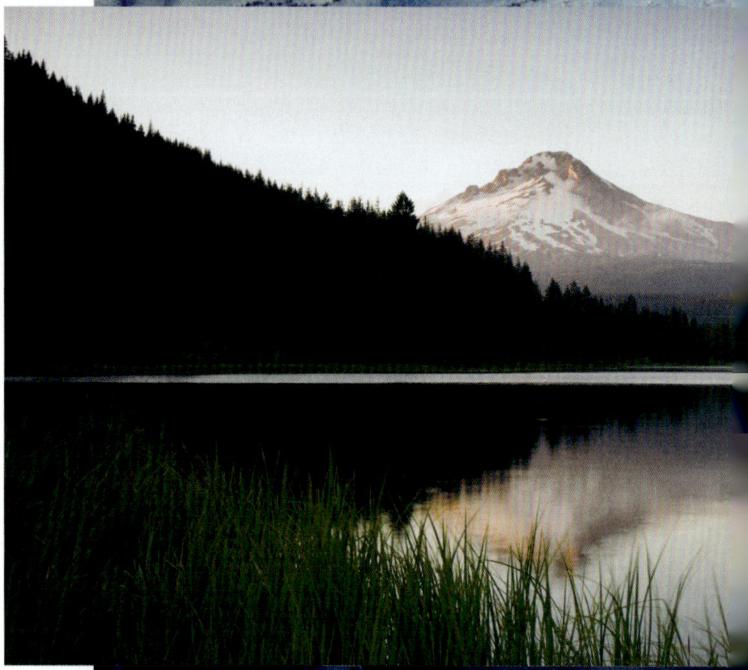

*The snow-covered summit of Mount Shasta rises above the surrounding bank of clouds in northern California (large picture). The active stratovolcano is the second highest peak in the Cascade Range.*

*A potentially active volcano: Mount Hood with Trillium Lake (small picture).*

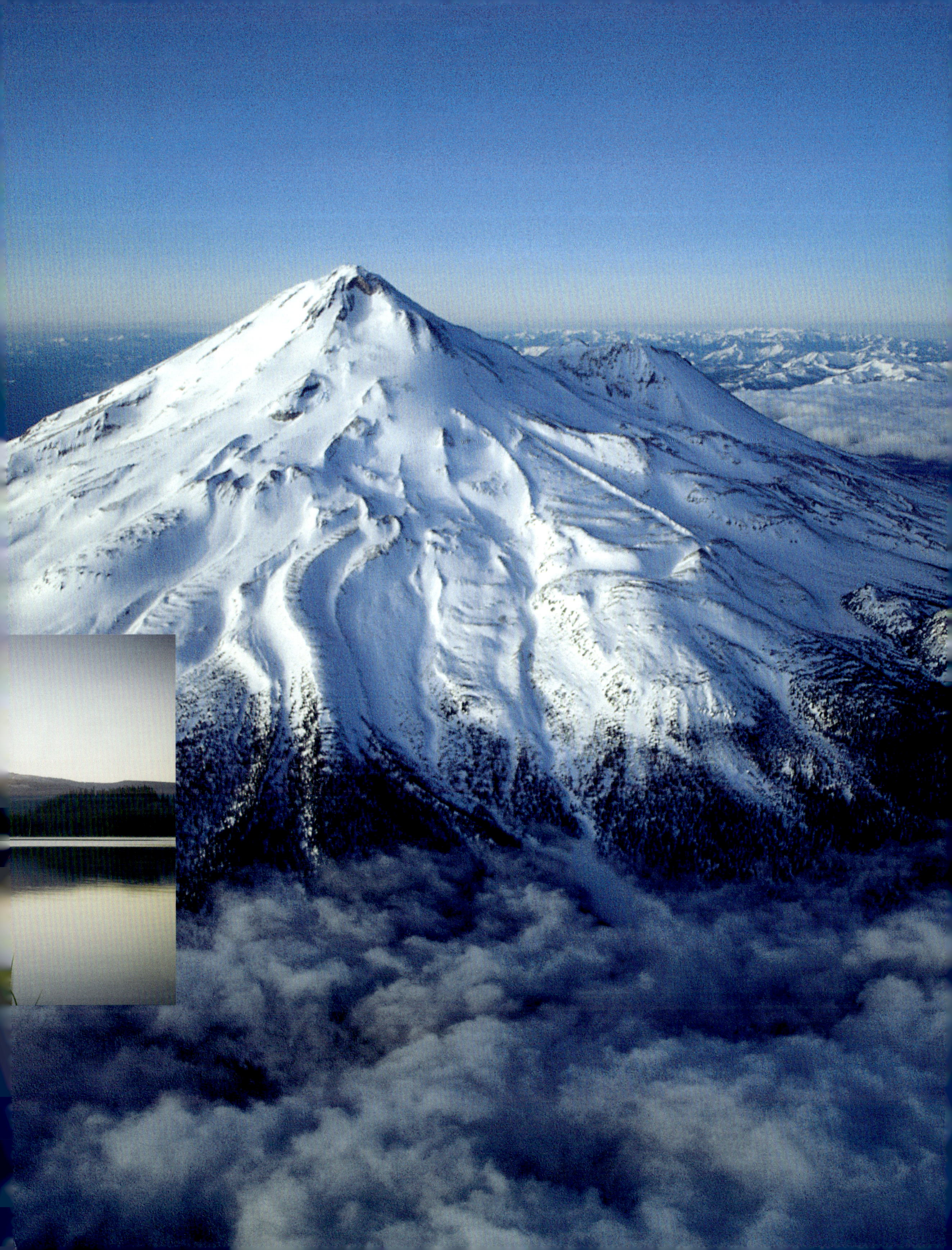

FOUR ELEMENTS

FIRE

# ANDES

The mountain range with the most volcanoes in the world

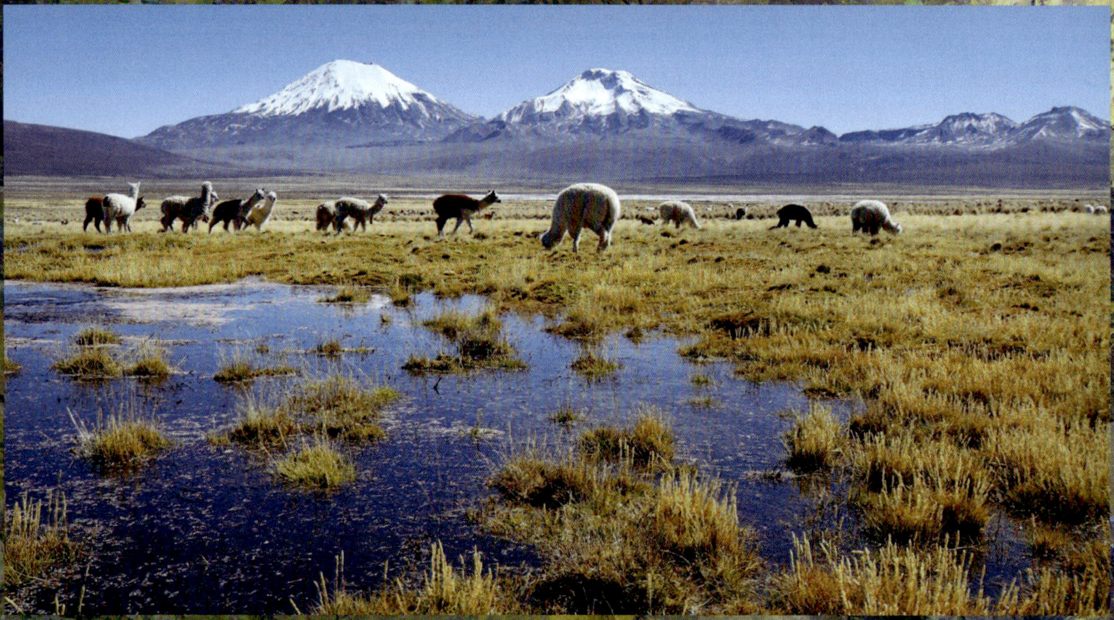

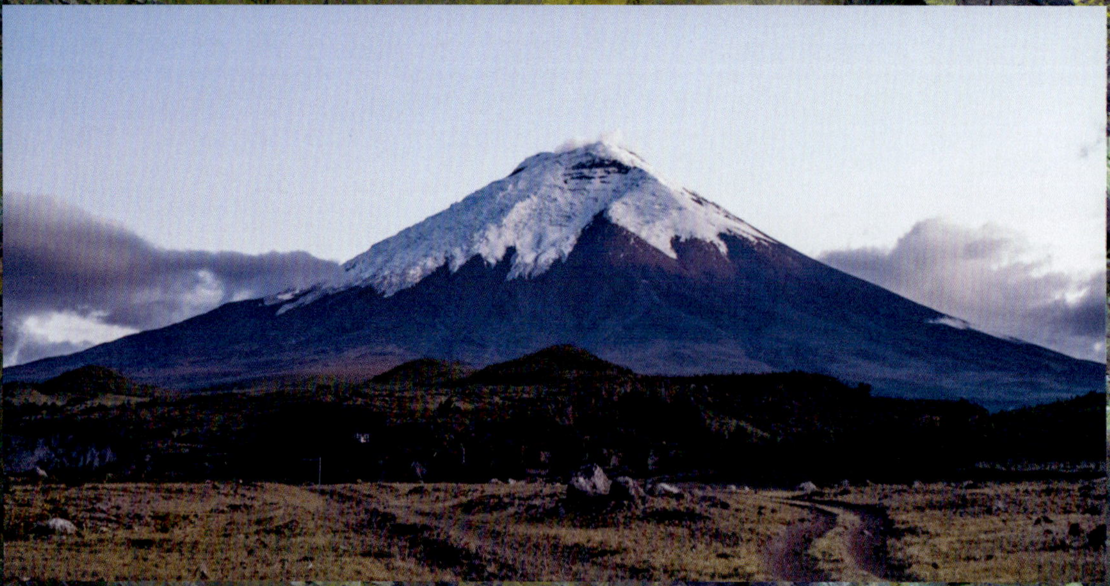

*From above: The snow-covered volcanoes, Pomerape and Parinacota, with llamas in the foothills, and the active volcano, Cotopaxi, in Ecuador, a popular mountaineering destination.*

*Large picture: A fascinating view from space of the volcanic landscape in Chile.*

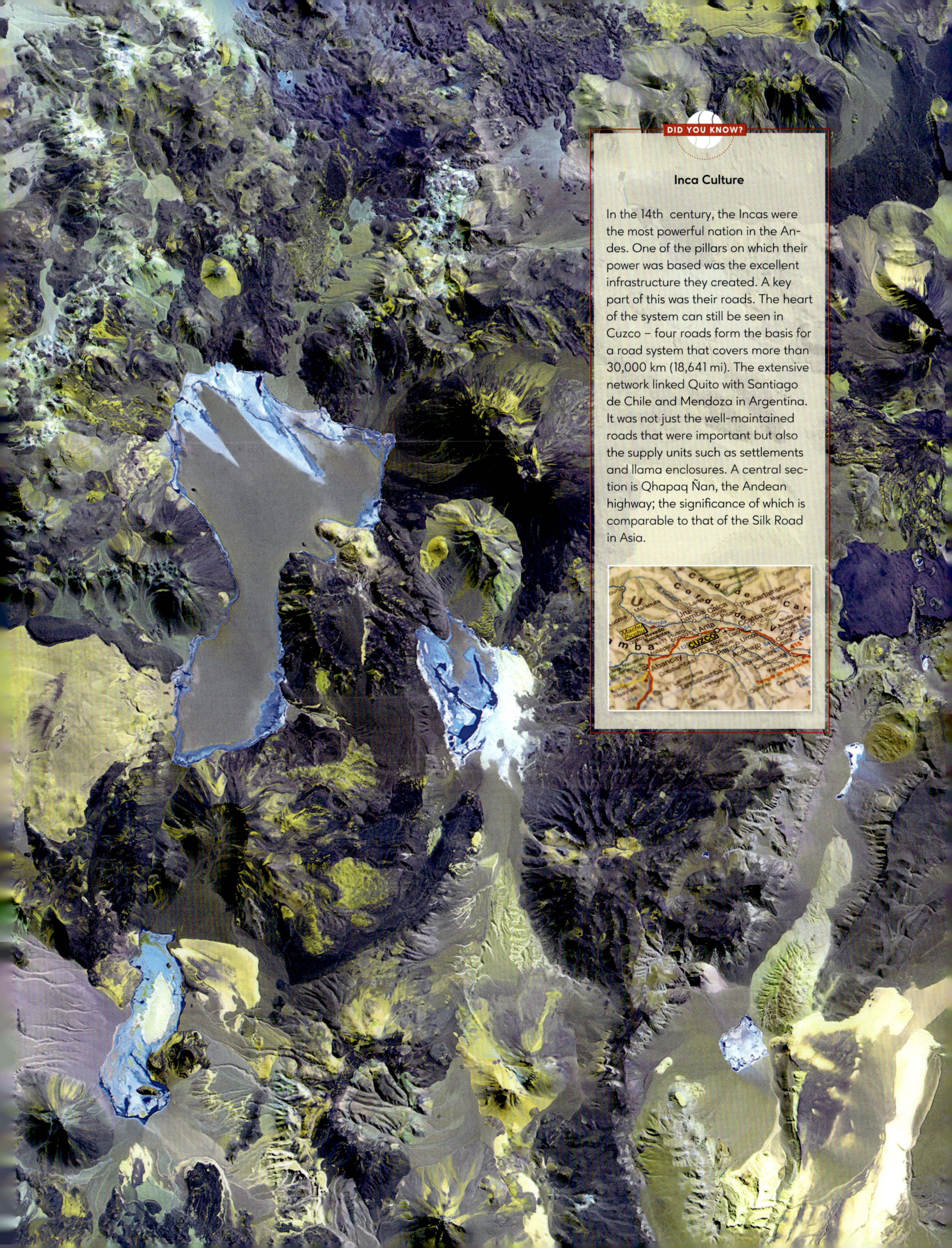

DID YOU KNOW?

### Inca Culture

In the 14th century, the Incas were the most powerful nation in the Andes. One of the pillars on which their power was based was the excellent infrastructure they created. A key part of this was their roads. The heart of the system can still be seen in Cuzco – four roads form the basis for a road system that covers more than 30,000 km (18,641 mi). The extensive network linked Quito with Santiago de Chile and Mendoza in Argentina. It was not just the well-maintained roads that were important but also the supply units such as settlements and llama enclosures. A central section is Qhapaq Ñan, the Andean highway; the significance of which is comparable to that of the Silk Road in Asia.

FOUR ELEMENTS

FIRE

# FIRE RITUALS

The most beautiful fire rituals

## BIIKEBRENNEN

Frisians light their traditional fire to banish winter and wake the spirits of spring. On Sylt, *biike* means 'fire signal' in Frisian, and this is impressive nature rituals is not only celebrated on Germany's most northerly North Sea island. Whether on an island, a hallig or on the mainland, the fire is traditionally lit on 21 February.

## SACRED HEARTS FIRES

This ritual may well be the loveliest time to visit Tyrol. The people light bonfires in a type of formation in the mountains. The fires originate from the time of a historical battle against the French, when the Tyroleans asked for divine help. They had an unexpected victory against the enemy, so the burning fires have remained an important tradition and are lit every year for the Sacred Hearts Festival on the third Saturday after Whitsun.

## WALPURGIS NIGHT

Walpurgis Night is shrouded in legends. The town of Harz, Germany is well-known for its large festivals with witches' dances and bonfires, but the tradition can in fact be traced back to the Celts. Beltane, as this festival is also called, is widely celebrated in Ireland. On the night of 30 April, students in Sweden and Finland take the opportunity to dance and rollick around the bonfires.

## WAKAKUSA YAMAYAKI, JAPAN

The burning mountain is one of the most sensational festivals in Japan. On the fourth Saturday of January, the dried grass on Mount Wakakusa is lit, turning the slopes into a fiery sea. The ritual dates back to the 12th century. The fire is still lit by monks; a procession sets out and visits several shrines to pray, before the monks set fire to the slopes.

## FALLAS, VALENCIA

Huge puppets fashioned from wood and papier maché form the centrepiece of this festival in Valencia in Spain. Similar to carnival time, the figures satirize the current political situation. They are first erected and then they dramatically go up in flames. Only one falla – one figure – remains intact; this is exhibited as a miniature in the town's Fallas Museum. The fallas hark back to a time when the carpenters would burn the old cloths and wood in their workshops once a year. The festivals take place around 15 March.

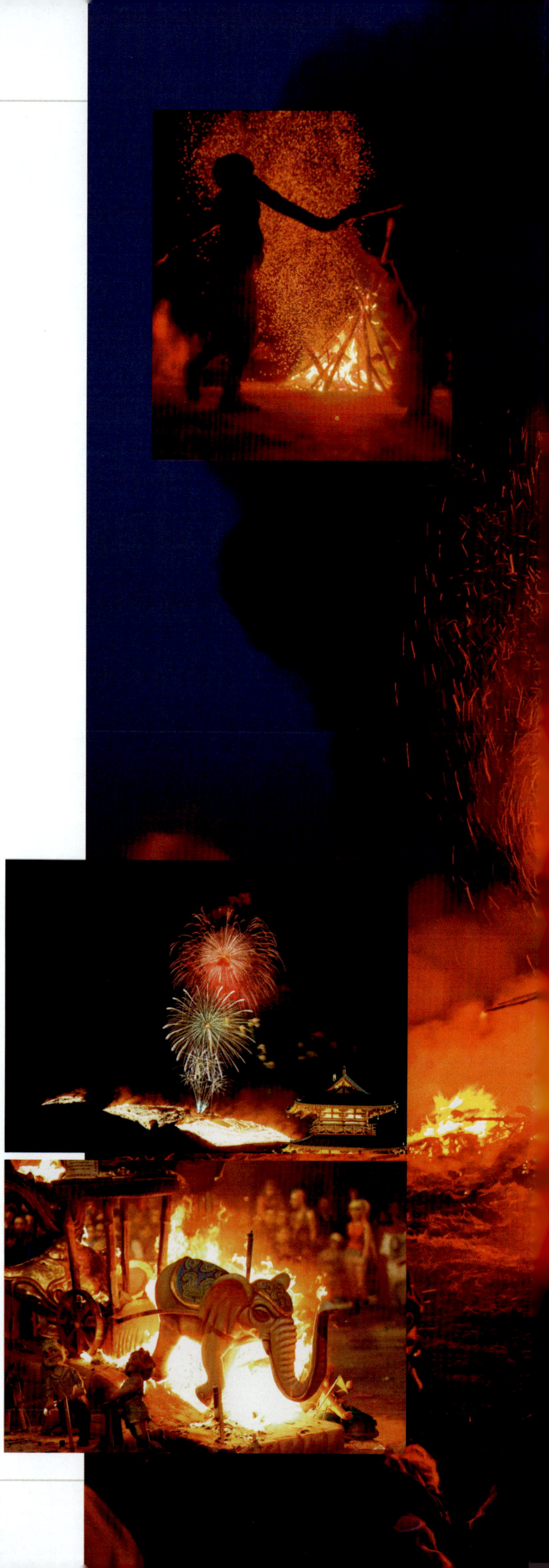

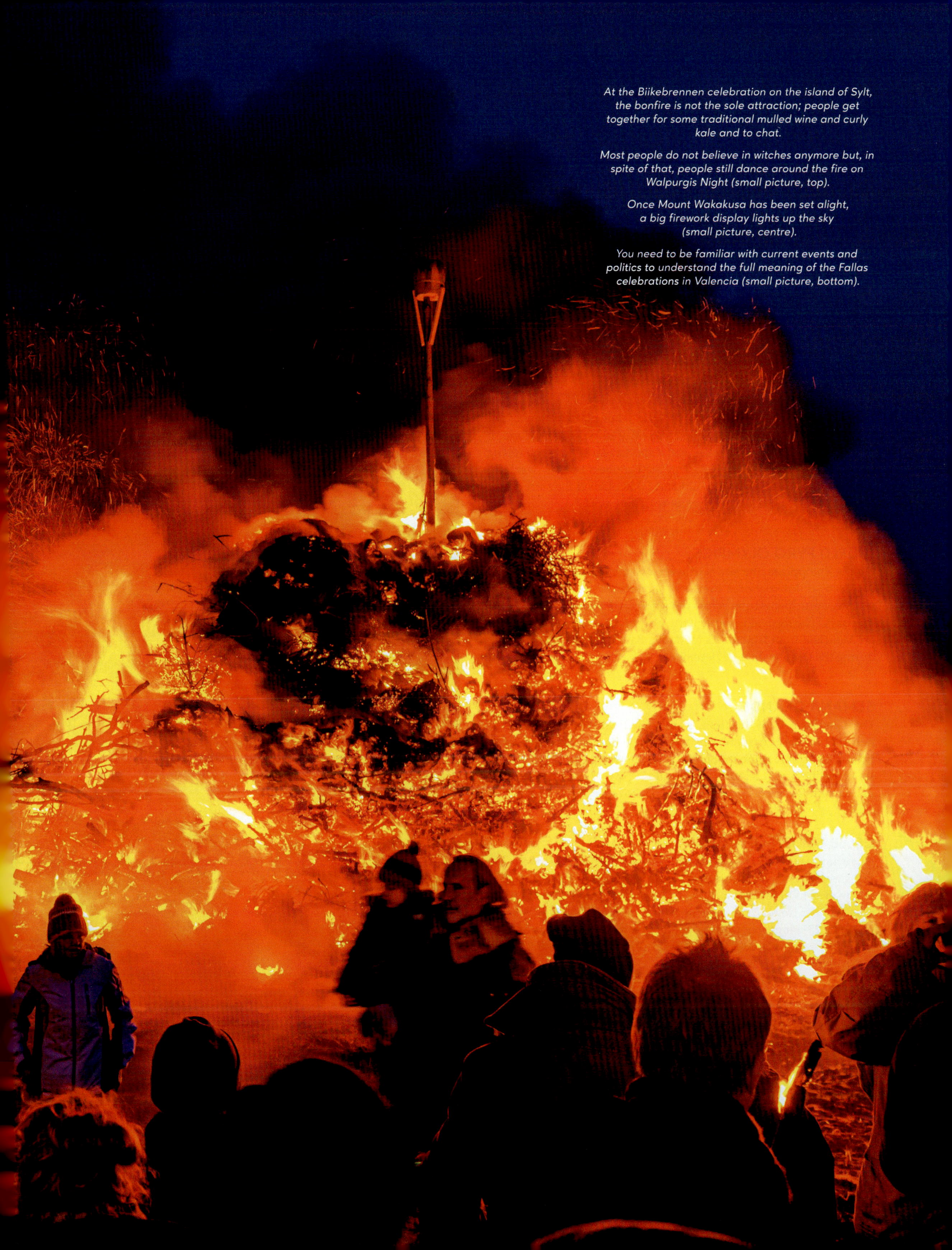

At the Biikebrennen celebration on the island of Sylt, the bonfire is not the sole attraction; people get together for some traditional mulled wine and curly kale and to chat.

Most people do not believe in witches anymore but, in spite of that, people still dance around the fire on Walpurgis Night (small picture, top).

Once Mount Wakakusa has been set alight, a big firework display lights up the sky (small picture, centre).

You need to be familiar with current events and politics to understand the full meaning of the Fallas celebrations in Valencia (small picture, bottom).

FOUR ELEMENTS

FIRE

# FIREWORKS

The five best annual firework displays around the world

## YONSHAKUDAMA FIREWORKS

The largest firework display in the world takes place in Japan. Yonshakudama Fireworks is the largest festival of its type in the world. It is hosted in the town of Katakai, and a spectacular sight is the 420 kg ignition ball, which even appears in the Guinness book of world records. The festival is 400 years old and takes place at the start of September, setting the skies there aglow with fireworks.

## WORLD PYRO OLYMPICS

On the Philippines, the sky is lit with a vast array of colours during this festival. International fireworks manufacturers compete at this festival. Participants from across the world spar against each other for five days. The festival is particularly beautiful because the fireworks are launched from large boats in Manila Bay. The festival takes place during the last few days of December.

## SEMANA GRANDE

The 'Grand Week', as this fiesta is known in English, starts with a cannon shot in San Sebastian. This town in the Basque region in Spain hosts pyrotechnic competitions for a whole week. The fireworks are set off in the bay. It features a marvellous play of colours in the sky, paired with the Spanish zest for life and, above all, lots of music.

## CELEBRATION OF LIGHT

Huge luminous balls of light in pink, showers of golden sparkles – the sky above Vancouver glows with the most beautiful blaze of sparks for two weeks at the end of July - beginning of August. Pyrotechnics artists gather from all over the world for the Celebration of Lights to present a 25-minute show to the public, accompanied by music. A particularly attractive aspect is that it takes place in the English Bay – directly above the water.

## INTERNATIONAL FIREWORK FESTIVAL MALTA

This little Mediterranean island loves fireworks. The sky there is often lit with artificial light, particularly in summer. At the end of April, the top pyrotechnics experts in the world come to Malta to mesmerize the public with their shows. The fireworks are set off from boats in the port of Valetta, and the buildings are often bathed in extraordinary light – a truly magical light show.

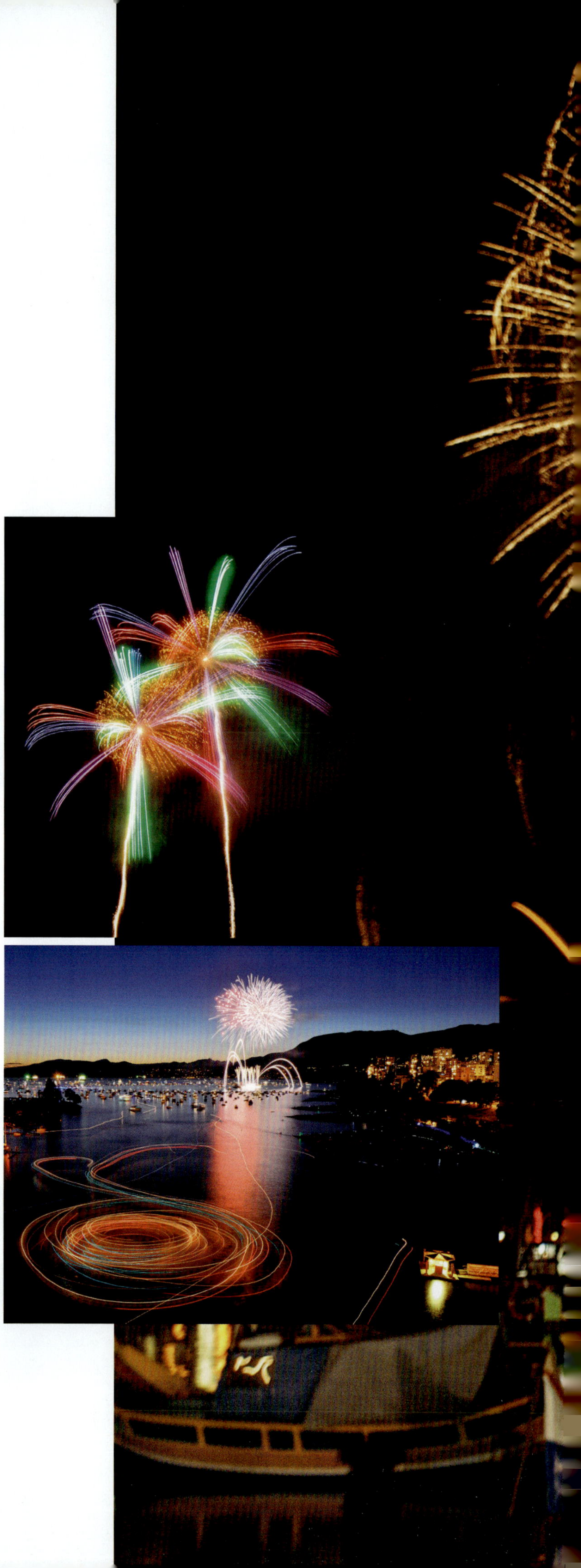

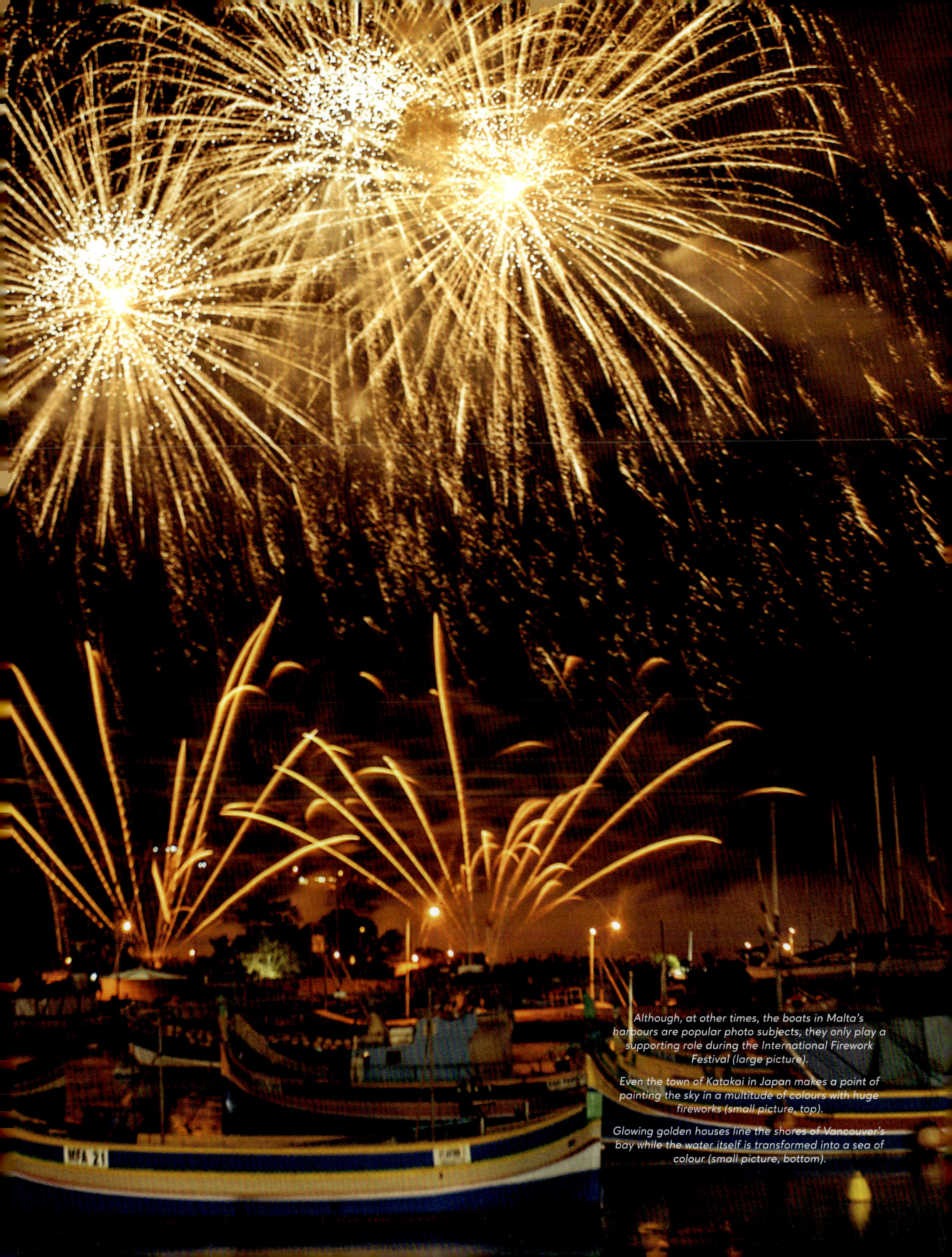

Although, at other times, the boats in Malta's harbours are popular photo subjects, they only play a supporting role during the International Firework Festival (large picture).

Even the town of Katakai in Japan makes a point of painting the sky in a multitude of colours with huge fireworks (small picture, top).

Glowing golden houses line the shores of Vancouver's bay while the water itself is transformed into a sea of colour (small picture, bottom).

FOUR ELEMENTS

FIRE

# SOLAR SYSTEM

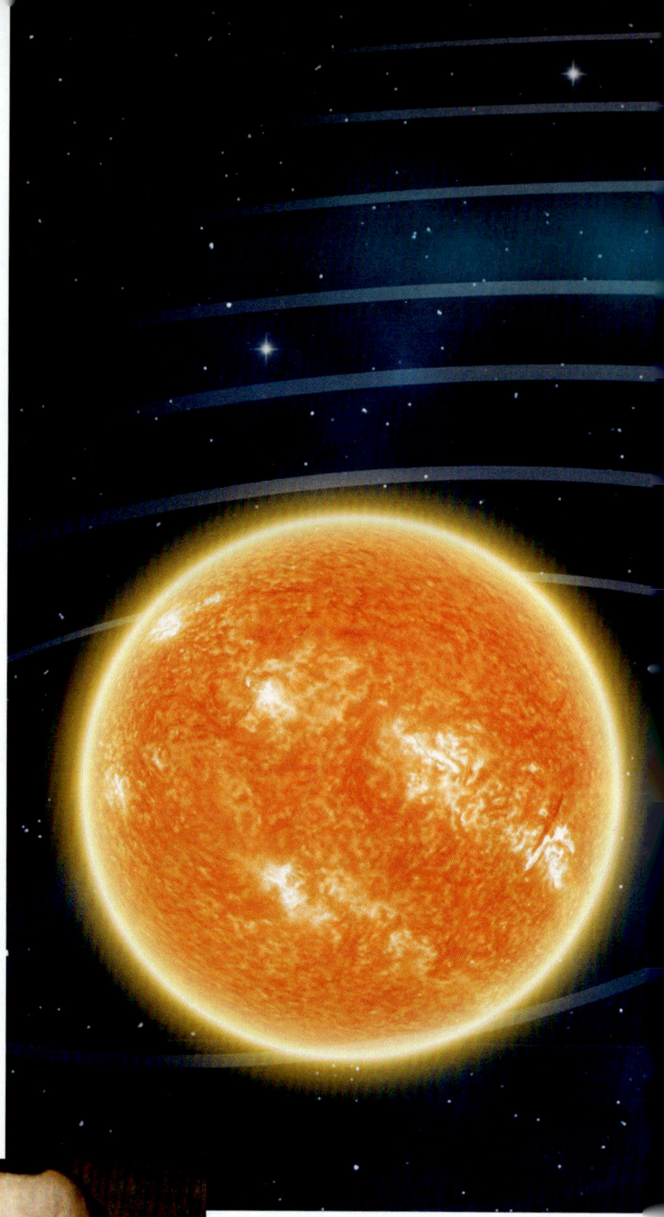

Our solar system, with the sun at its centre, has a diameter of around 14 billion km (8.7 billion mi)

Our galaxy, the Milky Way, measures 100,000 light years across, which is approximately 950 trillion km (590.3 trillion mi). Our sun, with its eight attendants – the planets Mercury, Venus, Earth, Mars, Jupiter, Saturn, Uranus and Neptune – is around 26,000 light years from the centre of the galaxy. The central star, the sun, accounts for almost the entire mass of the solar system, i.e. 99.86 per cent. Two-thirds of the remaining mass is taken up by the huge gas giant, Jupiter. The remaining 0.05 per cent is shared between the seven other planets and their moons in addition to the asteroids and minor and dwarf planets which also orbit the sun.

### The largest bodies in the solar system

Diameter at equator

1. **Sun**
   1,392,000 km (864,949 mi)
2. **Jupiter**
   142,984 km (88,846 mi)
3. **Saturn**
   120,536 km (74,898 mi)
4. **Uranus**
   51,118 km (31,763 mi)
5. **Neptune**
   49,528 km (30,775 mi)
6. **Earth**
   12,756 km (7,926 mi)
7. **Venus**
   12,104 km (7,521 mi)
8. **Mars**
   6,792 km (4,220 mi)
9. **Ganymed (Jupiter's moon)**
   5,262 km (3,270 mi)
10. **Titan (Saturn's moon)**
    5,150 km (3,200 mi)
11. **Mercury**
    4,879 km (3,032 mi)

### The heaviest bodies in the solar system

Earth's mass = 1

1. **Sun**
   332,270
2. **Jupiter**
   318
3. **Saturn**
   95
4. **Neptune**
   17
5. **Uranus**
   14.5
6. **Earth**
   1
7. **Venus**
   0.8
8. **Mars**
   0.1
9. **Mercury**
   0.06
10. **Ganymede**
    0.0025
11. **Titan**
    0.0023

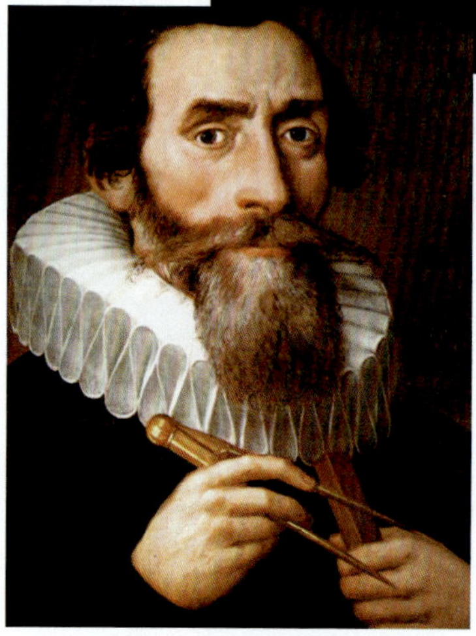

### JOHANNES KEPLER
**(1571–1630)**

Johannes Kepler, born in Weil der Stadt in Germany, discovered the laws of planetary motion which still bear his name today. He was able to show that the planetary orbits were not circles, as Copernicus had thought, but ellipses, and that the sun was always in one of the foci of the ellipse. Together with Galileo, Kepler is considered the founder of modern natural science.

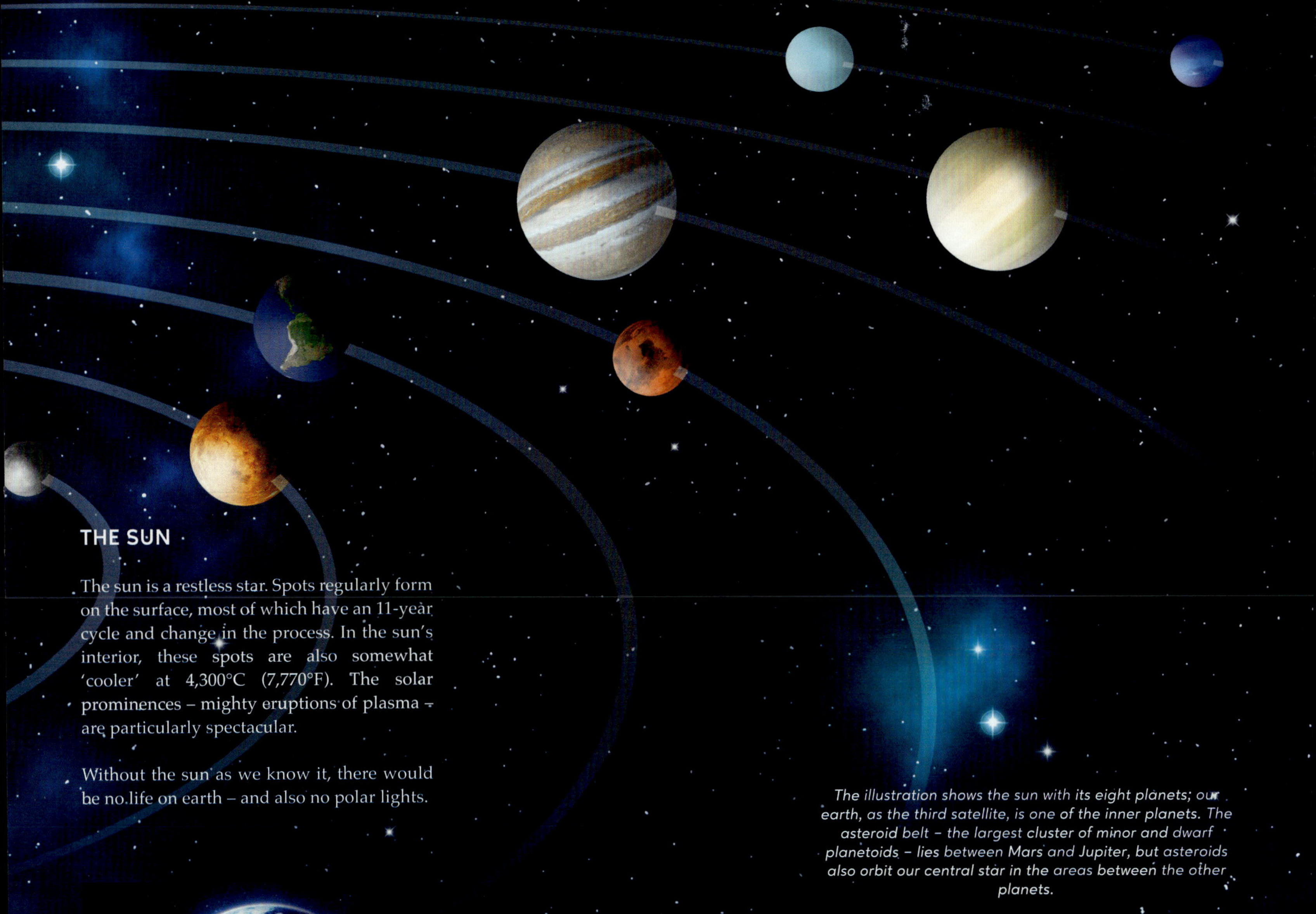

## THE SUN

The sun is a restless star. Spots regularly form on the surface, most of which have an 11-year cycle and change in the process. In the sun's interior, these spots are also somewhat 'cooler' at 4,300°C (7,770°F). The solar prominences – mighty eruptions of plasma – are particularly spectacular.

Without the sun as we know it, there would be no life on earth – and also no polar lights.

*The illustration shows the sun with its eight planets; our earth, as the third satellite, is one of the inner planets. The asteroid belt – the largest cluster of minor and dwarf planetoids – lies between Mars and Jupiter, but asteroids also orbit our central star in the areas between the other planets.*

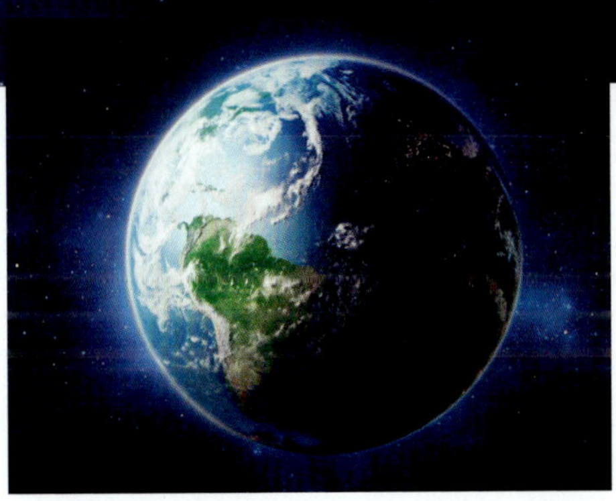

## MERCURY
**58 million km (36,039,530 mi)**

Mercury's mass is only 6 per cent of earth's mass. The thin atmosphere provides no protection from extreme temperatures of +425°C (+797°F) on its day side and -170°C (-274°F) on its night side.

## VENUS
**108 million km (67,108,090 mi)**

The completely opaque atmosphere consists primarily of carbon dioxide; the temperature is +460°C (+860°F) at ground level.

## EARTH
**150 million km (93,205,680 mi)**

The earth is one of the inner planets which are also classed as terrestrial, or rocky, planets, in contrast to the outer gas planets. One light second away, only one satellite, the moon, orbits the earth.

## MARS
**228 million km (141,672,630 mi)**

The fourth planet, Mars, looks red when seen from the earth. Mars is also one of the planets considered similar to earth, but it is significantly smaller than the earth.

## JUPITER
**778 million km (483,426,790 mi)**

Jupiter is by far the largest planet; it consists mainly of hydrogen and helium. Numerous satellites accompany it; Galileo had already discovered the four largest in 1610.

## SATURN
**1,427 million km (886,696,690 mi)**

Saturn is a gas giant; of all the planets, it has the lowest medium density. It is mainly composed of hydrogen and has a system of bands and zones. Its most distinctive feature are its pronounced rings.

## URANUS
**2,870 million km (1,783,335,320 mi)**

In 1986, Voyager 2 flew past the cold giant, and what was visible seemed almost structureless. It was possible, however, to confirm its ring system.

## NEPTUNE
**4,497 million km (2,794,306,250 mi)**

Neptune was discovered with the aid of a telescope in 1846. The gas planet has a somewhat higher proportion of methane; its temperature has been ascertained as -218°C (-360°F).

FOUR ELEMENTS

FIRE

# SUN

The hottest temperature of the sun

For astronomers, the sun is an average star. On the surface, the temperature reaches up to +5,500°C (+9,930°F); its interior temperature is +15 million°C (+27 million°F). Our central star consists of 70 per cent hydrogen and 28 per cent helium; heavy elements, mainly oxygen and carbon, only account for 1.5 to 2 per cent. The process by which the sun generates energy had long been a mystery. It could not be a chemical process such as the burning of carbon because then its energy stores would only have lasted 100,000 years. The question was resolved only after the discovery of nuclear energy. In 1938, the German-American physicist Hans Bethe described the proton-proton chain, which explained how hydrogen is 'burned' to make helium in the sun. The energy produced originates from the fact that the larger nuclei arising from nuclear fusion have less mass than the original nuclei; this difference, known as the mass defect, becomes available as energy according to Einstein's equation $e=m^*c^2$. It is due to the factor $c^2$ – the square of the speed of light, a very high value. However, the process only proceeds at a very slow pace. Even in the sun's interior, with its extremely high temperatures, only one proton in 100 million reaches the speed required for a collision. However, the sun contains so many protons that every second approximately 5 million tonne (5.5 tons) of mass is converted into energy. This has been a continuing process for 4.5 billion years, and it has not used even 4 per cent of the available hydrogen. The supply should therefore last for at least several billion years more.

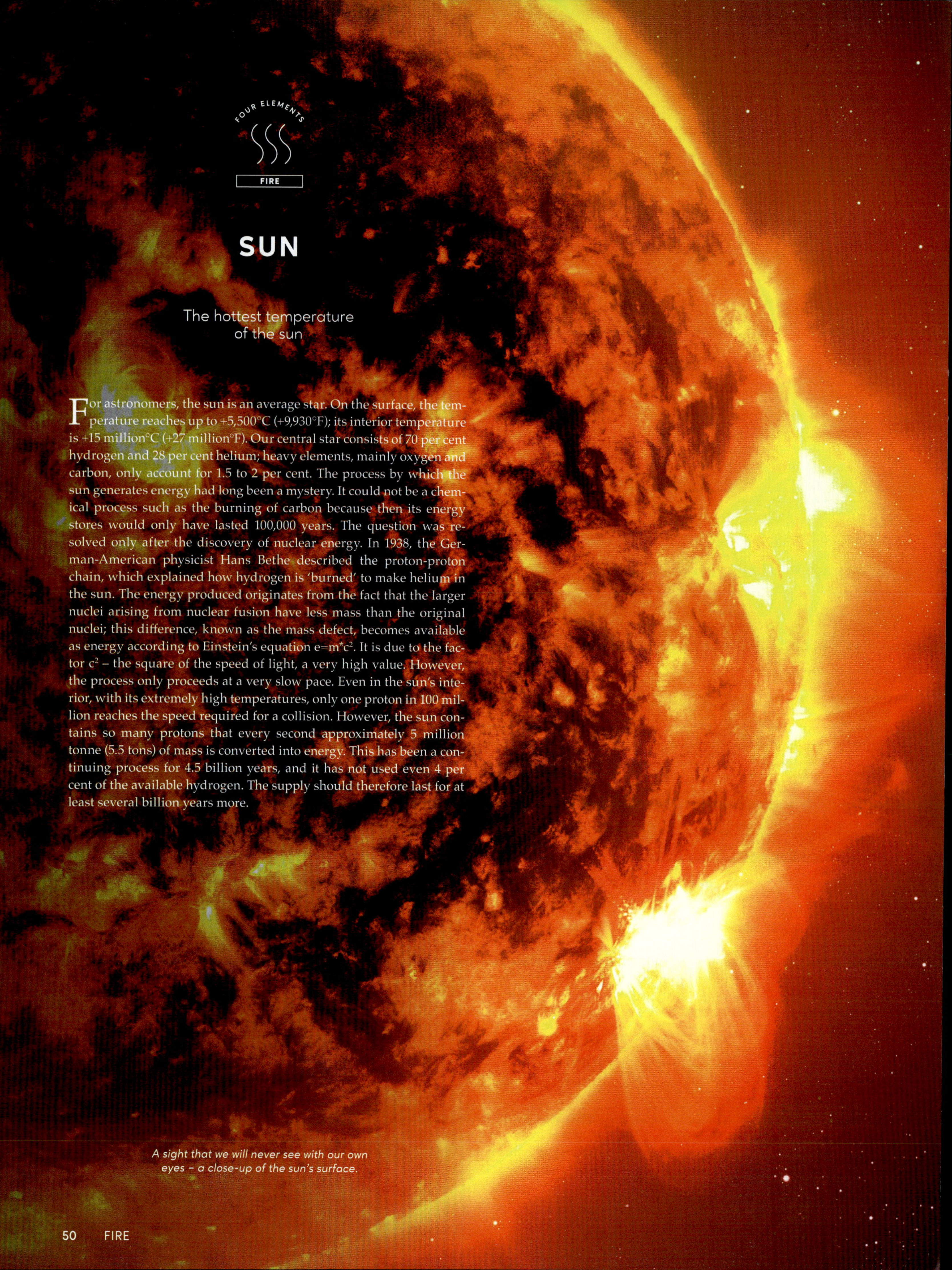

*A sight that we will never see with our own eyes – a close-up of the sun's surface.*

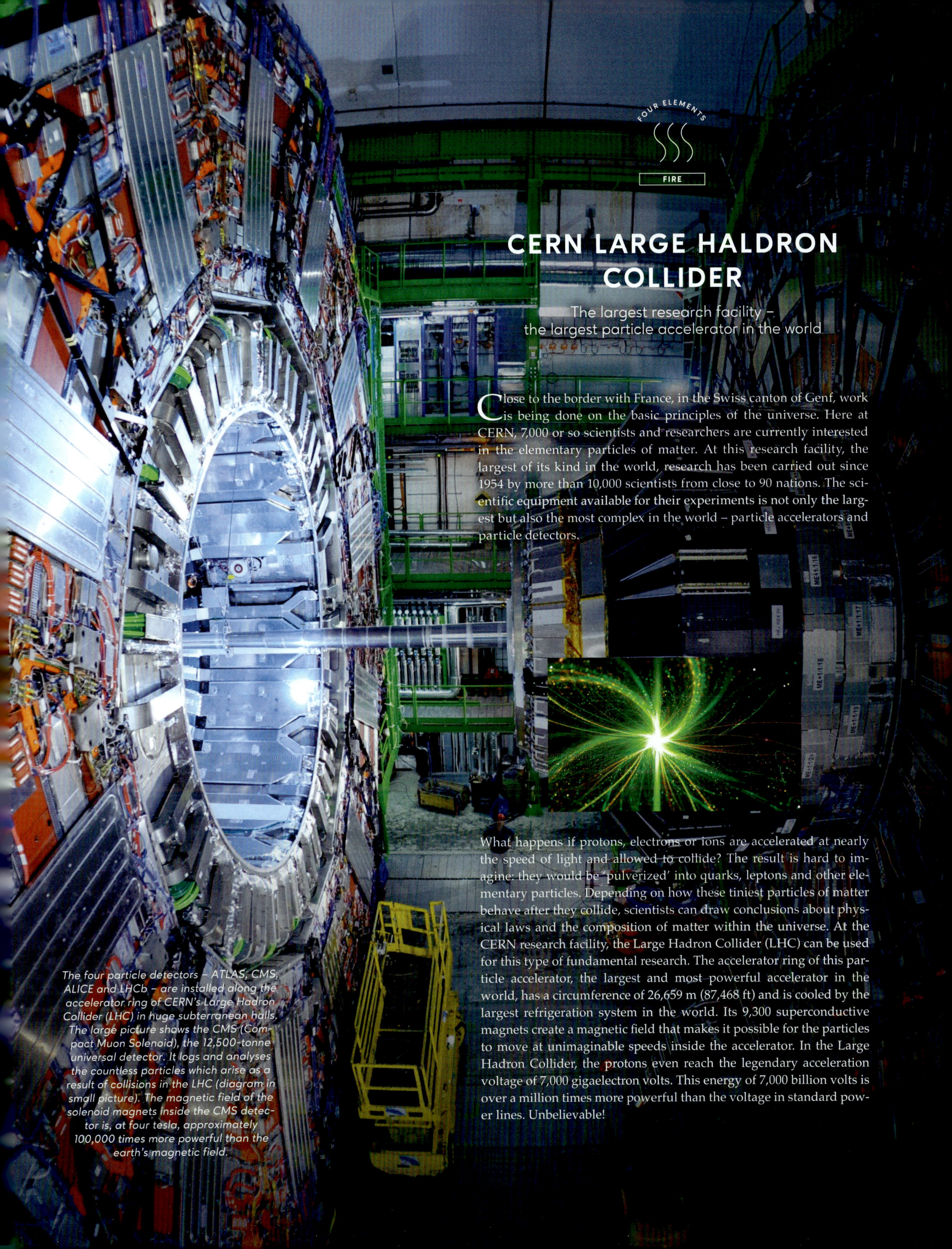

FOUR ELEMENTS

FIRE

# CERN LARGE HALDRON COLLIDER

The largest research facility –
the largest particle accelerator in the world

Close to the border with France, in the Swiss canton of Genf, work is being done on the basic principles of the universe. Here at CERN, 7,000 or so scientists and researchers are currently interested in the elementary particles of matter. At this research facility, the largest of its kind in the world, research has been carried out since 1954 by more than 10,000 scientists from close to 90 nations. The scientific equipment available for their experiments is not only the largest but also the most complex in the world – particle accelerators and particle detectors.

What happens if protons, electrons or ions are accelerated at nearly the speed of light and allowed to collide? The result is hard to imagine: they would be 'pulverized' into quarks, leptons and other elementary particles. Depending on how these tiniest particles of matter behave after they collide, scientists can draw conclusions about physical laws and the composition of matter within the universe. At the CERN research facility, the Large Hadron Collider (LHC) can be used for this type of fundamental research. The accelerator ring of this particle accelerator, the largest and most powerful accelerator in the world, has a circumference of 26,659 m (87,468 ft) and is cooled by the largest refrigeration system in the world. Its 9,300 superconductive magnets create a magnetic field that makes it possible for the particles to move at unimaginable speeds inside the accelerator. In the Large Hadron Collider, the protons even reach the legendary acceleration voltage of 7,000 gigaelectron volts. This energy of 7,000 billion volts is over a million times more powerful than the voltage in standard power lines. Unbelievable!

*The four particle detectors – ATLAS, CMS, ALICE and LHCb – are installed along the accelerator ring of CERN's Large Hadron Collider (LHC) in huge subterranean halls. The large picture shows the CMS (Compact Muon Solenoid), the 12,500-tonne universal detector. It logs and analyses the countless particles which arise as a result of collisions in the LHC (diagram in small picture). The magnetic field of the solenoid magnets inside the CMS detector is, at four tesla, approximately 100,000 times more powerful than the earth's magnetic field.*

# DESERTS

Many regions on earth have such extremes of aridity or cold that generally only dryland plants are able to develop, not plants that are specific to a particular landscape. Dry or hot deserts are frequently found in the zones of subtropical high-pressure regions, on coasts with upwelling cold deep water and in closed mountain basins. In full or core deserts, where very few plants grow, the annual precipitation level is under 100 mm (3.9 in).

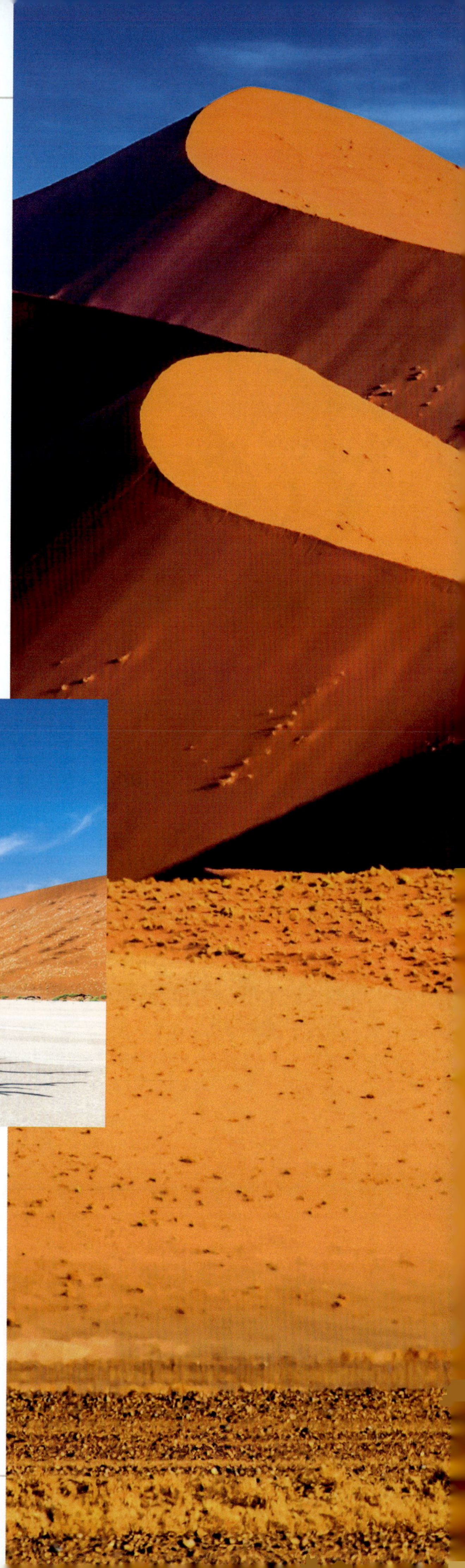

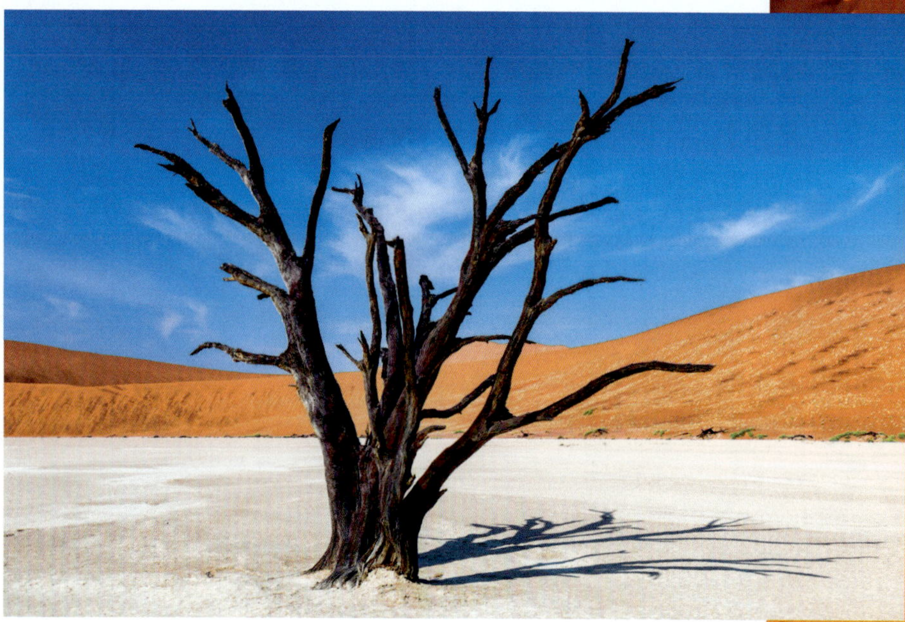

*Sand dunes, which change colour depending on the sun's altitude and humidity levels, characterize the Namib-Naukluft National Park in the Namib Desert in Namibia, the oldest desert in the world (large picture).*

*Above: Like a sculptural form, the bare black branches of the dead camelthorn trees protrude from the desert floor in Sossuvlei, Namibia.*

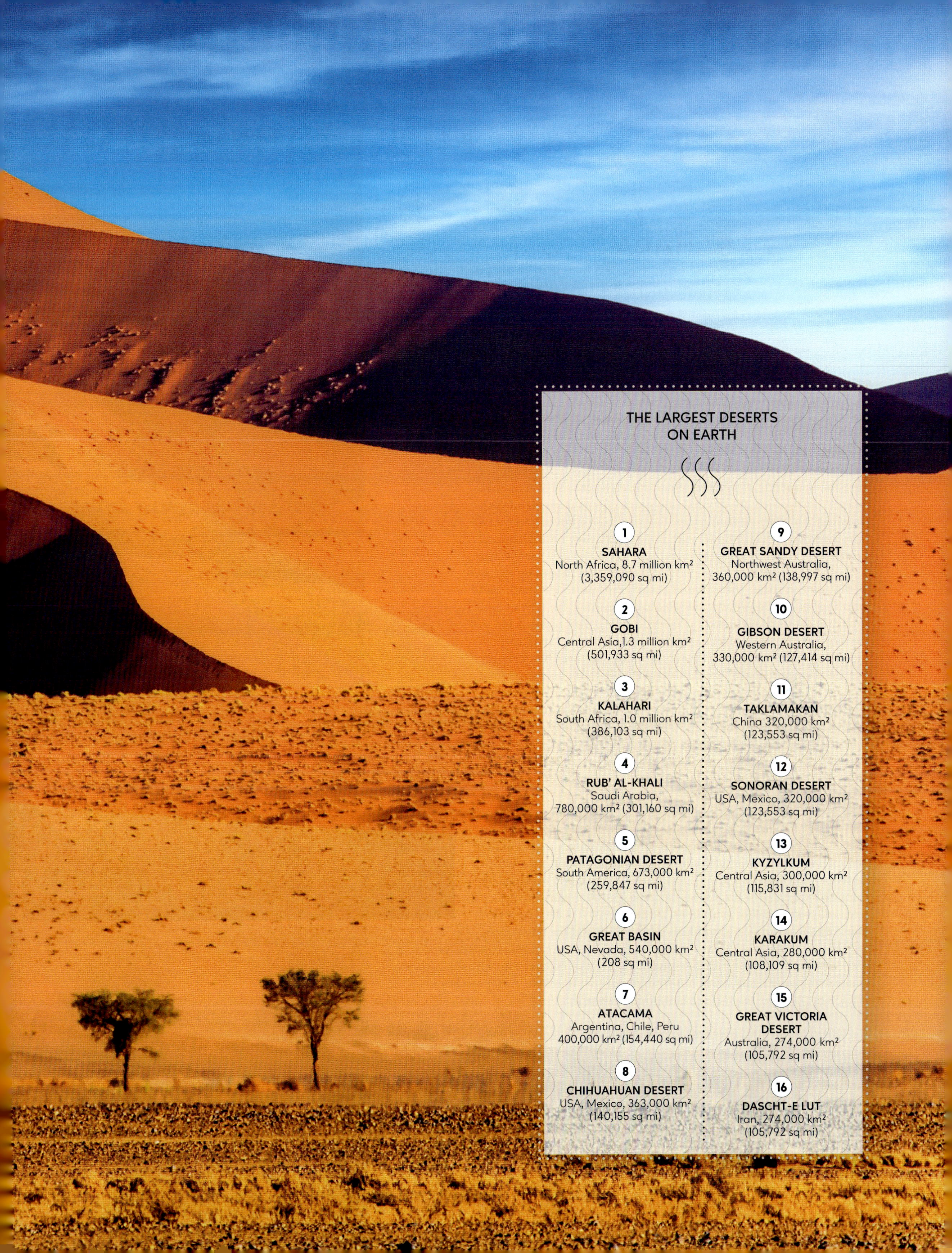

## THE LARGEST DESERTS ON EARTH

1 **SAHARA**
North Africa, 8.7 million km² (3,359,090 sq mi)

2 **GOBI**
Central Asia,1.3 million km² (501,933 sq mi)

3 **KALAHARI**
South Africa, 1.0 million km² (386,103 sq mi)

4 **RUB' AL-KHALI**
Saudi Arabia, 780,000 km² (301,160 sq mi)

5 **PATAGONIAN DESERT**
South America, 673,000 km² (259,847 sq mi)

6 **GREAT BASIN**
USA, Nevada, 540,000 km² (208 sq mi)

7 **ATACAMA**
Argentina, Chile, Peru 400,000 km² (154,440 sq mi)

8 **CHIHUAHUAN DESERT**
USA, Mexico, 363,000 km² (140,155 sq mi)

9 **GREAT SANDY DESERT**
Northwest Australia, 360,000 km² (138,997 sq mi)

10 **GIBSON DESERT**
Western Australia, 330,000 km² (127,414 sq mi)

11 **TAKLAMAKAN**
China 320,000 km² (123,553 sq mi)

12 **SONORAN DESERT**
USA, Mexico, 320,000 km² (123,553 sq mi)

13 **KYZYLKUM**
Central Asia, 300,000 km² (115,831 sq mi)

14 **KARAKUM**
Central Asia, 280,000 km² (108,109 sq mi)

15 **GREAT VICTORIA DESERT**
Australia, 274,000 km² (105,792 sq mi)

16 **DASCHT-E LUT**
Iran, 274,000 km² (105,792 sq mi)

FOUR ELEMENTS

FIRE

# SAHARA

The largest desert on earth

The largest desert region on earth covers an expanse of roughly 8.7 million km² (3,359,090 sq mi). This immense tableland, covered in sand and scree interspersed with small oases, stretches over approximately 6,000 km (3,728 mi) in an east-west direction from the Atlantic to the Red Sea and over 2,000 km (1,243 mi) in a north-south direction into the Sudan region. Only approximately 20 per cent of the Sahara is sandy desert; the remaining areas consist of scree and rocky basins, which are divided into gravel plains (serirs) and rock plateaus (hamadas). At their edges, the basins slope down into wadis – dry riverbeds which only fill with water after heavy rainfall. Inland depressions, covered in mighty sand seas (ergs) and salt flats (shatts or chotts), are typical of large parts of the northern Sahara up to the Mediterranean coast. These inland basins continue to the south and form the Ténéré Desert and the sections containing the Libyan and Arabian Deserts. Punctuated by the Nile valley, they reach as far as the Red Sea. In the Central Sahara, inselbergs and uplands such as Tademait, Tassili and Djado spring up, while high mountain ranges like the Ahaggar, Tibesti, Aïr, Iforas and Ennedi tower over them. The only permanent body of water in the Sahara is the Nile. The largest desert in the world has an extreme climate: the temperatures during the day can climb to over 50°C (122°F), but, in summer, they can drop to below freezing at night.

## TIBESTI AND AHAGGAR: THE HIGHEST MOUNTAINS IN THE SAHARA

The Ahaggar, or Hoggar Mountains, covering approximately 300,000 km² (115,831 sq mi), form the largest mountain region in the Central Sahara. The heart of this spectacular mountainous area, which developed 600 million years ago and has been constantly reshaped by volcanic activity, is the bizarre basalt landscape of the Atakor volcanic field, where the peak of Tahat rises 2,918 m (9,574 ft). The Tassili plateaus, which sprung up 250 million years ago, lie on the outskirts of the Ahaggar to the north and east. On these uplands, the sandstone layers were eroded by earlier river courses; further sand and wind erosion have transformed them into unusual forms. The vast Tassili n'Ajjer in Algeria, on which the Adrar reaches a height of 2,158 m (7,080 ft), resembles a fantastic moonscape.

The Tibesti Mountains, covering approximately 100,000 km² (38,610 sq mi) with their steep ravines and bizarre weathering effects, are one of the most impressive landscapes on the continent. These deeply cleft high mountains rise up out of the surrounding plain like a wedge. Layers of granite and schist that unfolded in the Quaternary and Tertiary periods form the base of the Tibesti Mountains. Volcanoes developed during this tectonic activity, and their massive, solidified lava outflows now sit atop this base as peaks.

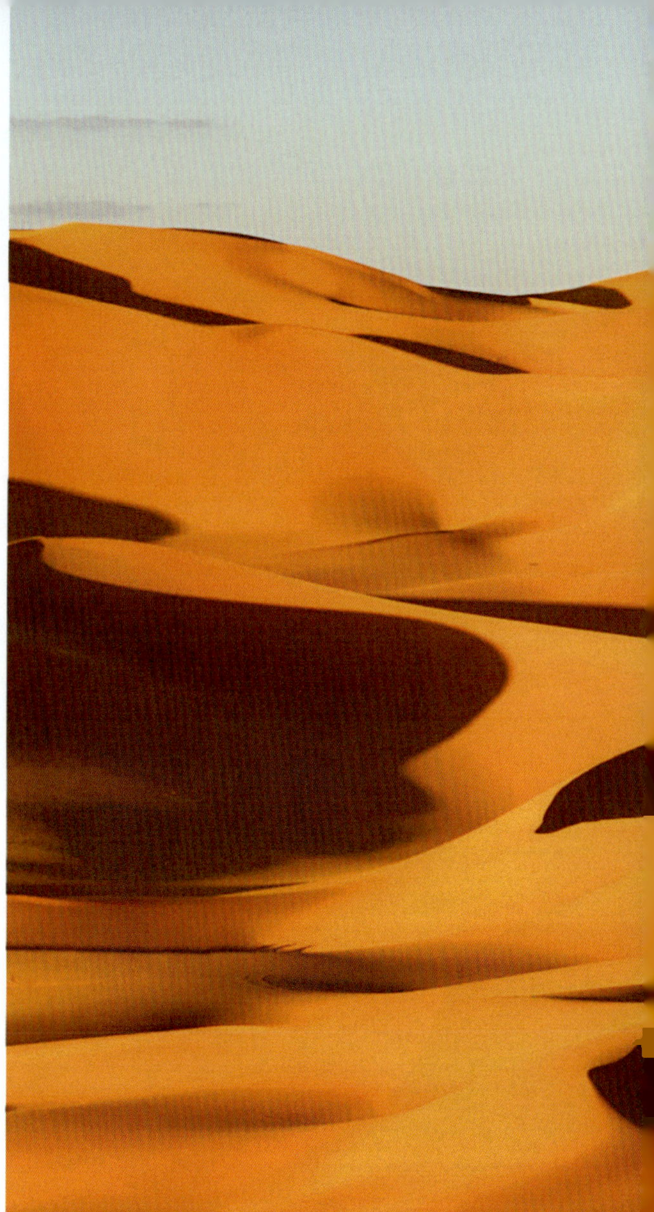

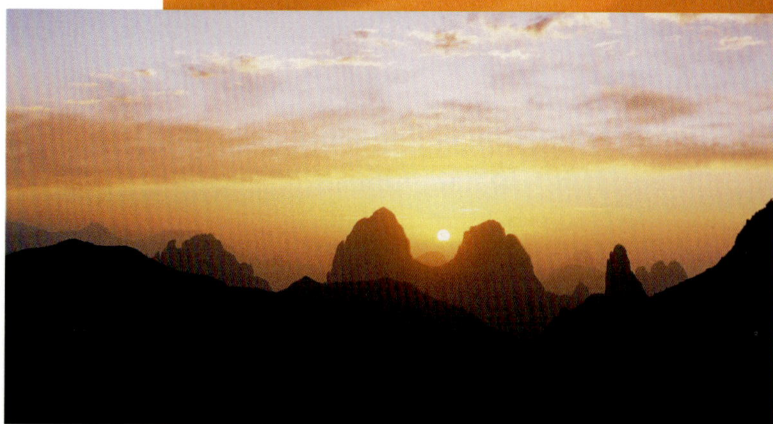

*Shaped by years and years of erosion and weathering, remarkable formations made up of stone columns and dune landscapes (above and right) leave their mark on the Ahaggar and Tibesti mountains in the Sahara.*

*The wide variety of different reliefs make the Sahara a multifaceted landscape where the mighty sand dunes are only a small part of the whole.*

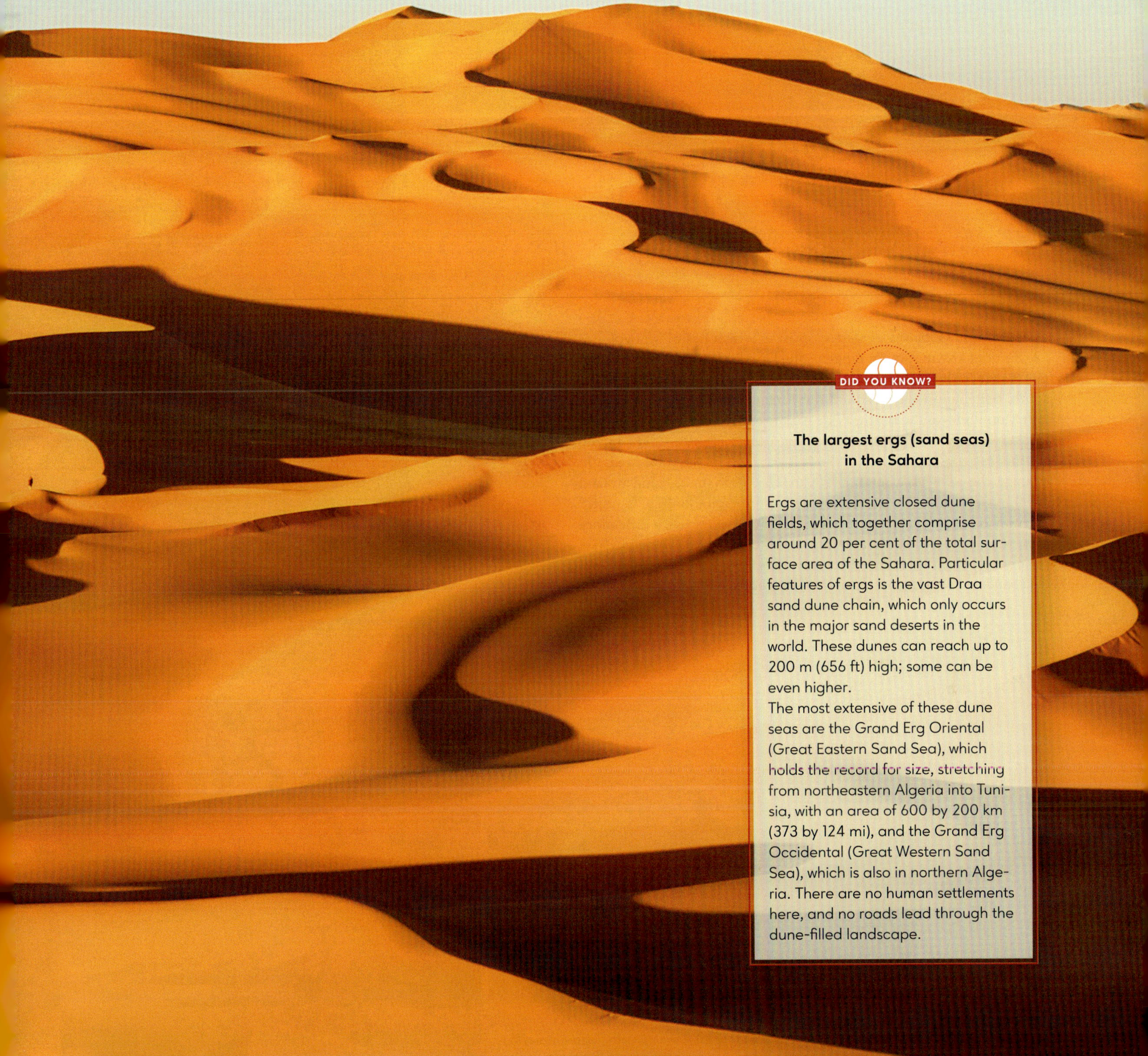

DID YOU KNOW?

**The largest ergs (sand seas) in the Sahara**

Ergs are extensive closed dune fields, which together comprise around 20 per cent of the total surface area of the Sahara. Particular features of ergs is the vast Draa sand dune chain, which only occurs in the major sand deserts in the world. These dunes can reach up to 200 m (656 ft) high; some can be even higher.

The most extensive of these dune seas are the Grand Erg Oriental (Great Eastern Sand Sea), which holds the record for size, stretching from northeastern Algeria into Tunisia, with an area of 600 by 200 km (373 by 124 mi), and the Grand Erg Occidental (Great Western Sand Sea), which is also in northern Algeria. There are no human settlements here, and no roads lead through the dune-filled landscape.

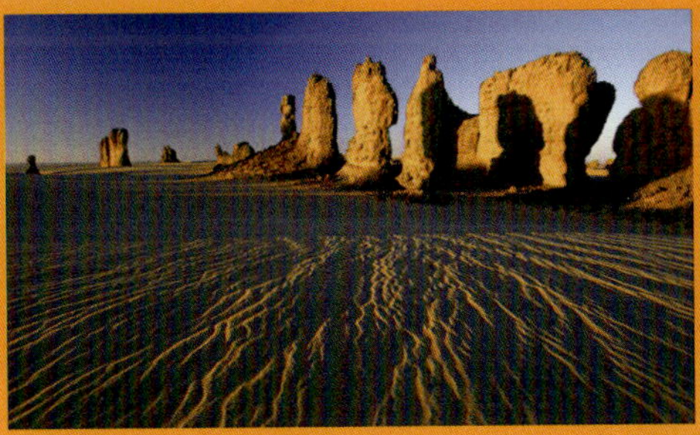

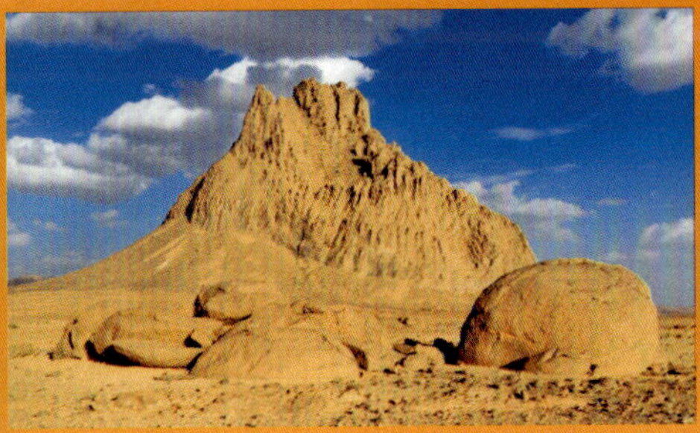

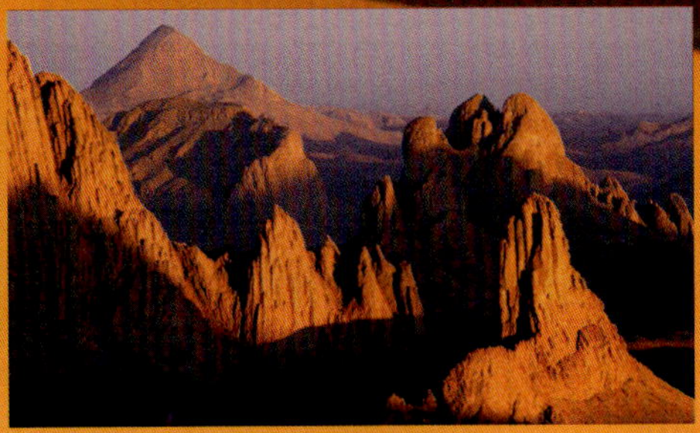

# GOBI

The highest sand dunes
in Asia

The central Gobi desert steppe, most of which lies in Mongolia, is bordered by mountain ranges – the Greater Khingan or Da Hinggan Range to the east and southeast, Altun Shan and Nan Shan in the south, Bei Shan in the west and the Khangai and Yablonovy (Yablonoi) Mountains to the north. Wide expanses are covered in scree and loess, cliffs and even lakes. The Gobi desert is only sandy in the south-western section that is about 3 per cent of the total area. On average, the Gobi has a height of 900 m (2,953 ft) in the east and around 1,500 m (4,921 ft) in the west. The annual precipitation rate is under 200 mm (8 in); only the border areas of the desert receive a little more rain, with some of this land being under cultivation. The climate is continental and has very cold winters and hot summers; in spring, cold sandstorms or snowstorms can even occur.

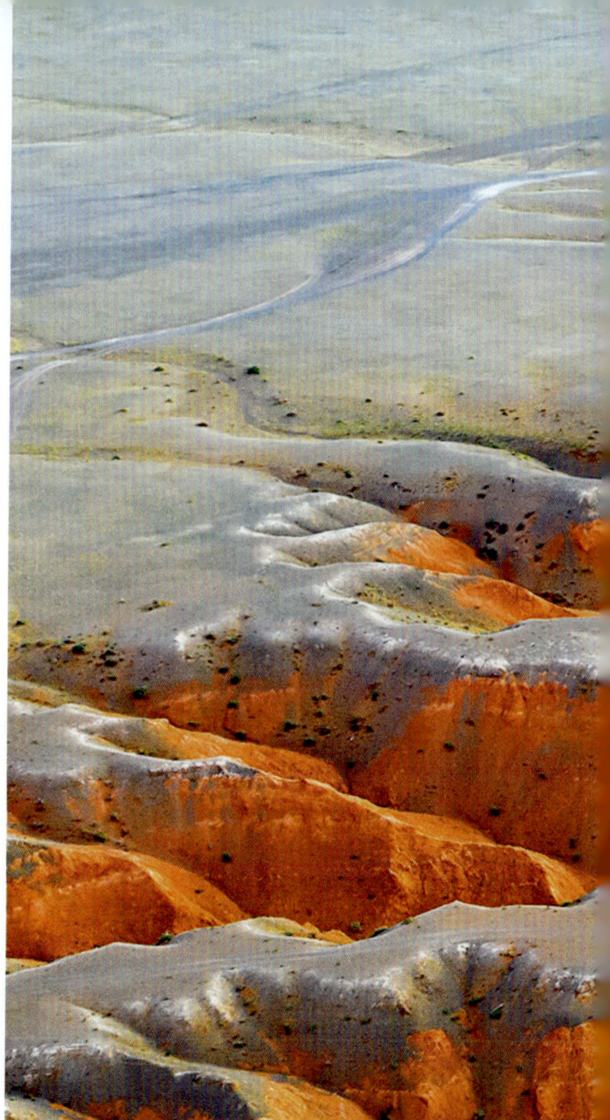

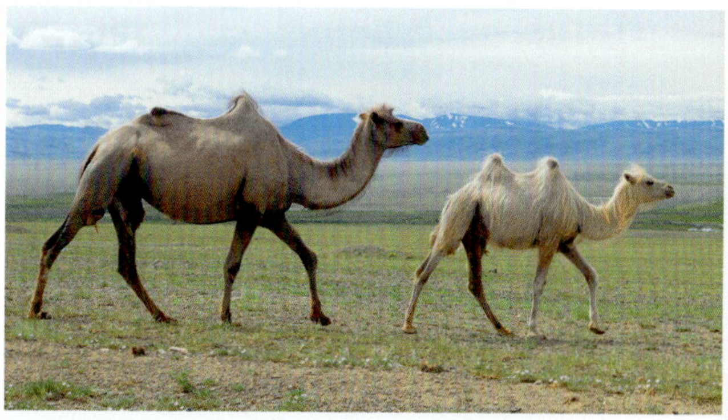

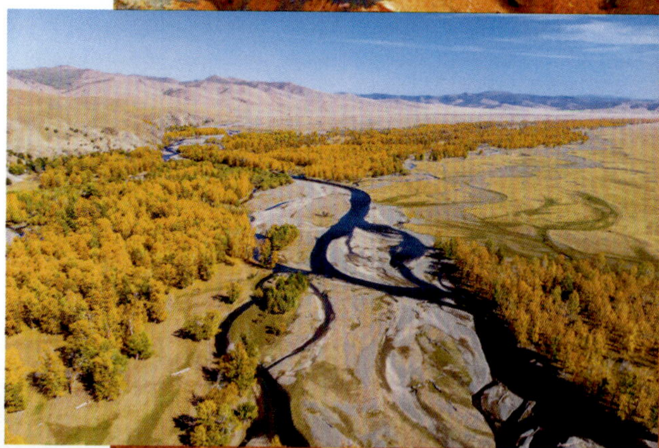

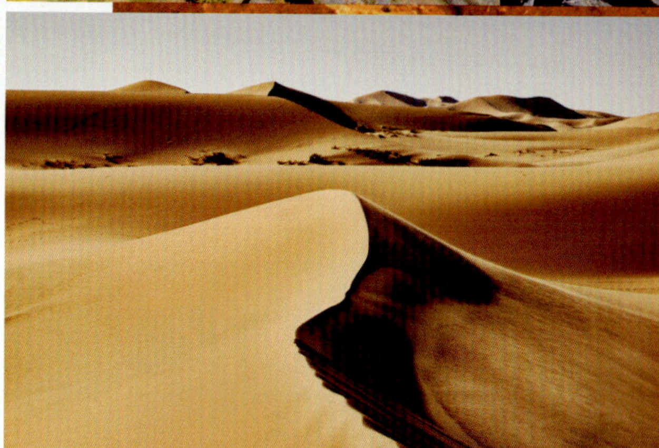

*Large picture: The Bayanzag, or the Flaming Cliffs, glow yellow and orange at sunset. The view from above across this stunning landscape always amazes.*

*Camels are at home in Mongolia (above).*

*Pictures on the right, from the top: The Gobi Desert is not hostile to life; in some regions, it is even dotted with fertile oases. The striking dune fields of Khongoryn Els are the largest dune fields in Mongolia.*

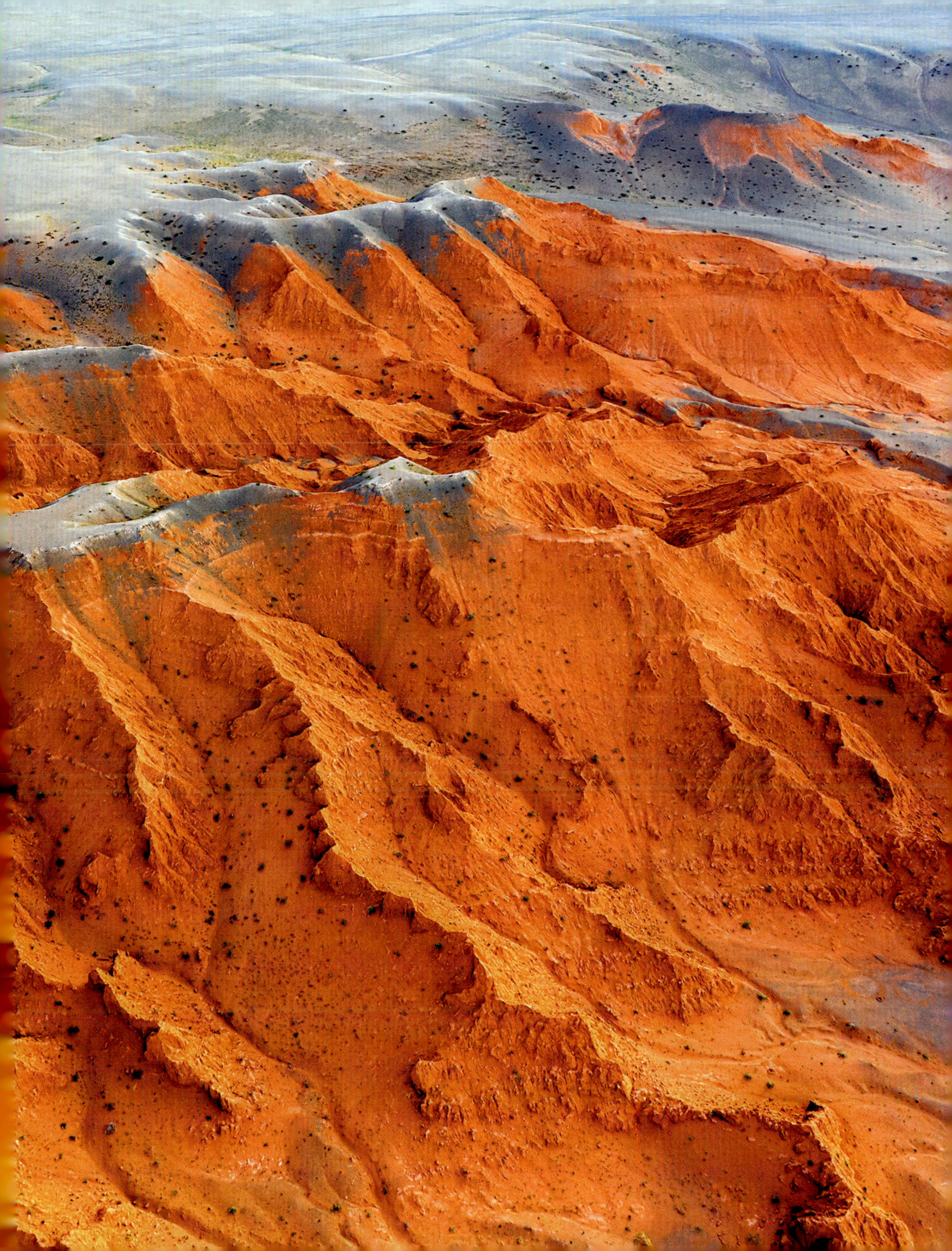

FOUR ELEMENTS

FIRE

# AFRICA

The Sahara is the largest desert in Africa and in the entire world. But this continent also features other extensive deserts and semi-deserts

## KALAHARI

**1,000,000 km² (386,102 sq mi)**

The dry basin landscape of the Kalahari Desert has no outlets and consists of vast upland areas which form the largest continuous expanse of sand on the planet. Some of the extensive dune fields, which are more prominent in the west, date from the ice age. Rocky hills are the only elevation in the otherwise endlessly flat, arid savannah which lies 900 m (2,953 ft) above sea level. In addition, the Kalahari has salt pans – the remains of ice age lakes – among which are the Makgadikgadi Pan and the Etosha Pan, two of the largest salt pans in the world. The marshy area of the Okavango basin stretches across the north. Unlike the dunes in the Namib desert, those in the Kalahari run parallel to each other in long, narrow ribs. These are fossil dunes whose foundations long ago turned into stone and which now do not change. In the dune valleys, water can be found near the surface; the underground network of this water connects to the inland delta of the Okavango river. This makes the Kalahari relatively fertile. Giraffes, oryx, zebras and springboks graze in the valleys.

## EASTERN DESERT

**220,000 km² (84,942 sq mi)**

The Eastern Desert, a dry, hilly region in eastern Egypt, is not to be confused with the Arabian Desert, also known as Rub' al-Khali. The Eastern Desert, together with the Nubian desert, forms the most eastern part of the Sahara Desert in Africa. It stretches from the Mediterranean in the north, the Red Sea and the Gulf of Suez in the east, the Nubian Desert in the south and the Nile in the west. Cliffs rise from the Nile valley here. Further to the east, it transforms into a deeply fissured upland plateau and finally slopes upwards to become craggy volcanic mountains on the Red Sea, only to drop suddenly at the coast. The city of Hurghada, the largest tourist centre in Egypt on the Red Sea, is here. The desert starts right behind the city. The highest elevation in the Eastern Desert is Gabal Sha'ib El Banat, standing 2,187 m (7,716 ft) above sea level. The desert registers occasional rainfall and is intersected by several wadis (dry riverbeds which fill on a seasonal basis). Completely cut off from the rest of Egypt, the desert is rich in natural resources such as crude oil (which is mined onshore as well as in the Gulf of Suez), phosphate rock, nitrate, uranium and gold.

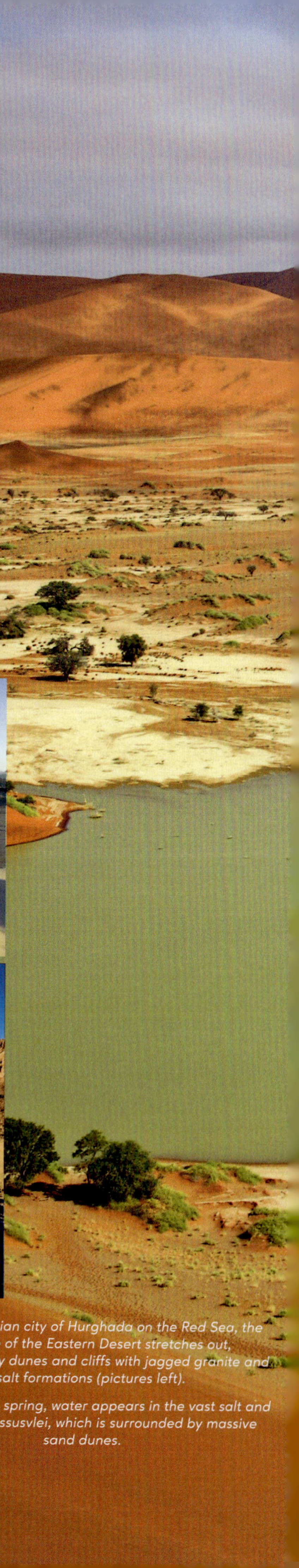

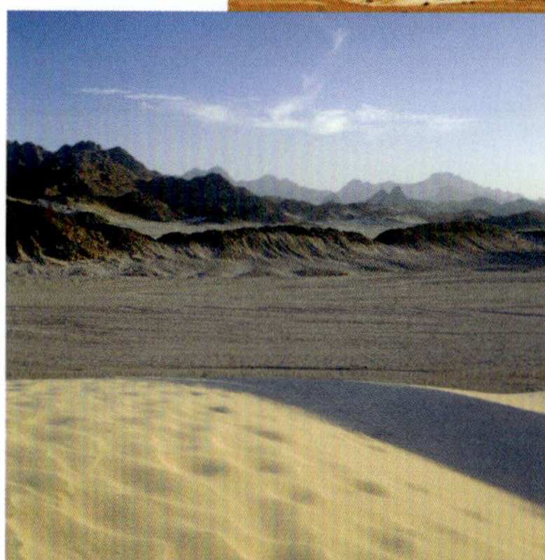

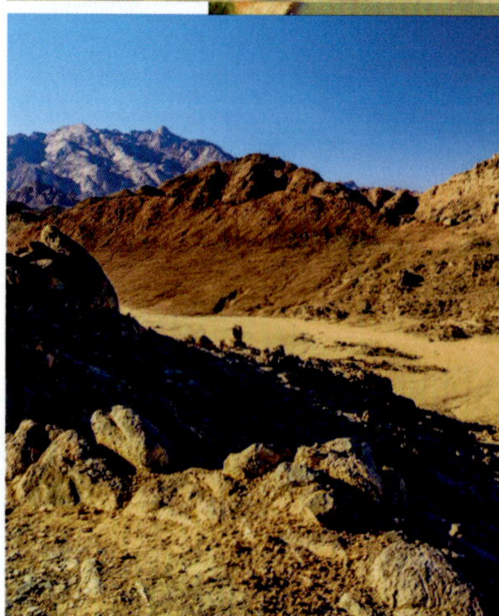

*Near the Egyptian city of Hurghada on the Red Sea, the landscape of the Eastern Desert stretches out, characterized by dunes and cliffs with jagged granite and basalt formations (pictures left).*

*Large picture: In spring, water appears in the vast salt and clay pan of Sossusvlei, which is surrounded by massive sand dunes.*

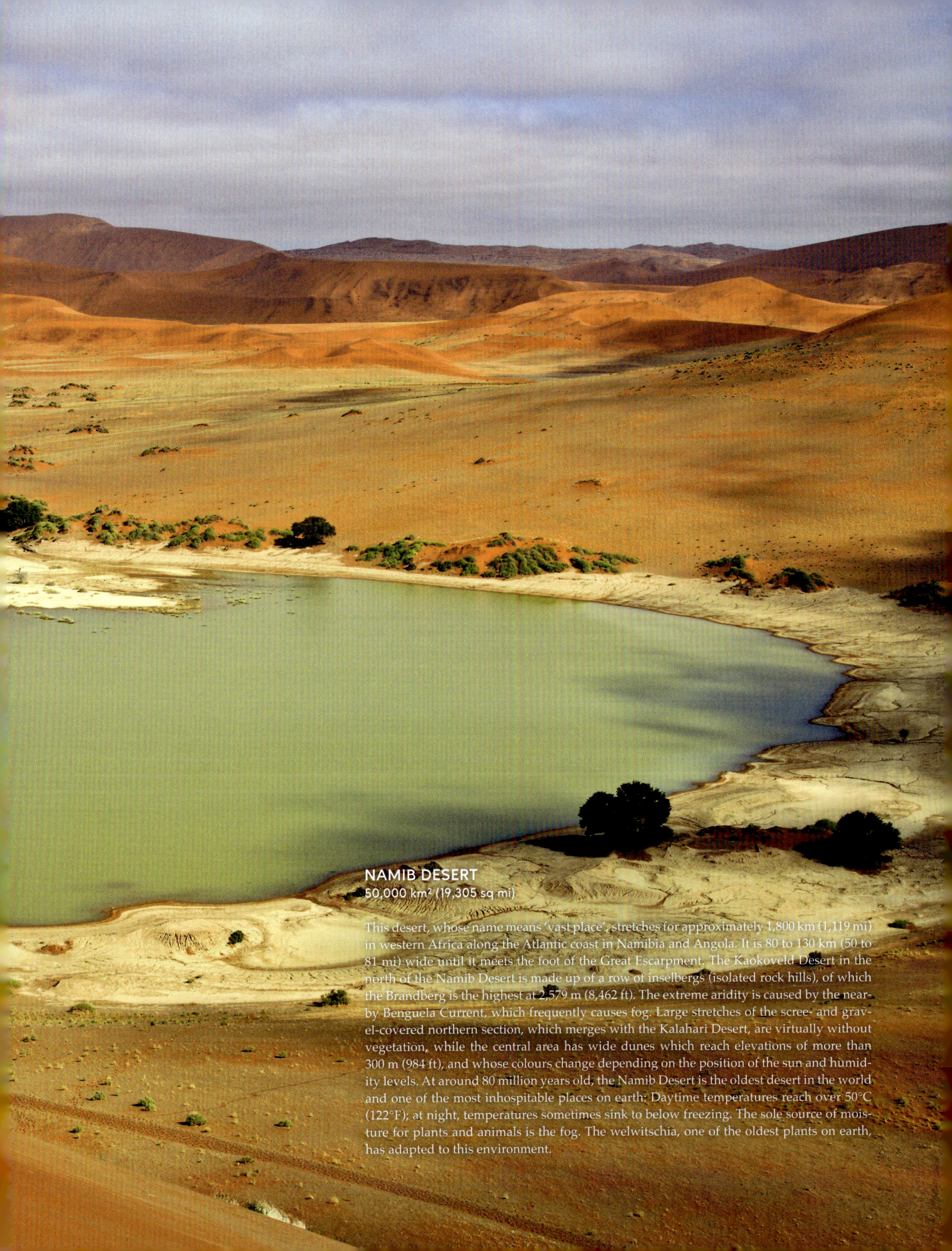

## NAMIB DESERT

50,000 km² (19,305 sq mi)

This desert, whose name means 'vast place', stretches for approximately 1,800 km (1,119 mi) in western Africa along the Atlantic coast in Namibia and Angola. It is 80 to 130 km (50 to 81 mi) wide until it meets the foot of the Great Escarpment. The Kaokoveld Desert in the north of the Namib Desert is made up of a row of inselbergs (isolated rock hills), of which the Brandberg is the highest at 2,579 m (8,462 ft). The extreme aridity is caused by the nearby Benguela Current, which frequently causes fog. Large stretches of the scree- and gravel-covered northern section, which merges with the Kalahari Desert, are virtually without vegetation, while the central area has wide dunes which reach elevations of more than 300 m (984 ft), and whose colours change depending on the position of the sun and humidity levels. At around 80 million years old, the Namib Desert is the oldest desert in the world and one of the most inhospitable places on earth: Daytime temperatures reach over 50°C (122°F); at night, temperatures sometimes sink to below freezing. The sole source of moisture for plants and animals is the fog. The welwitschia, one of the oldest plants on earth, has adapted to this environment.

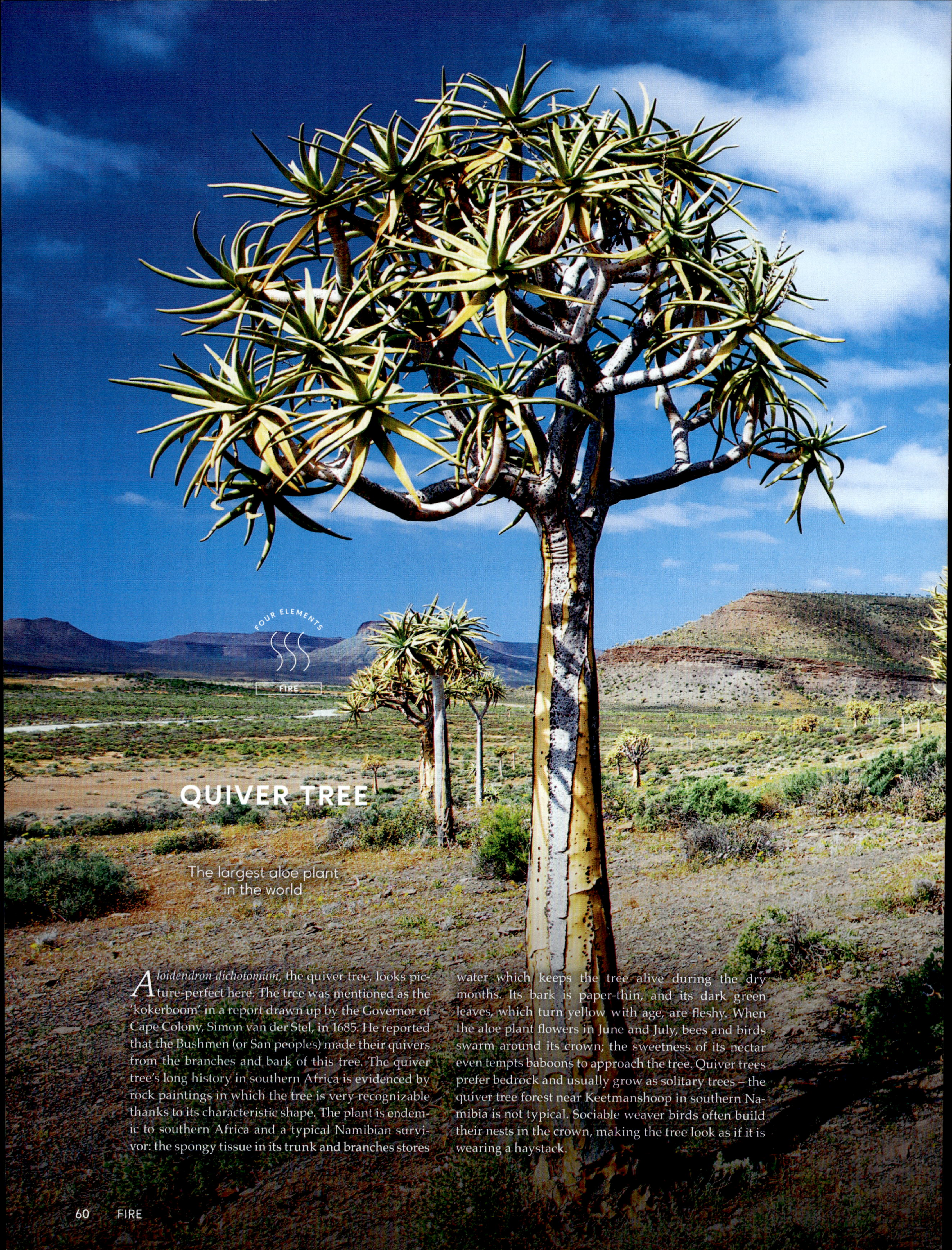

FOUR ELEMENTS
FIRE

# QUIVER TREE

The largest aloe plant in the world

*Aloidendron dichotomum*, the quiver tree, looks picture-perfect here. The tree was mentioned as the 'kokerboom' in a report drawn up by the Governor of Cape Colony, Simon van der Stel, in 1685. He reported that the Bushmen (or San peoples) made their quivers from the branches and bark of this tree. The quiver tree's long history in southern Africa is evidenced by rock paintings in which the tree is very recognizable thanks to its characteristic shape. The plant is endemic to southern Africa and a typical Namibian survivor: the spongy tissue in its trunk and branches stores water which keeps the tree alive during the dry months. Its bark is paper-thin, and its dark green leaves, which turn yellow with age, are fleshy. When the aloe plant flowers in June and July, bees and birds swarm around its crown; the sweetness of its nectar even tempts baboons to approach the tree. Quiver trees prefer bedrock and usually grow as solitary trees – the quiver tree forest near Keetmanshoop in southern Namibia is not typical. Sociable weaver birds often build their nests in the crown, making the tree look as if it is wearing a haystack.

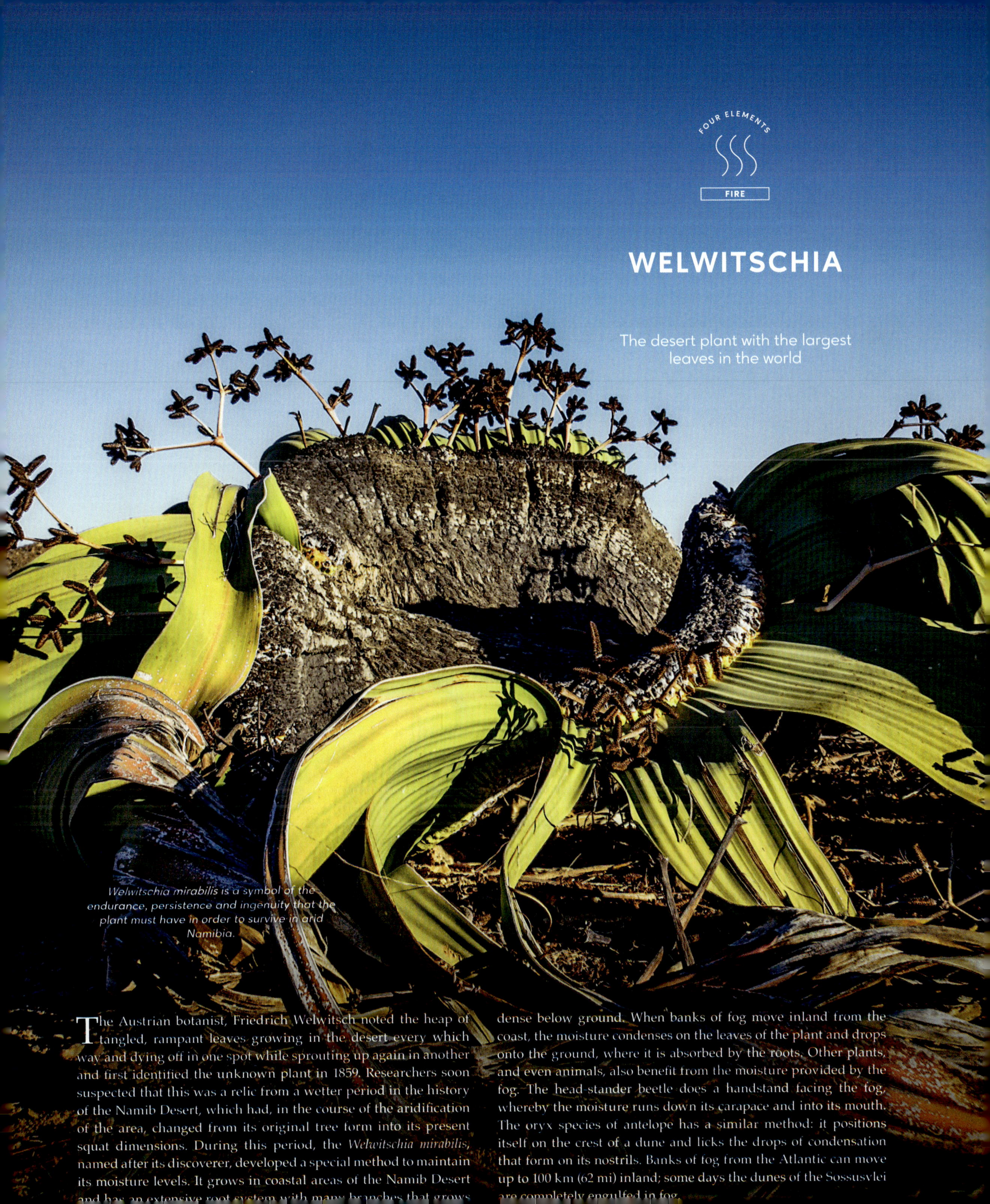

FOUR ELEMENTS
FIRE

# WELWITSCHIA

The desert plant with the largest leaves in the world

*Welwitschia mirabilis is a symbol of the endurance, persistence and ingenuity that the plant must have in order to survive in arid Namibia.*

The Austrian botanist, Friedrich Welwitsch noted the heap of tangled, rampant leaves growing in the desert every which way and dying off in one spot while sprouting up again in another and first identified the unknown plant in 1859. Researchers soon suspected that this was a relic from a wetter period in the history of the Namib Desert, which had, in the course of the aridification of the area, changed from its original tree form into its present squat dimensions. During this period, the *Welwitschia mirabilis*, named after its discoverer, developed a special method to maintain its moisture levels. It grows in coastal areas of the Namib Desert and has an extensive root system with many branches that grows dense below ground. When banks of fog move inland from the coast, the moisture condenses on the leaves of the plant and drops onto the ground, where it is absorbed by the roots. Other plants, and even animals, also benefit from the moisture provided by the fog. The head-stander beetle does a handstand facing the fog, whereby the moisture runs down its carapace and into its mouth. The oryx species of antelope has a similar method: it positions itself on the crest of a dune and licks the drops of condensation that form on its nostrils. Banks of fog from the Atlantic can move up to 100 km (62 mi) inland; some days the dunes of the Sossusvlei are completely engulfed in fog.

FOUR ELEMENTS

FIRE

# ASIA

The Rub' al Khali, the Taklamakan and the Gobi deserts are among the most famous in the world. Some of the largest deserts on earth lie on the continent of Asia.

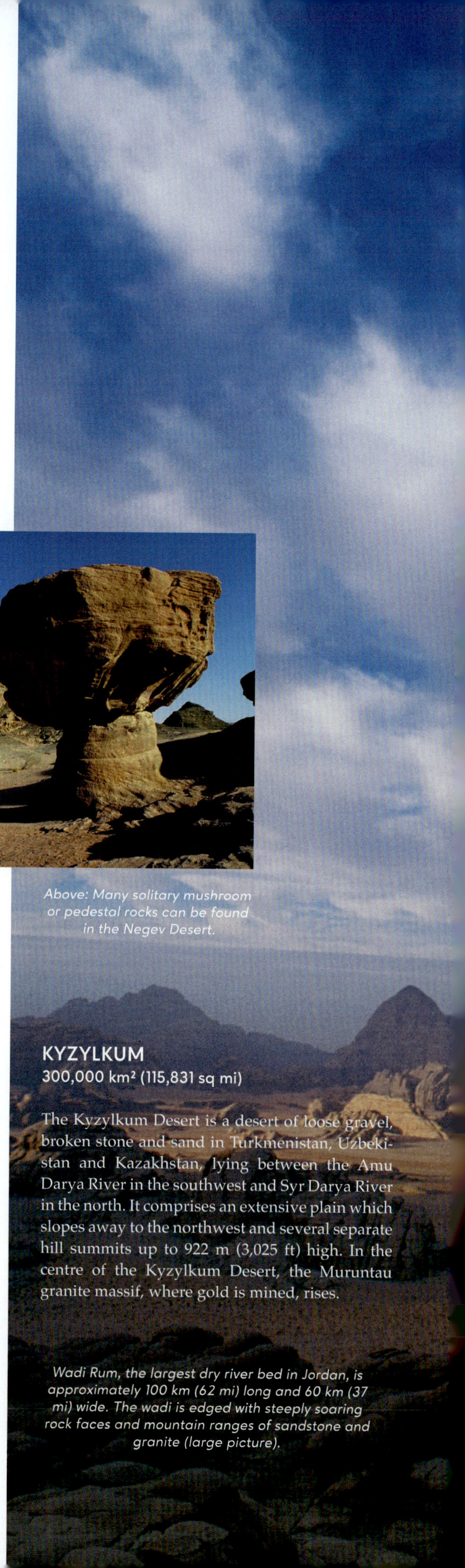

## NEGEV DESERT

**12,000 km² (4,633 sq mi)**

The Negev Desert covers around 60 per cent of the surface of Israel and forms part of the desert belt which stretches from the Atlantic Ocean to India. An unvarying dusty plain covers the northern and western sections of the desert; in the south, the region is covered in mountain ranges and eroded craters. After it rains, the dried-up wadis, or riverbeds, transform into raging rivers, and, for a brief time, the desert blooms.

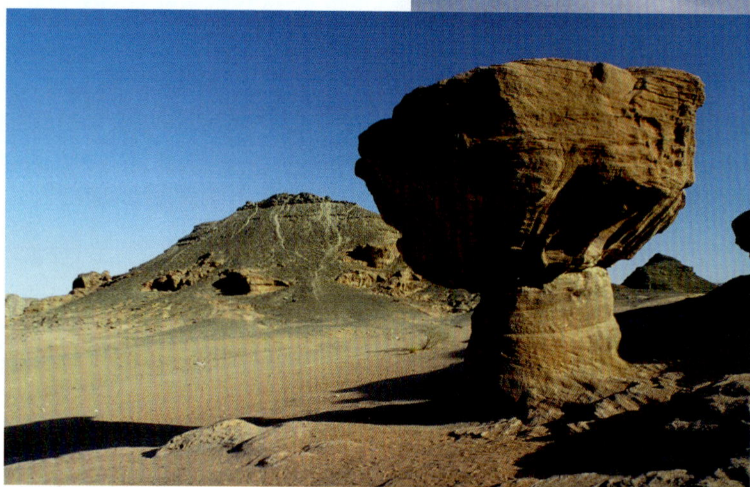

*Above: Many solitary mushroom or pedestal rocks can be found in the Negev Desert.*

## NEFUD DESERT

**104,000 km² (40,155 sq mi)**

The Nefud Desert in Saudi Arabia, with its red sand, is well-known for its sudden violent winds, which lead to the formation of 180-m (590-ft) high dunes. Some of the desert's sandstone cliffs are eroded into bizarre shapes. In some of the oases, fruit, vegetable and grains can be grown. In the southeast, the Lesser Nefud merges into the Rub' al Khali Desert.

## AD-DAHNA

**104,000 km² (40,155 sq mi)**

The Ad-Dahna Desert, an arc-shaped corridor of sand and gravel approximately 1,300 km (808 mi) long and up to 50 km (31 mi) wide, connects the two major deserts in Saudi Arabia: the Rub' Al Khali in the south and the Nefud Desert in the north. The dunes of the Ad-Dahna Desert appear mainly reddish to deep red in colour because of the high levels of iron oxide in the sand. Important public highways run through the desert, linking Kuwait with Riyadh and Riyadh with Al-Hasa.

## RUB' AL-KHALI

**780,000 km² (301,160 sq mi)**

Rub' Al Khali (the Empty Quarter), or Arabian Desert, is the largest continuous expanse of sand in the world and covers the southern third of the Arabian Peninsula, from Najd in Saudi Arabia in the north to Hadhramaut in Yemen in the south and the United Arab Emirates in the east.

## KARAKUM

**280,000 km² (108,109 sq mi)**

The Karakum Desert ('black sand'), together with the neighbouring Kyzylkum Desert ('red sand'), forms a large part of the Turan Lowland in central Asia. The Karakum lies almost completely within Turkmenistan; a small section is in Uzbekistan.

## KYZYLKUM

**300,000 km² (115,831 sq mi)**

The Kyzylkum Desert is a desert of loose gravel, broken stone and sand in Turkmenistan, Uzbekistan and Kazakhstan, lying between the Amu Darya River in the southwest and Syr Darya River in the north. It comprises an extensive plain which slopes away to the northwest and several separate hill summits up to 922 m (3,025 ft) high. In the centre of the Kyzylkum Desert, the Muruntau granite massif, where gold is mined, rises.

*Wadi Rum, the largest dry river bed in Jordan, is approximately 100 km (62 mi) long and 60 km (37 mi) wide. The wadi is edged with steeply soaring rock faces and mountain ranges of sandstone and granite (large picture).*

## SYRIAN DESERT

**260,000 km² (100,387 sq mi)**

The Syrian Desert, a steppe- and semi-desert cretaceous limestone plateau, lies in the north of the Arabian Peninsula. It stretches over northern Saudi Arabia, northeastern Jordan, southeastern Syria, from which it gets its name, and western Iraq. The dry zone is bordered by the plains of the Fertile Crescent to the north, the River Euphrates in the northeast and the volcanic massif of the Hauran in the southeast. Massive pipelines cross the Syrian Desert, connecting the oilfields of Saudi Arabia and Iraq with the ports on the Mediterranean. Along the old caravan routes through the steppe and desert landscape, the ruins of several historic cities originally built on the sites of oases lie, such as Tudmur, dating from the first century AD. Because the Syrian Desert only receives the average equivalent of 130 mm (5 in) rainfall per year, any land under cultivation has to be intensively irrigated.

## THAR

**260,000 km² (100,387 sq mi)**

The Thar Desert is a desert region on the Indian subcontinent, two-thirds of which lies in Rajasthan. The remainder is in Haryana,

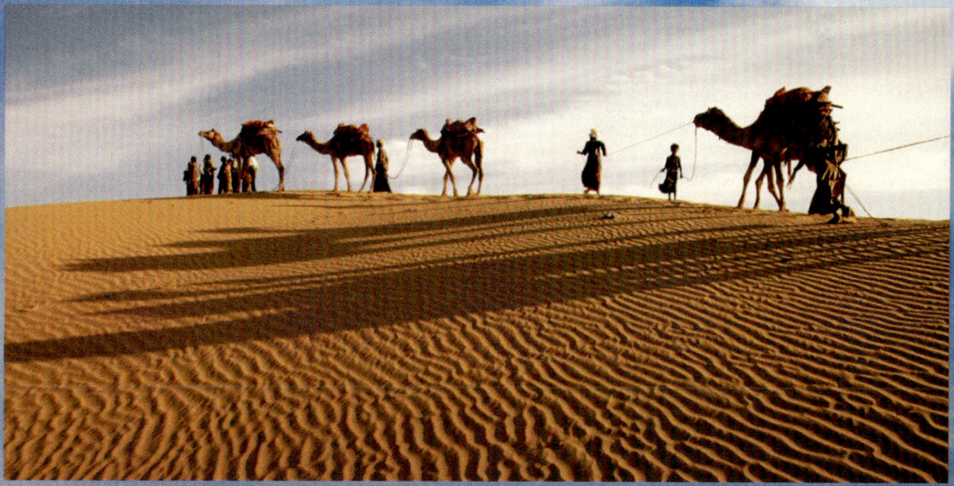

Punjab and Gujarat. It comprises fixed and shifting sand dunes, cliffs and salt pans. Extreme temperature fluctuations from 0°C (32°F) in winter to 50°C (122°F) in summer are a feature of the climate. The annual rainfall is less than 150 mm (6 in) in the west and 350 mm (14 in) in the east.

DID YOU KNOW?

**Lut Desert 70°C (158°F) – the hottest place on earth, 274,000 km² (105,792 sq mi)**

Dascht-e Lut, southwest of the Zagros Mountains, is the largest desert in Iran. In the north, the country's second largest desert, the Kavir, connects with it. In Lut Desert, there are several very high dunes, which are otherwise only found in the ergs in the Sahara Desert. With temperatures up to 70.7°C (160°F), the Lut Desert is the hottest place in the world. No evidence has yet been found of earlier settlements or fossils, and today the desert is still unpopulated.

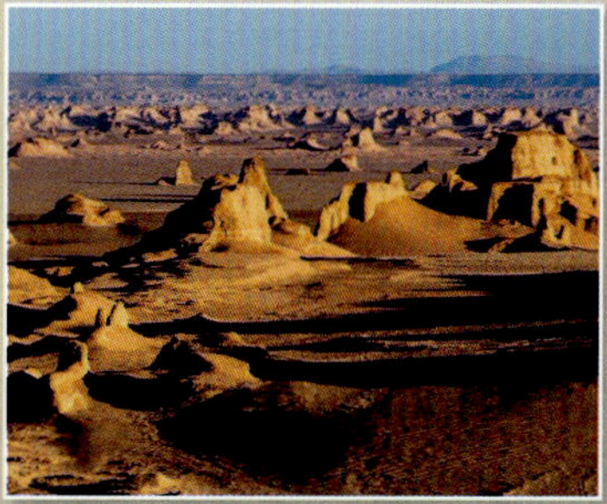

## TAKLAMAKAN

**320,000 km² (123,553 sq mi)**

The Taklamakan Desert in central Asia is the second largest sand desert in the world. It extends from Xinjiang in China through the Tarim Basin to Highway 218 (to the east of this highway lies the Lop Desert). The Taklamakan mainly consists of dunes over 100 m (328 ft) high, which swiftly migrate due to the strong winds. With less than a 30-mm (1.2-in) rainfall a year, the Taklamakan Desert is classed as extremely arid.

FOUR ELEMENTS

FIRE

# NORTH AND CENTRAL AMERICA

The deserts and semi-deserts stretch from Texas to the mountains of southern California and from Oregon to Mexico, all with very different types of flora.

## GREAT BASIN

**540,000 km² (208,495 sq mi)**

The Great Basin is an extensive desert upland with no outlet to the sea and consists of around 100 sub-basins. It lies between the Colombia Plateau in the north, the Mojave Desert in the south, the Sierra Nevada in the west and the Wasatch Range, a cordillera of the Rocky Mountains, in the east. Most of the Great Basin lies within the US state of Nevada; there are also smaller sections in Oregon, Idaho, Utah, Arizona and California. One of the highest points is the Humboldt Range, with a 3,677-m (12,064-ft) high summit. The national park is famous for the Great Basin bristlecone pine trees, the oldest known living organisms in the world.

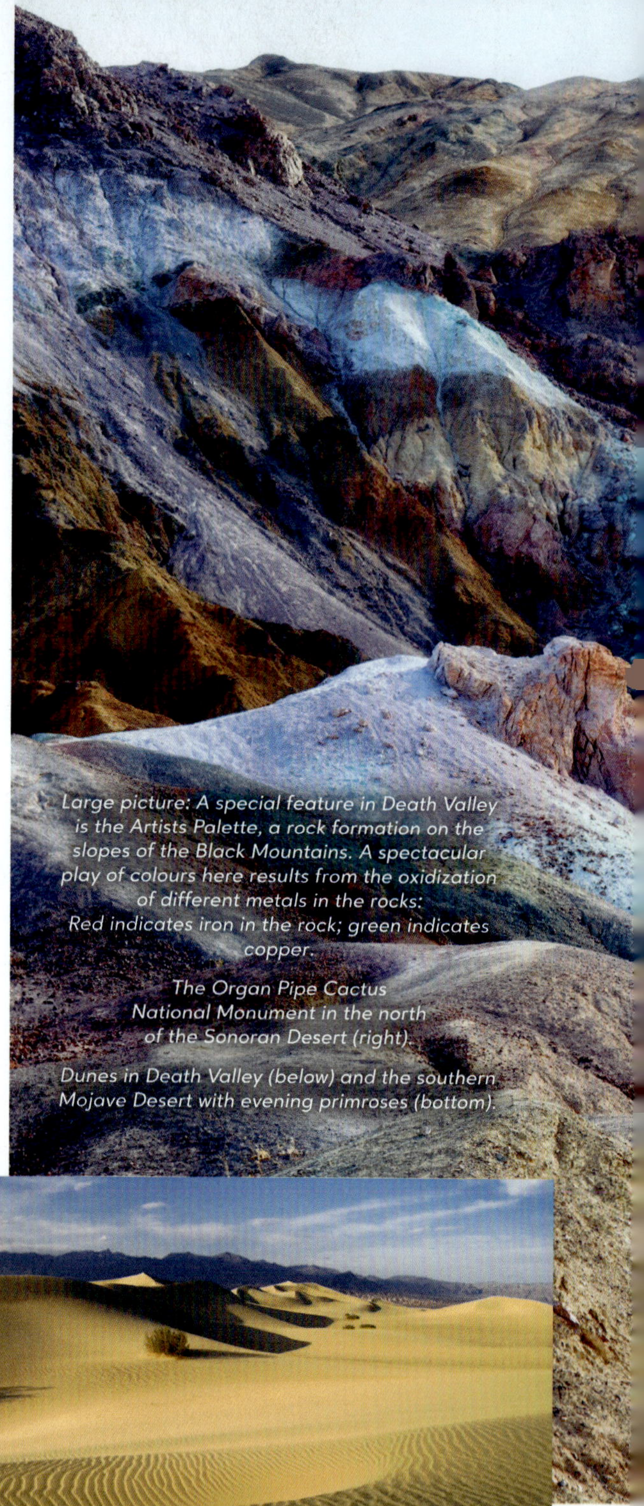

*Large picture: A special feature in Death Valley is the Artists Palette, a rock formation on the slopes of the Black Mountains. A spectacular play of colours here results from the oxidization of different metals in the rocks: Red indicates iron in the rock; green indicates copper.*

*The Organ Pipe Cactus National Monument in the north of the Sonoran Desert (right).*

*Dunes in Death Valley (below) and the southern Mojave Desert with evening primroses (bottom).*

## DEATH VALLEY

**8,500 km²(3,282 sq mi)**

Death Valley, named so because its climate is reputedly hostile to life, is a valley in the Great Basin, surrounded by several mountains. It is accessed via the Badwater Basin, a saline pool that is the deepest point in the western hemisphere (85.5 m (281 ft) below sea level). Although Death Valley is only a few hundred kilometres from the Pacific Ocean, it is the driest place on earth. The heat record here stands at 56.7°C (134°F).

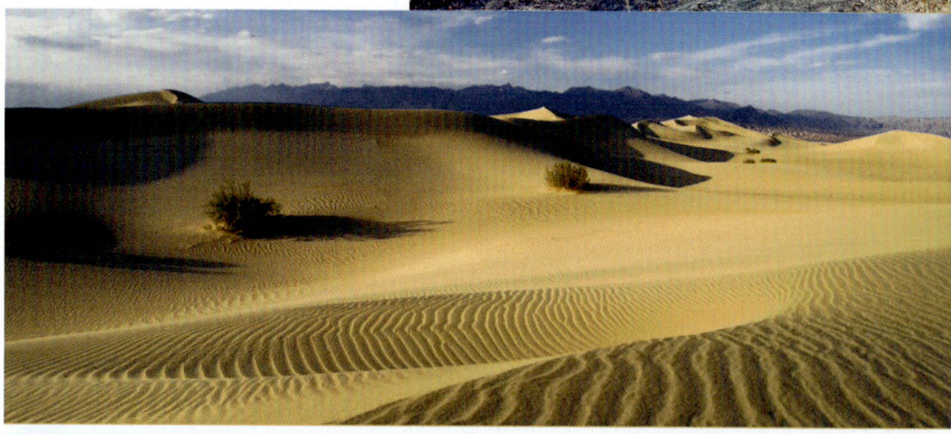

## MOJAVE DESERT

**40,000 km² (15,444 sq mi)**

The Mojave Desert lies in the US states of California, Utah, Nevada and Arizona and is a desert basin with no outlet to the sea, surrounded by several mountain ranges. The desert contains various ghost towns which cropped up at the time of the silver rush and the construction of Route 66. Many different reptiles live here, including rattlesnakes. The Joshua tree is only found here.

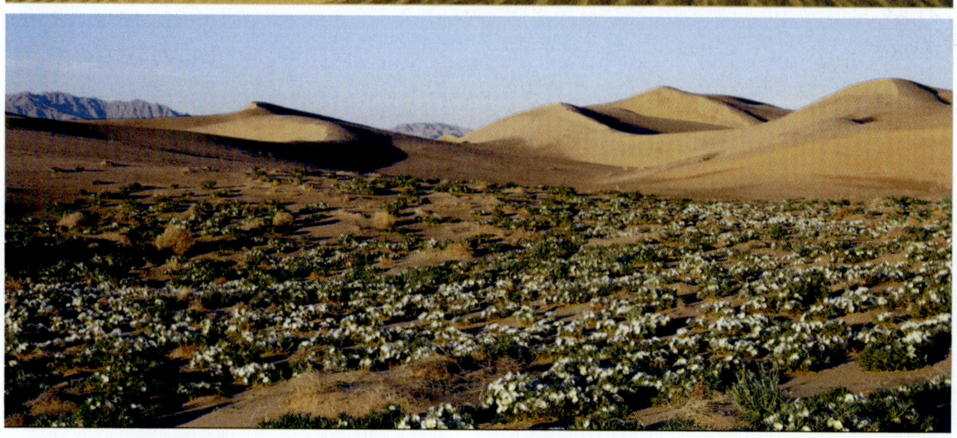

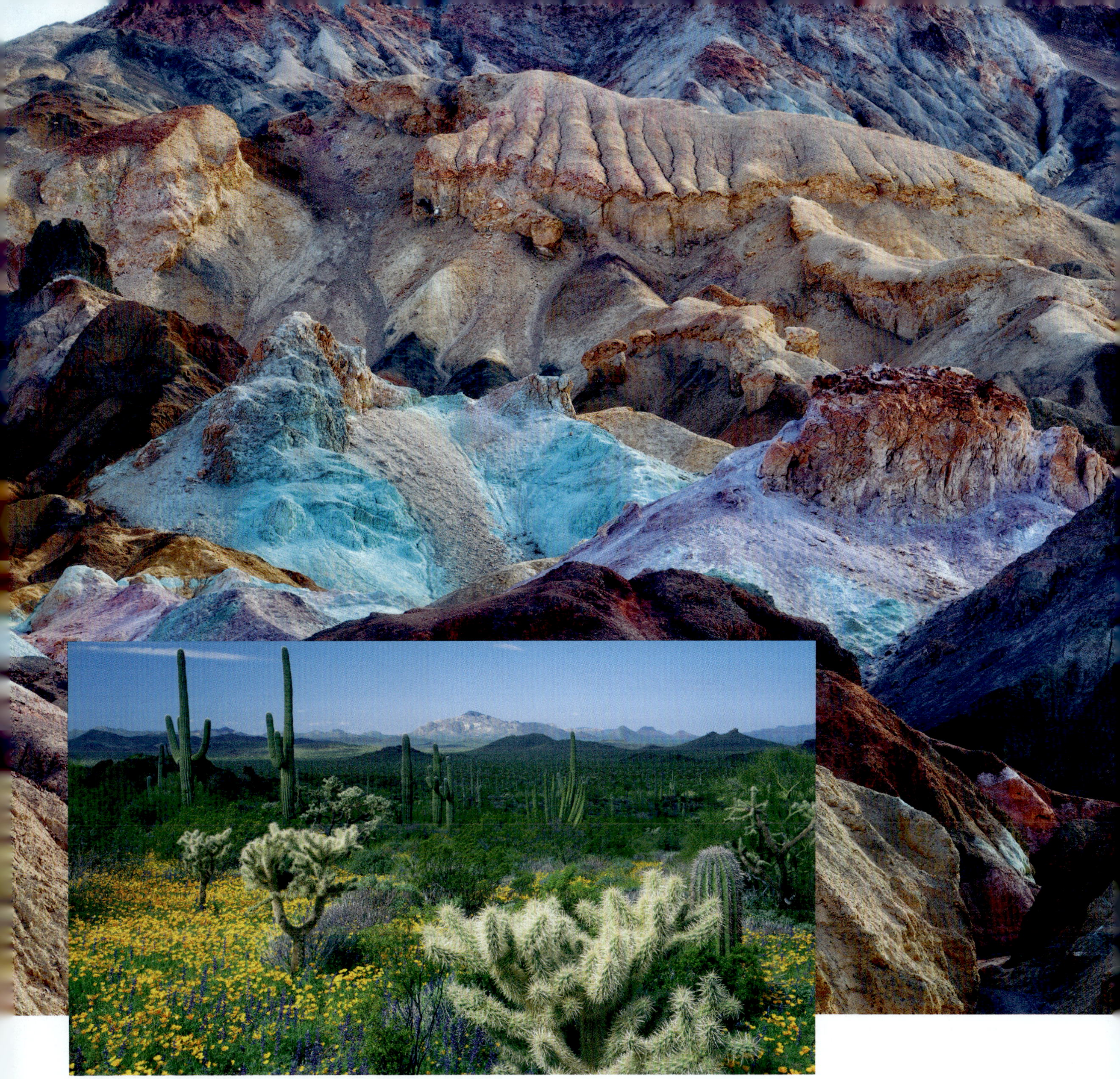

## SONORAN DESERT
**320,000 km² (123,553 sq mi)**

The parallel mountain ranges with their wide valleys and the North Pacific coastal plains in the Mexican state of Sonora were formed during the middle and late Tertiary periods. In the north, this dry landscape begins in the Sonoran Desert, which stretches to California and southwest Arizona. It is bordered by the Mojave Desert in the north and by the Chihuahuan Desert in the east. The Sonoran Desert is one of the most varied and biodiverse desert regions in the world. Many reptile and bird species live here as well as many different types of cacti, which are a particular attraction at the Organ Pipe Cactus National Monument.

## CHIHUAHUAN DESERT
**363,000 km² (140,155 sq mi)**

The Chihuahuan Desert extends to the US states of New Mexico in the south and Texas in the west as well as to the Mexican states of Chihuahua and Coahuila. In the Sierra del Carmen, there are coniferous forests. The Chihuahuan Desert has typical desert vegetation – agaves, cacti, peyote cacti and grasses. The most striking feature of the Chihuahuan Desert is White Sands National Park, with dunes formed by white gypsum crystals. When it rains, the gypsum is washed from the surrounding mountains into Lake Lucero, which has no outlet; the lake dries up in summer, and the gypsum crystals remain and are then shaped into dunes by the wind.

FOUR ELEMENTS

FIRE

# JOSHUA TREES AND SAGUARO CACTI

The largest desert plants in America

## JOSHUA TREES – THE LARGEST AGAVE PLANTS IN THE WORLD

Like silent guardians, the Joshua trees – Joshua yuccas, in fact – stand out against the mostly clear skies in the Joshua Tree National Park, east of Palm Springs. These cactus-like plants are related to yuccas and can grow up to 12 m (39 ft) high and live to be 900 years old. The plants are an important element of the Mojave Desert's ecosystem and provide many desert creatures with food and shelter. The fibres of the yucca are processed to make ropes and cables.

## SAGUAROS – THE LARGEST CACTUS ON EARTH

The nature conservation area of Saguaro National Park extends over southern Arizona in the Tucson Mountains and the Sonoran Desert. The park holds a particularly rich and beautiful collection of saguaro cacti. The botanical name of the plant is *Carnegiea gigantea*, in honour of the American industrialist and philanthropist, Andrew Carnegie.

The mighty succulents (fleshy plants) reach a height of up to 15 m (49 ft) and an average age of 85 years; individual plants can even live to be 200 years. A full-grown saguaro can weigh as much as 15 tonne (16.5 tons). Each of the long-tubed flowers is pollinated by bats, hummingbirds and woodpeckers, and they bloom in May for just one day, shortly before the rainy season. The fruits, which are red inside and out, are 6 to 9 cm (2.3 to 3.5 in) long. The saguaro cactus grows extremely slowly – only 4 cm (1.6 in) a year. The biggest threat to the plants is from desert hares, bush rats and bighorn sheep, who all like to eat them. Woodpeckers make their breeding burrows in the trunks of the cacti, and larger birds, such as falcons, build their eyries on the plants. The delicious fruits are used in dried form by the Papago Indians (or Tohono O'odham Nation) as a foodstuff.

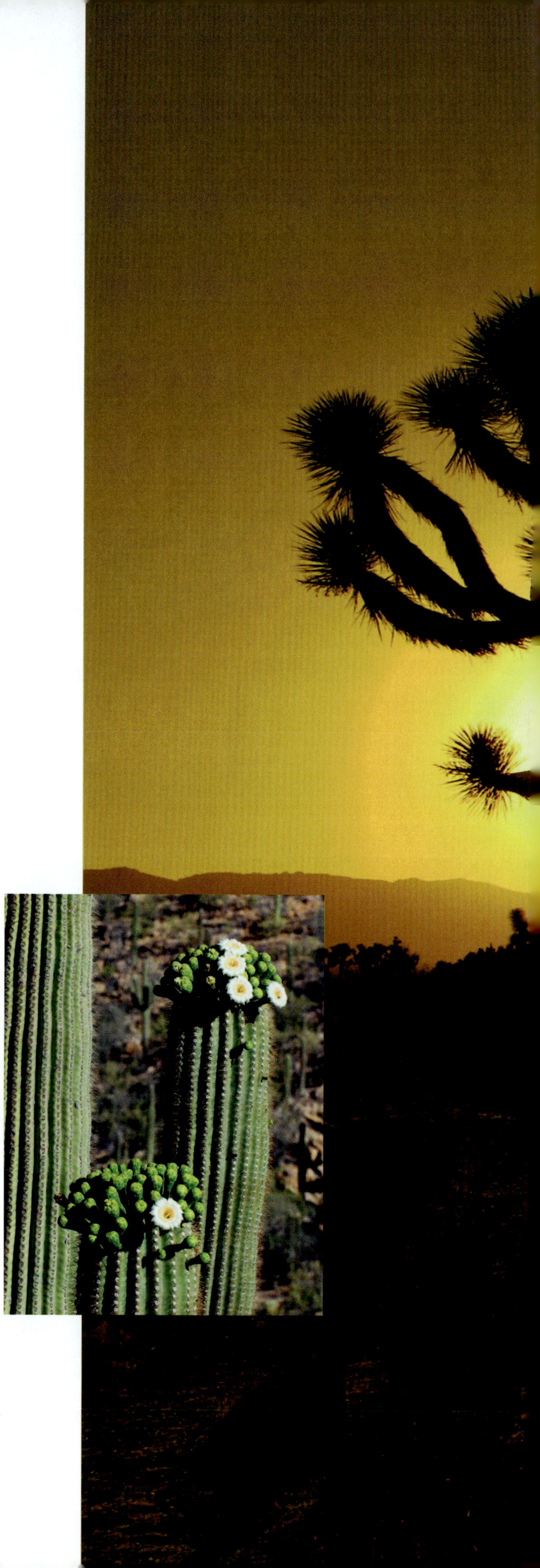

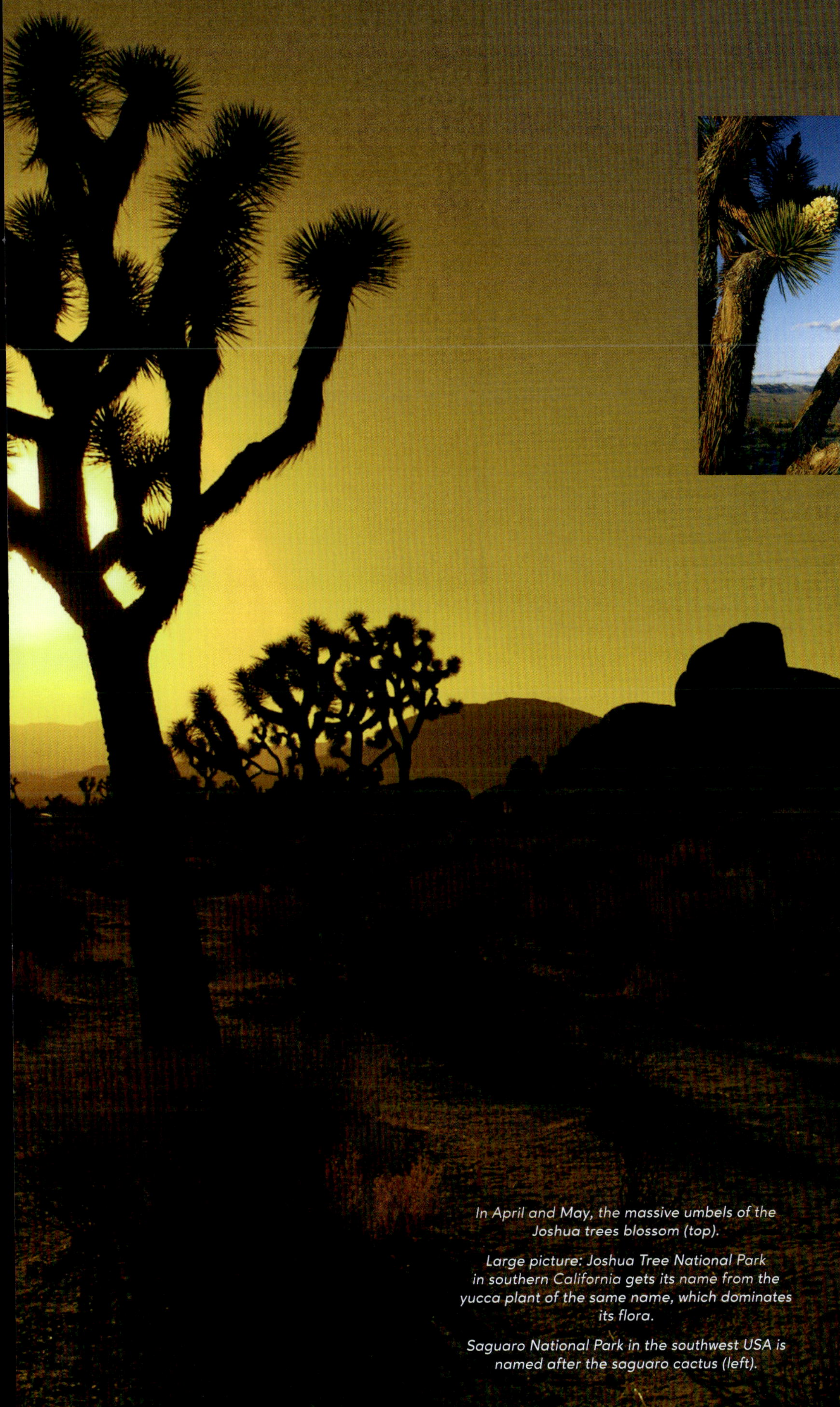

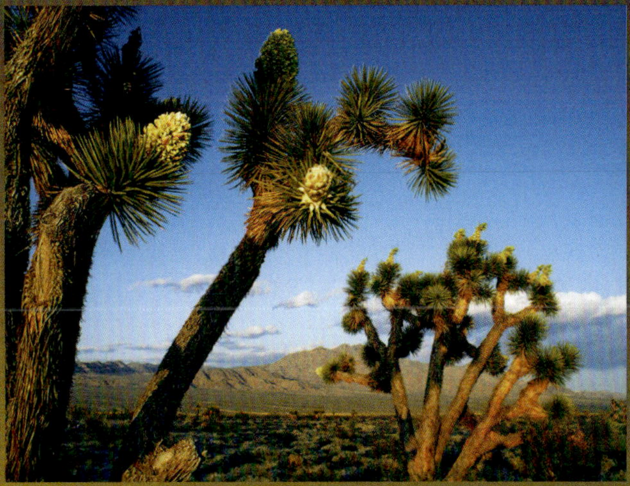

*In April and May, the massive umbels of the Joshua trees blossom (top).*

*Large picture: Joshua Tree National Park in southern California gets its name from the yucca plant of the same name, which dominates its flora.*

*Saguaro National Park in the southwest USA is named after the saguaro cactus (left).*

DID YOU KNOW?

**Coachella Festival**

Hippie flair with a desert backdrop – this festival is famous for its spectacular scenery. A Ferris wheel in the middle of the desert. Indie, pop and electronic music is just as likely to be played here as hip-hop. Coachella Valley, which gives the festival its name, lies in the Colorado Desert; since 1999, the festival has been held here. Every year, in mid April, the time comes, and famous stars gather here. Rihanna was a guest artist here as were Beyoncé and Paul McCartney. The festival runs over three days and is famous for the multi-coloured crocheted fashion and flower-power outfits as well as the increasingly important role that sustainability plays in it.

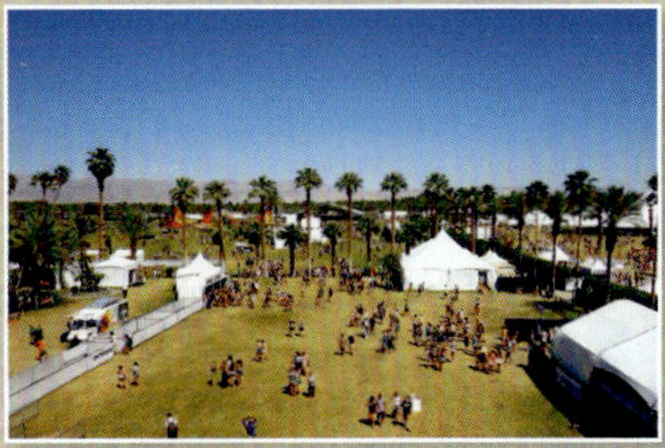

FOUR ELEMENTS

FIRE

# AUSTRALIA

Around two-thirds of Australia, the driest continent on earth, is covered in deserts or semi-deserts, which feature many impressive geological formations

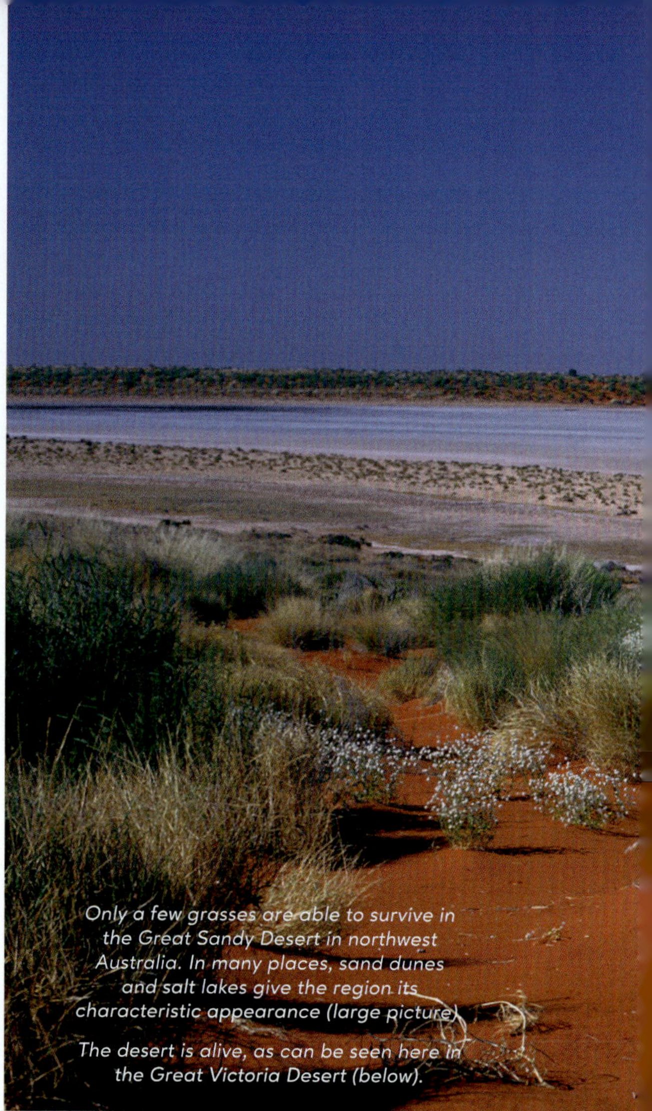

Only a few grasses are able to survive in the Great Sandy Desert in northwest Australia. In many places, sand dunes and salt lakes give the region its characteristic appearance (large picture).

The desert is alive, as can be seen here in the Great Victoria Desert (below).

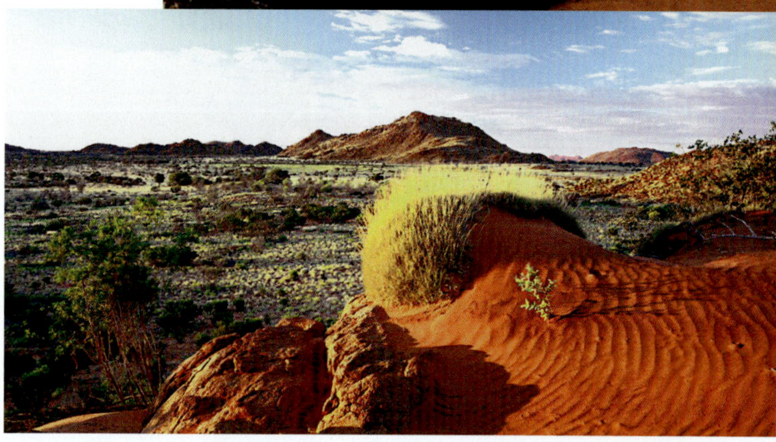

## THE GREAT SANDY DESERT

**388,000 km² (149,808 sq mi)**

In Western Australia, in the northwest of the continent, the Great Sandy Desert lies in the Canning Basin, a plain between the Pilbara and Kimberley Mountains. The desert has several salt lakes, and only several groups of indigenous Australians live there. In the coastal area and around the Kimberley Mountains, the annual rainfall is 300 mm (12 in) although this falls at very irregular intervals. In the rest of the desert, the rainfall is around 250 mm (10 in) annually. This is relatively high for a desert although the heat means that evaporation rates are high. In summer, the temperature soars to around 38°C (100°F); in the short winters, it never falls below 25°C (77°F). Given these conditions, only a few grasses grow here. In spite of the inhospitable environment, the area is home to the princess parrot, a species of parrot adapted to arid zones.

## GIBSON DESERT

**330,000 km² (127,414 sq mi)**

The Gibson Desert in Western Australia lies south of the Great Sandy Desert and north of the Great Victoria Desert, between Kumpupintil Lake (formerly Lake Disappointment) and Lake Macdonald on the Tropic of Capricorn. In geographical terms, Gibson Desert is part of Australia's Western Plateau. It consists of wide expanses of undulating sand and dune fields, low rocky ridges (the highest elevation is only 500 m (1,640 ft) above sea level) and areas with numerous laterite formations. Several saltwater lakes lie in the centre of the Gibson Desert. Only a few roads cross the desert, and the singular notable settlement is Warburton at the edge of the dry zone. The climate of the Gibson Desert is hot and dry, with an annual rainfall between 200 and 250 mm (7.9 - 9.8 in) (with a high evaporation rate). Temperatures are over 40°C (100°F) in the summer months and a maximum of 18°C (65°F) to 6°C (43°F) in the winter.

## GREAT VICTORIA DESERT

**274,000 km² (105,792 sq mi)**

The Great Victoria Desert is a semi-desert in southern Australia in the states of South and West Australia. The sand dunes, which reach an average height of 15 m (49 ft), are in part stabilized by the grass cover; others are shifting sand dunes. Between them lie grasslands and salt lakes. The Great Victoria Desert is known for its crescent-shaped dunes. These form on the leeward edges of lakes which are only rarely filled with water and are composed of sediment from the lake bed. The average temperatures are between 18°C (65°F) and 40°C (100°F) with occasional frosts in winter. In some areas of the desert, it is damp enough for mulga, a variety of acacia native to Australia, to grow. Reptiles, such as geckos and monitor lizards, inhabit the desert.

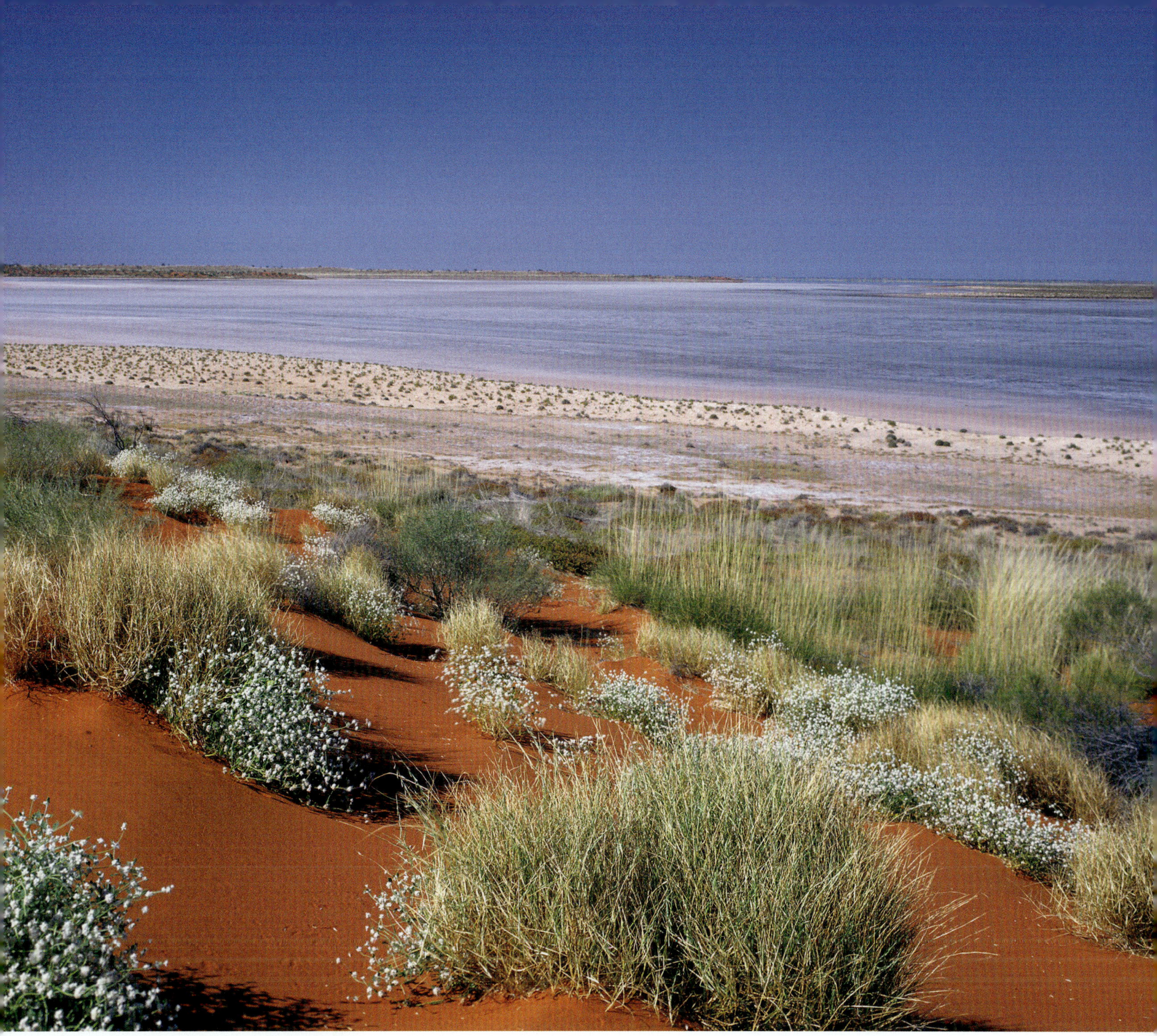

## TANAMI DESERT

**180,000 km² (69,498 sq mi)**

Tanami Desert, the most northerly desert in Australia, lies in the Northern Territory, east of the Great Sandy Desert and northwest of Alice Springs. This far north in the continent, the rainfall is greater than in other regions: The Tanami Desert registers up to 400 mm (15.7 in) per year. However, due to the heat, the water evaporates very quickly. The daytime temperature in summer is around 38°C (100°F); in winter, it still reaches about 25°C (77°F) with occasional night frosts. Vegetation here is only sparse. The popular Tanami Track, 1,000 km (621 mi) long, runs across the wide red sand plain of the Tanami Desert, which is interrupted in places by the low hill ranges; the track is the shortest connecting route between the Kimberley region and the centre of Australia. This natural track is in good condition, but, in summer's rainy season, it sometimes floods and becomes unpassable.

## SIMPSON DESERT

**250,000 km² (96,526 sq mi)**

The Simpson Desert also extends into the Northern Territory: In the west, Finke River and Mabel Range border the desert; Adam Range lies to the north with the Georgina and Diamantina Rivers in the west, and Lake Eyre is to the south. There is a large aquifer under the desert; however, this threatens to run dry, since it has been over-exploited in recent years. In the interior of the Simpson Desert, there are extensive shifting sand dunes, and stark rock formations stand at its edges. The region is largely uninhabited and protected by nature conservation areas, among them Simpson Desert National Park and Witjira National Park. Because there is very little rainfall in the Simpson Desert in comparison to other deserts in Australia (less than 150 mm (6 in) annually), only bush vegetation grows by its edges. Deeper in the desert's interior, the bush is increasingly replaced by spinifex grass, a type of sweet grass.

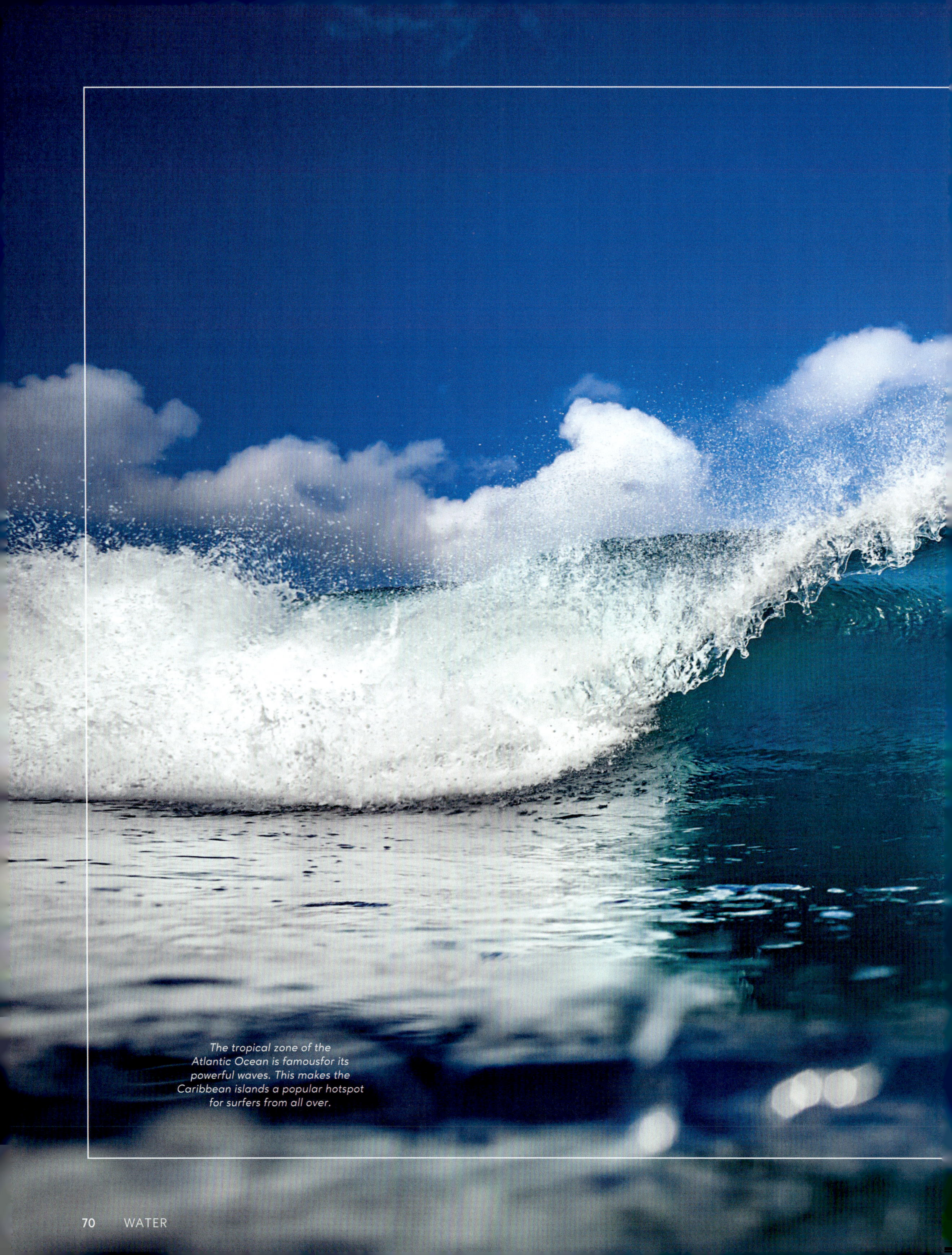

The tropical zone of the Atlantic Ocean is famousfor its powerful waves. This makes the Caribbean islands a popular hotspot for surfers from all over.

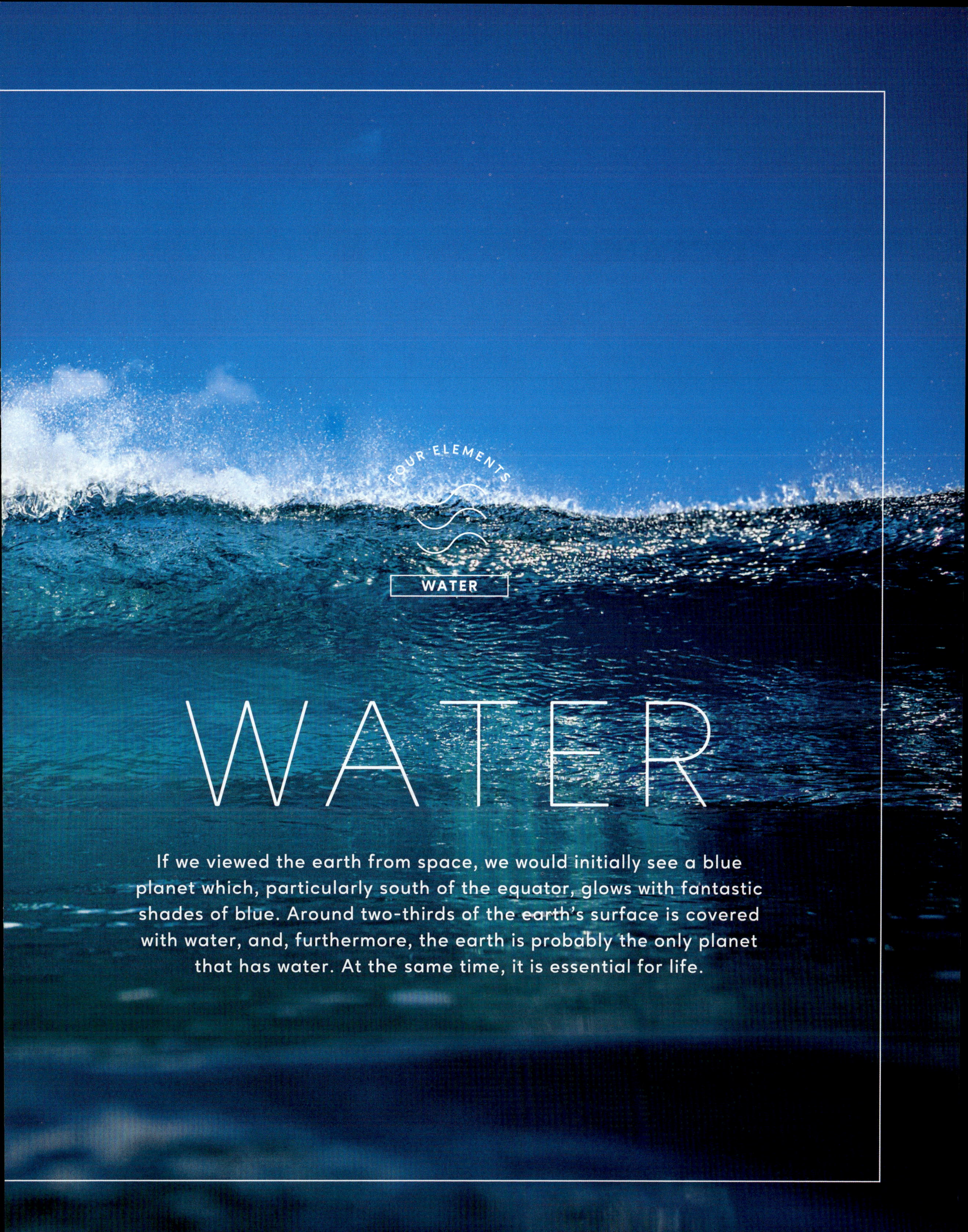

FOUR ELEMENTS

WATER

# WATER

If we viewed the earth from space, we would initially see a blue planet which, particularly south of the equator, glows with fantastic shades of blue. Around two-thirds of the earth's surface is covered with water, and, furthermore, the earth is probably the only planet that has water. At the same time, it is essential for life.

FOUR ELEMENTS

WATER

# OCEANS

Oceans cover 71 per cent of our planet and, in an expanse of 361 million km² (13,899,678 sq mi), form a contiguous body of water that makes the earth appear like a 'blue planet' when seen from space.

The three largest oceans – the Pacific, Atlantic and Indian Oceans – account for around 90 per cent of this area; the remaining 10 per cent is divided between the Arctic and Antarctic Oceans. The oceans contain 97.5 per cent of the earth's water; only 2.5 per cent is freshwater on the mainland. The ocean salinity averages 3.5 per cent. The average depth of the seabed ranges between 3,000 m (9,843 ft) and 4,500 m (14,764 ft). The transitions to continents and islands vary. Depending on tectonic events, there may be shelf seas less than 200 m (656 ft) deep and continental breakups or ocean trenches which reach down to depths of up to 11,022 m (36,163 ft).

## OCEANS ACCORDING TO SURFACE AREA AND AVERAGE DEPTH

1
**PACIFIC OCEAN**
Surface area 181,340,000 km² (70.015,765 sq mi)
Average depth 4,188 m (13,740 ft)

2
**ATLANTIC OCEAN**
Surface area 106,400,000 km² (41,081,270 sq mi)
Average depth 3,339 m (10,955 ft)

3
**INDIAN OCEAN**
Surface area 73,556,000 km² (28,400,130 sq mi)
Average depth 3,400 m (11,155 ft)

4
**ANTARCTIC OCEAN**
Surface area 20,327,000 km² (7,848,299 sq mi)
Average depth 4,500 m (14,764 ft)

5
**ARCTIC OCEAN**
Surface area 14,090,000 km² (5,440.179 sq mi)
Average depth 1,500 m (4,92 ft)

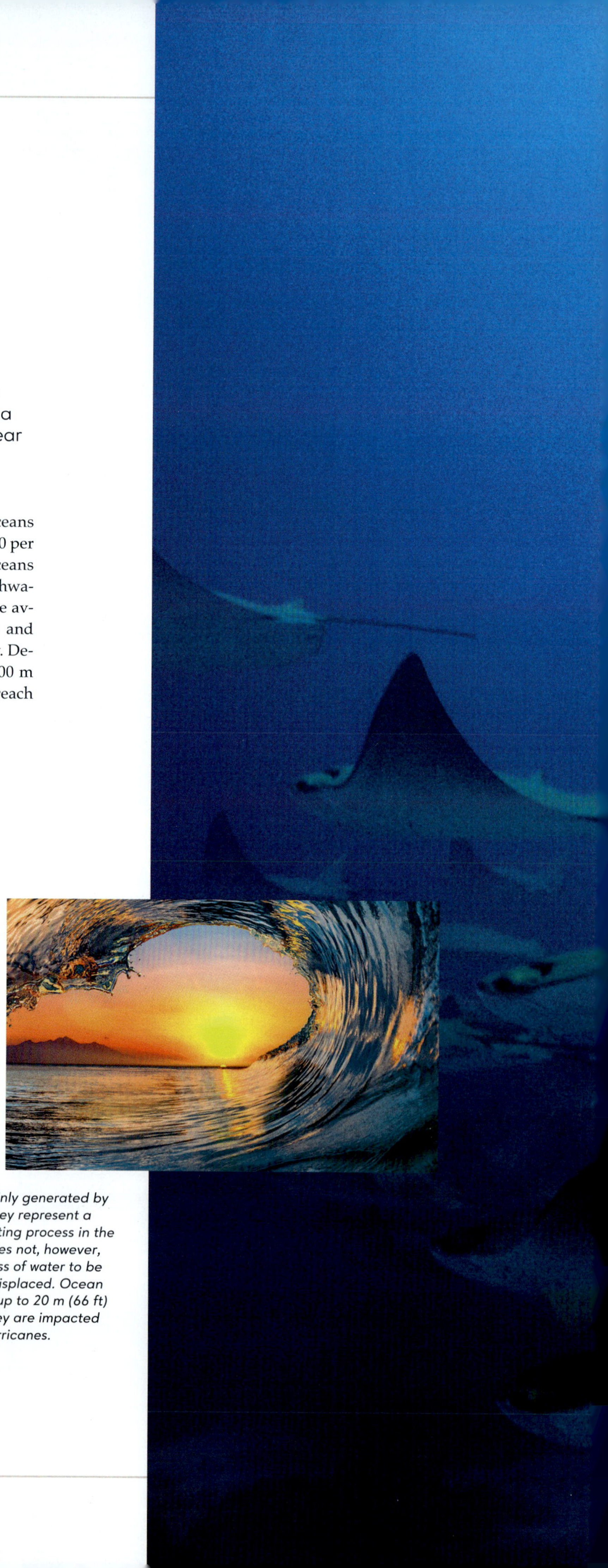

*Waves are mainly generated by the wind. They represent a dynamic oscillating process in the water that does not, however, cause the mass of water to be significantly displaced. Ocean waves can be up to 20 m (66 ft) high when they are impacted by hurricanes.*

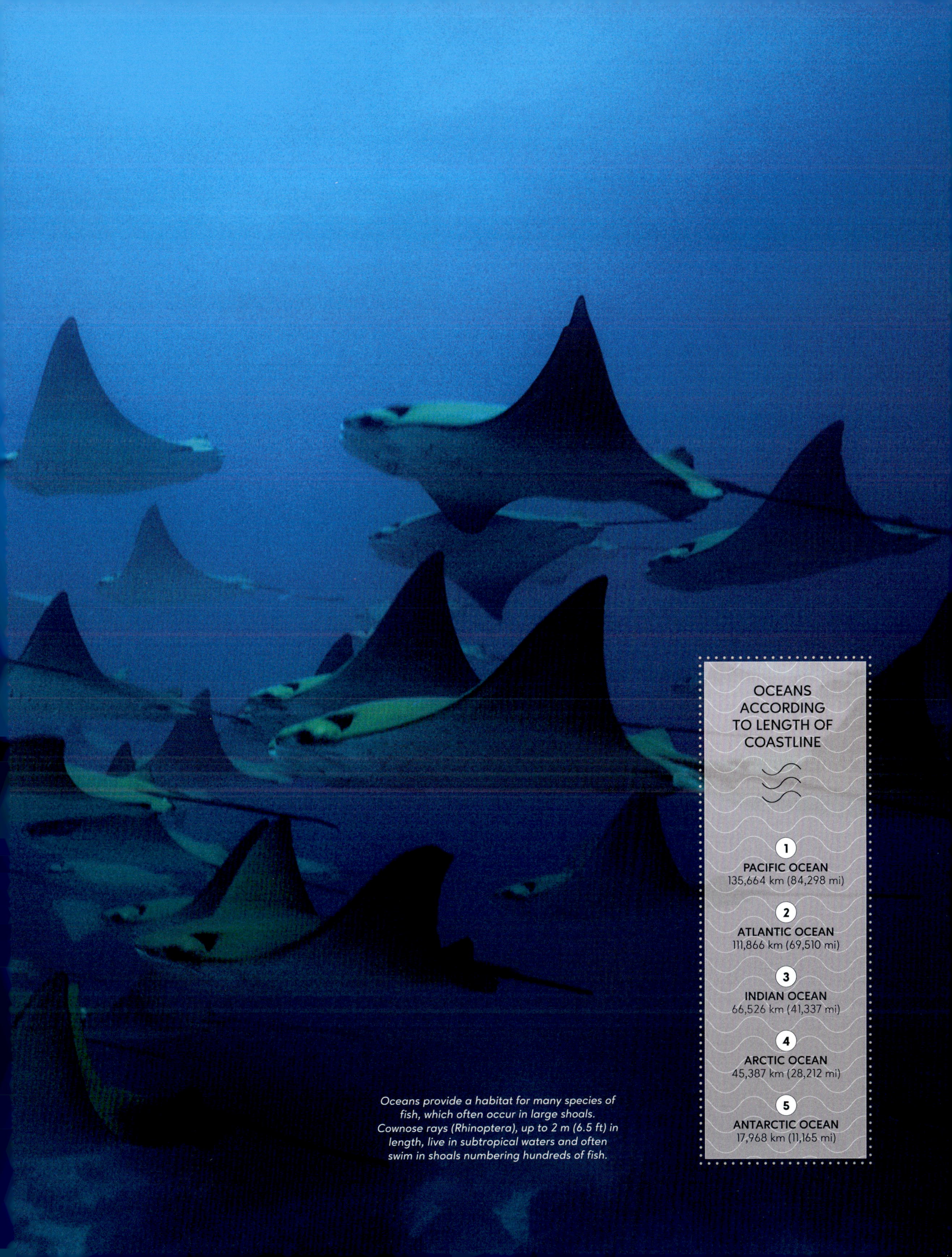

## OCEANS ACCORDING TO LENGTH OF COASTLINE

1. **PACIFIC OCEAN**
   135,664 km (84,298 mi)
2. **ATLANTIC OCEAN**
   111,866 km (69,510 mi)
3. **INDIAN OCEAN**
   66,526 km (41,337 mi)
4. **ARCTIC OCEAN**
   45,387 km (28,212 mi)
5. **ANTARCTIC OCEAN**
   17,968 km (11,165 mi)

*Oceans provide a habitat for many species of fish, which often occur in large shoals. Cownose rays (Rhinoptera), up to 2 m (6.5 ft) in length, live in subtropical waters and often swim in shoals numbering hundreds of fish.*

# PACIFIC OCEAN

Along with its marginal seas, the Pacific Ocean covers 181,340,000 km² (70,015,765 sq mi), which is a little over a third of the earth's total surface.

Almost all its marginal seas, delimited by islands and island chains, are located in the western part of the ocean. The ocean trenches and underwater ridges of the seabed are particularly varied. The Challenger Deep in the Mariana Trench is the deepest point of all the world's oceans. The seabed in the eastern Pacific Ocean is not as varied. Apart from the elongated East Pacific Rise, which runs from Lower California to the Antarctic Ocean, it only has a series of fault zones and a few islands.

## THE AUSTRALASIAN MEDITERRANEAN SEA – THE LARGEST MARGINAL SEA OF THE PACIFIC OCEAN

The Australasian Mediterranean Sea includes all the marginal seas of the Pacific Ocean. Islands mark its border with the Indian Ocean, and there are wide straits on the border to the Pacific. It is a tropical body of water with a correspondingly rich variety of flora and fauna. Coral reefs occur in all parts of the Australasian Mediterranean Sea in varying distributions. The massive shallow continental shelves are mainly found in the western part of the sea; in the east, deep basins prevail, descending to depths of 7,440 m (24,409 ft).

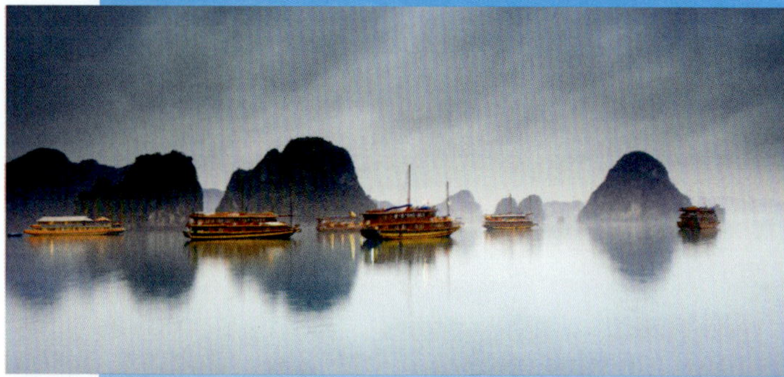

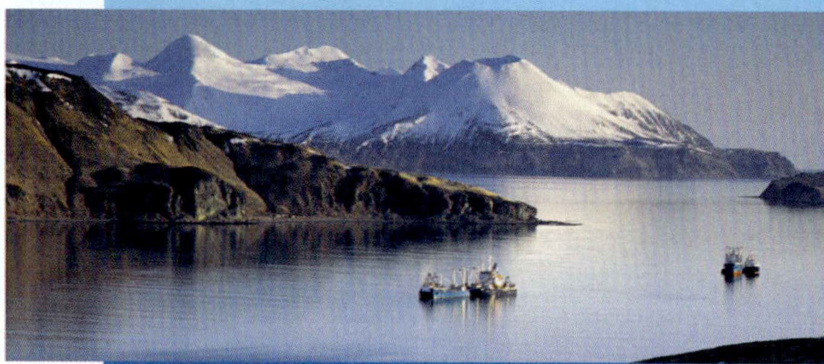

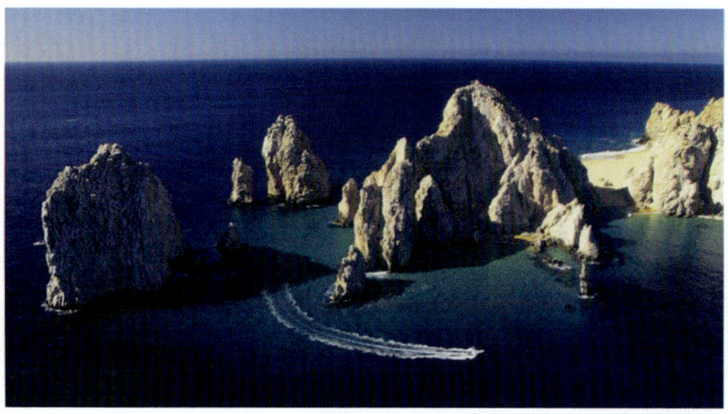

*Junks – wooden Chinese boats – sail in Halong Bay in the Gulf of Tonkin in northern Vietnam (above, top).*

*The Aleutian Islands, an extremely mountainous island chain at the southern edge of the Bering Sea (above, bottom).*

*Cabo San Lucas, the southernmost tip of the Baja California peninsula, or Lower Californian Peninsula (above).*

### THE LARGEST MARGINAL SEAS

| | |
|---|---|
| 1 **AUSTRALASIAN MEDITERRANEAN SEA** Western Pacific 9,080,000 km² (3,505,808 sq mi) | 6 **GULF OF ALASKA** North Pacific 1,533,000 km² (591,895 sq mi) |
| 2 **PHILIPPINE SEA** Western Pacific approx. 5,000,000 km² (1,930,510 sq mi) | 7 **SEA OF OKHOTSK** North Pacific 1,530,000 km² (590,736 sq mi) |
| 3 **CORAL SEA** Southwest Pacific 4,791,000 km² (1,849,815 sq mi) | 8 **EAST CHINA SEA** Western Pacific 1,250,000 km² (482,628 sq mi) |
| 4 **BERING SEA** North Pacific 2,300,000 km² (885,035 sq mi) | 9 **SEA OF JAPAN** Western Pacific 1,049,000 km² (405,021 sq mi) |
| 5 **TASMAN SEA** Southwest Pacific 2,300,000 km² (885,035 sq mi) | 10 **GULF OF CALIFORNIA** Eastern Pacific 160,000 km² (61,776 sq mi) |

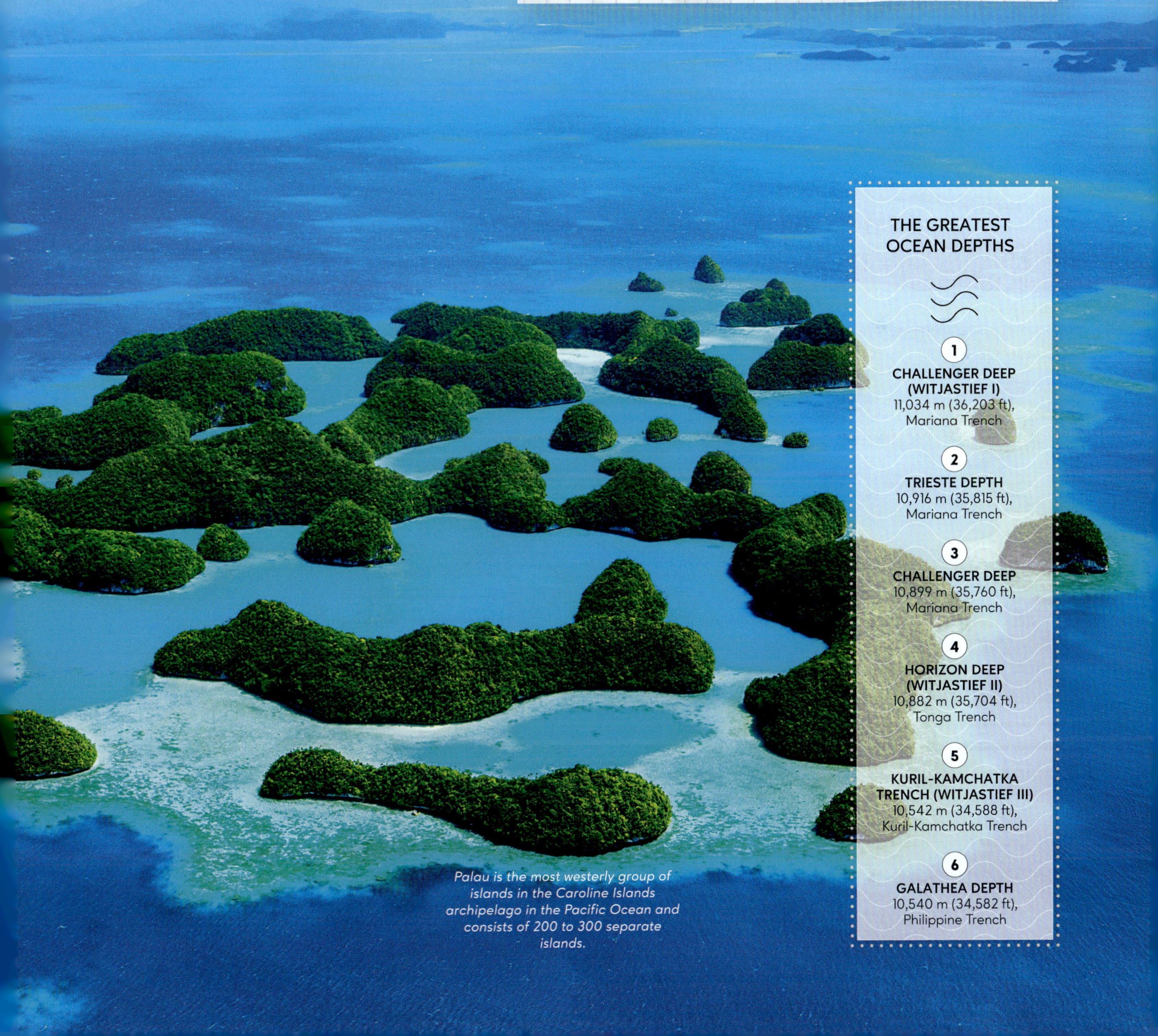

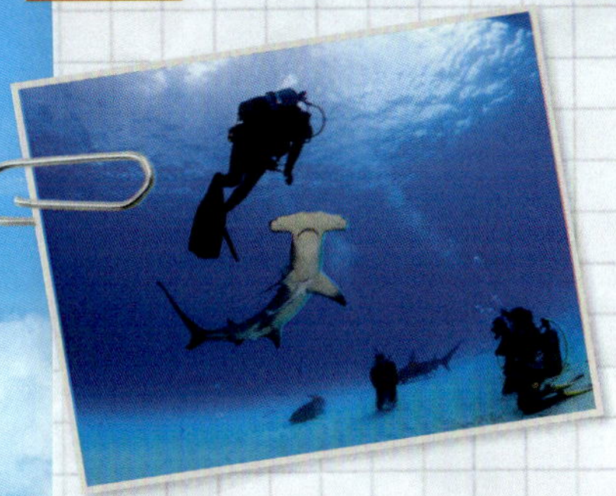

## TIP

**DIVE WITH HAMMERHEAD SHARKS OFF ISLA DE COCO IN AUGUST**

It is considered the island of hammerhead sharks and draws divers from all over the world – the reef off the Isla de Coco is well known for its abundance of underwater creatures. They move through the waters off the coast in huge schools of up to a hundred fish and are not disturbed by the presence of divers. Isla de Coco is part of Costa Rica and observes stringent protection measures, as it is also a UNESCO World Heritage natural site. Travel to the island requires a special permit. It lies approximately 500 km (310 mi) off the mainland and is accessed via a lengthy boat trip. Only nature rangers live on the island; there are consequently no overnight stays for tourists.

*Palau is the most westerly group of islands in the Caroline Islands archipelago in the Pacific Ocean and consists of 200 to 300 separate islands.*

## THE GREATEST OCEAN DEPTHS

1. **CHALLENGER DEEP (WITJASTIEF I)**
   11,034 m (36,203 ft), Mariana Trench
2. **TRIESTE DEPTH**
   10,916 m (35,815 ft), Mariana Trench
3. **CHALLENGER DEEP**
   10,899 m (35,760 ft), Mariana Trench
4. **HORIZON DEEP (WITJASTIEF II)**
   10,882 m (35,704 ft), Tonga Trench
5. **KURIL-KAMCHATKA TRENCH (WITJASTIEF III)**
   10,542 m (34,588 ft), Kuril-Kamchatka Trench
6. **GALATHEA DEPTH**
   10,540 m (34,582 ft), Philippine Trench

# AS IF AN ELEPHANT WERE STANDING ON YOUR THUMB

FOUR ELEMENTS
WATER

DID YOU KNOW?

### How do creatures adapt to the extreme conditions in deep oceans?

The permanent darkness is one of the biggest challenges of living in the ocean depths. To be able to live in these dark zones in spite of this, these creatures have developed various strategies. Some of them, the colossal squid for example, have eyes that are extremely large and highly sensitive, so that they can absorb the small amount of available light and make optimal use of it. Other species, such as the pelican eel, have tiny eyes but, on the other hand, have a light-emitting organ (photophore) at the end of their tails.
Researchers who have penetrated the depths of the oceans report bluish flashes of light emitted by the organisms living there. Around 90 per cent of all deep-sea dwellers produce their own light. They use it for camouflage, to lure their prey or to signal to potential reproductive partners. Because the food available in the extreme depths is scarce, the organisms living here need to expend as little energy as possible. They tend not to hunt their prey: instead, they lie in wait for it. During the seasons when there is even less food available, many deep-sea dwellers go into a kind of hibernation.

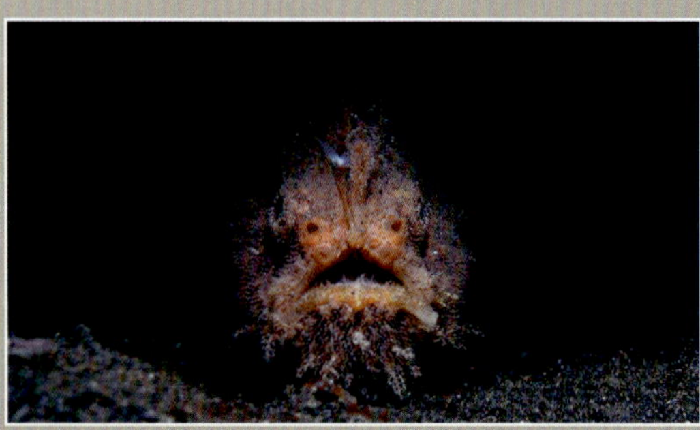

0 m

OCEAN SUNFISH (MOLA MOLA)
Depth below SL: up to 800 m (2,624 ft)
Length: over 4 m (13 ft)
Weight: over 3 tonne (3.3 ton)

- 1,000 m (3,280 ft)

COLOSSAL SQUID
Depth below SL: up to 1,000 m (3,280 ft)
Length: up to 20 m (65 ft)
Weight: up to 50 kg (110 lb)

- 2,000 m (6,562 ft)

BEAKED WHALE
Depth below SL: almost 3,000 m (9,843 ft)
Length: up to 13 m (43 ft)
Weight: up to 15 tonne (16.5 ton)

- 3,000 m (9,843 ft)

- 4,000 m (13,123 ft)

- 5,000 m (16,404 ft)

- 6,000 m (19,685 ft)

- 7,000 m (22,966 ft)

DEEP-SEA ANGLERFISH
Depth below SL: believed to be up to 8,000 m (26,247 ft)
Length: Females up to 120 cm (47 in) corresponding males up to 12 cm (4.7 in)

- 8,000 m (26,247 ft)

COMMON SEASNAIL
Depth below SL: over 8,000 m (26,247 ft)
Length: up to 70 cm (27.5 in)

- 9,000 m (29,528 ft)

Mariana Trench

- 10,000 m (32,808 ft)

TRIESTE DEPTH, MARIANA TRENCH
Depth below SL: 10,916 m (35,813 ft)
First dive: 23 January 1960
Persons: 2

CHALLENGER DEEP (WITJASTIEF I) MARIANA TRENCH
Depth below SL: 11,034 m (36,201 ft)
First dive: None
Persons: None

- 11,000 m (36,089 ft)

- 12,000 m (39,370 ft)

## WHICH CREATURES LIVE IN THE DEEP SEA?

'Deep sea' refers to any water deeper than 800 m (2,625 ft). From that point, there is total darkness and the water pressure is immense. At a depth of 10,000 m (32,810 ft), a weight of around one tonne weighs down on each square centimetre.

No life can exist here, or so explorers thought up until the middle of the 19th century. Doubts were raised when a telegraph cable encrusted with underwater creatures was hauled up from a depth of 1,800 m (5,906 ft) under the ocean's surface. A 1869 expedition southwest of Ireland by the HMS Porcupine took ground samples from a depth of 4,000 m (13,124 ft) and finally provided the proof that there was life even in these depths of the ocean. Indeed, the number of underwater creatures decreases the deeper the water, but the variety of species is astounding.

Creatures such as the vampire squid, batfish or anglerfish not only look bizarre: they have also developed unusual hunting techniques in their search for prey.

On the seabed, 200,000 species have already been identified; it is surmised that the number of different species could run into the millions. Even in the sea mud of the deepest trenches, there are crustaceans, molluscs, bristle worms and sea cucumbers. At a depth of 4,000 m (13,124 ft), these account for around half of the mass of all organisms, and, at 8,500 m (27,888 ft) deep, the figure rises to 90 per cent.

Tube worms of several feet or single-cell organisms measuring up to 25 cm (9.8 in) have become specialized in the art of living next to extinct deep sea volcanoes. Of particular interest to researchers are the xenophyophores (literally, 'bearers of foreign bodies') with casings made of dead plankton. In turn, these casings are inhabited by various tenants, such as different species of sea stars and isopods.

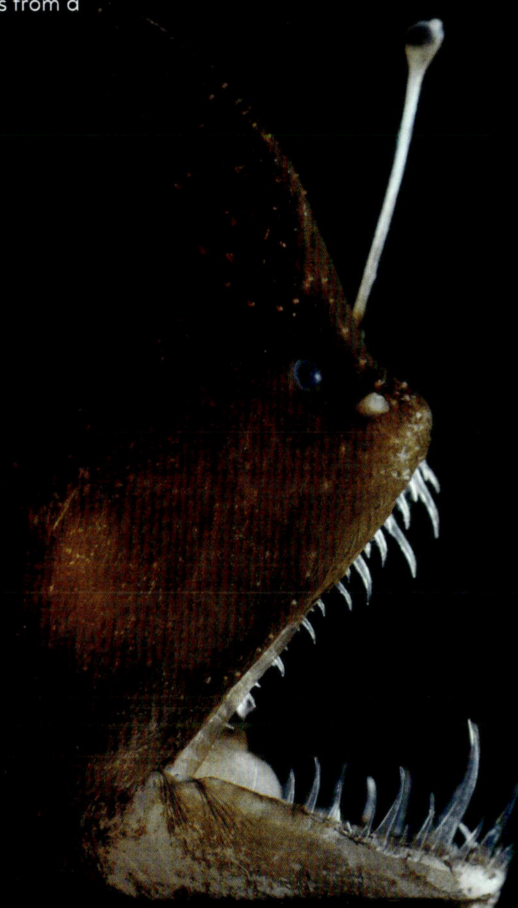

*The mating behaviour of some deep-sea anglerfish is very curious: the very small males fuse themselves temporarily or permanently to the much larger females. In the process, the skin and blood circulation of the two fish grow together.*

## THE DEEPEST CREWED DIVE IN THE WORLD

In the Trieste, a bathyscaphe (a submersible vessel specially constructed for deep sea exploration), Jacques Piccard from Switzerland and Don Walsh from the USA succeeded in diving 10,910 m (35,796 ft) into the Mariana Trench in January 1960 (picture below). The bathyscaphe is a buoyancy body, which is filled with seawater when it dives. The submersible vessel's crew are housed in a small white spherical pressure chamber underneath the body of the vessel which is built to withstand a pressure of over one tonne per cm². At the end of the dive, the location of the dive was named the Trieste Depth.

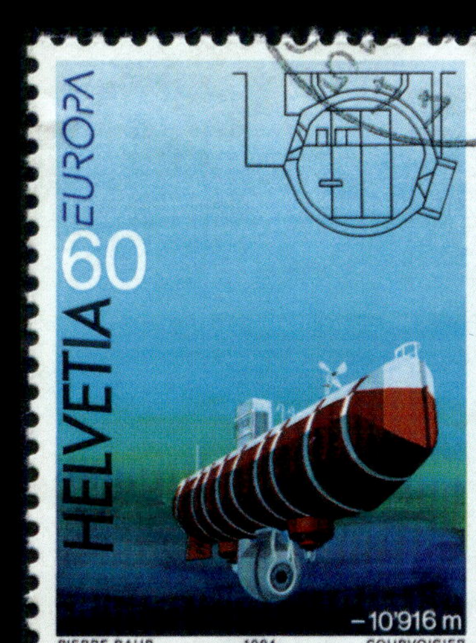

FOUR ELEMENTS

WATER

# CORALS

The smallest architects in the world

They look like underwater plants and enthrall divers with their multitude of different colours. But they are not underwater flowering plants; they are fixed colony-building cnidarians that feed on microplankton by filtering them from the flowing seawater with their tentacles. Reefs are built by hard or stony corals that form a skeleton through calcification. Living corals constantly grow over dead skeleton material, leading to the construction of complex reefs. Clear water with a maximum depth of around 60 m (197 ft) and water temperatures between 18 (65°F) and 35°C (95°F) are prerequisites for their existence. These conditions only exist in tropical and subtropical waters between the 25° north and 25° south latitudes. The largest and most famous coral reef of this type is the Great Barrier Reef in eastern Australia. In addition to the reef-building stony corals, there is also a wide variety of soft corals that do not build reefs and occur in deeper waters.

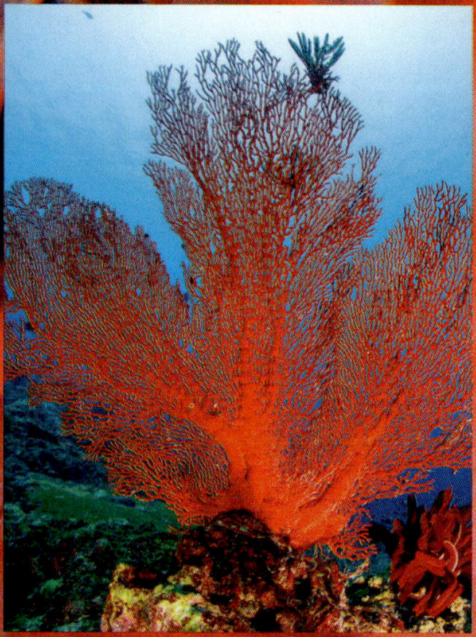

*Soft corals are a subgroup of the anthozoa class which comprises around 6,000 different species. These four pictures from the Red Sea and the western Pacific provide a brief glimpse of the enormous biodiversity in the coral world.*

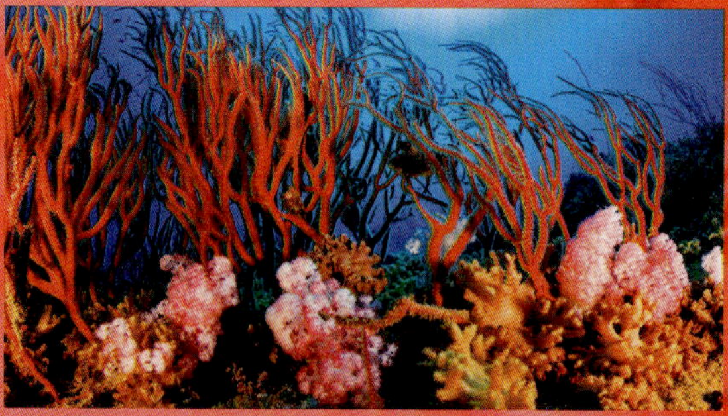

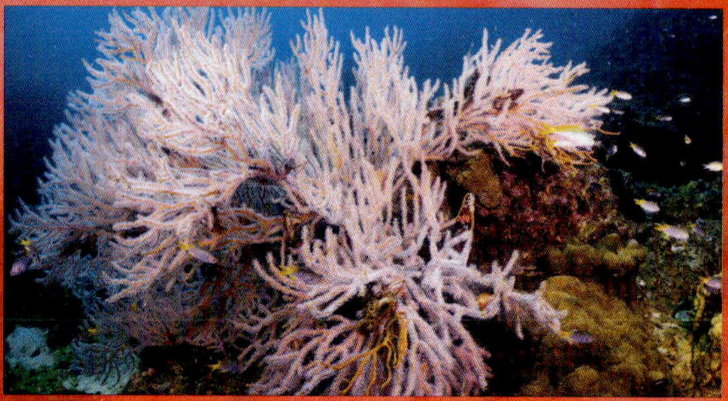

*Large picture: Anthomastus ritteri is a deep-sea coral which lives off the coast of California and Mexico at a depth of between 400 m (1,312 ft) and 1,200 m (3,937 ft).*

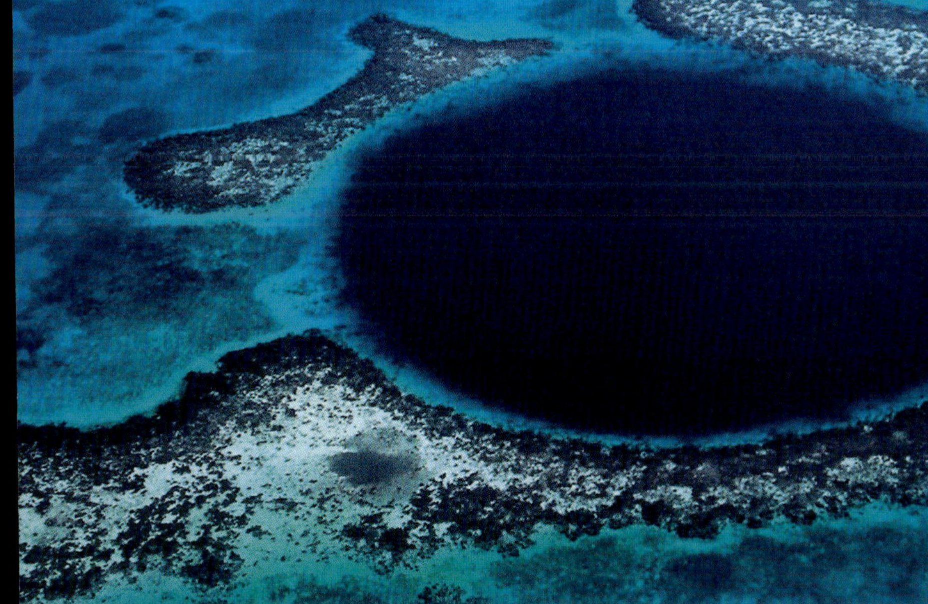



# CORAL REEFS

The largest 'structures' in nature

### THE LARGEST CORAL REEFS ON EARTH

1. **GREAT BARRIER REEF**
Australia, 348,700 km² (134,634 sq mi)

2. **NEW CALEDONIAN BARRIER REEF**
New Caledonia, 15,743 km² (6,078 sq mi)

3. **ANDROS BARRIER REEF**
Bahamas, 6,000 km² (2,317 sq mi)

4. **BELIZE BARRIER REEF**
Belize, 963 km² (372 sq mi)

5. **EAST RENNELL**
Solomon Islands, 370 km² (143 sq mi)

6. **TUBBATAHA REEF**
Philippines, 332 km² (128 sq mi)

Coral reefs are often compared to the tropical rainforests in terms of their biodiversity and significance. Two statistics give some idea of their importance: They occupy only 0.015 per cent of the ocean surface, yet they account for over 25 per cent of all marine life. In this, it is not only the corals themselves that, with their wide range of varieties and colours, contribute to the complexity of this marine ecosystem: the algae, fish, crustaceans, molluscs and other organisms occurring in this biotope in a living community with the coral also play a role. They live with and from each other. Life is finely balanced in this symbiotic system – the multitude of colours is used to lure, delude or camouflage in order to eat or avoid being eaten. There is hunting, stinging and scrounging. Each species has developed its own survival and reproduction strategies, lending complexity to the whole system but also making it very sensitive to any disruptions. Causes for concern are the rising sea levels and the increase in water temperatures, as these can lead to the dreaded coral bleaching which signifies nothing less than the death of the coral reefs.

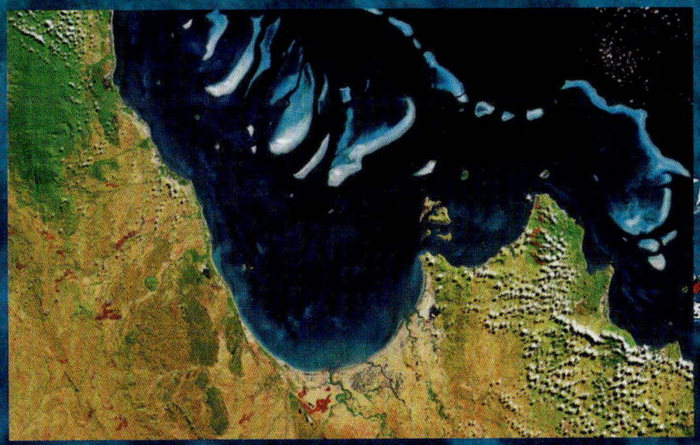

*Large picture: With vertically-inclined reefs along its edge, the Great Blue Hole is an underwater sinkhole (chimney- or funnel-shaped sink) which, with its associated caverns, forms an extraordinary phenomenon on the Belize Barrier Reef.*

*Right: A satellite image shows a section of the Great Barrier Reef north of Princess Charlotte Bay and Cape Melville.*

FOUR ELEMENTS

WATER

# ATLANTIC OCEAN

With 106,400,000 km² (41,081,270 sq mi), the Atlantic Ocean is the second largest ocean in the world; it stretches from the Arctic Sea to the Antarctic Sea and covers around one-fifth of the earth's surface.

The Mid-Atlantic Ridge runs more or less down the middle of the ocean, from north to south. Extending for approximately 11,000 km (6,835 mi), it is the longest continuous cordillera on earth. Elevations visible above sea level are Jan Mayen Island, Iceland and the Azores in the northern hemisphere as well as the relatively small Ascension, Tristan da Cunha, Gough and Bouvet Islands in the southern hemisphere.

The Mid-Atlantic Ridge was originally just a small fissure which divided the supercontinent of Pangea. Over the course of the last 200 million years, plate tectonics have caused the continents to drift apart, leading to the widening of the Atlantic Ocean by as much as 5,000 km (3,107 mi). These processes are still ongoing and very active. The underwater mountain range is traversed by a trench 25 to 50 km (15 to 31 mi) wide, which, as a result of emergent magma, pushes the Atlantic Ocean roughly 2 to 3 cm (0.8 to 1.2 in) apart in a west-east direction every year in the northern hemisphere and roughly 4 cm (1.6 in) apart in the southern hemisphere.

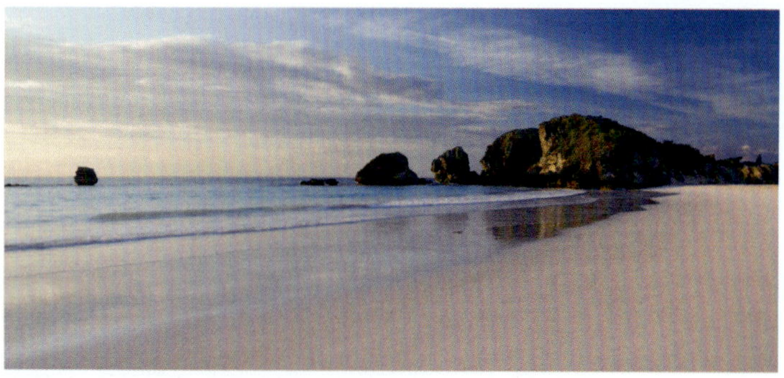

## THE SARGASSO SEA – THE LARGEST MARGINAL SEA IN THE ATLANTIC

Warm ocean currents mean that the waters in the Sargasso Sea spin in a clockwise direction as though they were in a gigantic whirlpool. The rapid shift from calm, smooth seas to high waves means that this area, which is also part of the Bermuda Triangle, is not popular with seafarers.

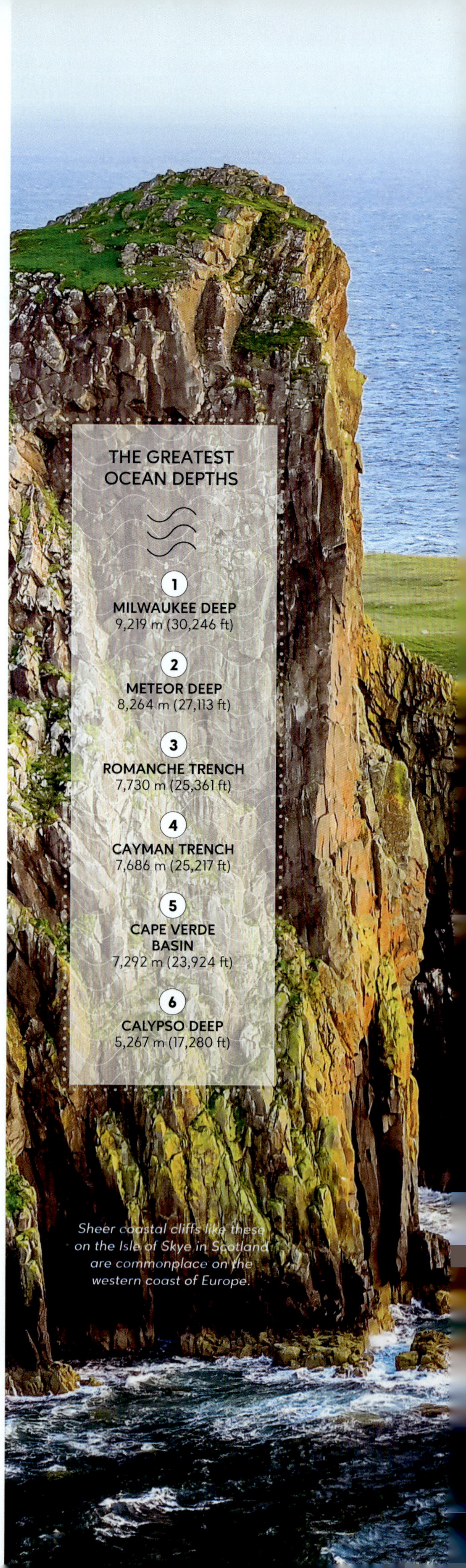

### THE GREATEST OCEAN DEPTHS

1. **MILWAUKEE DEEP** 9,219 m (30,246 ft)
2. **METEOR DEEP** 8,264 m (27,113 ft)
3. **ROMANCHE TRENCH** 7,730 m (25,361 ft)
4. **CAYMAN TRENCH** 7,686 m (25,217 ft)
5. **CAPE VERDE BASIN** 7,292 m (23,924 ft)
6. **CALYPSO DEEP** 5,267 m (17,280 ft)

*Sheer coastal cliffs like these on the Isle of Skye in Scotland are commonplace on the western coast of Europe.*

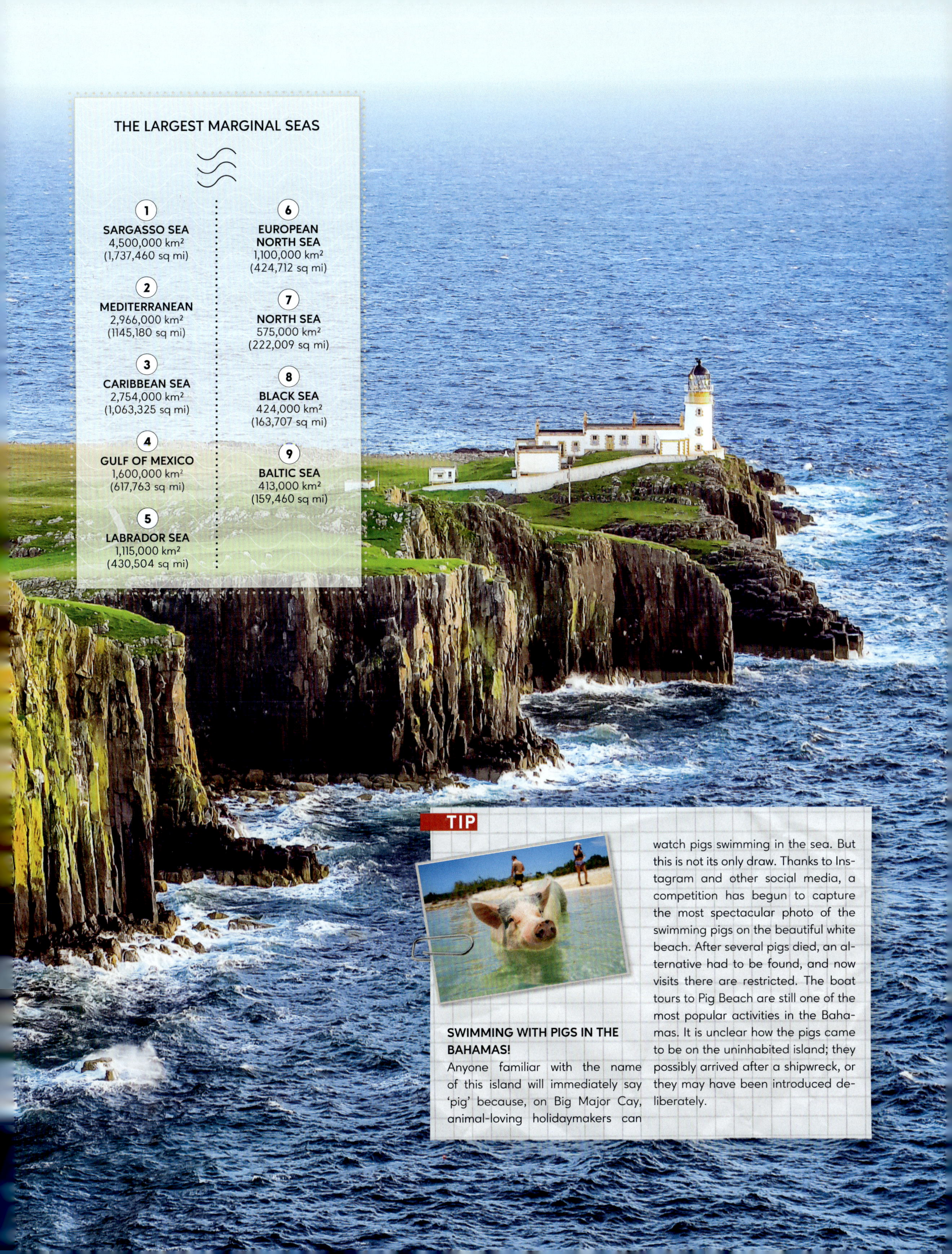

## THE LARGEST MARGINAL SEAS

1. **SARGASSO SEA**
   4,500,000 km²
   (1,737,460 sq mi)
2. **MEDITERRANEAN**
   2,966,000 km²
   (1145,180 sq mi)
3. **CARIBBEAN SEA**
   2,754,000 km²
   (1,063,325 sq mi)
4. **GULF OF MEXICO**
   1,600,000 km²
   (617,763 sq mi)
5. **LABRADOR SEA**
   1,115,000 km²
   (430,504 sq mi)
6. **EUROPEAN NORTH SEA**
   1,100,000 km²
   (424,712 sq mi)
7. **NORTH SEA**
   575,000 km²
   (222,009 sq mi)
8. **BLACK SEA**
   424,000 km²
   (163,707 sq mi)
9. **BALTIC SEA**
   413,000 km²
   (159,460 sq mi)

**TIP**

**SWIMMING WITH PIGS IN THE BAHAMAS!**

Anyone familiar with the name of this island will immediately say 'pig' because, on Big Major Cay, animal-loving holidaymakers can watch pigs swimming in the sea. But this is not its only draw. Thanks to Instagram and other social media, a competition has begun to capture the most spectacular photo of the swimming pigs on the beautiful white beach. After several pigs died, an alternative had to be found, and now visits there are restricted. The boat tours to Pig Beach are still one of the most popular activities in the Bahamas. It is unclear how the pigs came to be on the uninhabited island; they possibly arrived after a shipwreck, or they may have been introduced deliberately.

# I AM KING OF THE WORLD!

### THE LARGEST CRUISE SHIPS IN THE WORLD
(according to gross tonnage (GT))

1. **SYMPHONY OF THE SEAS**
Royal Caribbean
228,081

2. **HARMONY OF THE SEAS**
Royal Caribbean
226,963

3. **COSTA SMERALDA**
Costa Crociere
185,010

4. **MSC MERAVIGLIA**
MSC Cruises
171,598

5. **SPECTRUM OF THE SEAS**
Royal Caribbean
169,379

And the race to achieve the highest ranks continues. Virtually every year bigger cruise ships are being built to sail on all the oceans of the world.

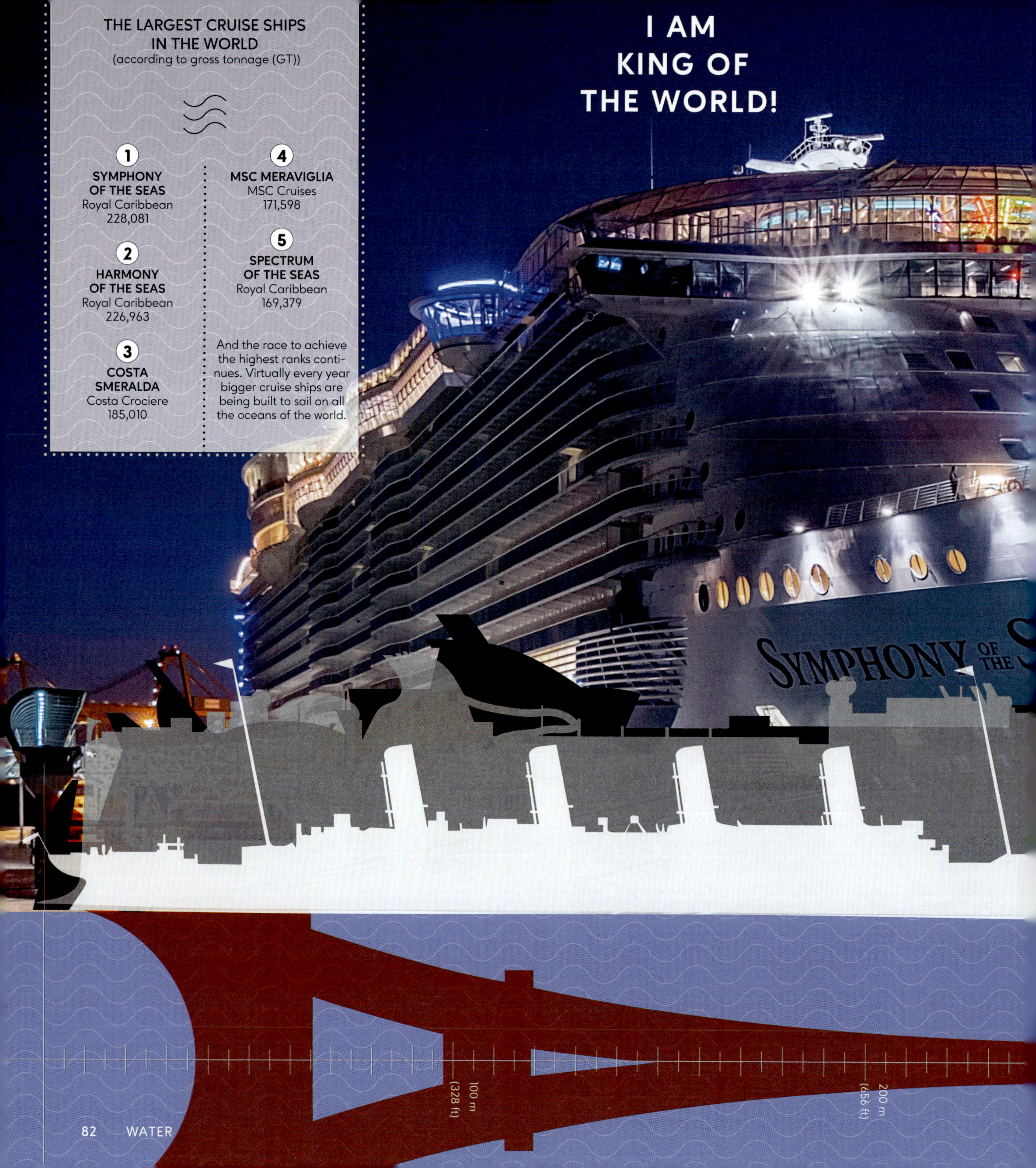

DID YOU KNOW?

### Higher, bigger, wider, more eccentric

Until the 1990s, cruises were a luxury that only a few could afford. Now, however, operators offer trips to suit every taste and budget. Special expedition, adventure, fun and family cruises have been introduced alongside the classic cruise. Increasing numbers of people are interested in going on a cruise. The market is booming. Ocean liners can accommodate thousands of passengers. They are often compared to floating cities, offering every type of amusement and diversion imaginable. The offers on board the mega-liners are not merely limited to swimming pools, wellness zones, entertainment areas with theatres, jazz clubs and dance lounges for the whole family to enjoy; there are also a great variety of indoor and outdoor activities, such as an ice skating rink, mini-golf, a climbing wall, a surf park or a volleyball court. Restaurants and cafés cater to every taste and attend to the guests' physical well-being.

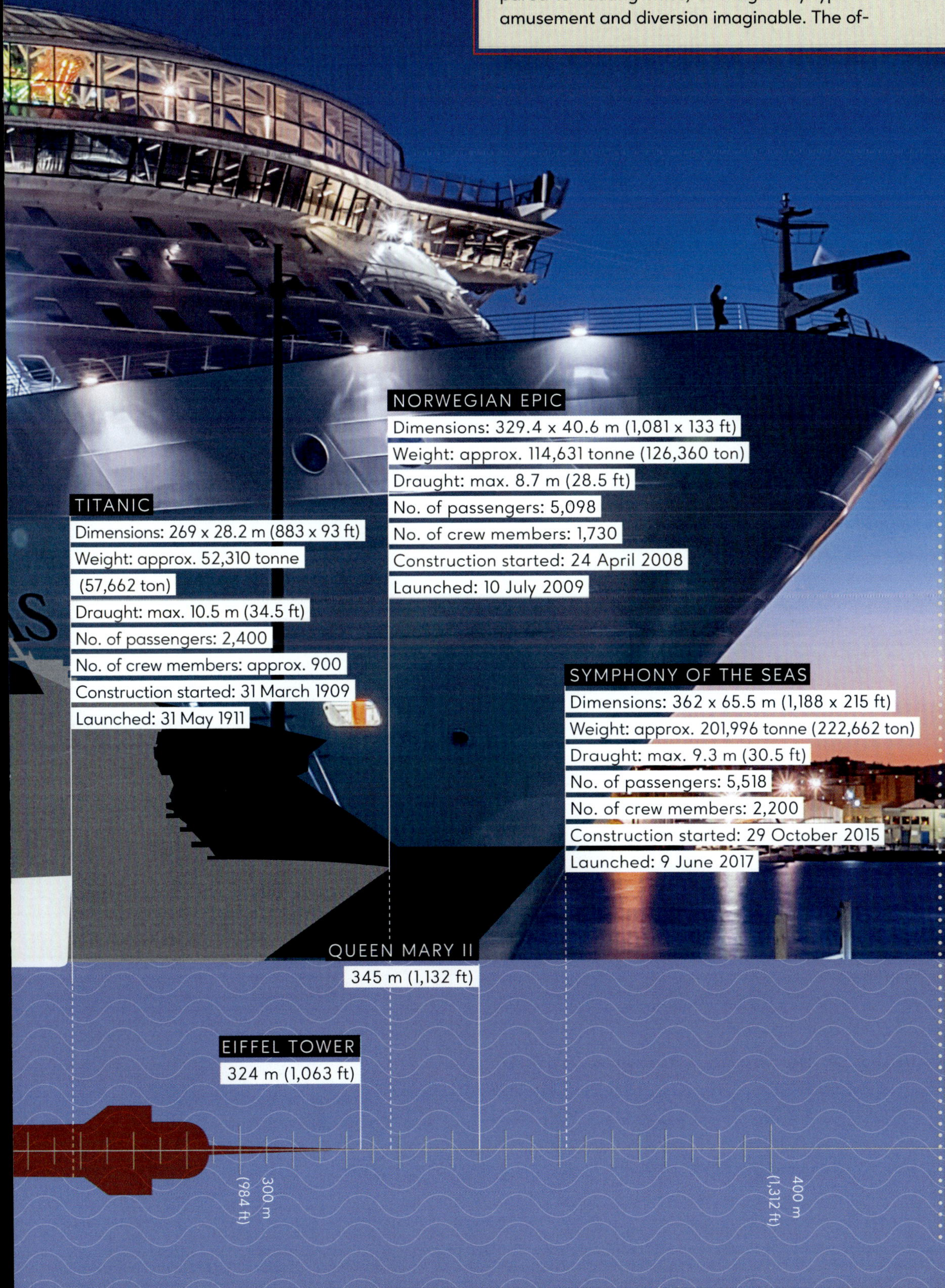


## CRUISE PORTS AND THEIR PASSENGER NUMBERS

AS OF: 2018

1. **MIAMI** 5.3 million
2. **PORT CANAVERAL** 4.5 million
3. **COZUMEL** 4.1 million
4. **PORT EVERGLADES** 3.8 million
5. **BARCELONA** 2.7 million
6. **NASSAU** 2.6 million
7. **CIVITAVECCHIA** 2.2 million

FOUR ELEMENTS

WATER

# INDIAN OCEAN

The Indian Ocean ranks third largest of oceans of the world. Its seabed has a complex structure – the fault zones of the Mid-, Southwest and Southeast Indian Ridges subdivide into huge basins which in turn are divided by plateaus and trenches

The ocean currents in the Indian Ocean are very changeable. In winter, the monsoon winds propel the currents north of the equator into the coast of Africa, whereas, in summer, they are propelled to the Indian coast, accompanied by heavy rainfall. South of the equator, following the almost windless zone of the horse latitudes, the Roaring Forties lie in the West Wind Drift area between latitudes 40° and 50° south and are characterized by strong, stormy, westerly winds.

## THE ARABIAN SEA – THE LARGEST MARGINAL SEA IN THE INDIAN OCEAN

The Arabian Sea lies between the Arabian Peninsula, the Horn of Africa and the Indian subcontinent and covers an area of around 3.86 million km$^2$ (1,490,354 sq mi). The temperature difference between the open sea and the mainland make it a 'weather kitchen' in the summer months for the southwest monsoon, providing regions in India with the highest rainfall in the world.

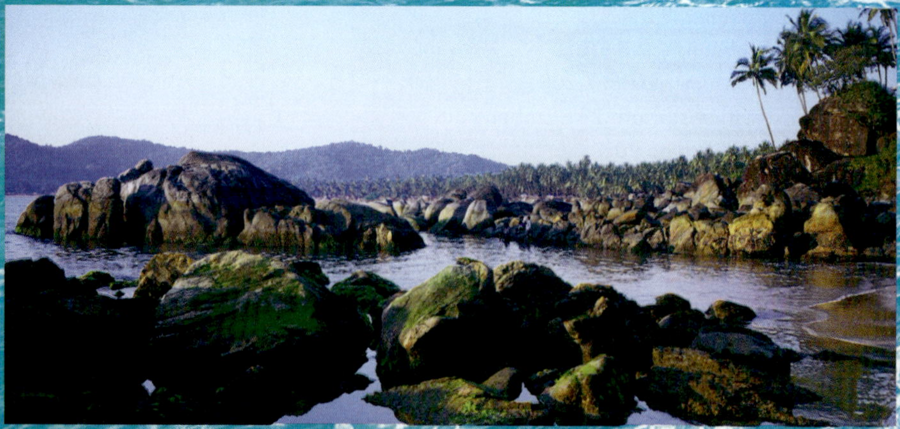

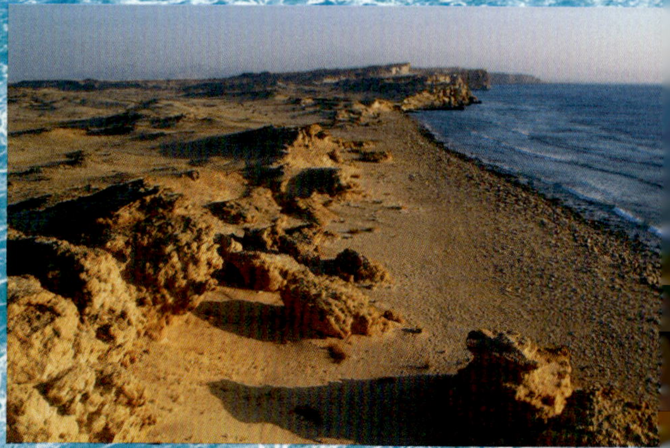

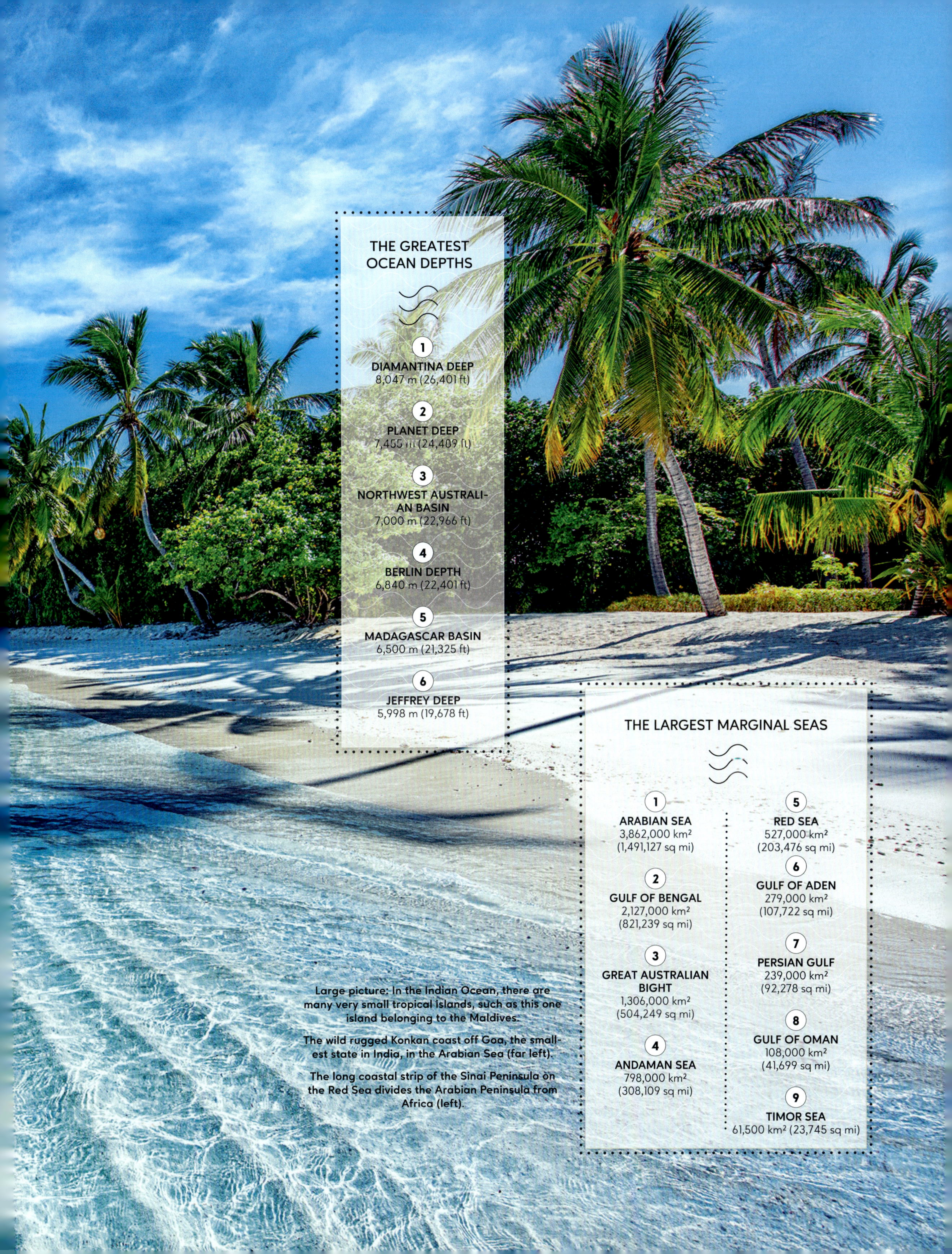

## THE GREATEST OCEAN DEPTHS

1. **DIAMANTINA DEEP**
   8,047 m (26,401 ft)
2. **PLANET DEEP**
   7,455 m (24,409 ft)
3. **NORTHWEST AUSTRALIAN BASIN**
   7,000 m (22,966 ft)
4. **BERLIN DEPTH**
   6,840 m (22,401 ft)
5. **MADAGASCAR BASIN**
   6,500 m (21,325 ft)
6. **JEFFREY DEEP**
   5,998 m (19,678 ft)

Large picture: In the Indian Ocean, there are many very small tropical islands, such as this one island belonging to the Maldives.

The wild rugged Konkan coast off Goa, the smallest state in India, in the Arabian Sea (far left).

The long coastal strip of the Sinai Peninsula on the Red Sea divides the Arabian Peninsula from Africa (left).

## THE LARGEST MARGINAL SEAS

1. **ARABIAN SEA**
   3,862,000 km² (1,491,127 sq mi)
2. **GULF OF BENGAL**
   2,127,000 km² (821,239 sq mi)
3. **GREAT AUSTRALIAN BIGHT**
   1,306,000 km² (504,249 sq mi)
4. **ANDAMAN SEA**
   798,000 km² (308,109 sq mi)
5. **RED SEA**
   527,000 km² (203,476 sq mi)
6. **GULF OF ADEN**
   279,000 km² (107,722 sq mi)
7. **PERSIAN GULF**
   239,000 km² (92,278 sq mi)
8. **GULF OF OMAN**
   108,000 km² (41,699 sq mi)
9. **TIMOR SEA**
   61,500 km² (23,745 sq mi)

FOUR ELEMENTS

WATER

# ANTARCTIC OCEAN

It sounds strange, but this ocean has only been in existence since the year 2000. It was in this year that the International Hydrographic Organization decided to consolidate the area south of the 60th parallel with the Antarctic Ocean. Prior to this, the various areas of the ocean were part of the Atlantic, Pacific and Indian Oceans respectively. Measuring 20,327 million km² (7,848,298 sq mi), it is the second smallest ocean in the world and completely surrounds the fifth largest and coldest continent, Antarctica.

## THE LARGEST MARGINAL SEAS

1. **WEDDELL SEA** 2,800,000 km² (1,081,086 sq mi)
2. **BELLINGSHAUSEN SEA** 1,500,000 km² (579,153 sq mi)
3. **ROSS SEA** 960,000 km² (370,658 sq mi)
4. **D'URVILLE SEA** 950,000 km² (366,797 sq mi)
5. **DAVIS SEA** 900,000 km² (347,492 sq mi)
6. **AMUNDSEN SEA** 770,000 km² (297,299 sq mi)
7. **COOPERATION SEA** 230,000 km² (88,803 sq mi)

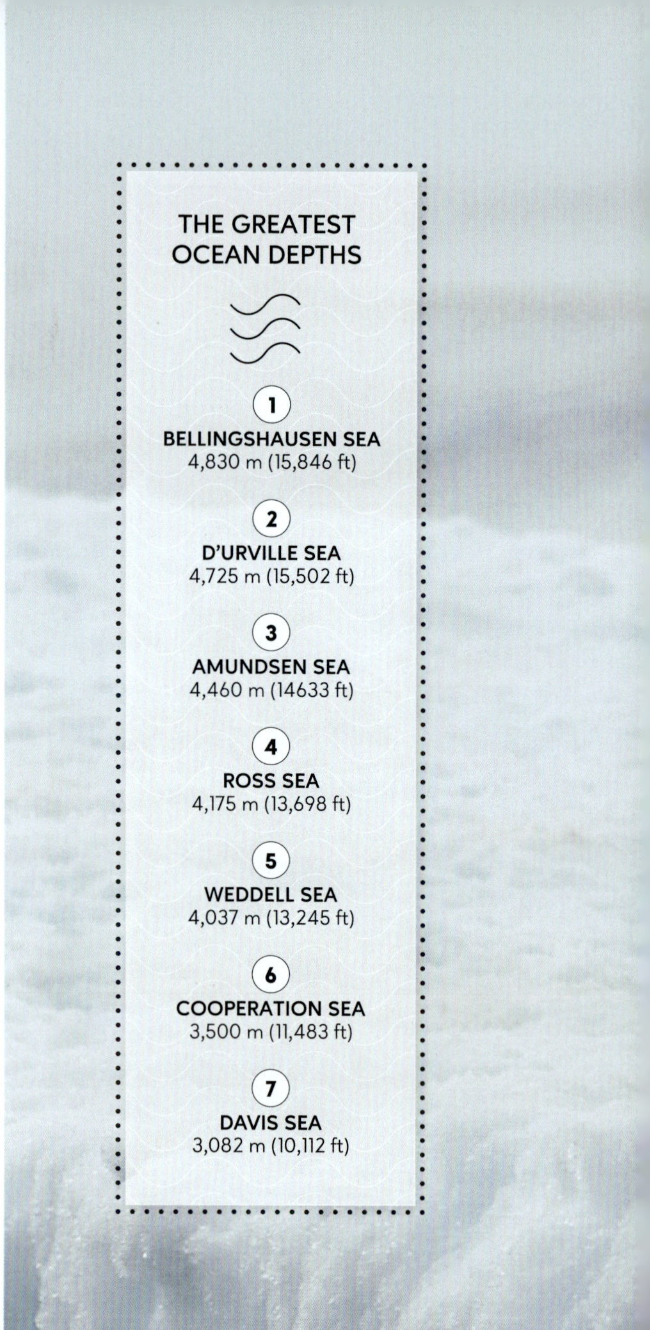

## THE GREATEST OCEAN DEPTHS

1. **BELLINGSHAUSEN SEA** 4,830 m (15,846 ft)
2. **D'URVILLE SEA** 4,725 m (15,502 ft)
3. **AMUNDSEN SEA** 4,460 m (14633 ft)
4. **ROSS SEA** 4,175 m (13,698 ft)
5. **WEDDELL SEA** 4,037 m (13,245 ft)
6. **COOPERATION SEA** 3,500 m (11,483 ft)
7. **DAVIS SEA** 3,082 m (10,112 ft)

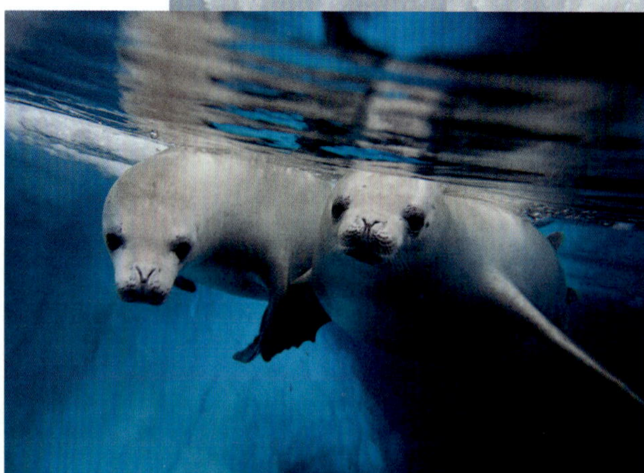

*Large picture: Icebergs, in particular table icebergs formed from shelf ice, are typical of the Antarctic Ocean. For penguins, they are a launch site but can also be a barrier in their search for food at times.*

*Above: Inquisitive seals in Weddell Sea.*

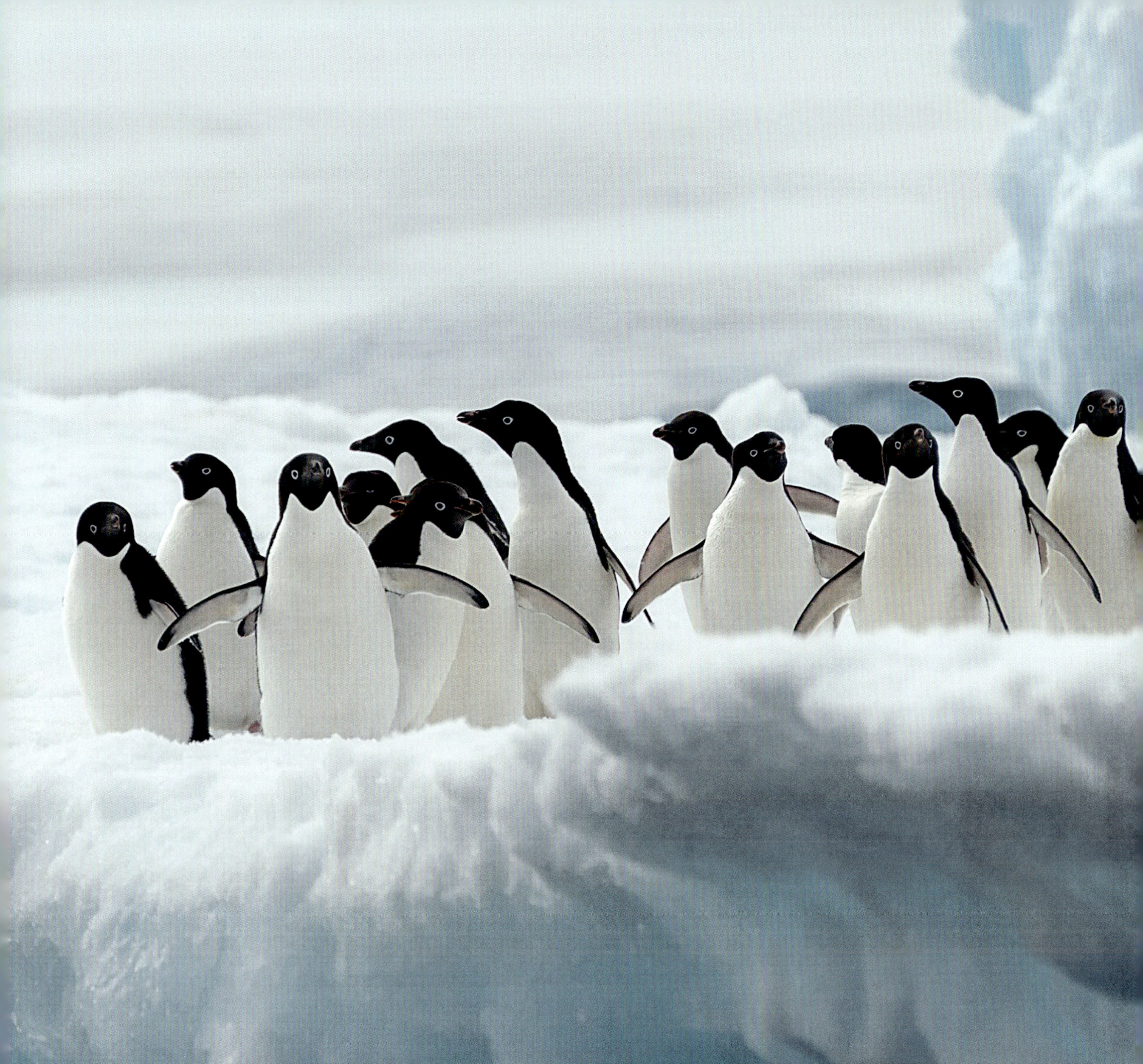

## THE WEDDELL SEA – THE LARGEST MARGINAL SEA IN THE ANTARCTIC OCEAN

The Weddell Sea, covering an area of 2.8 million $km^2$ (1,081,086 sq mi), is the largest marginal sea in Antarctica. It extends 2,000 km (1,243 mi) west to east between the Antarctic Peninsula and the coastal region of Coats Land in the eastern part of the ocean. Up to 4,037 m (13,245 ft) deep, the sea borders the Filchner-Ronne Ice Shelf in the south. Large areas of the sea are permanently covered in pack ice, in particular on the lee side of the Antarctic Peninsula, which benefits from the unusual behaviour of the currents in the Weddell Sea – the clockwise-rotating eddies push ice from the north initially to the south and then west onto the peninsula.

FOUR ELEMENTS

WATER

# ARCTIC OCEAN

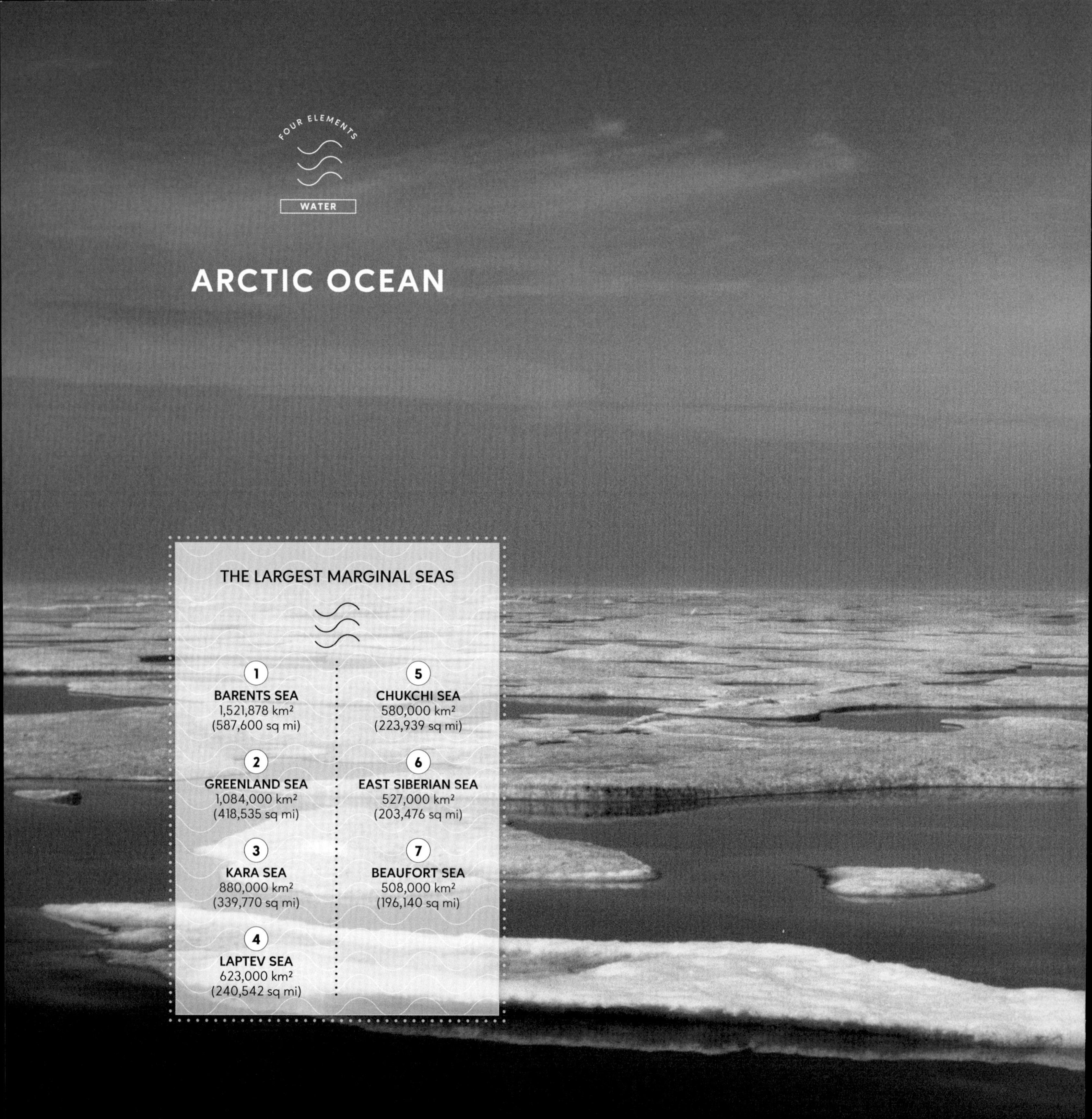

## THE LARGEST MARGINAL SEAS

1. **BARENTS SEA**
1,521,878 $km^2$
(587,600 sq mi)

2. **GREENLAND SEA**
1,084,000 $km^2$
(418,535 sq mi)

3. **KARA SEA**
880,000 $km^2$
(339,770 sq mi)

4. **LAPTEV SEA**
623,000 $km^2$
(240,542 sq mi)

5. **CHUKCHI SEA**
580,000 $km^2$
(223,939 sq mi)

6. **EAST SIBERIAN SEA**
527,000 $km^2$
(203,476 sq mi)

7. **BEAUFORT SEA**
508,000 $km^2$
(196,140 sq mi)

The Arctic Ocean appears ice-covered, inhospitable and hostile to life; with an area of 14,090,000 $km^2$ (5,440,179 ft), it is the smallest of the world's five oceans. Almost completely surrounded by the mainlands of northern Asia, North America and Europe, the most northerly section of the Arctic Ocean is an ice cap, which remains between 2 and 4 m (6.5 and 13 ft) thick throughout the year. The ice is not fixed; it drifts and is constantly in motion. Ocean currents and the rotation of the earth are the main causes of this movement. On the outer margins of the ice cap, bizarre pack ice landscapes are pushed together by the wind and waves; further south, these break up in the considerably more open drift-ice zones with their drifting icebergs. These processes are subject to strong seasonal variations.

With the narrow Bering Strait between North America and Asia, the Arctic Ocean has a link to the Pacific Ocean as well as a link to the Atlantic Ocean via the Norwegian Sea. Large, flat oceanic shelf areas, particularly off the Siberian coast, are a feature of large areas of the seabed.

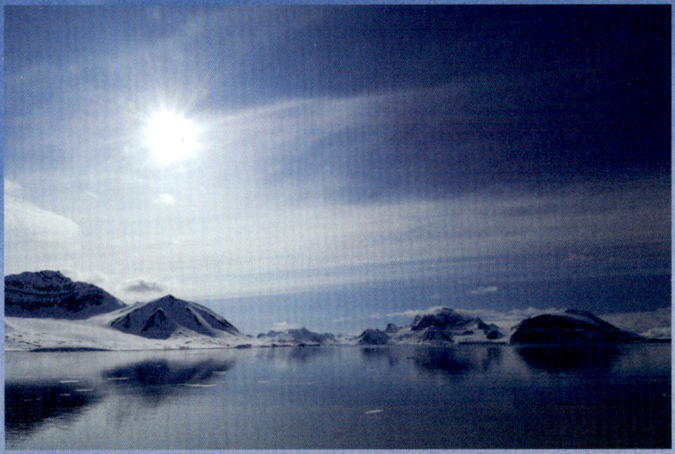

*The Hornsundfjord in Spitzenbergen (Svalbard) in Sweden is pristine and peaceful.*

## THE BARENTS SEA – THE LARGEST MARGINAL SEA IN THE ARCTIC OCEAN

The Barents Sea is bordered by the Spitzbergen archipelago, Franz Joseph Land and Novaya Zemlya as well as by the European mainland. The wide passage to the Norwegian Sea between the North Cape of Norway and the Spitzbergen archipelago allows the tail of the warm gulf stream to penetrate this shelf sea. As a result, the Barents Sea is largely ice-free, and ports such as Murmansk on the north coast of Europe remain accessible to shipping even in winter. Murmansk is therefore also the major base for the Russian North Sea fleet.

The relatively high temperature of the ocean waters compared to other parts of the Arctic Ocean lead to rapid production of phytoplankton in spring, subsequently greatly increasing the zooplankton. These are the main food source for cod, which are caught here in large numbers by the Russian and Norwegian fleets.

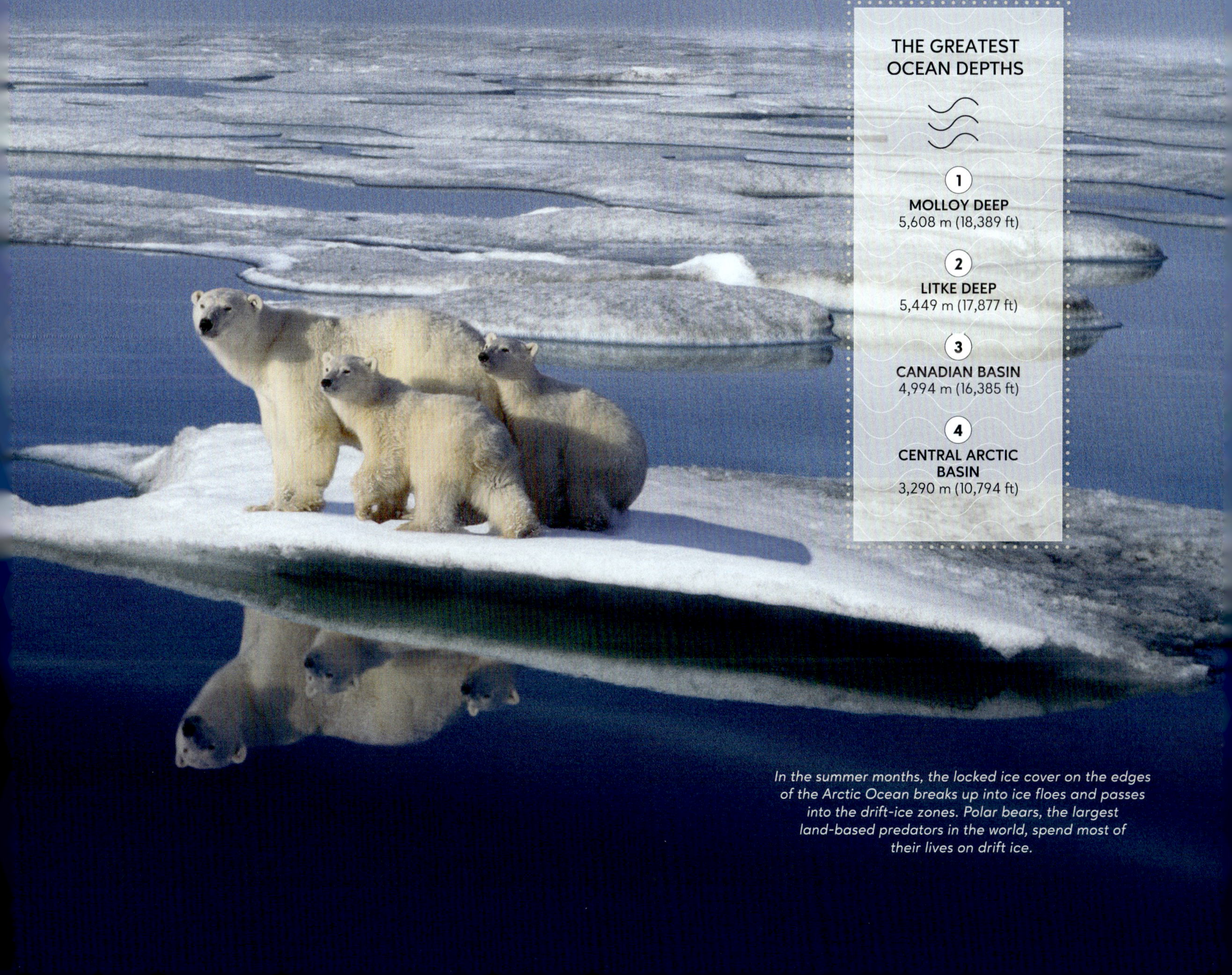

### THE GREATEST OCEAN DEPTHS

1. **MOLLOY DEEP**
5,608 m (18,389 ft)

2. **LITKE DEEP**
5,449 m (17,877 ft)

3. **CANADIAN BASIN**
4,994 m (16,385 ft)

4. **CENTRAL ARCTIC BASIN**
3,290 m (10,794 ft)

*In the summer months, the locked ice cover on the edges of the Arctic Ocean breaks up into ice floes and passes into the drift-ice zones. Polar bears, the largest land-based predators in the world, spend most of their lives on drift ice.*

FOUR ELEMENTS

WATER

# NORTH POLE

The Geographic North Pole is the northernmost point on earth

No survey point marks this location in the Arctic Ocean because it does not lie on the mainland but on a layer of ice, 2 to 3 m (6.5 - 9.8 ft) thick, floating on an ocean approximately 4,000 m (13,124 ft) deep, that is constantly in motion. It is the northernmost endpoint of the earth's rotational axis. The sightlines to the west and east are not valid here – all directions point south. The Magnetic North Pole is in direct relation to the Geographic North Pole and lies in northern Canada. At the North Pole, six months of the year are polar days (21 March - 23 September) with a short transitional period. In simpler terms: the sun never sets. The other six months of the year are polar nights in which the sun does not rise.

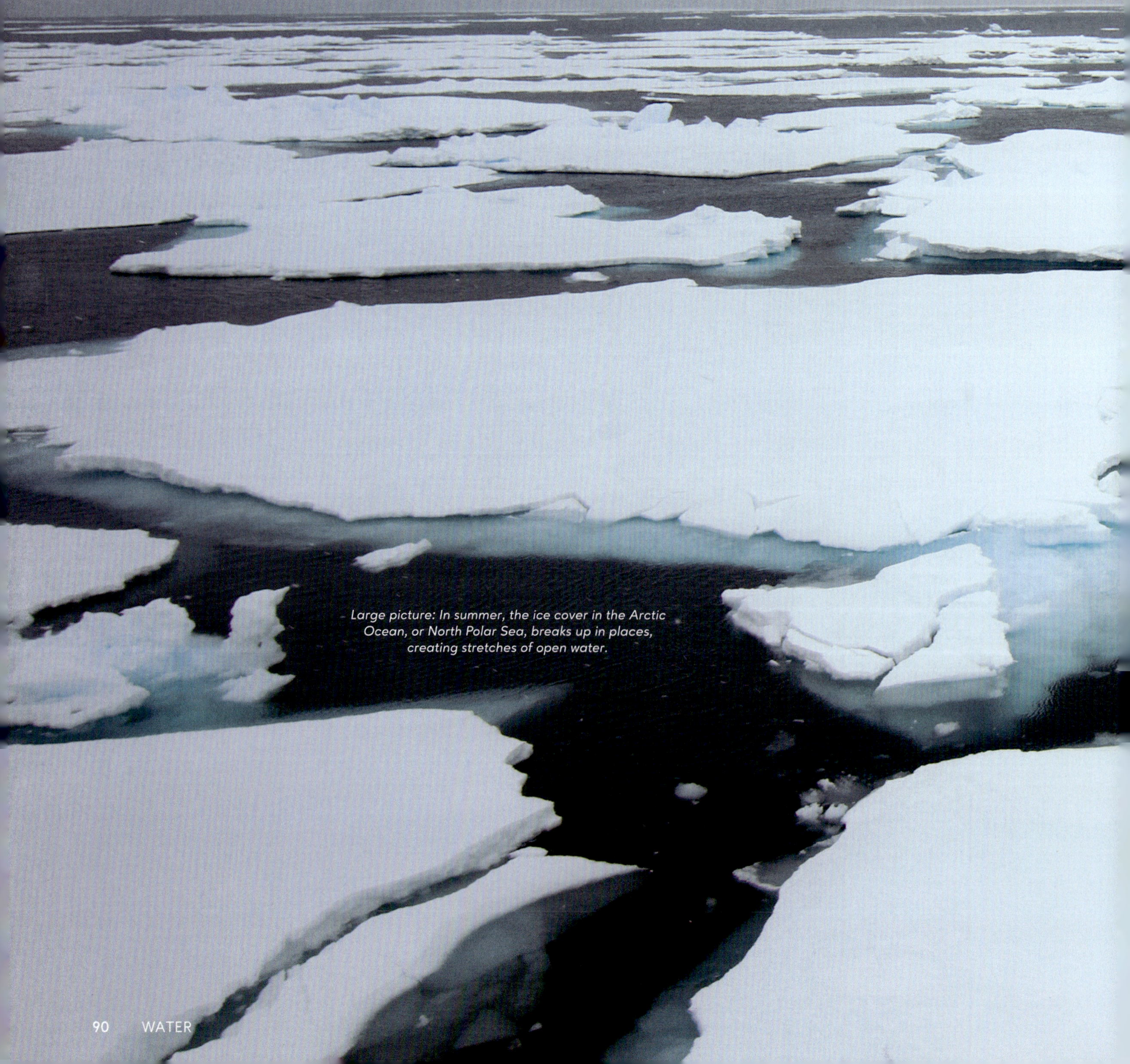

*Large picture: In summer, the ice cover in the Arctic Ocean, or North Polar Sea, breaks up in places, creating stretches of open water.*

The polar explorer Robert Edwin Peary

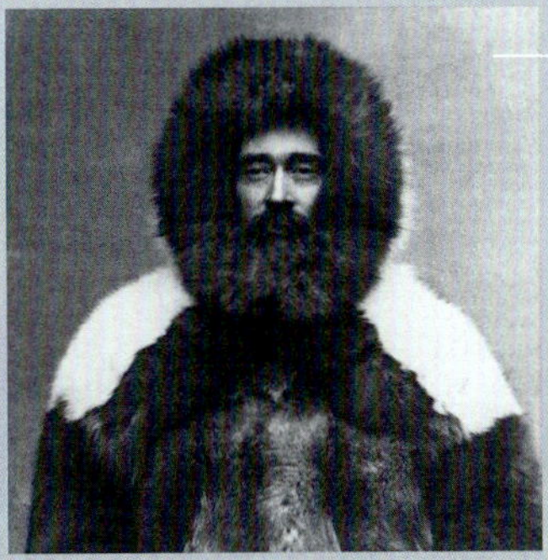

Peary's expedition to the North Pole

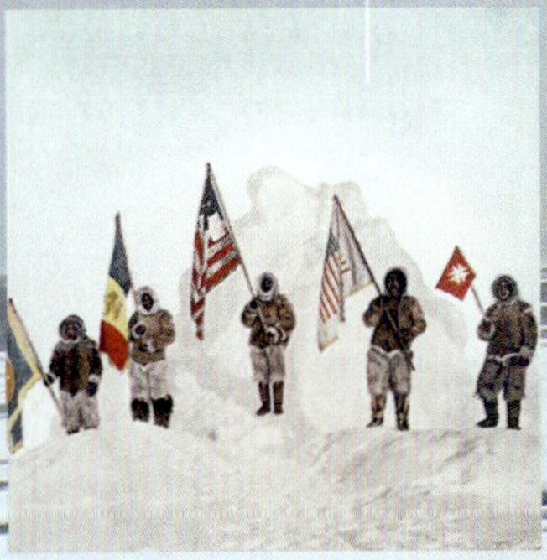

The polar explorer Matthew A. Henson

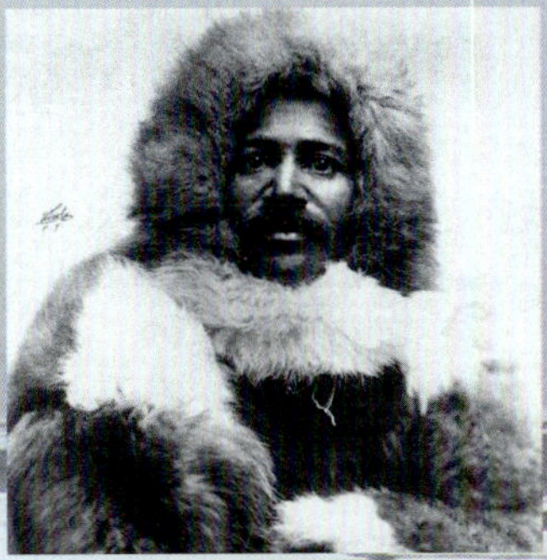

DID YOU KNOW?

**The Race to the Pole**

There is a long history of expeditions to the North Pole. Obsessed with the idea of being the first to reach the North Pole, explorers and adventurers engaged in a race which was full of intrigue and drama.

There were grave doubts in scientific circles as to whether Robert E. Peary (1856 - 1920), together with Matthew A. Henson, had in fact been the first persons to reach the North Pole on 6 April 1909. The explorer Frederick Cook (1865 - 1940) claimed that he had already reached the North Pole a year earlier, which was also considered very unlikely. However, it was verified that the flight of the airship *Norge*, undertaken by Umberto Nobile, Roald Amundsen and Lincoln Ellsworth in May 1926, did pass over the Geographic North Pole. This at least confirmed victory in the air.

In 1958, the USS *Nautilus* became the first submarine to pass under the closed ice cover of the Geographic North Pole. This finally proved that there was no Arctic continent.

The first person to reach the North Pole on foot, however, was the British explorer Walter William Herbert in 1969.

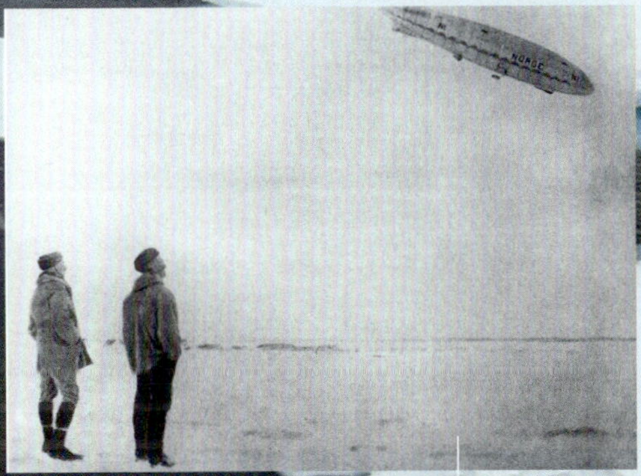

Amundsen and Ellsworth watch the *Norge*.

A submarine surfaces between ice floes at the North Pole

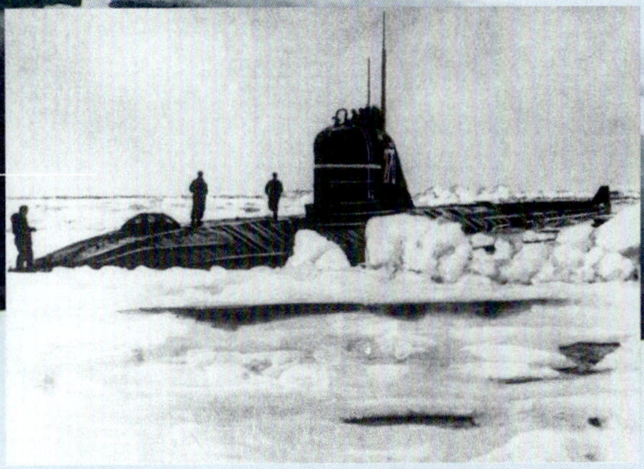

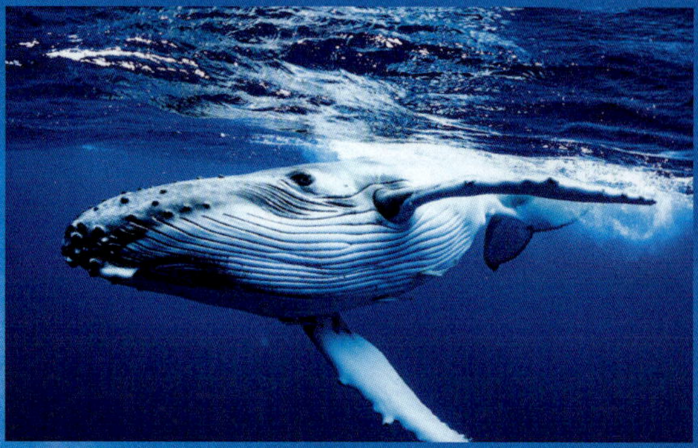

# WHALES

The largest mammals

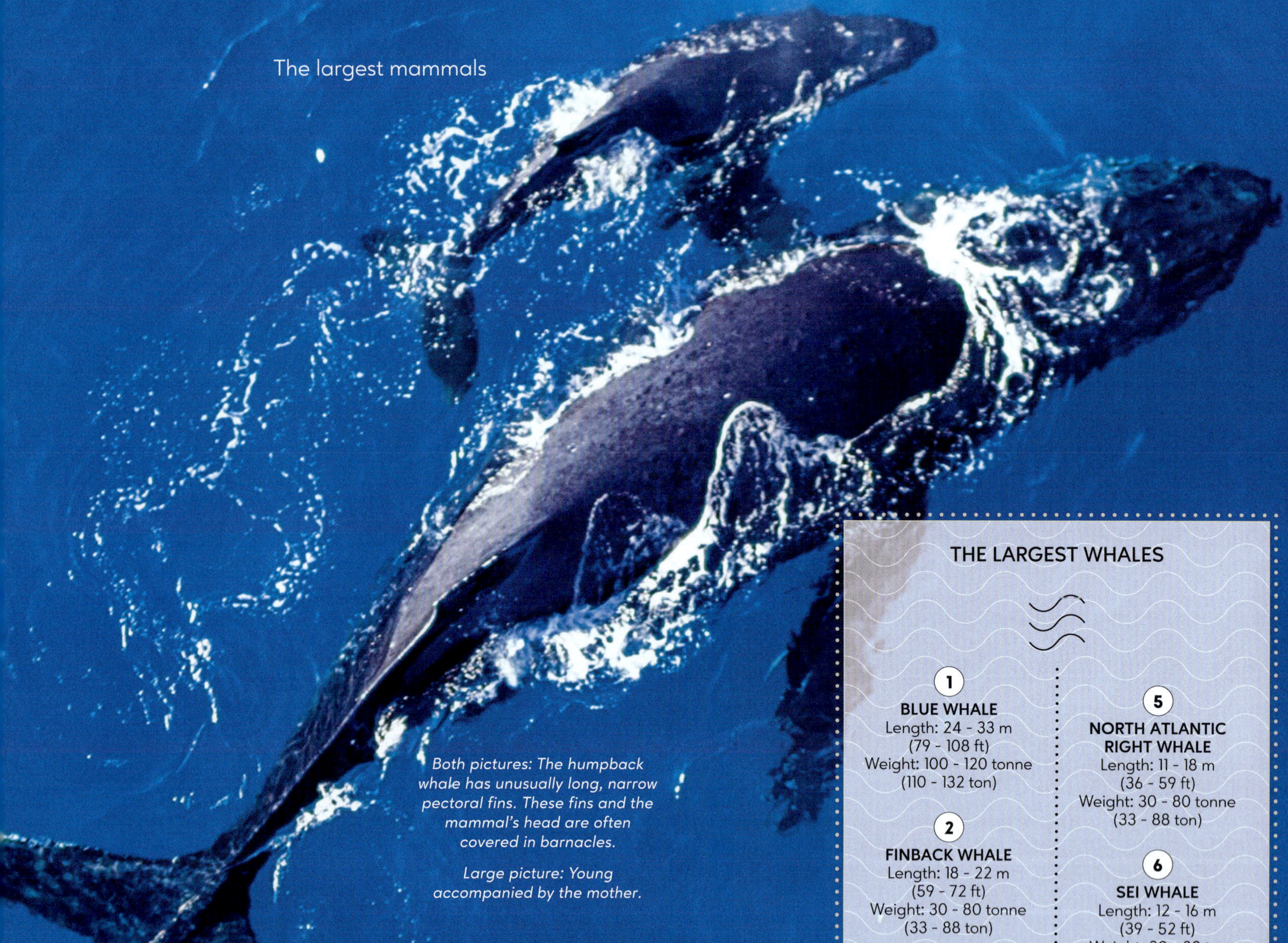

*Both pictures: The humpback whale has unusually long, narrow pectoral fins. These fins and the mammal's head are often covered in barnacles.*

*Large picture: Young accompanied by the mother.*

The ninety or more existing whale species are divided into two main groups – the baleen whales and the toothed whales. The latter group also includes dolphins. The baleen whales (*Mysticeti*) include the largest whales, such as the blue whale, which grows up to 33 m (108 ft) long and is the largest animal in the world. By means of their baleen, these whales filter small crustaceans from the sea: these are their primary food source. The toothed whales (*Odontoceti*) use their teeth to catch prey, but they do not use their teeth to tear or chew it. The main source of food for toothed whales is fish. All whales are mammals; they bear fully developed calves and breathe air. They have highly developed social behaviours and live in groups. Whales communicate with each other by means of the mysterious whale song.

## THE LARGEST WHALES

1. **BLUE WHALE**
Length: 24 - 33 m (79 - 108 ft)
Weight: 100 - 120 tonne (110 - 132 ton)

2. **FINBACK WHALE**
Length: 18 - 22 m (59 - 72 ft)
Weight: 30 - 80 tonne (33 - 88 ton)

3. **SPERM WHALE**
Length: 11 - 18 m (36 - 59 ft)
Weight: 20 - 50 tonne (22 - 55 ton)

4. **BOWHEAD WHALE**
Length: 14 - 18 m (46 - 59 ft)
Weight: 60 - 100 tonne (66 - 110 ton)

5. **NORTH ATLANTIC RIGHT WHALE**
Length: 11 - 18 m (36 - 59 ft)
Weight: 30 - 80 tonne (33 - 88 ton)

6. **SEI WHALE**
Length: 12 - 16 m (39 - 52 ft)
Weight: 20 - 30 tonne (22 - 33 ton)

7. **GRAY WHALE**
Length: 12 - 15 m (39 - 49 ft)
Weight: 14 - 35 tonne (15 - 38 ton)

8. **HUMPBACK WHALE**
Length: 11 - 15 m (36 - 49 ft)
Weight: 25 - 30 tonne (28 - 33 ton)

# WHALE SHARKS, BASKING SHARKS, GREAT WHITE SHARKS

The largest fish

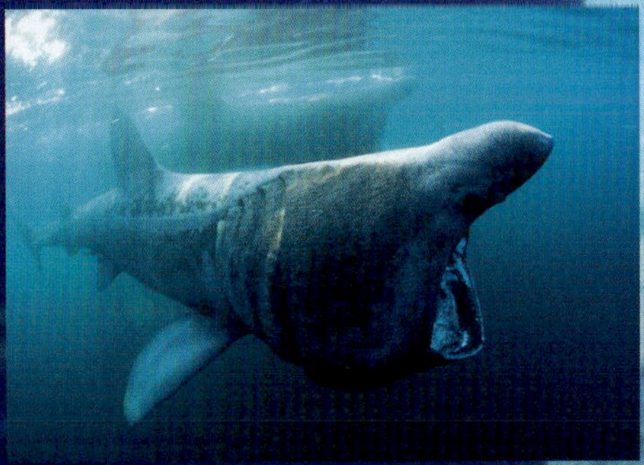

## BASKING SHARK

**Length: up to 15 m (49 ft)**
**Weight: up to 7 tonne (7.7 ton)**

In spite of its terrifying size, basking sharks (*Cetorhinus maximus*) are not ocean predators. Nor do they pose any threat to humans. After whale sharks, they are the second largest fish species in the world. They feed on plankton which they ingest passively by allowing water to pass through their open mouths and over their gills, where plankton is filtered out (large picture). In one hour, 2,000 tonne (2,205 ton) of water can pass through their mouths. This requires extremely small gill slits which almost completely span the head and are a characteristic feature of this species of shark. Basking sharks are mainly found in the cool to warm waters of the Atlantic and Pacific Oceans. They are not solitary animals; they live in large groups of over a hundred or more. The basking shark population is threatened by hunting. Hunters are mainly interested in the shark's liver, which can constitute up to 30 per cent of the animal's body weight, corresponding to 1,500 litres (396.25 gallons) of cod liver oil.

## WHALE SHARKS

**Length: up to 18 m (59 ft)**
**Weight: up to 13 tonne (14.3 ton)**

The whale shark (*Rhincodon typus*) is the largest fish in the world – a titan among sea creatures. It lives in pods in tropical and subtropical regions of the oceans north and south of the equator. With 300 rows of teeth, each containing hundreds of individual teeth, it has the most teeth of any animal on earth. But just why they need so many teeth remains a mystery because, like the basking sharks, they only eat plankton – the only difference is that they actively suck the water in through their open mouths. From time to time, they close their mouths in order to be able to swallow their prey. To do this, they often adopt a vertical, upright position. Whale sharks can live to be up to 100 years old; they are able to reproduce when they reach the age of ten. Very little is known about how they reproduce. They are viviparous – they bear live young. They do not seem to have specific mating seasons. It is believed that they are able to bear up to 300 young, which they can carry inside their bodies at various stages of development. Their existence is threatened by shipping and fishing.

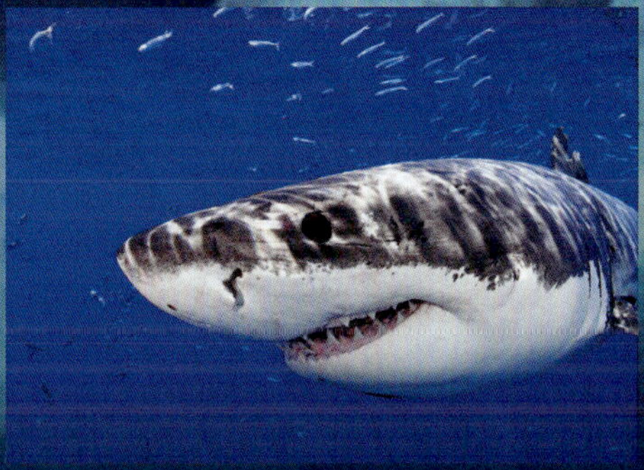

## GREAT WHITE SHARK

**Length: up to 8 m (26 ft)**
**Weight: up to 3 tonne (3.3 ton)**

Dramatically portrayed in a film by Steven Spielberg, no other predatory fish arouses so much fear and terror as the great white shark (*Carcharodon carcharias*). Surfers and swimmers look very much like seals to these sharks, and seals are their natural prey. That is why there are frequent 'shark attacks' although these have nothing to do with any evil intent on the part of the fish. They are just following their natural instincts for hunting. They are found in all the world's oceans and live in small groups, demonstrating complex social behaviours which have still not been sufficiently studied. They are also a

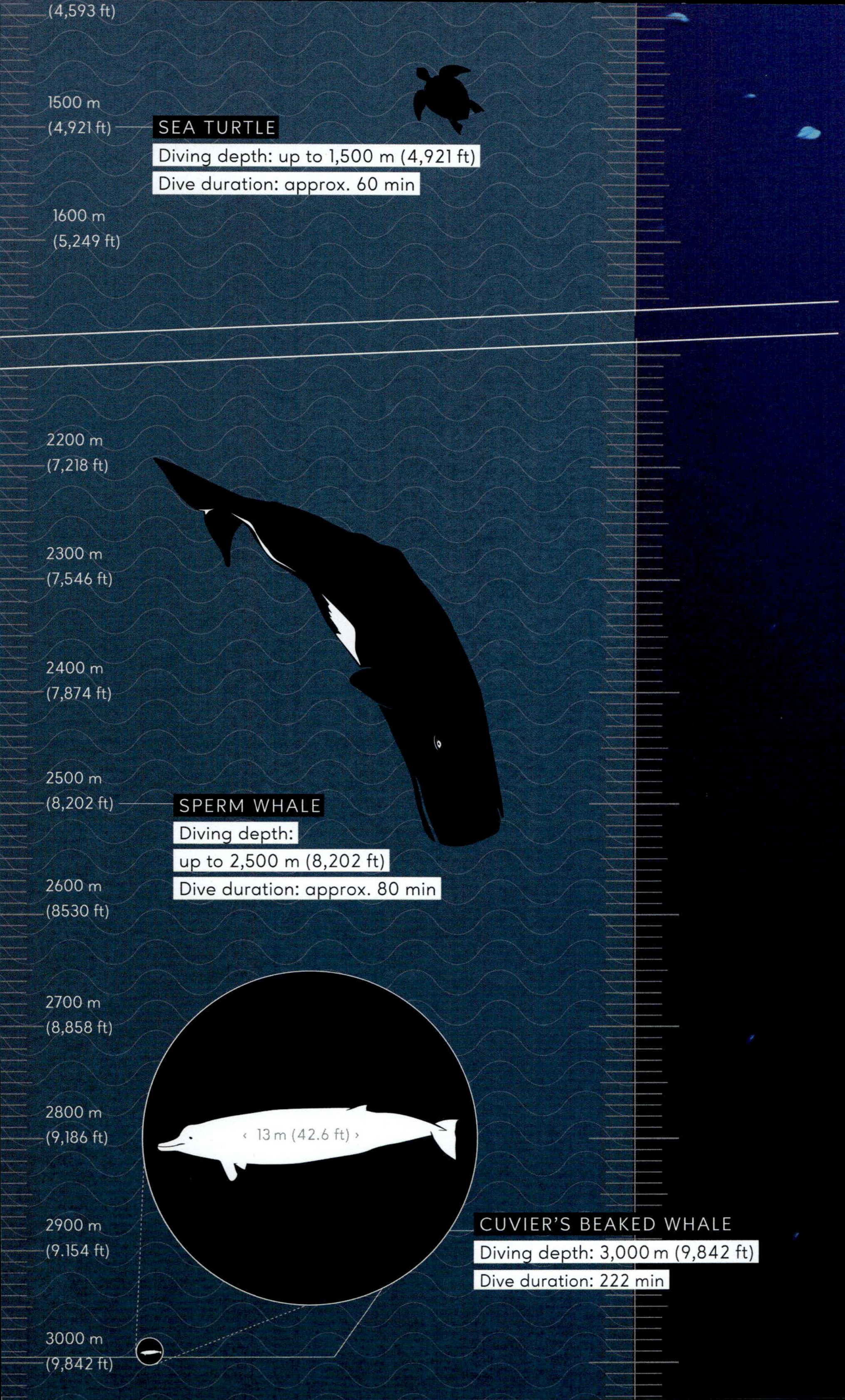


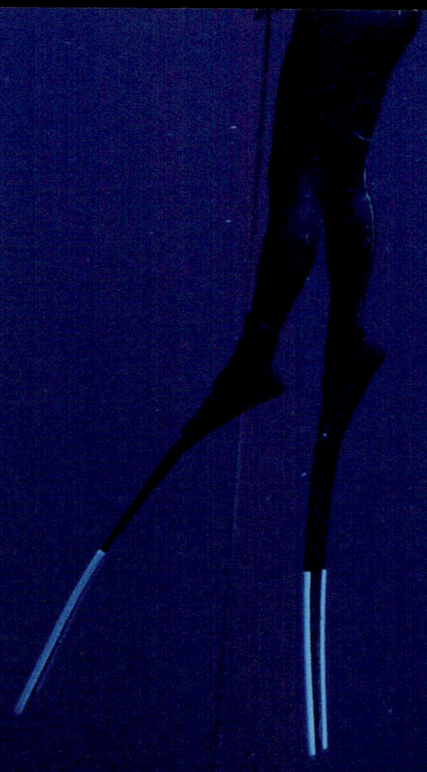

## HUMAN

Diving depth: 125 m (410 ft)

Dive duration: 4 mins 39 seconds

No-one has yet been able to dive to this depth without an external air supply.

The Russian Alexey Molchanov carried out a dive in Dean's Blue Hole on Long Island, one of the islands in the Bahamas, and achieved a new world record in apnea (free diving). The sportsman remained suspended at a depth of 125 m (410 ft) before making his way back up to the surface. For this feat, he had to hold his breath for an astounding 4 minutes and 39 seconds.

## ANIMALS

Diving depth: 3,000 m (9,842 ft)

Dive duration: 222 min

A Cuvier's beaked whale, which researchers fitted with a depth-recorder, currently holds the deep-sea diving record: 3,000 m (9,843 ft). During this, the animal stayed underwater for 222 minutes. It is highly likely that whales can dive to even greater depths. However, to verify this, the animals would have to be fitted beforehand with a measuring device.

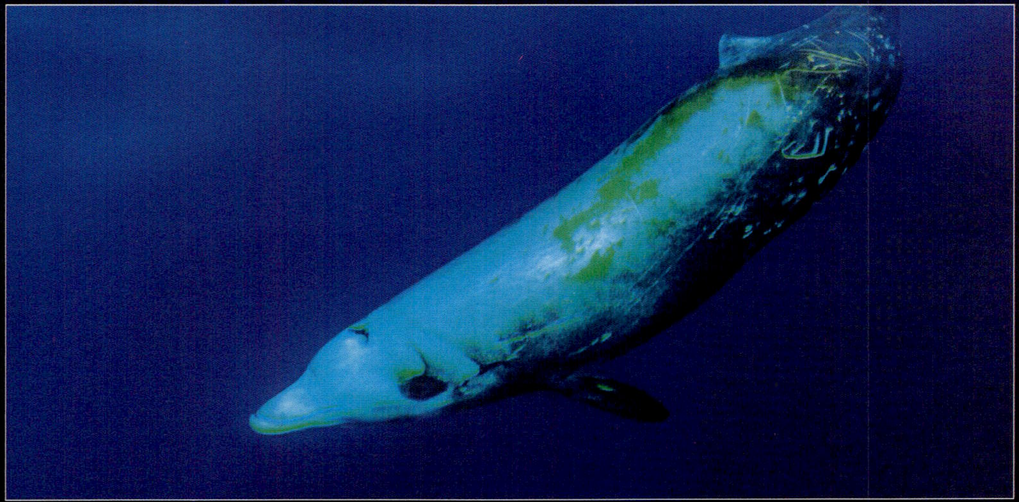

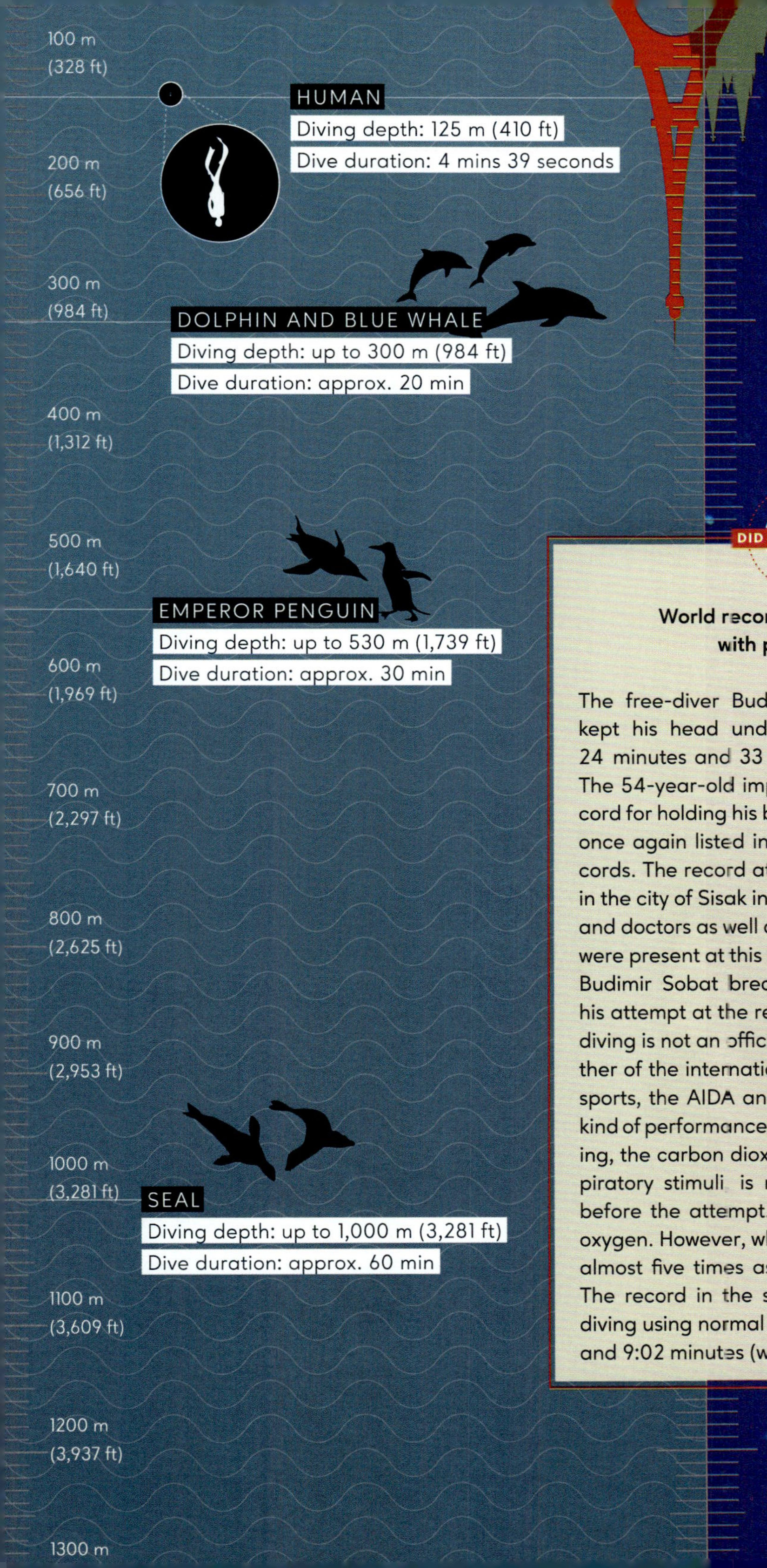


FOUR ELEMENTS

WATER

# NOW HOLD YOUR BREATH!

DID YOU KNOW?

**World record for timed diving with pure oxygen**

The free-diver Budimir Sobat, from Croatia, kept his head underwater for an incredible 24 minutes and 33 seconds – a world record. The 54-year-old improved on his own best record for holding his breath by 22 seconds and is once again listed in the Guinness Book of Records. The record attempt took place in a pool in the city of Sisak in Croatia. Emergency divers and doctors as well as reporters and spectators were present at this unbelievable performance. Budimir Sobat breathed pure oxygen before his attempt at the record. This version of timed diving is not an officially recognized sport of either of the international associations for diving sports, the AIDA and the CMAS – it is more a kind of performance. In this type of breath-holding, the carbon dioxide, which triggers the respiratory stimuli is nearly completely exhaled before the attempt. Normal air is 21 per cent oxygen. However, when pure oxygen is inhaled, almost five times as much reaches the lungs. The record in the sport of static apnea (free diving using normal air) is 11:54 minutes (men's) and 9:02 minutes (women's)

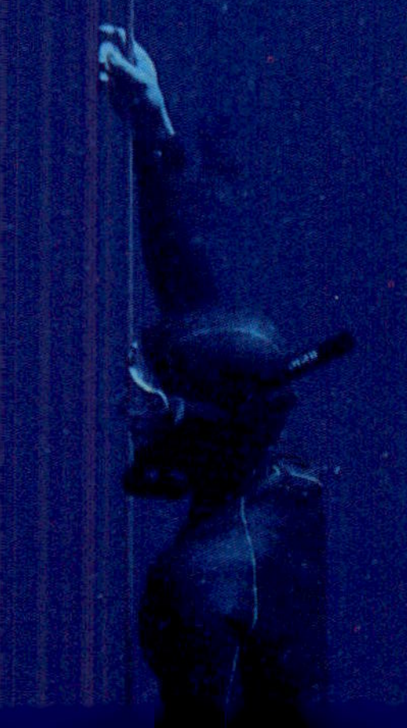

FOUR ELEMENTS

WATER

# ORCAS

## The largest dolphins

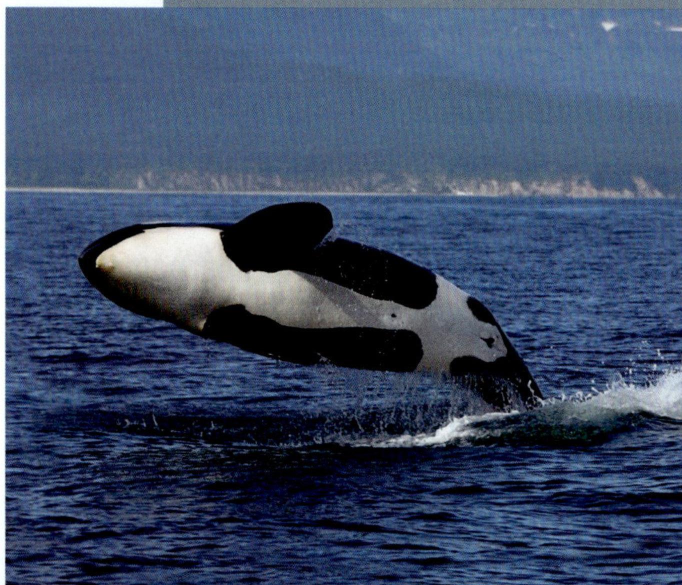

Orcas (*Orcinus orca*) are the largest species in the dolphin family. They belong to the toothed whale group, weighing between 5 and 9 tonne (5.5 and 10 ton) and with a length of 6 to 8 m (20 to 26 ft). These animals, also called killer whales because of their high dorsal fins, can be found in every ocean, but particularly in the cooler sea areas. They live to be around 50 to 80 years old and are identified as 'resident' or 'transient' orcas. Resident orcas stay in their traditional sea areas, often in pods of more than 50 animals, and feed primarily on fish. Transient orcas live in smaller pods of five to ten animals that move around in the world's oceans, and, because they are fast swimmers and ingenious hunters, they tend to specialize in hunting larger prey. They are not afraid to hunt blue whales, which are considerably larger than themselves. The hunting techniques employed by the orcas operating in these pods are astounding. In the Antarctic, they ensnare seals and penguins from ice floes; in Patagonia, they throw themselves onto the sand in order to catch sea lions. In the waters off New Zealand, they also kill sharks with a powerful blow from their tail fins. These bold hunting techniques have earned the orcas the name 'killer whales'.

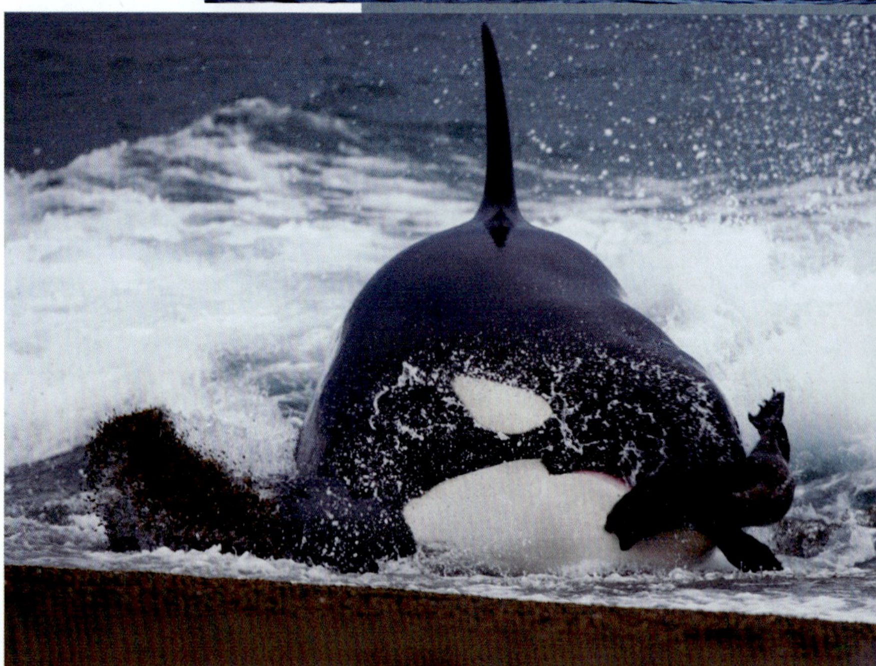

Orcas live in family pods with a matriarchal structure. After reaching sexual maturity, they calve every five to six years. Following a gestation period of 15 months, the calves are raised by the female leader and trained in a wide range of hunting techniques. The training of these highly intelligent animals to become bold, successful hunters lasts several years. To communicate acoustically, they use high-frequency clicks and whistles, which are generally referred to as 'whale song'.

During the mating season, the bulls leave their group but return to their families. Procreation within their own family is taboo. Among the orcas, the bulls do not seem to indulge in fighting each other. In the course of evolution, they seem to have learned the art of resolving conflicts without violence and instead seem to cooperate with each other.

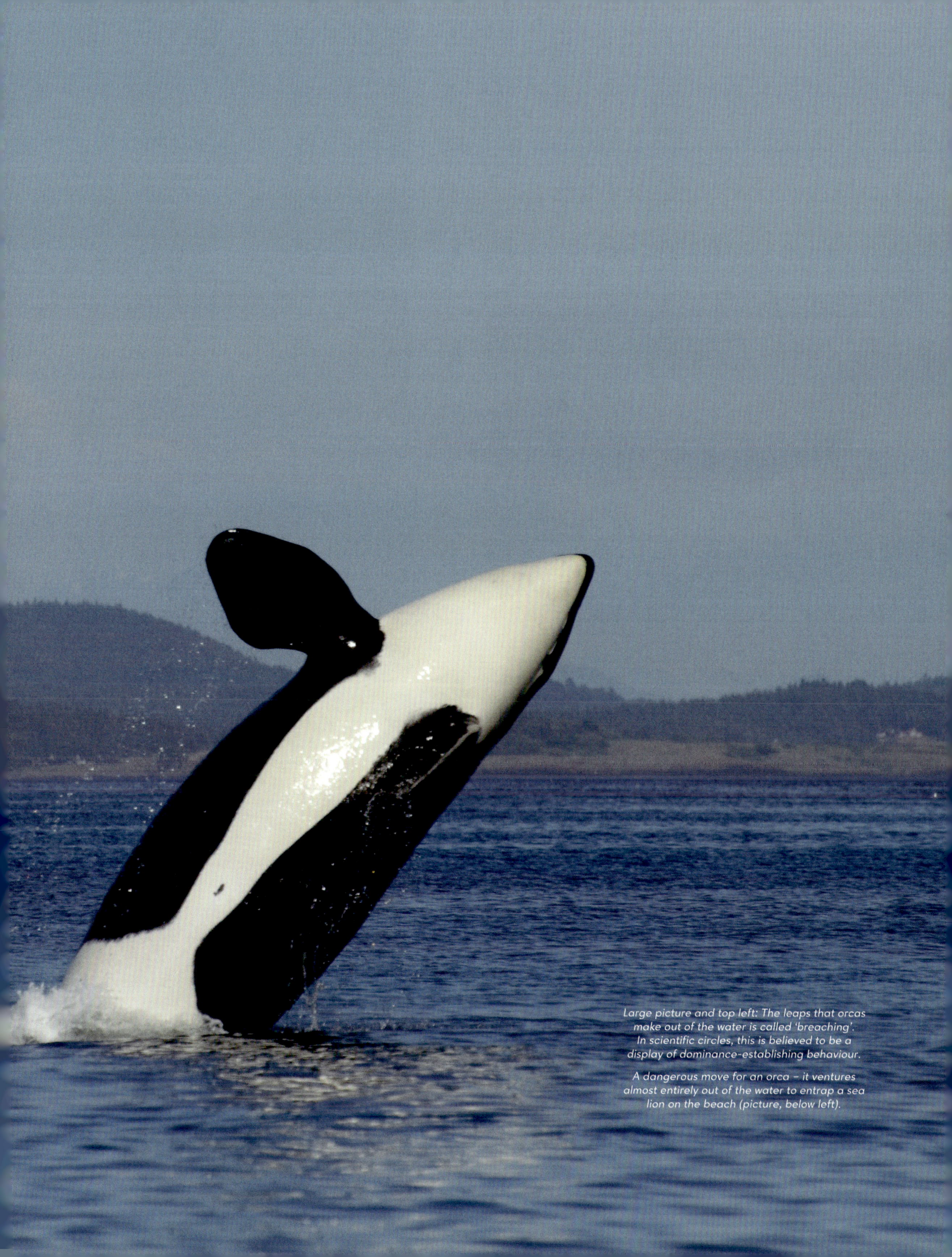

*Large picture and top left: The leaps that orcas make out of the water is called 'breaching'. In scientific circles, this is believed to be a display of dominance-establishing behaviour.*

*A dangerous move for an orca – it ventures almost entirely out of the water to entrap a sea lion on the beach (picture, below left).*

FOUR ELEMENTS

WATER

# RIVERS

Rivers shape our landscapes. Over the course of millions of years, they have cut deep valleys through mountains, brought water to plains and transformed bare desert regions into fertile oases

When rivers join together to form currents that drain into the oceans, they then become a vital part of the global hydrologic cycle.

The longest rivers do not necessarily have the most water. The Nile, as an ephemeral river, loses so much water through evaporation en route to its river mouth that it only discharges 2,800 $m^3$ (98,881 cubic ft) per second into the Mediterranean. Rivers much shorter in length often carry volumes several times this amount. But this pales beside the rivers which have the highest flows in the world, particularly the river with, by far, the highest flow – the Amazon.

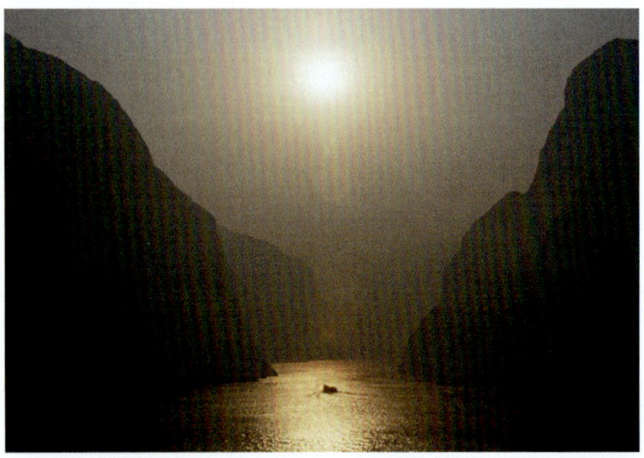

*The Qutang Gorge on the Yangtze Kiang in the Three Gorges region in China (above).*

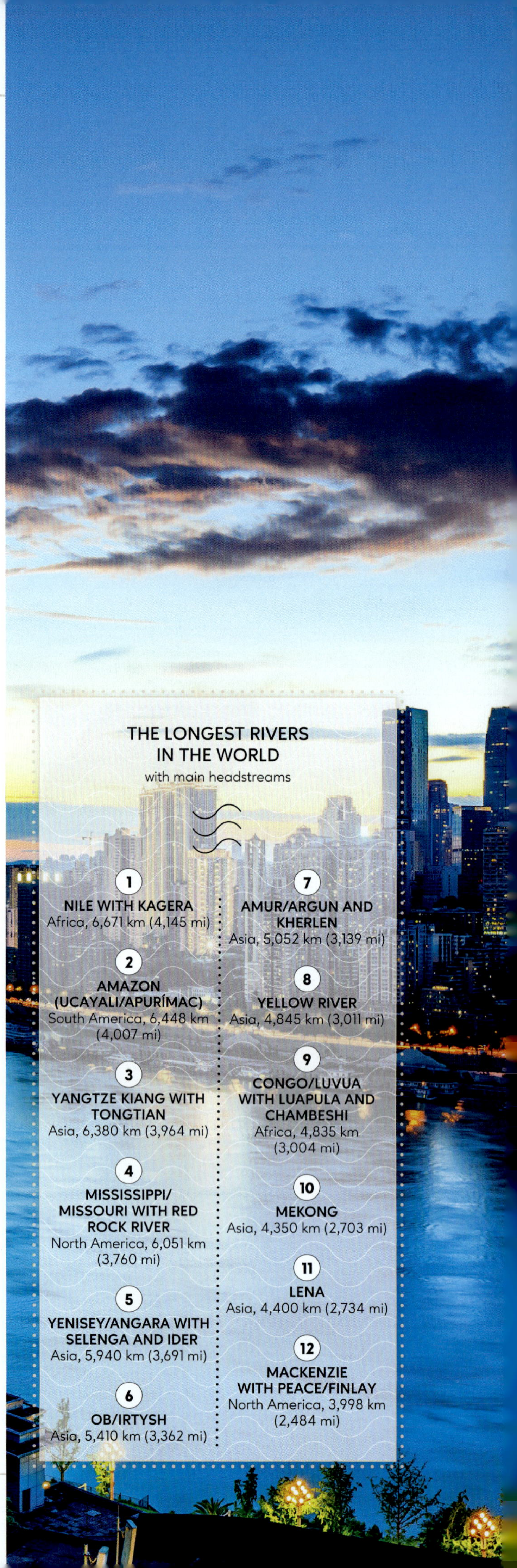

## THE LONGEST RIVERS IN THE WORLD

with main headstreams

1. **NILE WITH KAGERA**
Africa, 6,671 km (4,145 mi)

2. **AMAZON (UCAYALI/APURÍMAC)**
South America, 6,448 km (4,007 mi)

3. **YANGTZE KIANG WITH TONGTIAN**
Asia, 6,380 km (3,964 mi)

4. **MISSISSIPPI/ MISSOURI WITH RED ROCK RIVER**
North America, 6,051 km (3,760 mi)

5. **YENISEY/ANGARA WITH SELENGA AND IDER**
Asia, 5,940 km (3,691 mi)

6. **OB/IRTYSH**
Asia, 5,410 km (3,362 mi)

7. **AMUR/ARGUN AND KHERLEN**
Asia, 5,052 km (3,139 mi)

8. **YELLOW RIVER**
Asia, 4,845 km (3,011 mi)

9. **CONGO/LUVUA WITH LUAPULA AND CHAMBESHI**
Africa, 4,835 km (3,004 mi)

10. **MEKONG**
Asia, 4,350 km (2,703 mi)

11. **LENA**
Asia, 4,400 km (2,734 mi)

12. **MACKENZIE WITH PEACE/FINLAY**
North America, 3,998 km (2,484 mi)

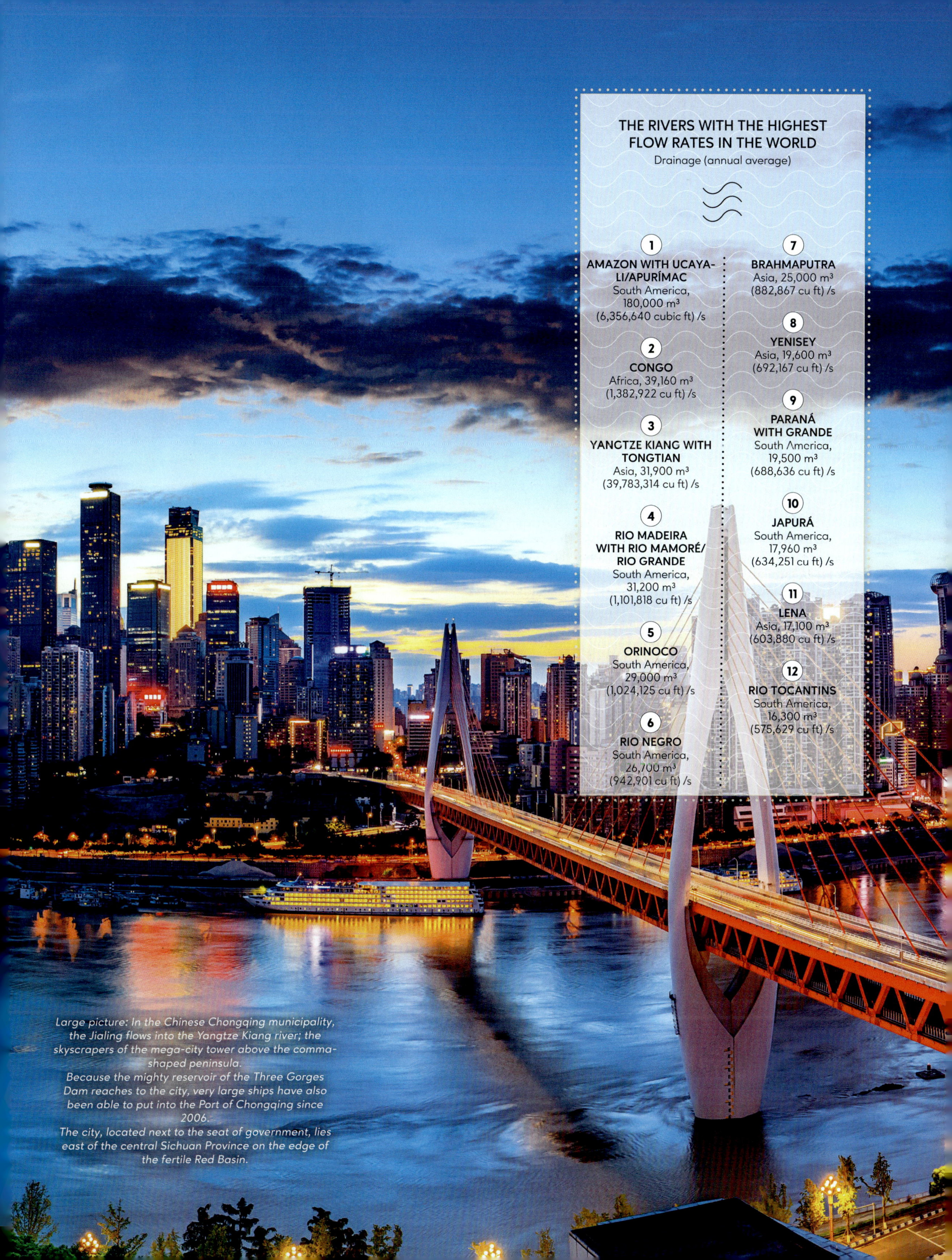

## THE RIVERS WITH THE HIGHEST FLOW RATES IN THE WORLD

Drainage (annual average)

**1**
**AMAZON WITH UCAYALI/APURÍMAC**
South America, 180,000 m³ (6,356,640 cubic ft) /s

**2**
**CONGO**
Africa, 39,160 m³ (1,382,922 cu ft) /s

**3**
**YANGTZE KIANG WITH TONGTIAN**
Asia, 31,900 m³ (39,783,314 cu ft) /s

**4**
**RIO MADEIRA WITH RIO MAMORÉ/ RIO GRANDE**
South America, 31,200 m³ (1,101,818 cu ft) /s

**5**
**ORINOCO**
South America, 29,000 m³ (1,024,125 cu ft) /s

**6**
**RIO NEGRO**
South America, 26,700 m³ (942,901 cu ft) /s

**7**
**BRAHMAPUTRA**
Asia, 25,000 m³ (882,867 cu ft) /s

**8**
**YENISEY**
Asia, 19,600 m³ (692,167 cu ft) /s

**9**
**PARANÁ WITH GRANDE**
South America, 19,500 m³ (688,636 cu ft) /s

**10**
**JAPURÁ**
South America, 17,960 m³ (634,251 cu ft) /s

**11**
**LENA**
Asia, 17,100 m³ (603,880 cu ft) /s

**12**
**RIO TOCANTINS**
South America, 16,300 m³ (575,629 cu ft) /s

*Large picture: In the Chinese Chongqing municipality, the Jialing flows into the Yangtze Kiang river; the skyscrapers of the mega-city tower above the comma-shaped peninsula.*
*Because the mighty reservoir of the Three Gorges Dam reaches to the city, very large ships have also been able to put into the Port of Chongqing since 2006.*
*The city, located next to the seat of government, lies east of the central Sichuan Province on the edge of the fertile Red Basin.*

FOUR ELEMENTS

WATER

# NILE

The longest river in the world

The Nile is both a lifeline and a legend. Its source was only discovered at the end of the 19th century. It is considered to be an ephemeral river because it draws its water from the wet regions of its headwaters and acquires very little water before reaching its river mouth.

The drainage area of the longest river on our planet - 6,671 km (4,145 mi) - covers three million km$^2$ (1,158,306 sq mi). That is more than 10 per cent of the entire continent of Africa. The headstream is the White Nile which is supplied by the Kagera River. These two headstreams rise in Burundi and Rwanda; the longer of the two rises at Luvironza in Burundi, which has the highest summit in the small African country, standing at 2,700 m (8,859 ft), and is therefore considered to be the actual headstream. At the Tanzanian-Ugandan border, the Kagera flows into Lake Victoria, into which many other rivers also flow. It therefore serves as the water reservoir for the entire Nile. As the Albert Nile, it discharges its mass of water over the Murchison Falls and flows on northwards through Uganda and Sudan. At Khartoum, the Blue Nile joins the White Nile, and, a little further north, the Atbara also joins; both come from the Ethiopian uplands. With no further rivers joining it, the river winds its way through the Nubian Desert and finally on to Egypt, where it empties into the Mediterranean at its wide delta north of Cairo.

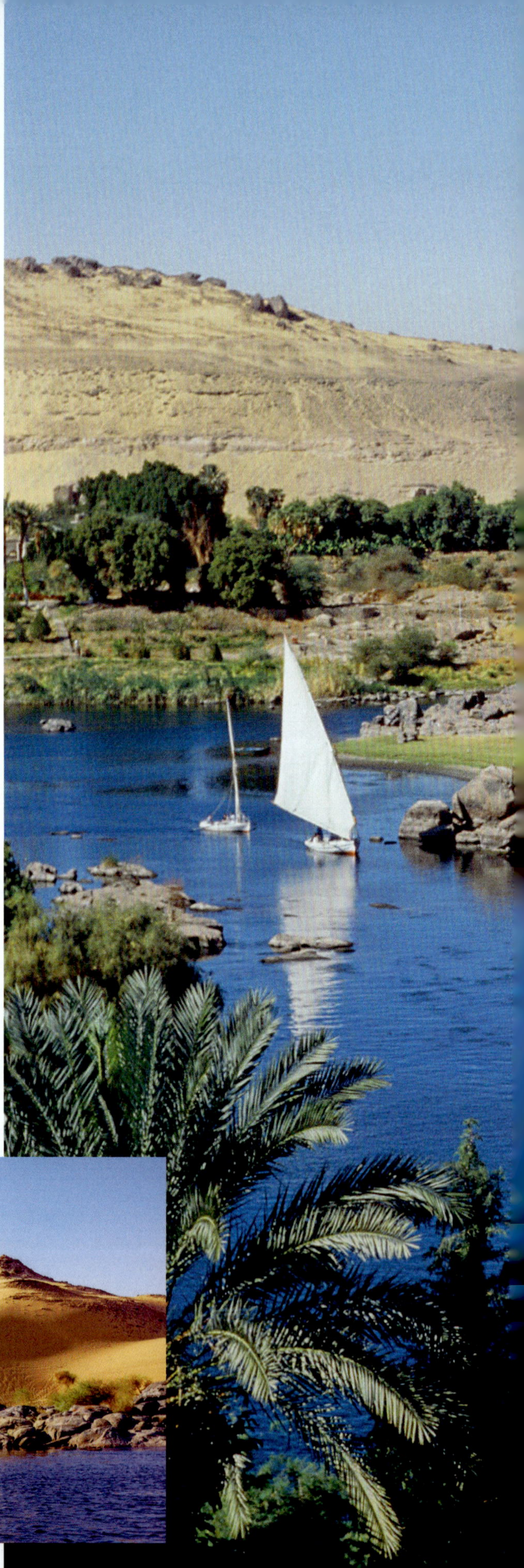

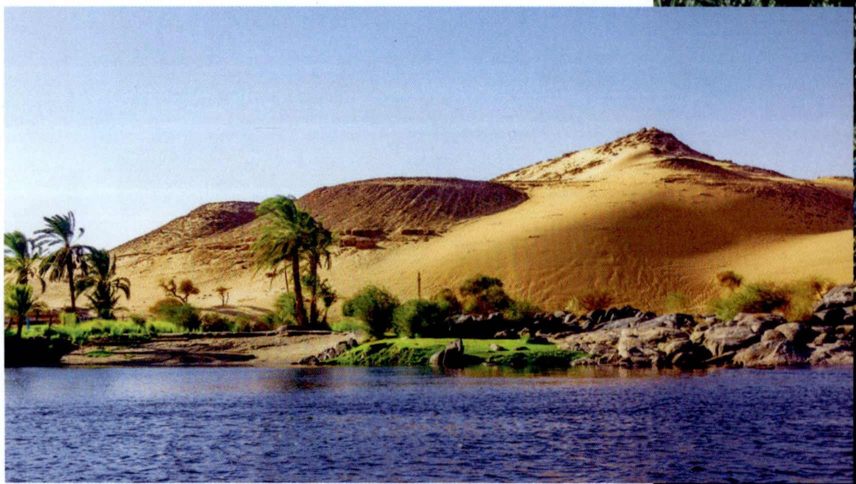

*The Nile at Aswan; in the background, the end of the Aga Khan Mausoleum, built in the 1950s, is visible. The river banks are an oasis; directly behind them, the barren desert begins.*

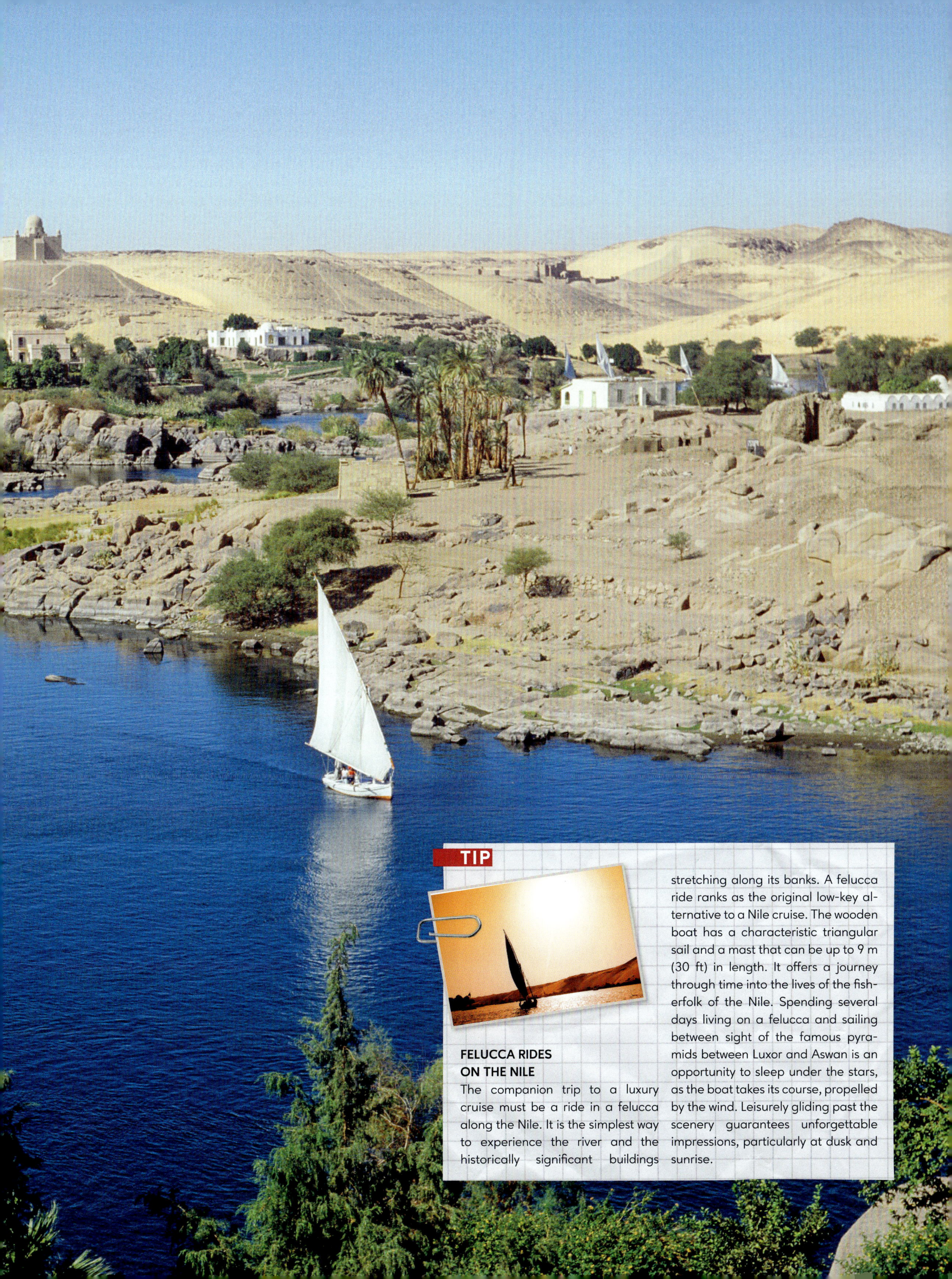

TIP

**FELUCCA RIDES ON THE NILE**

The companion trip to a luxury cruise must be a ride in a felucca along the Nile. It is the simplest way to experience the river and the historically significant buildings stretching along its banks. A felucca ride ranks as the original low-key alternative to a Nile cruise. The wooden boat has a characteristic triangular sail and a mast that can be up to 9 m (30 ft) in length. It offers a journey through time into the lives of the fisherfolk of the Nile. Spending several days living on a felucca and sailing between sight of the famous pyramids between Luxor and Aswan is an opportunity to sleep under the stars, as the boat takes its course, propelled by the wind. Leisurely gliding past the scenery guarantees unforgettable impressions, particularly at dusk and sunrise.

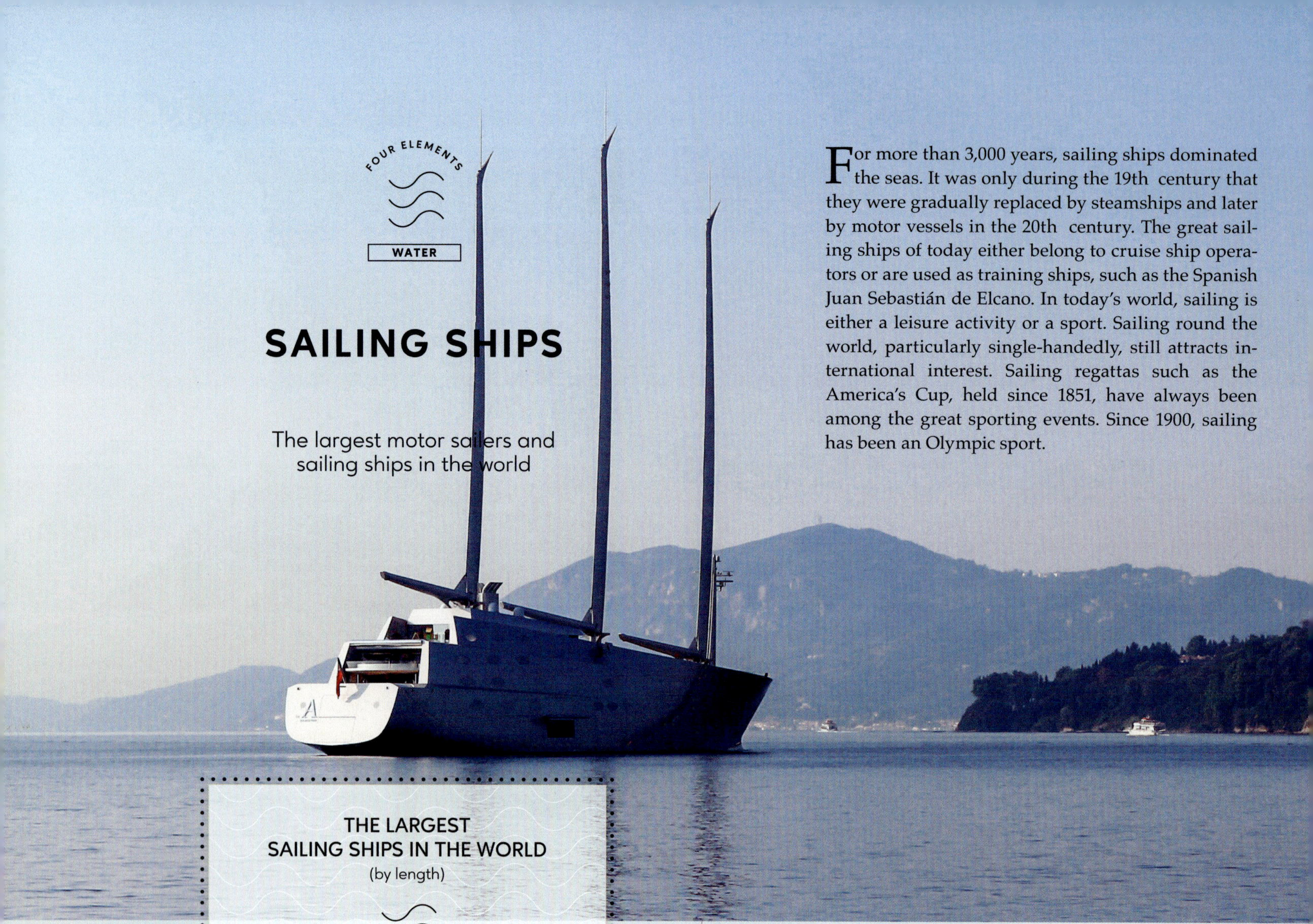

FOUR ELEMENTS

WATER

# SAILING SHIPS

The largest motor sailers and sailing ships in the world

For more than 3,000 years, sailing ships dominated the seas. It was only during the 19th century that they were gradually replaced by steamships and later by motor vessels in the 20th century. The great sailing ships of today either belong to cruise ship operators or are used as training ships, such as the Spanish Juan Sebastián de Elcano. In today's world, sailing is either a leisure activity or a sport. Sailing round the world, particularly single-handedly, still attracts international interest. Sailing regattas such as the America's Cup, held since 1851, have always been among the great sporting events. Since 1900, sailing has been an Olympic sport.

## THE LARGEST SAILING SHIPS IN THE WORLD

(by length)

1. **SAILING YACHT A** Germany, 142.8 m (469 ft)
2. **ROYAL CLIPPER** Malta, 132.2 m (434 ft)
3. **SEDOV** Russia, 117.5 m (385 ft)
4. **STAR CLIPPER/STAR FLYER** Malta, 115.5 m (379 ft)
5. **UNIÓN** Peru, 115 m (377 ft)
6. **KRUZENSHTERN** Russia, 114.5 m (376 ft)
7. **JUAN SEBASTIÁN ELCANO** Spain, 113 m (371 ft)
8. **ESMERALDA** Chile, 113 m (371 ft)
9. **MIR** Russia, 109 m (358 ft)
10. **NADEZHA** Russia, 109 m (358 ft)

## SAILING YACHT A – SY A THE LARGEST SAILING SHIP IN THE WORLD

The construction of *Sailing Yacht A* is reported to have cost 400 million euro; it was commissioned by the Russian entrepreneur and millionaire Andrei Melnitschenko. The construction of the *White Pearl*, as it was called before its completion, was carried out to a large extent in secrecy in a boatyard in Rendsburg, located in Schleswig-Holstein in Germany, following a design drawn up by the French designer Philippe Starck. When *Sailing Yacht A* left the boatyard at the start of 2017, it became the largest sailing ship in the world. Its total length was a massive 142.81 m (469 ft), and it had eight decks to call its own. Three masts rose 90 m (295 ft) into the sky; the largest sail had a surface area of more than 1,760 m² (18,945 sq ft). On the underside of the keel, curved glass had been used so that the underwater world could be observed. Overall, *Sailing Yacht A* can sail at almost 40 km (25 mi) per hour, or it can use the diesel or electric motors for support. The use of teak for the construction was controversial, as it reportedly came from illegal plantations.

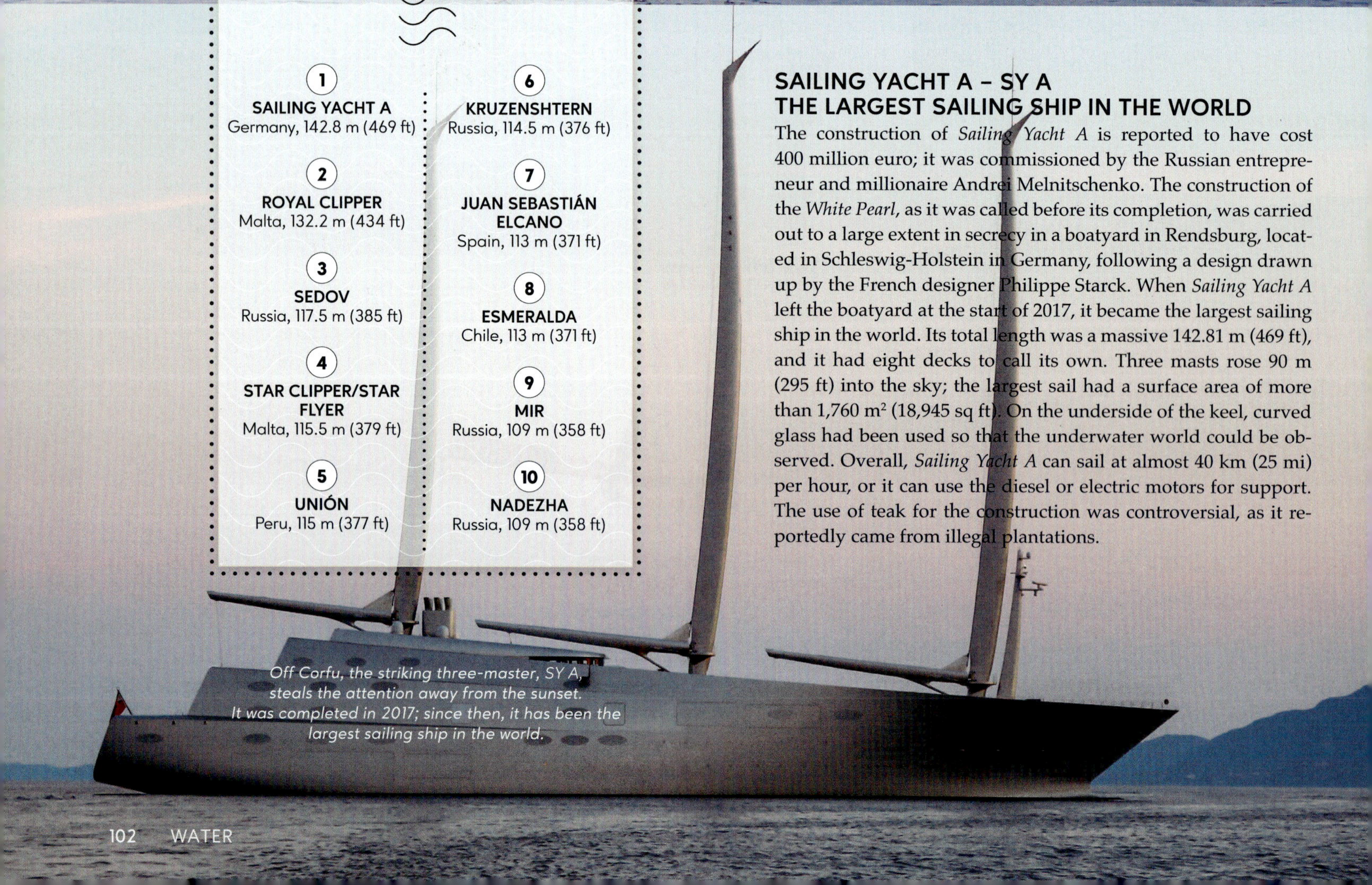

*Off Corfu, the striking three-master, SY A, steals the attention away from the sunset. It was completed in 2017; since then, it has been the largest sailing ship in the world.*

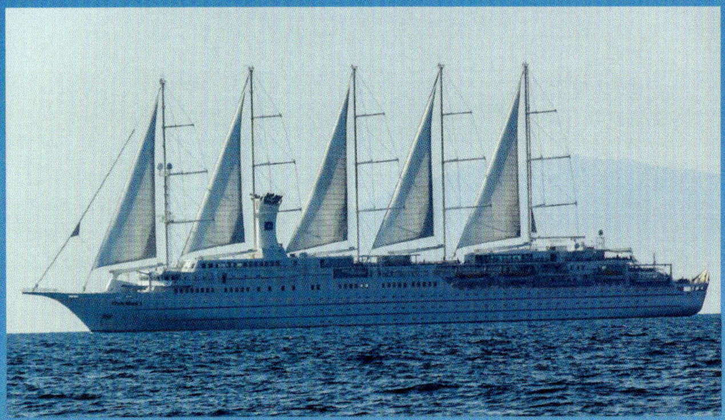

## WIND SURF AND CLUB MED 2 – THE LARGEST MOTOR SAILERS IN THE WORLD

The *Fünfmaster* made trips as *Club Med 1* until 1998. It now makes crossings for the American sailing cruise operator, Windstar Cruises, in the Caribbean in winter and in the Mediterranean and the Baltic Sea in summer. Modern computer technology not only supports the ship's navigation, it also controls the furling and unfurling of the seven sails, which have a total area of 2,600 $m^2$ (27,986 sq ft). *Wind Surf* is fitted with two electromotors, so that, even in calm waters, it can move along at a speed of up to 12 knots. Alongside *Club Med 2* (left), it is the largest motor sailer in the world (2021).

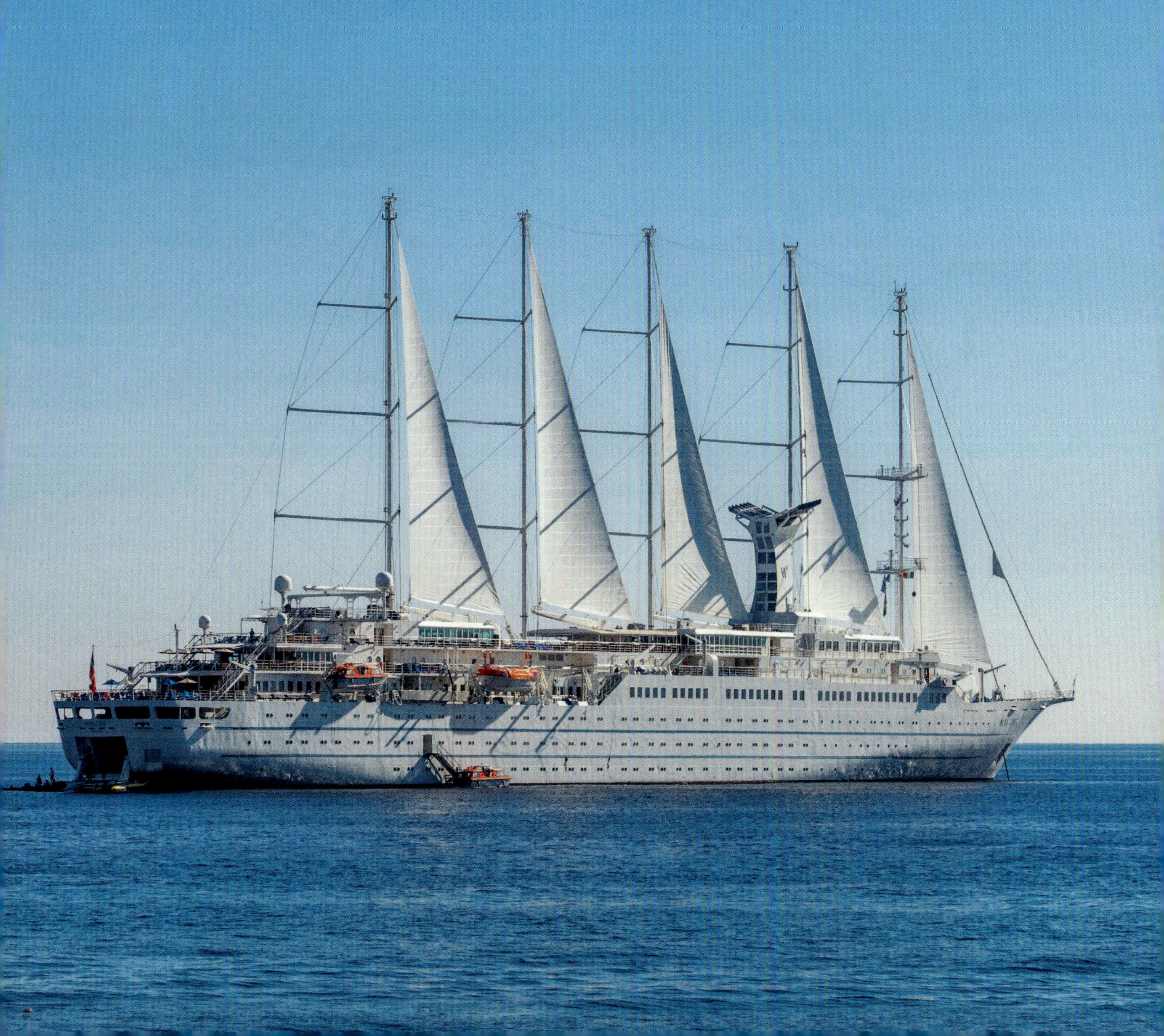

# RIVERS OF AFRICA

The lengths of the rivers in Africa which rank after the Nile is not always undisputed fact. This is due not least to the extreme variations in the quantity of water during the rainy and dry seasons. The data for this continent therefore vary widely.

## Niger
**4,184 km (2,600 mi)**

The third longest river on the African continent rises in Guinea, near the border with Sierra Leone. It flows in an arc eastwards through Mali, passes the city of Timbuktu, bends sharply to the southeast further on, flows through Niger, touches the border with Benin and finally drains in Nigeria after leaving the Kainji Reservoir via a wide delta in the Gulf of Guinea.

## Congo and Luvua
**with Luapula and Chambeshi**
**4,835 km (3,004 mi)**

The second longest river in Africa has the highest flow rate in the world, after the Amazon. Its drainage basin, covering 3.7 million $m^2$ (39,826,468 sq ft), is even greater than that of the Nile. A great number of rivers form part of the river system of the Congo. The Chambeshi, as the headwater of the Luapula, is the headwater situated furthest away from the mouth of the river. It rises in northern Zambia close to the border with Tanzania, west of Lake Malawi, and flows on southwest through the Bangweulu Wetlands, which are then drained by the Luapula. At the point where the Congo turns towards the southwest, it forms the border between the Democratic Republic of the Congo and the Republic of the Congo – both of the capitals, Brazzaville and Kinshasa, lie opposite each other on the Pool Malebo, a widening in the river which resembles a lake.

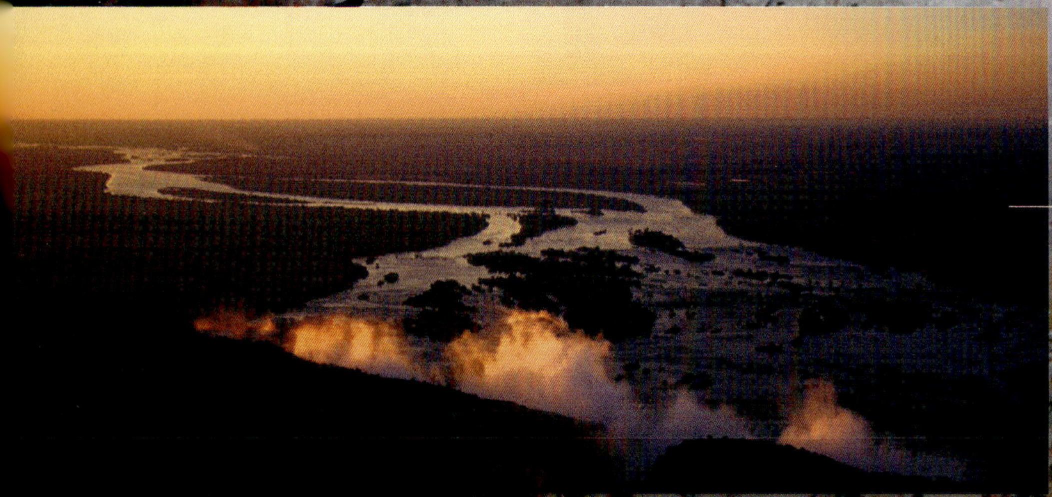

## Zambezi
**2,574 km (1,599 mi)**

The Zambezi is the largest African river that drains in the east. It rises in the northwest of Zambia at a height of 1,500 m (4,921 ft), flows through Angola and Zambia, is the border river to Zimbabwe and empties into the Indian Ocean in Mozambique in a vast delta covering 880 $km^2$ (9,472 sq ft). The Victoria Falls on the Zambezi River between Zimbabwe and Zambia are a UNESCO World Heritage site.

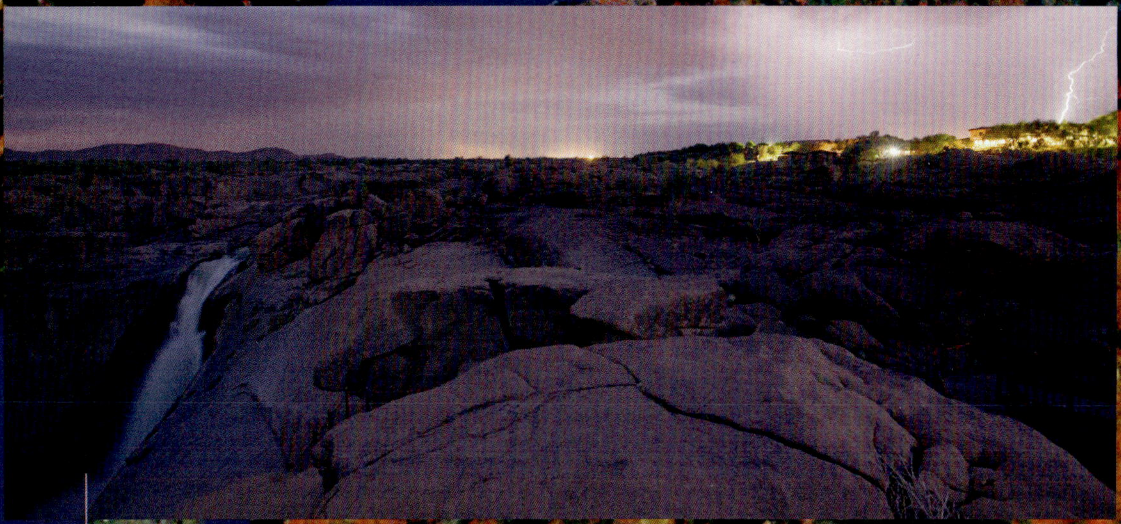

## Oranje
2,160 km (1,342 mi)

The Oranje, as the Orange River is called in Afrikaans, is the largest river in the south of Africa. It flows through Lesotho and South Africa and, in its lower course, forms the border river with Namibia. Its source lies in the uplands of Lesotho; from there, it flows westward through South Africa and discharges into the Atlantic Ocean at its river mouth. The Orange River is a UNESCO World Heritage.

## Jubba
1,658 km (1,030 mi)

The headwaters of the Jubba, or Juba, are in the Ethiopian uplands; they join together at Dolow on the border with Somalia to become the Jubba, which then flows southwards through Somalia before emptying into the Indian Ocean at Kismaayo. The Jubba Basin is a savannah which the river has transformed into fertile land fit for cultivation. The Jubba often floods as a result of heavy rainfall.

## Okavango
1,800 km (1,118 mi)

The Okavango River, known as the Cubango in its upper reaches, originates in the uplands of Bié in Angola in southwest Africa. It flows south through the uplands and, for 400 km (249 mi), forms the border river between Namibia and Angola. It reaches the northwestern regions of Botswana on the far side of Andara. There, the water seeps away into the swampy Okavango delta, the inland delta covering 15,000 $km^2$ (5,791 sq mi) in the northeastern section of the Kalahari Desert.

## Volta with the Black Volta
1,600 km (994 mi)

The Volta River flows in West Africa and is fed by three headwaters which rise in Burkina Faso (formerly known as Upper Volta): the Black Volta (Mouhoun), Red Volta (Nazinoun) and White Volta (Nakambé). After the confluence of the rivers, the flow is dammed at Salaga in Ghana by the Akosombo Dam, the second largest water reservoir in the world after Lake Victoria. The Volta then flows southeast and empties into the Gulf of Guinea.

## Limpopo
1,750 km (1,087 mi)

The Limpopo River is the most northerly border river in South Africa. After the Zambezi, it is the second longest river in Africa that drains into the Indian Ocean. Its source lies north of Johannesburg in the Witwatersrand mountain ridge, at a height of around 1,800 m (5,906 ft), where it is known as the Crocodile. It flows southeast in an arc through Mozambique and empties into the Indian Ocean north of Mabuto. Its largest tributary is the Olifants River, which flows through Kruger National Park. Since 2005, it has been officially known as the Lepelle.

## Chari
with Ouham, 1,740 km (1,081 mi)

The Chari (or Shari) River is the main tributary of Lake Chad, which has no outlet to the sea, and lies on the southern edge of the Sahara in the area bordering Chad, Cameroon, Nigeria and Niger. The lake receives ninety per cent of its waters from the Chari, which is why the water level is dependent on the variable rainfall levels in the headwater region. The main headwater, the Ouham River, rises in the Central African Republic and then flows through Chad before emptying at Sarh into the Chari, which then flows into Lake Chad further to the northwest.

## Senegal
with Bafing, 1,430 km (889 mi)

The Bafing River, the headwater of the Senegal River, rises in the Fouta Djallon highlands in Mali, close to the border with Guinea. The Bafing flows northwards, is dammed at Manantali, and then, through its confluence with the Bakoyé River, forms the Senegal River. The Senegal River flows further westwards as the border river between the countries of Mauritania and Senegal and drains into the Atlantic at the Island of Saint-Louis, a UNESCO World Heritage site. In the estuary area, there is a national park where many of the European white storks spend the winter months.

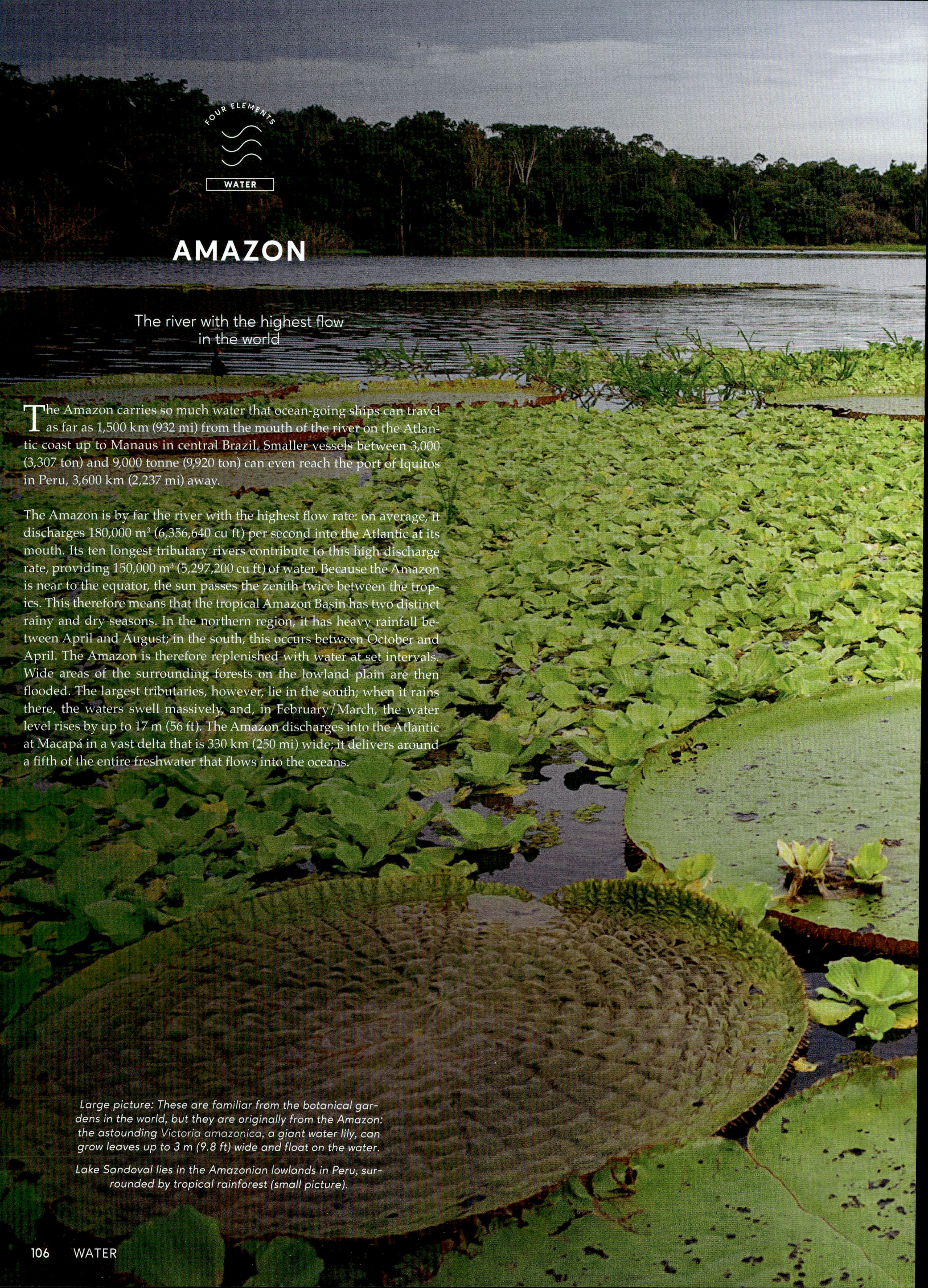

FOUR ELEMENTS

WATER

# AMAZON

The river with the highest flow in the world

The Amazon carries so much water that ocean-going ships can travel as far as 1,500 km (932 mi) from the mouth of the river on the Atlantic coast up to Manaus in central Brazil. Smaller vessels between 3,000 (3,307 ton) and 9,000 tonne (9,920 ton) can even reach the port of Iquitos in Peru, 3,600 km (2,237 mi) away.

The Amazon is by far the river with the highest flow rate: on average, it discharges 180,000 $m^3$ (6,356,640 cu ft) per second into the Atlantic at its mouth. Its ten longest tributary rivers contribute to this high discharge rate, providing 150,000 $m^3$ (5,297,200 cu ft) of water. Because the Amazon is near to the equator, the sun passes the zenith twice between the tropics. This therefore means that the tropical Amazon Basin has two distinct rainy and dry seasons. In the northern region, it has heavy rainfall between April and August; in the south, this occurs between October and April. The Amazon is therefore replenished with water at set intervals. Wide areas of the surrounding forests on the lowland plain are then flooded. The largest tributaries, however, lie in the south; when it rains there, the waters swell massively, and, in February/March, the water level rises by up to 17 m (56 ft). The Amazon discharges into the Atlantic at Macapá in a vast delta that is 330 km (250 mi) wide; it delivers around a fifth of the entire freshwater that flows into the oceans.

*Large picture: These are familiar from the botanical gardens in the world, but they are originally from the Amazon: the astounding Victoria amazonica, a giant water lily, can grow leaves up to 3 m (9.8 ft) wide and float on the water.*

*Lake Sandoval lies in the Amazonian lowlands in Peru, surrounded by tropical rainforest (small picture).*

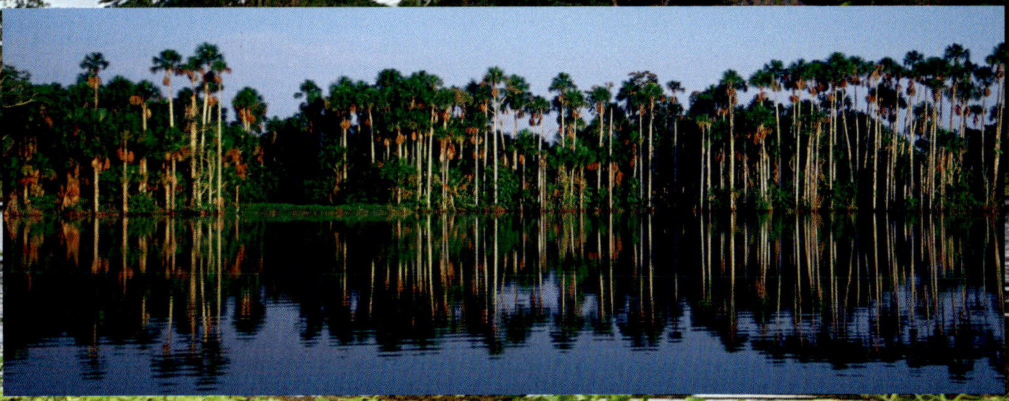

**TIP**

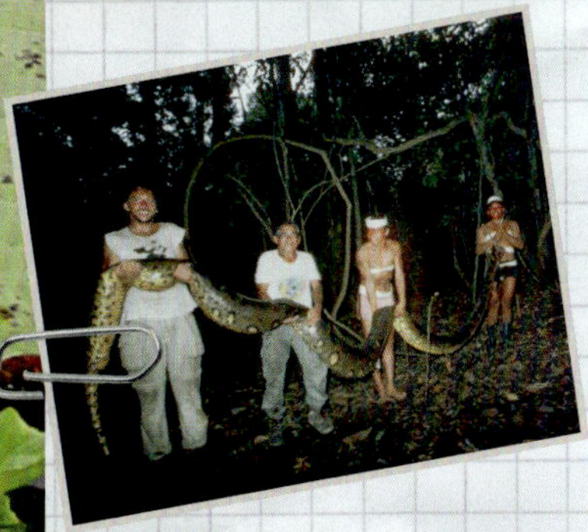

**SURVIVAL TRIP WITH TARANTULAS AND ANACONDAS**

The largest snake in the world is awesomely terrifying, not least because it can grow to be 9 m (30 ft) long. Its head can be as big as that of a mid-sized dog. The snake, a member of the boa family, lives in the waters of the rainforest and lurks there, waiting for prey; it then entwines itself around its prey, before strangling it. It feeds on large animals, such as deer or hogs, which it swallows whole. To do this, it unhooks its jaws; then it can survive up to 500 days without food. The snake can reach up to 250 kg (551 lb) in weight and possesses vestigial limbs. Boas give birth to live young. There are many operators offering jungle trips like these, where you can exchange your hotel bed for adventure and outdoor experiences and observe anacondas, tarantulas and piranha fish at close range. You also learn how to forage for food.

# RIVERS OF ASIA

The longest rivers in Asia

Some of the world's mightiest rivers can be found in Asia, the largest continent on earth. In the Asian part of Russia, the majority of the rivers drain into the northern Arctic Ocean, whilst the rivers in South and East Asia empty into oceans lying to the south and east.

## Yenisey
5,940 km (3,691 mi)

The Yenisey River flows from south to north through Siberia and discharges into the Kara Sea in the Arctic Ocean. Its total length includes the Angara with the Selenga, into which it issues as a headwater of the Ider, and the estuary bay. The source of the river system is located in the Khangai Mountains southwest of Lake Baikal. In fall, the first ice floes form; it remains completely frozen over until May.

## Yangtze Kiang
6,380 km (3,964 mi)

For the Chinese, the third longest river in the world is officially known simply as Chang Jiang, the 'long river'. Its source region lies at a height of 5,405 m (17,734 ft) in the sparsely populated northern uplands of Tibet. The source region covers several hundred $km^2$; its headwater is considered to be the Tuotuo He. In the upper reaches, the river is known as Tongtian He – 'the river that crosses to heaven'. It flows in a southerly direction, but, at Shigu, the mass of water hits a limestone massif head on. The river is then diverted northeast in a sharp curve. Without this 'Great Bend at Shigu', the Yangtze Kiang, like the neighbouring Mekong, would discharge to the south. After several hooks, the river then flows east.

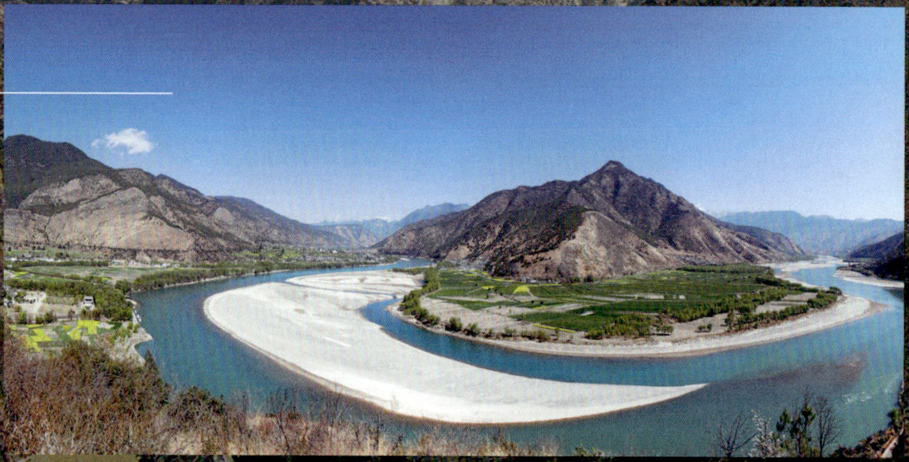

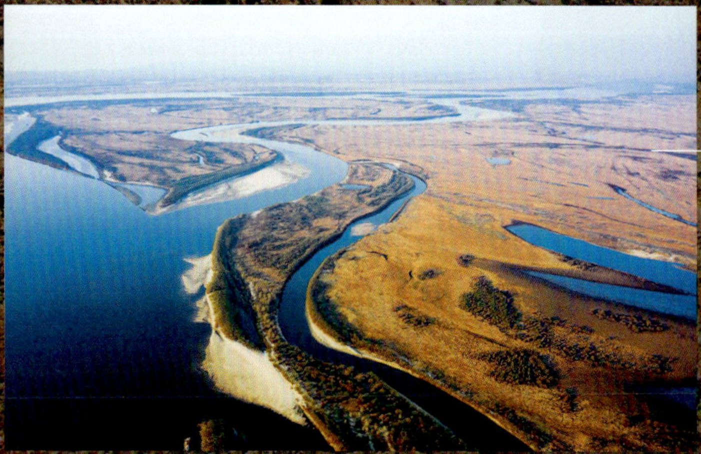

## Amur/Argun and Kherlen
5,052 km (3,139 mi)

The Kherlen River (also known as the Kerülen) rises in Mongolia north of Ulaanbaatar (previously Ulan-Bator) and then flows to China in the Hulun Nur, a steppe lake. In years with high levels of precipitation, this lake overflows and drains in the north into the Argun (or Ergune) River which forms the China-Russia border. The Argun joins the Amur River further east in the Russian city of Chabarowsk. This river flows further north and then empties in Nikolayevsk-on-the-Amur opposite Sakhalin Island in the Tatar Strait.

## Ob and Irtysh
5,410 km (3,362 mi)

The Irtysh, literally 'the White River', rises in the Mongolian Altai Mountains in Dzungaria in the Chinese province of Xinjiang. It flows through Kazakhstan, reaches Russia at a point southeast of Omsk and joins the Ob River at Khanty-Mansijsk. The Ob flows on northwards through western Siberia and drains into the Kara Sea in an estuary 800 km (497 mi) long and 70 km (43 mi) wide.

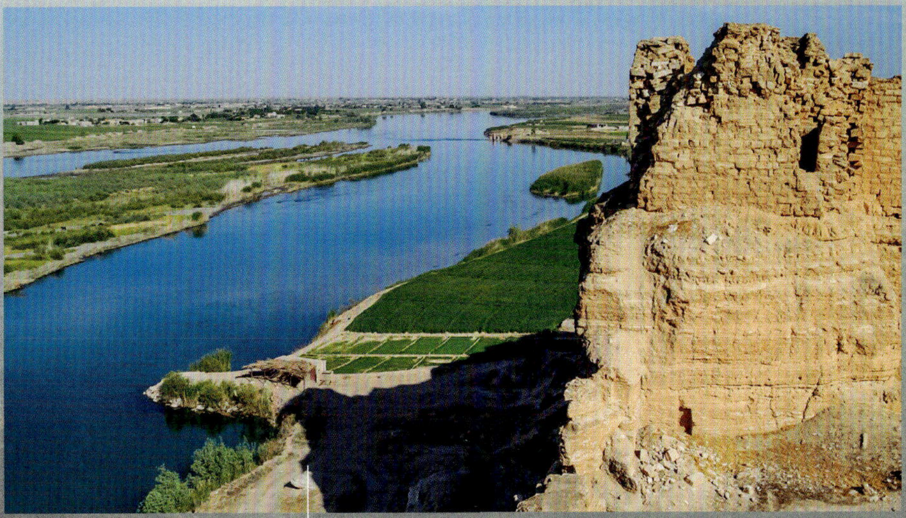

## Euphrates and Murat
3,380 km (2,100 mi)

The Murat, the source of the Euphrates River, rises in the vicinity of Ararat, the highest mountain in Turkey (5,165 m (16,945 ft)), in the uplands of Eastern Anatolia. It flows west and joins the Euphrates at Malatya, the capital of the province bearing the same name. The Euphrates subsequently flows through Syria and Iraq, where, at the Shatt al-Arab in Mesopotamia, it joins its twin river, the Tigris – the border river between Iraq and Iran. Mesopotamia has historically been considered the cradle of civilization.

## Yellow River
4,845 km (3.011 mi)

The Yellow River (Huang He) is the second longest river in China, after the Yangtze Kiang. It also rises in the Tibetan Plateau, in the Bayan Har Mountains in the province of Qinghai. It flows through nine Chinese provinces and discharges south of Beijing into the Gulf of Bohai on the Yellow Sea. Its name comes from the high levels of loess that it carries with it, which benefits agriculture but also leads to problems due to erosion and sedimentation.

## Mekong
4,500 km (2,796 mi)

The Mekong rises in the Tibetan Plateau – the exact sources are disputed. It flows south, forming the border river between Laos and Myanmar and between Laos and Thailand. The Mekong flows on through the capital of Laos, Vientiane (Viangchan). It then crosses Cambodia not far from the capital, Pnomh Penh. It flows over the border to Vietnam and empties into the South China Sea in a wide delta.

## Lena
4,400 km (2,734 mi)

The Lena is a river in Russia in the border area of central to eastern Siberia. Its source lies in the Baikal Mountains, west of Lake Baikal. It flows via Ust-Kut, from which point it is navigable, to Jakutsk, the capital of the Russian republic of Sakha. From here, it flows further north, before discharging its waters in a wide river delta into the Laptev Sea, a marginal sea of the Arctic Ocean. From October to June, the Lena River is frozen over.

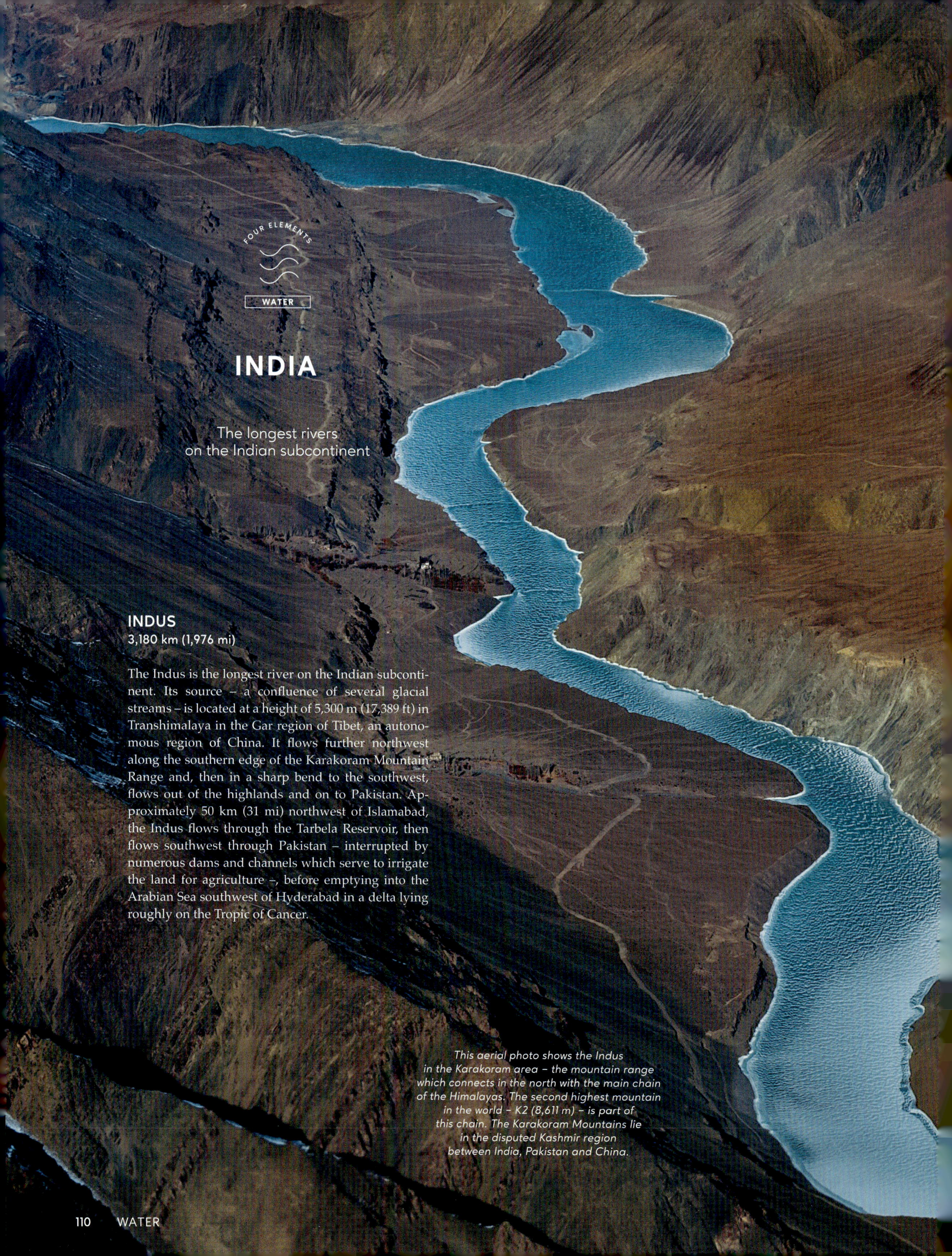

FOUR ELEMENTS

WATER

# INDIA

The longest rivers on the Indian subcontinent

## INDUS

**3,180 km (1,976 mi)**

The Indus is the longest river on the Indian subcontinent. Its source – a confluence of several glacial streams – is located at a height of 5,300 m (17,389 ft) in Transhimalaya in the Gar region of Tibet, an autonomous region of China. It flows further northwest along the southern edge of the Karakoram Mountain Range and, then in a sharp bend to the southwest, flows out of the highlands and on to Pakistan. Approximately 50 km (31 mi) northwest of Islamabad, the Indus flows through the Tarbela Reservoir, then flows southwest through Pakistan – interrupted by numerous dams and channels which serve to irrigate the land for agriculture –, before emptying into the Arabian Sea southwest of Hyderabad in a delta lying roughly on the Tropic of Cancer.

*This aerial photo shows the Indus in the Karakoram area – the mountain range which connects in the north with the main chain of the Himalayas. The second highest mountain in the world – K2 (8,611 m) – is part of this chain. The Karakoram Mountains lie in the disputed Kashmir region between India, Pakistan and China.*

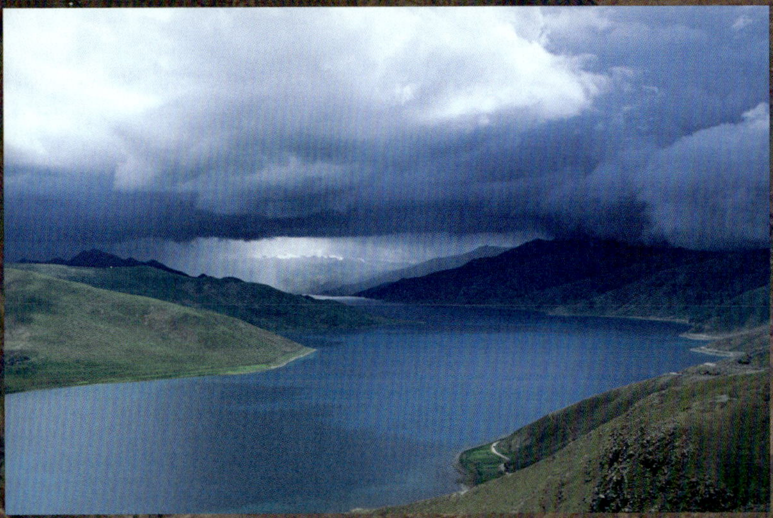

## BRAHMAPUTRA

**2,896 km (1,800 mi)**

The Brahmaputra River rises at a height of 6,000 m (19,686 ft) from a glacier on the northern side of the Himalayas, close to the holy Mount Kailash. Inside Tibet (left), it flows east for 1,500 km (932 mi), breaks through in the Yarlung Tsangpo Grand Canyon to head south and joins with the Ganges to discharge into the Gulf of Bengal.

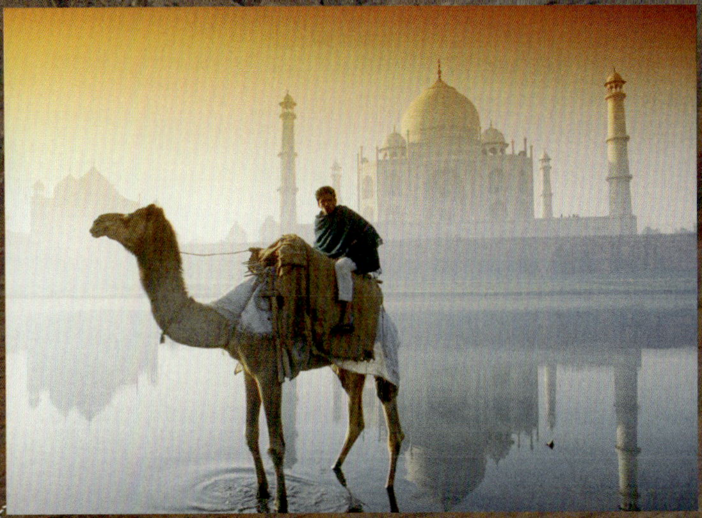

## GANGES

**2,511 km (1,560 mi)**

The Ganges starts as a confluence of two headwaters in the Garhwal district of India in the southwest Himalayas. It is the holy river of the Hindus. The river initially flows southeast through the Ganges lowlands. In Bangladesh, it joins the Brahmaputra and empties at the Ganges delta into the Gulf of Bengal. Picture above: The Ganges tributary: the River Yamuna at Agra.

**TIP**

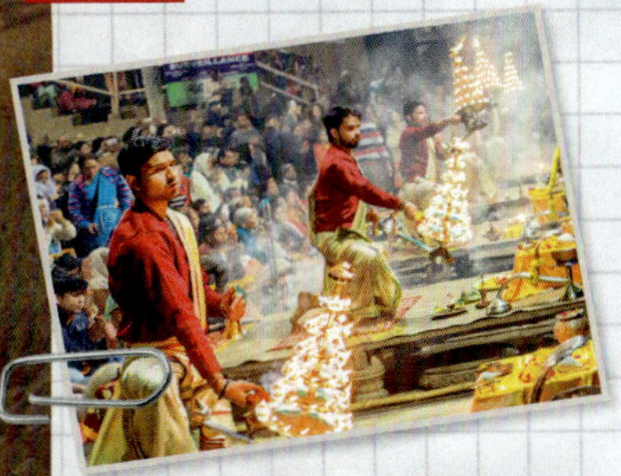

**GANGA AARTI, FIRE RITUAL ON THE GANGES**

The river is also called Mother Ganga because the Ganges is not just the most important river in India; for the Indian people, it is also the embodiment of the river goddess. They view it as the personification of the goddess Ganga, which is why it bears her name. It is therefore not surprising that, despite all the pollution in the water, bathing in the Ganges is still considered a holy act. Not only do people frequently take a dip in the Ganges, it is also a ceremonial site – for example, the Ganga Aarti, a fire ceremony, is held here. In the cities of Varanasi, Rishikesh and Haridwar, there are well-known rituals; in Rishikesh, believers light candles and float them on the water while praying and asking for a blessing and healing.

# NORTH AND CENTRAL AMERICA

The subcontinent of North America borders three of the world's oceans. In the north, the rivers of Canada and the USA drain into the Arctic Ocean, into the Pacific in the west and into the Atlantic in the east, including all the marginal seas of these oceans.

## MISSISSIPPI RIVER

**(with Missouri River and Red Rock River), 6,051 km (3,760 mi)**

The river system that includes the Mississippi begins in the Rocky Mountains in the midwestern states of the USA. In the state of Montana, Red Rock River rises west of Yellowstone Park, on the border with Idaho. It is the longer of the two headwaters of the Jefferson River, which in turn is the longest of the three headwaters which converge at Three Forks to form the Missouri River. The Missouri then runs further northwards through mountainous canyons, leaves the mountains east of Great Falls and then flows southeast until it joins the Mississippi at St Louis on the Illinois border. The source of the Mississippi lies west of Lake Ober in Minnesota. It flows southward, and wide stretches of it form the border river between adjoining US states. After joining the Missouri, it continues south and discharges in a wide delta into the Gulf of Mexico at New Orleans.

## MACKENZIE RIVER

**(with Peace River and Finlay River), 4,241 km (2,635 mi)**

The source of the Finlay River lies in western Canada east of the continental water divide in the Rocky Mountains. It flows southeastwards into Williston Lake, which is drained by Peace River. Further east, Peace River joins the Athabasca River, then flows northwest through the Great Slave Lake, before discharging into the Beaufort Sea as the Mackenzie River.

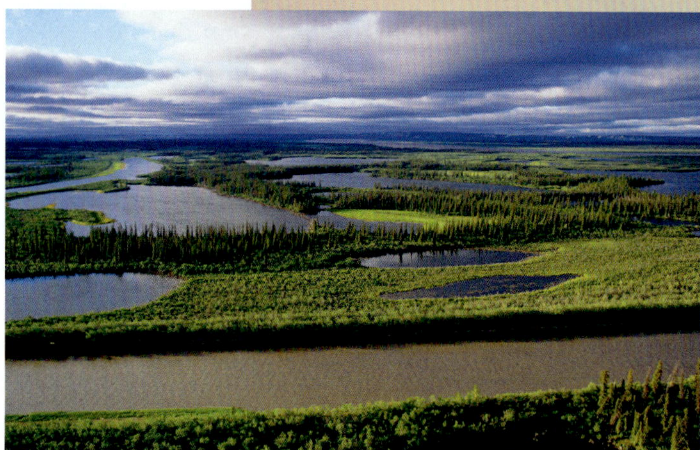

## ST LAWRENCE RIVER

**(with North River), 3,058 km (1,900 mi)**

If the waterfalls of the largest lakes are included in the reckoning, then the river that is furthest from the Atlantic is North River, a headwater of the St Louis River, which empties into Lake Superior at Duluth. The actual St Lawrence River starts as an outlet of Lake Ontario and drains into the Atlantic at Québec. The upper reaches flow between the USA and Canada (Picture, right).

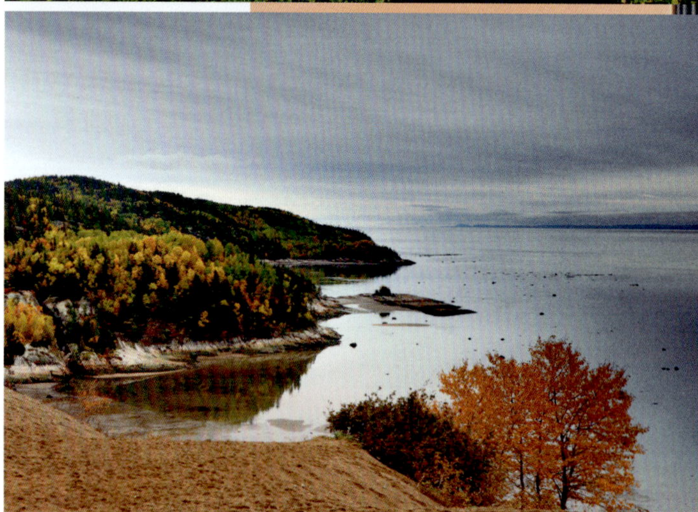

## RIO GRANDE

**3,034 km (1,885 mi)**

At a height of approximately 3,000 m (9,843 ft), the Rio Grande rises as a confluence of smaller headwaters at the foot of Canby Mountain in the US state of Colorado. From there, it flows southward through the San Luis Valley, then through New Mexico via Albuquerque to El Paso in Texas. Further south, it curves to the northeast and leaves the mountain area, flowing southeast to the Gulf of Mexico. After El Paso, the Rio Grande forms the border to Mexico.

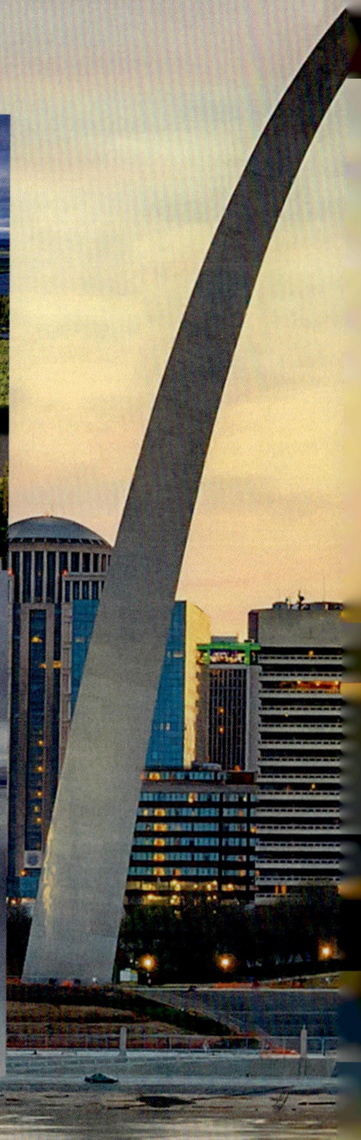

*Large picture: The Mississippi at the foot of the St Louis skyline in the US state of Missouri with the Gateway Arch, the city's famous landmark inaugurated in 1968, which stands 192 m (630 ft) high. At St Louis, the Missouri River joins the Mississippi (top left).*

## NELSON RIVER

**(with Saskatchewan River), 2,671 km (1,660 mi)**

The sources of both arms of the Saskatchewan River, the 'source rivers' of the Nelson River, are in the Rocky Mountains. The two arms join east of Prince Albert to form one river and flow into Lake Winnipeg. The Nelson River forms the outlet of the lake and discharges into Hudson Bay after 664 km (413 mi).

## COLORADO RIVER

**2,333 km (1,450 mi)**

The sources of the Colorado River lie in the Rocky Mountains National Park in the US state of Colorado. The Colorado River flows first through Colorado, then through Utah and Arizona and then bends southward when it reaches the elevation of Las Vegas. The river is famous for its dramatic canyons, including the Grand Canyon where it has cut through the rocks by eroding them.

## YUKON

**(with Teslin River and Nisutlin River), 3,185 km (1,979 mi)**

The gold prospectors made this river famous. Its headwater, the Nisutlin River, rises in the Canadian Pelly Mountains, then flows south into the Teslin Lake and so into the Teslin river, which joins the Yukon at Hootalinqua northwest of that point. It then flows on through Alaska and drains into the Bering Sea.

## COLUMBIA RIVER

**(with Snake River), 2,240 km (1,392 mi)**

The Colombia River is fed by Lake Colombia, which lies at a height of 820 m (2,690 ft) in the Canadian province of British Colombia. It flows south into the US state of Washington and bends off to the west ahead of the Blue Mountains. At about this height, Snake River, which rises in Yellowstone National Park, joins the Colombia River as its left tributary. West of Portland, Oregon, it finally discharges into the Pacific.

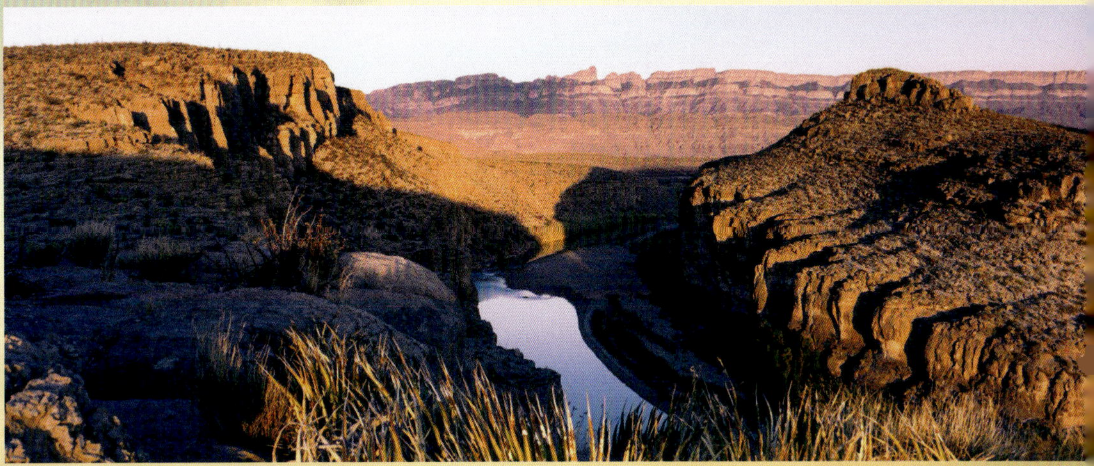

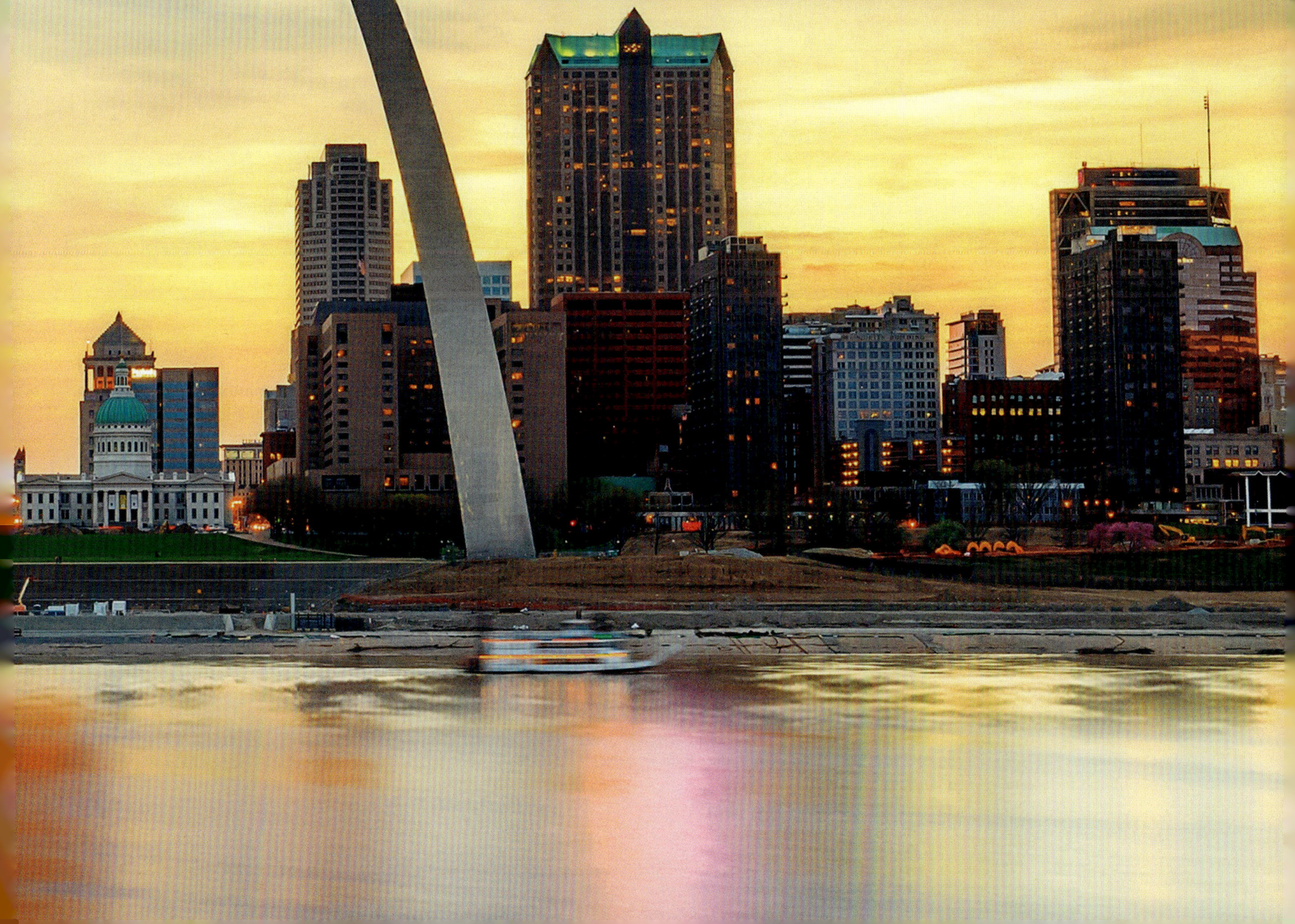

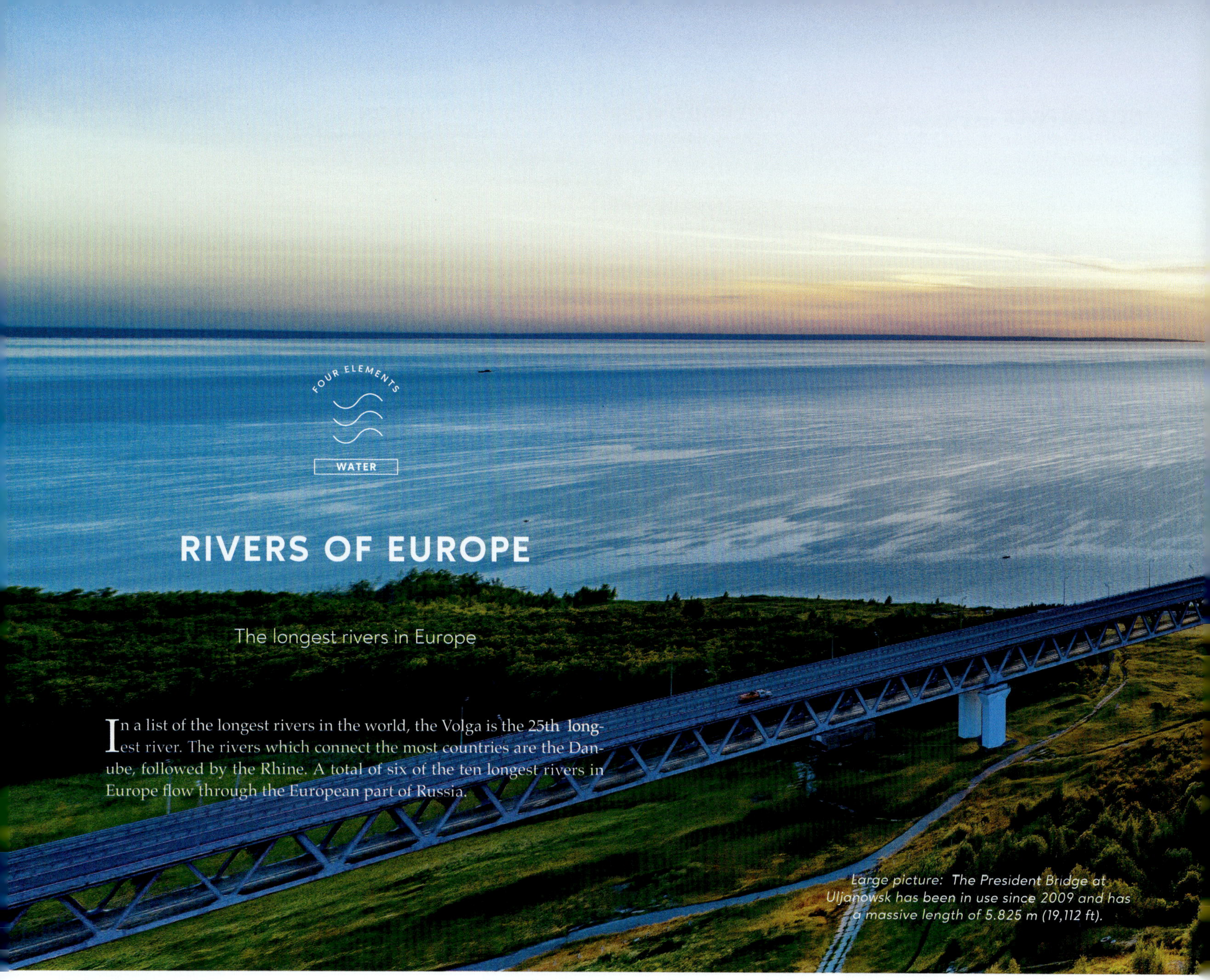

FOUR ELEMENTS

WATER

# RIVERS OF EUROPE

The longest rivers in Europe

In a list of the longest rivers in the world, the Volga is the 25th longest river. The rivers which connect the most countries are the Danube, followed by the Rhine. A total of six of the ten longest rivers in Europe flow through the European part of Russia.

*Large picture: The President Bridge at Uljanowsk has been in use since 2009 and has a massive length of 5.825 m (19,112 ft).*

## VOLGA

**3,534 km (2,196 mi)**

The longest river in Europe, the Volga, rises approximately 300 km (186 mi) northwest of Moscow, in the Valdai Hills. With a flow rate of 8,000 $m^3$ (282,517 cu ft) per second, it is also the river with the highest flow in the continent. The Volga first flows eastward and then, at Rybinsk bis Kasan, in a southeasterly direction. It continues through the steppe landscape of the Caspian Depression and finally empties into the Caspian Sea south of Astrakhan.

## DNIEPER

**2,285 km (1,420 mi)**

The Dnieper rises northeast of Smolensk, flows southwest past Smolensk through Belarus, then further south into Ukraine as far as Kyiv and Dnipro and finally sweeps in an arc east of Odessa to discharge into the Black Sea. In the Ukrainian section, the Dnieper has a series of vast reservoirs. The hydroelectric power plants here provide the heavily populated industrial region with electricity.

## URAL

**2,428 km (1,509 mi)**

The Ural Mountains and the eponymous river are considered the border between Europe and Asia – the western bank is in Europe and the eastern bank in Asia. The source of the Ural River lies in the southern Ural mountain range, around 150 km (93 mi) north of Magnitogorsk. It flows in a southerly direction until Orsk, then branches west at almost a right angle, before emptying into the Caspian Sea at Atyrau.

## DANUBE

**2,888 km (1,795 mi)**

The Danube rises in the Black Forest, flows east through six countries – and forms the border river for a further four – and discharges into the Black Sea in Ukraine, south of Odessa. The Breg river is now considered to be its headwater; it rises north of Furtwangen at a height of 1,078 m (3,537 ft). The Danube is navigable from Kelheim. The Main-Danube Canal makes maritime traffic possible from the North Sea through to the Black Sea.

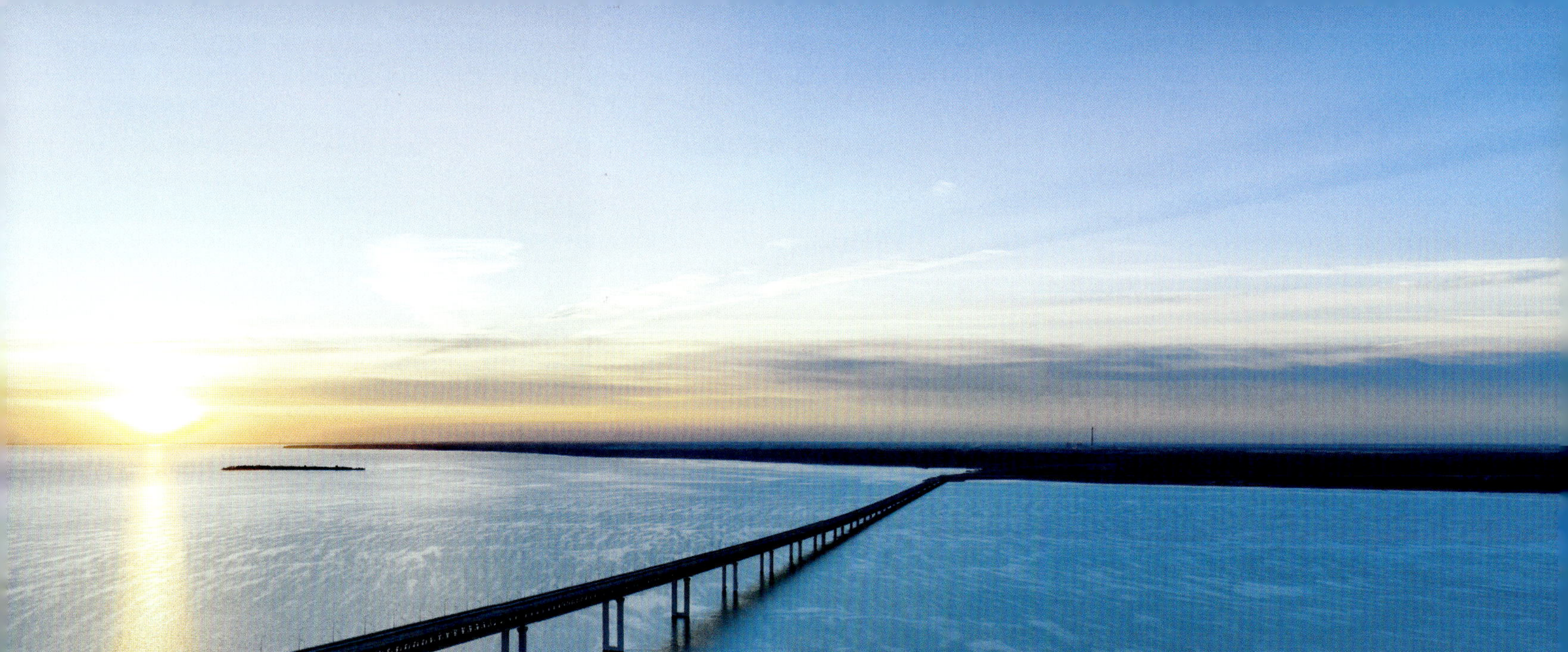

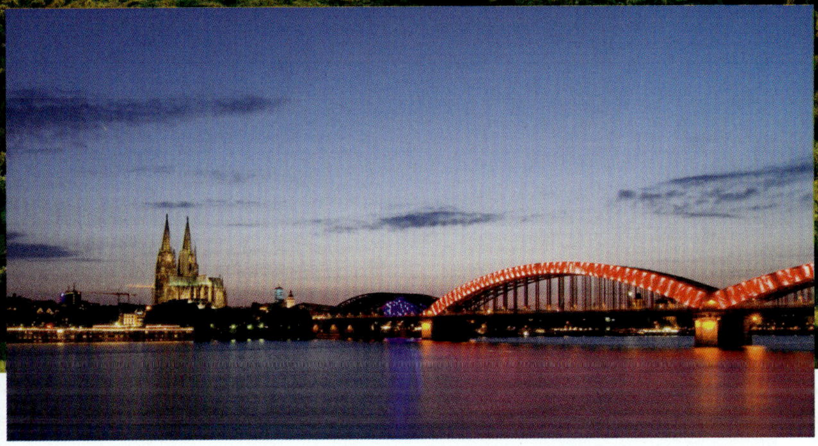

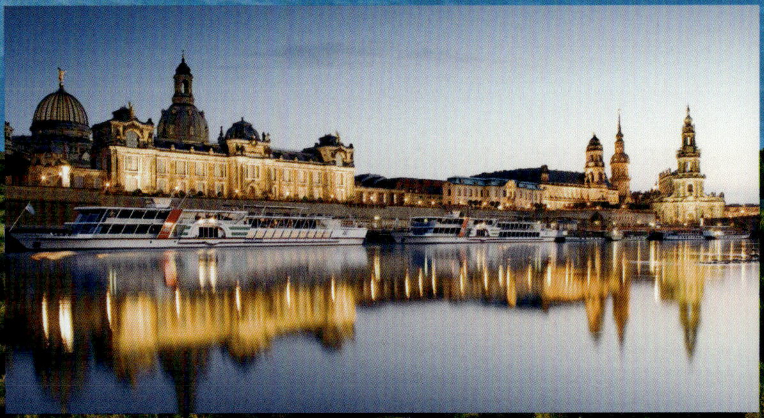

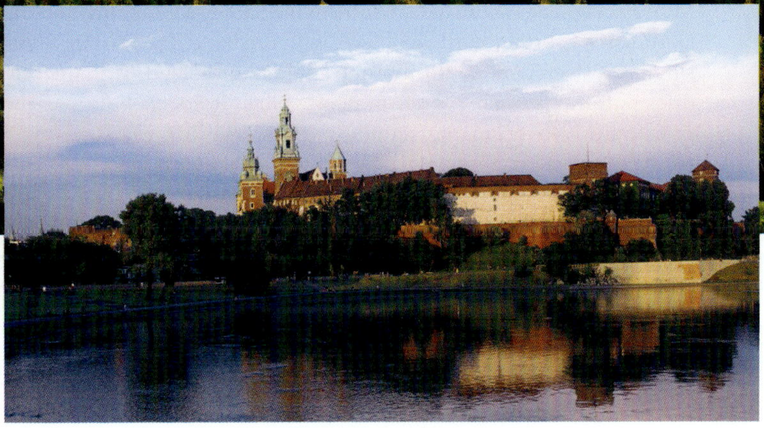

## RHINE

**1,320 km (820 mi)**

The headwater of the Rhine furthest from its mouth is the Rein da Medel in the Swiss cantons of Ticino (Tessin) and Grison (Graubünden). The rivers Aare, Moselle and Main are among its largest tributaries. The Rhine is one of the busiest waterways in the world. It is also the largest tributary of the North Sea. It flows through Switzerland, Austria, Germany and the Netherlands. It is a border river between France and Liechtenstein. Protection from the frequent flooding is provided by dykes in some areas. Picture above: The Rhine at Cologne – on the left, Cologne Cathedral; on the right, the Hohenzollern Bridge, with its festive lights in the twilight.

## ELBE WITH VLTAVA (MOLDAU)

**1,252 km (778 mi)**

The source of the Elbe lies in the Giant Mountains of the Czech Republic, at a height of 1,386 m (4,547 ft). The total length is arrived at when acknowledging the Vltava, which joins the Elbe 30 km (19 mi) north of Prague, as the headwater. The Elbe flows through the Czech Republic and Germany and is the only waterway that connects the Bohemians with the North Sea. It flows from the Elbe Sandstone Highlands to the tidal flats of the Wadden Sea. Picture, above right: Brühl's Terrace, 'The Balcony of Europe', on the Elbe in Dresden.

## VISTULA

**1,047 km (651 mi)**

The headwaters of the Vistula (Weichsel) River rise in the Silesian Beskids, at a height of 1,100 m (3,609 ft); after a few kilometres, it flows into the Zbiornik Czernianski reservoir lake. The river flows east through the old royal city of Krakow and, further north, through Warsaw, before finally discharging its waters into the Baltic Sea at Danzig. There are two rivers' mouths there: the older one to the west by Danzig and a more recent one further to the east. Picture above: The Vistula in Krakow with Wawel Castle

# AUSTRALIA/OCEANIA

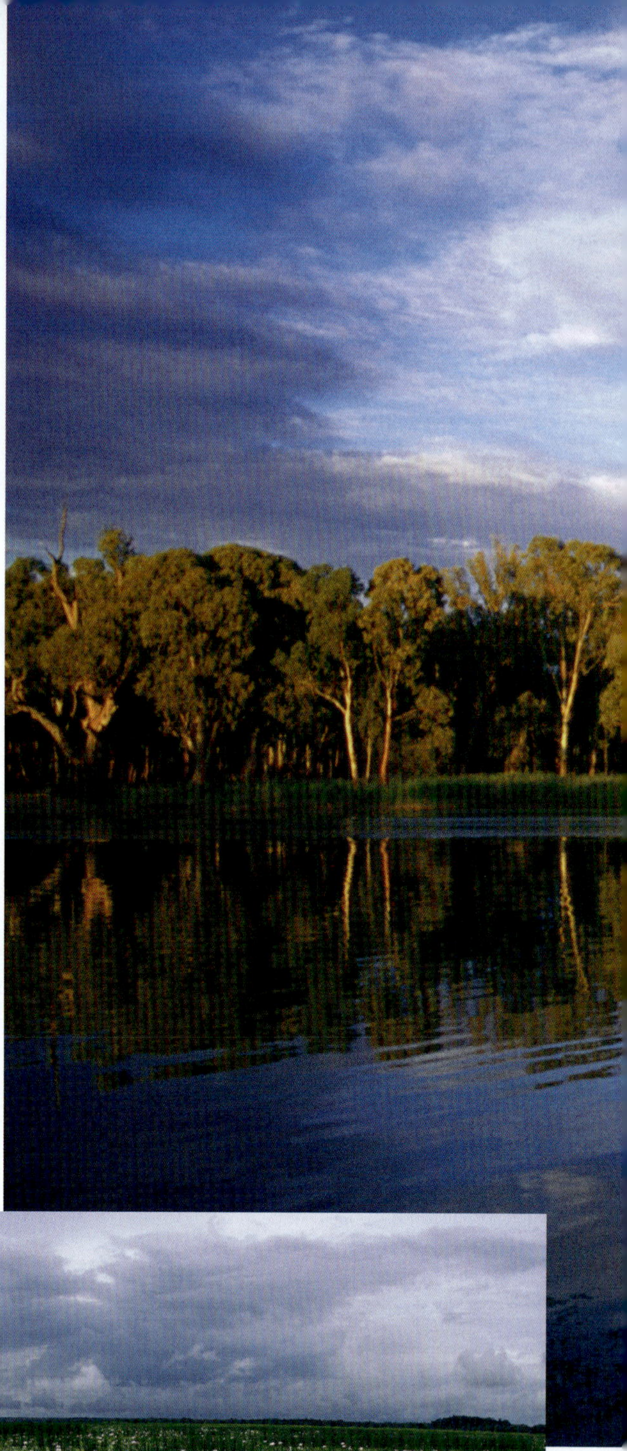

The fifth continent includes the offshore islands of Tasmania and Papua New Guinea as well as the islands of Oceania. The interior of Australia itself is very dry; as a result, not many rivers here are in flow all year round

## DARLING WITH MURRAY

**3,370 km (2,094 mi)**

The Darling River, at 2,739 km (1,702 mi) long, is the longest river in Australia. It joins the Murray River, which is 2,589 km (1,609 mi) long, in New South Wales at Wentworth, thus forming a river system covering 3,370 km (2,094 mi) from the source of the Darling to its estuary on the Indian Ocean. Both rivers rise in the western slopes of the Great Dividing Range, a mountainous area on Australia's west coast: the Murray originates south of Canberra in the Snowy Mountains, and the Darling rises further north on a level with Brisbane. The Darling flows sluggishly west and south through the dry grassland. It frequently dries out, leaving just a series of pools. When it floods, however, the river quickly overruns surrounding areas. The Murray flows in a westerly direction. It is navigable up to almost 2,000 km (1,243 mi); trips on houseboats are very popular here. Four reservoir lakes, among them Lake Hume, mean that only a little water reaches the sea; the estuary is at risk of silting up.

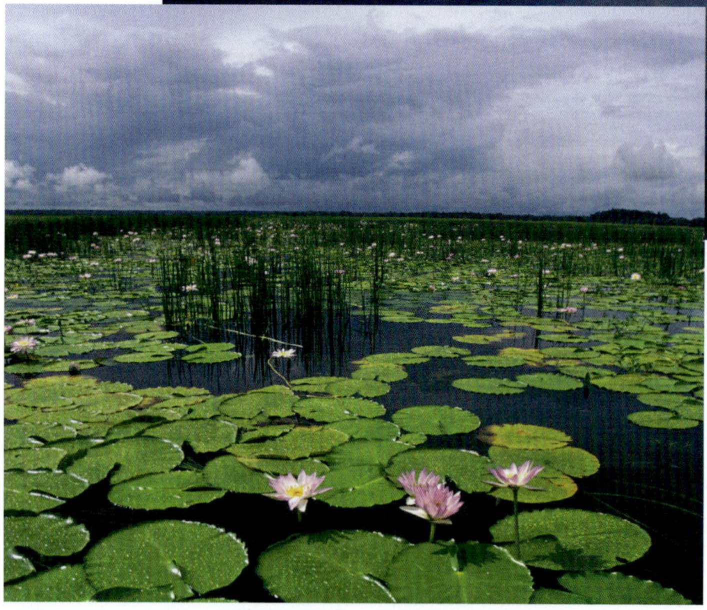

## MURRUMBIDGEE

**1,579 km (981 mi)**

The right-hand tributary of the Murray River in New South Wales rises on the northeastern slopes of the Snowy Mountains, close to Mount Kosciuszko, which, at 2,228 m (7,310 ft), is the highest summit on the Australian mainland. Murrumbidgee means 'big water' in Wiradjuri, one of the local Australian Aboriginal languages. From the source region, the river first flows north for a stretch, then west through grassland and finally joins the Murray north of Swan Hill. The Australian capital, Canberra, is on the Murrumbidgee.

## COOPER CREEK

**1,420 km (882 mi)**

In Australia, Cooper Creek rises south of Killarney Park and west of the mountains of the Carnarvon Range, a group of mountains situated before the Great Dividing Range in the west. In its first section, it is called the Barcoo River. It initially flows northwest, then southwest at Blackall where it joins Thomson Creek. Known as Cooper Creek from this point on, it flows on through the desert and empties into Lake Eyre, a salt lake, in the north. The river is only infrequently in flow; at these times, lilies and eucalyptus grow here (above).

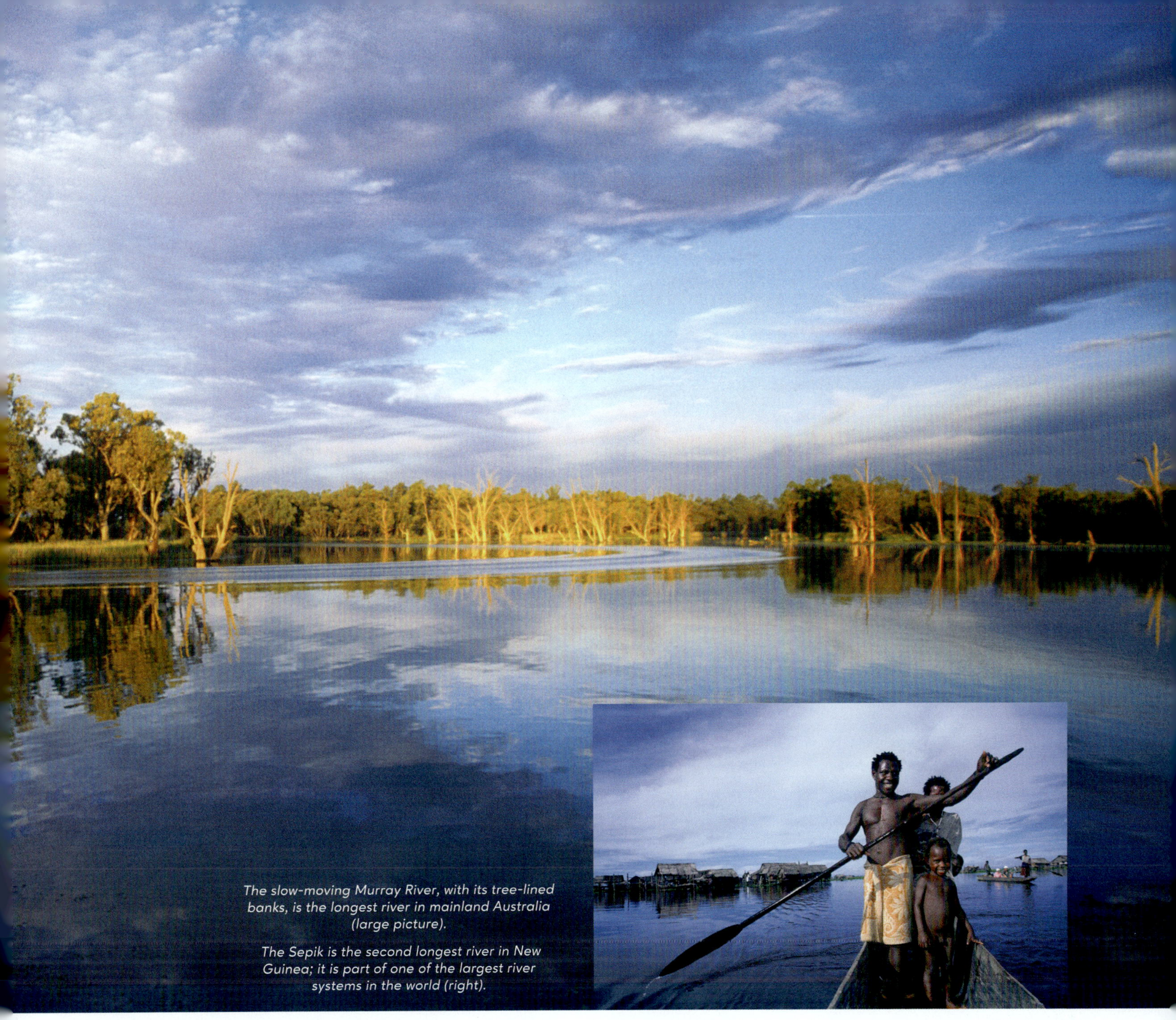

*The slow-moving Murray River, with its tree-lined banks, is the longest river in mainland Australia (large picture).*

*The Sepik is the second longest river in New Guinea; it is part of one of the largest river systems in the world (right).*

## MAMBERAMO WITH TARITATU AND SOBGER

**1,175 km (730 mi)**

The Sobger rises in the province of Papua on New Guinea, belonging to Indonesia. New Guinea is the second largest island in the world, after Greenland. Its two headwaters lie in the Sterren Range, west of the Antares Mountains. After roughly 130 km (81 mi), the Sobger flows northwest into the Taritatu/Idenburg River, flows through vast mangrove swamps and finally flows sharply north – first through a deep valley, then into the Pacific in a vast delta.

## FLY

**1,120 km (696 mi)**

After the Sepik, which discharges in the north, Fly River is the third largest river on the island of New Guinea. It rises on the southern side of the Victor Emanuel Range, which is part of the Maoke Mountains, and flows to the south. A small section of the river forms the western border to West New Guinea (or West Papua), the Indonesian part of the island. Fly River flows on further southeast, through sprawling swamplands, and finally discharges into the Gulf of Papua in an estuary 56 km (35 mi) wide. The Gulf contains many inhabited islands which are very fertile.

## SEPIK

**1,126 km (700 mi)**

The Sepik is the second longest river on the island of New Guinea and is located in the independent island state of Papua New Guinea, which occupies the eastern half of the island. The river rises in the Victor Emanuel Range, at a height of 2,170 m (7,120 ft), flows northwest. It then leaves the mountains and sweeps in a large arc to the north, before discharging into the Bismarck Sea. The drainage area is vast and characterized by swampland and tropical rainforest; only a few inhabitants, who are well-known for their practice of ritual scarification, live there.

# EYE OF TYRANNOS

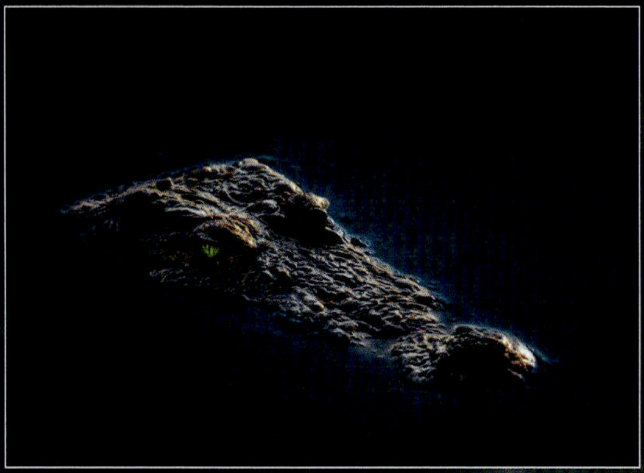

## PATIENCE PAYS OFF

Crocodiles are able to wait for the perfect opportunity for days on end; they stay quietly in one place, watching, with only their nostrils and their prominent eyes visible above the water. In terms of expending energy, they have perfected this technique so well that they can sometimes go for months without eating – until then, they do what they do best: they lurk.

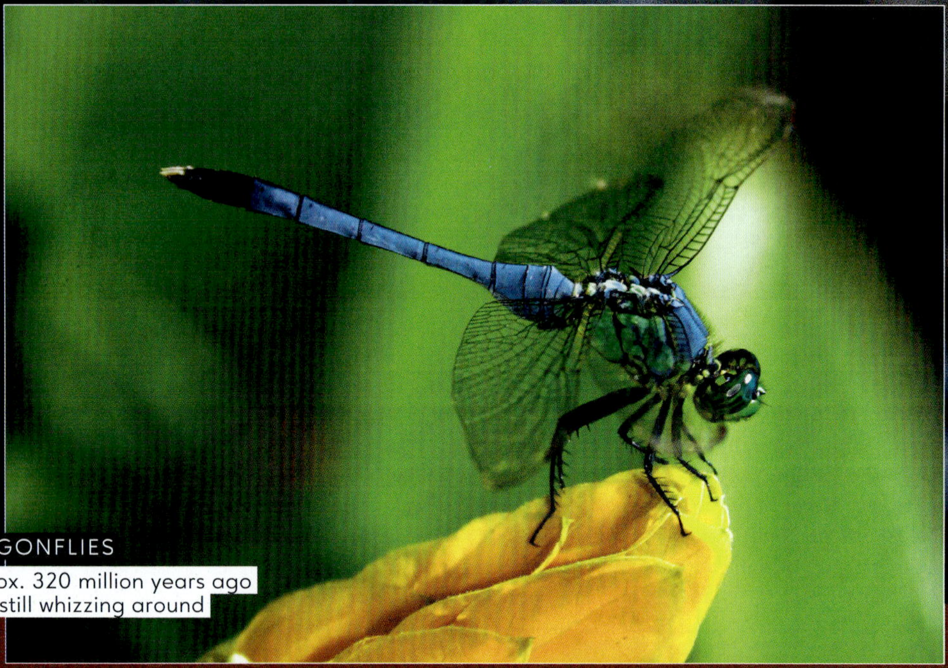

## WITNESSES FROM THE PREHISTORIC ERA

Thanks to fossils, we now know that as early as 320 million years ago, the swampy forests were alive with dragonflies – or rather, the ancestors of our current dragonflies. That is, more than 150 million years before the dinosaurs. However, at that time, they were much larger. Wing spans of 60 to 75 cm (23.5 to 29.5 in) were not unusual – in comparison, that of present-day dragonflies is between 2 and 15 cm (0.8 and 6 in). Apart from this, they have retained much of their original form: The body structure, behaviours and appearance generally reflect those of the species living 150 million years ago. And one myth persists even now – that dragonflies can sting. The graphic descriptions of them as 'horse stingers' or 'the Devil's darning needles' are unfounded because dragonflies do not have a stinger.

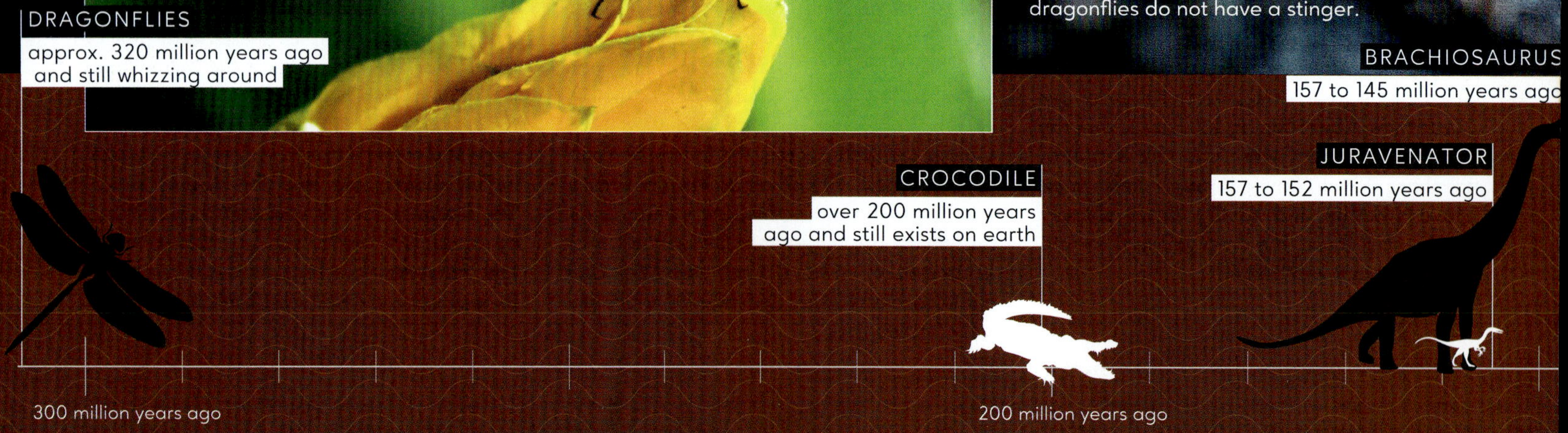

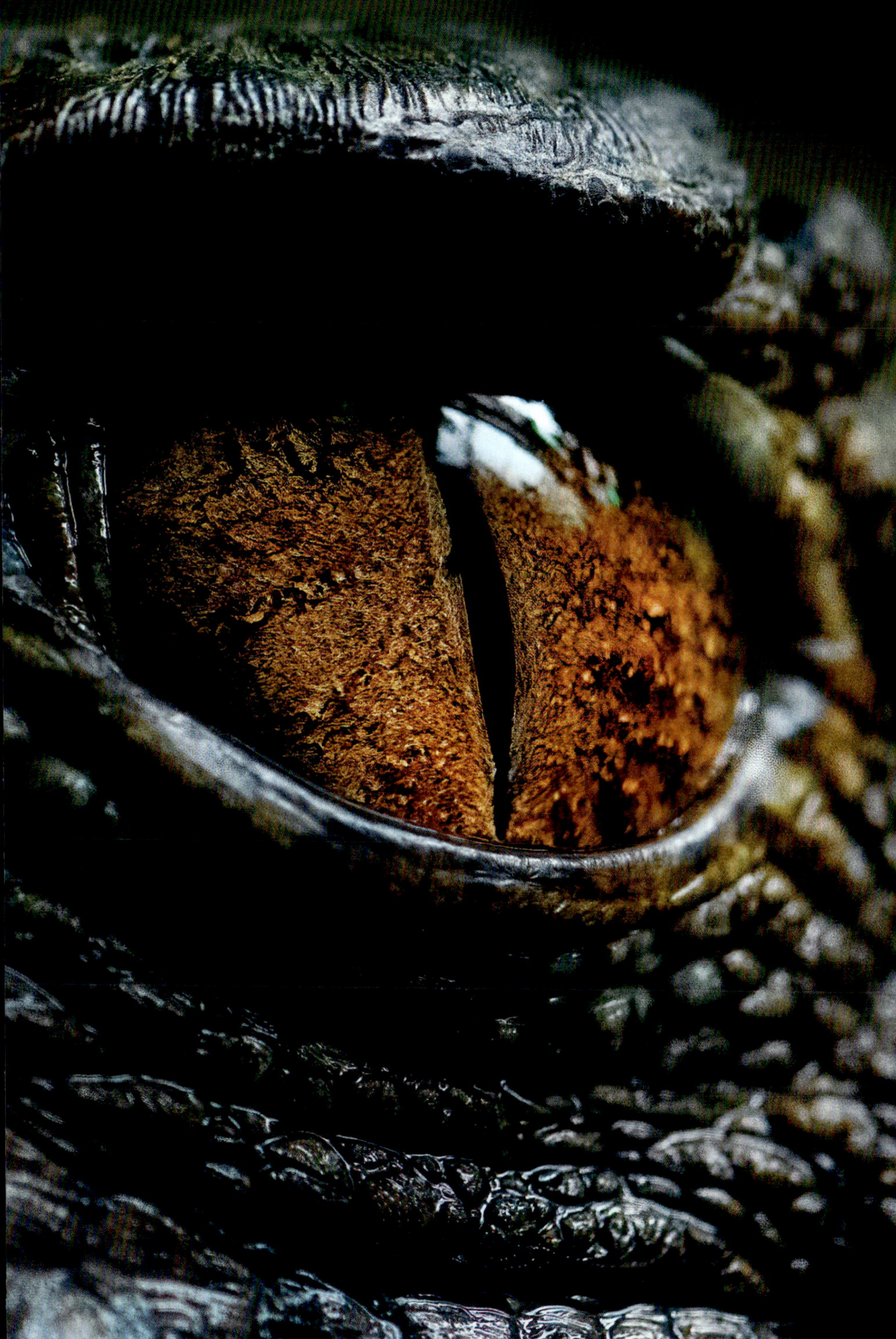

## WHEN THE EARTH HAD JUST ONE CONTINENT

Roughly 250 million years ago, satellites (if they had been around then) would have taken very different pictures of the world from above. Because, at that time, all the landmasses on earth were joined and surrounded by a single ocean. This supercontinent was called Pangea; it lasted from approximately 325 million years ago to 150 million years ago – from the late Carboniferous into the Jurassic period. About 200 million years ago, the continent began to break up, initially into two parts – Laurasia and Gondwana. It was not until later that both of these broke into smaller parts, in which the modern forms of North and South America, Asia, Africa and Europe were already recognizable, with one distinction – they were all much closer together.

And the fauna? The fauna at that time consisted of dinosaurs, among other creatures, who in turn also died out. But the crocodiles, who shared a habitat with the dinosaurs over a long period of time, are still living on earth now. It is only their geographical spread that has changed, i.e. they too are now extinct in Europe.

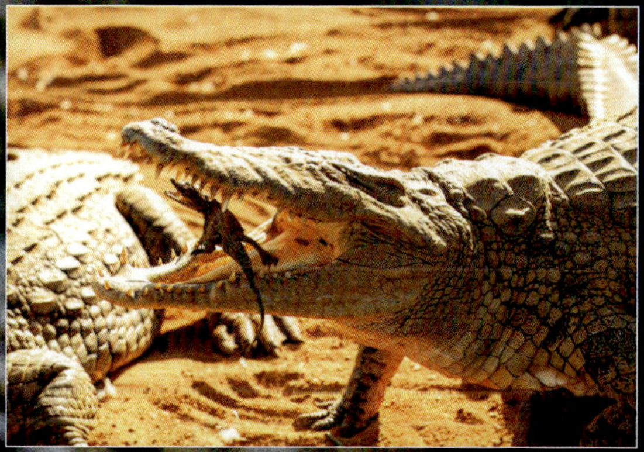

## MOTHER LOVE

Those not familiar with the animal may at first be alarmed at the sight of a huge crocodile carrying a newly hatched baby crocodile between its extremely sharp and numerous teeth. But fear not – the crocodile mother is exercising great caution, and is only carrying her charge to the water after it has broken its way out of its shell. Special nerves at the roots of her teeth ensure that she does not bite down too hard.

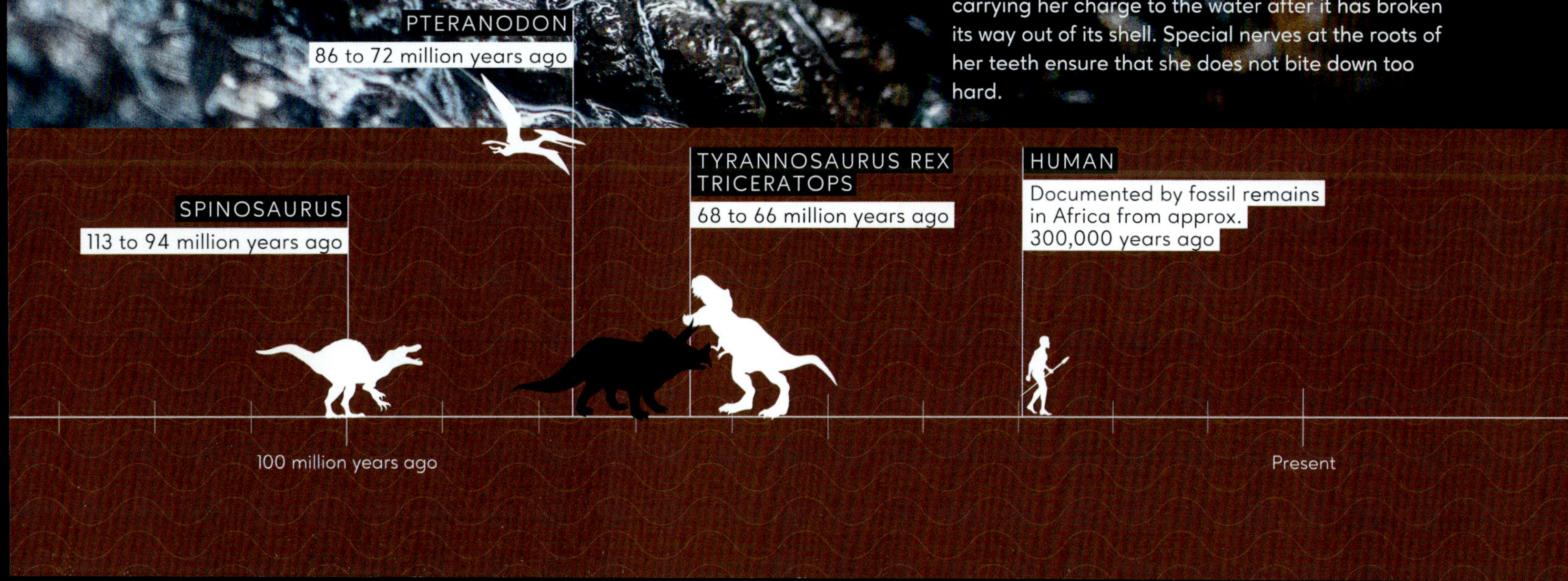

FOUR ELEMENTS

WATER

# WATERFALLS

At geological breakpoints or similarly formed substrata, the streams of flowing water often plunge down dozens or even hundreds of metres. This can occur in a single step or, less frequently, in a cascade down several steps

When the water flow is accelerated due to the gradient or narrowing and becomes a torrent, it is referred to as rapids. If the flow of swirling water is divided by rock faces, this is called a cataract. Rapids and waterfalls are among the most awe-inspiring natural spectacles on the planet.

### The most powerful rapids and waterfalls

1 **Chutes Inga**
Dem. Rep. Congo, 42,500 $m^3$ (1,500,873 cu ft) /s

2 **Chutes de Livingstone**
Dem. Rep. Congo, 35,000 $m^3$ (1,126,013 cu ft) /s

3 **Boyoma Falls**
Dem. Rep. Congo, 17,000 $m^3$ (600,349 cu ft) /s

4 **Chutes de Khone**
Laos, 11,600 $m^3$ (409,650 cu ft) /s

5 **Salto Pará**
Venezuela, 3,500 $m^3$ (123,601 cu ft) /s

6 **Cachoeira de Paulo Afonso**
Brazil, 2,800 $m^3$ (98,881 cu ft) /s

7 **Niagara Falls**
USA, 2,500 $m^3$ (88,287 cu ft) /s

8 **Igauzu Falls (Saltos do Iguaçu)**
Argentina/ Brazil, 1,750 $m^3$ (61,801 cu ft) /s

9 **Saltos dos Patos e Maribondo**
Brazil, 1,500 $m^3$ (52,972 cu ft) /s

10 **Victoria Falls**
Zimbabwe/ Zambia, 1,100 $m^3$ (38,846 cu ft) /s

11 **Virginia Falls**
Canada, 1,000 $m^3$ (35,315 cu ft) /s

12 **Churchill Falls**
Canada, 990 $m^3$ (34,961 cu ft) /s

### The highest waterfalls

1 **Salto Angel**
Venezuela, 979 m (3,212 ft)

2 **Tugela Falls**
South Africa, 948 m (3,110 ft)

3 **Cataratas las tres Hermanas**
Peru, 914 m (2,999 ft)

4 **Olo'upena Falls**
USA, 900 m (2,953 ft)

5 **Catarata Yumbilla**
Peru, 896 m (2,940 ft)

6 **Vinnufossen**
Norway, 860 m (2,822 ft)

7 **Baläifossen**
Norway, 850 m (2789 ft)

8 **Pu'uka'oku Falls**
USA, 850 m (2,789 ft)

9 **James Bruce Falls**
Canada, 840 m (2,756 ft)

10 **Browne Falls**
New Zealand, 836 m (2,743 ft)

11 **Strupenfossen**
Norway, 820 m (2,690 ft)

12 **Ramnefjellsfossen**
Norway, 818 m (2,684 ft)

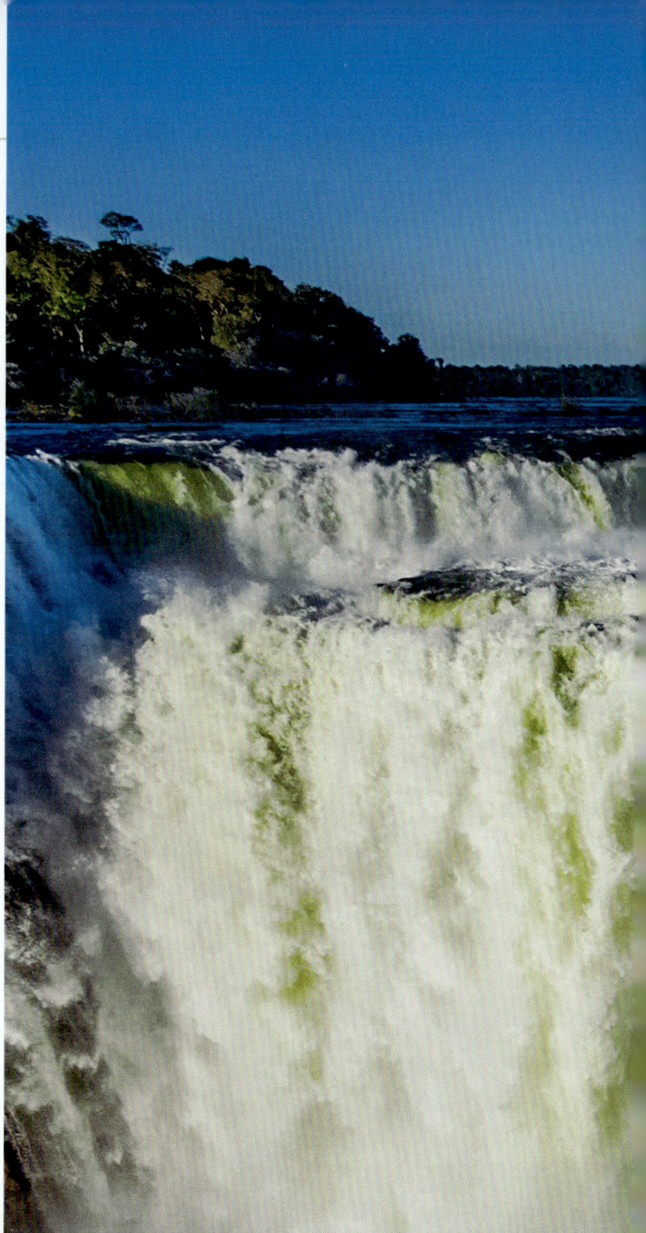

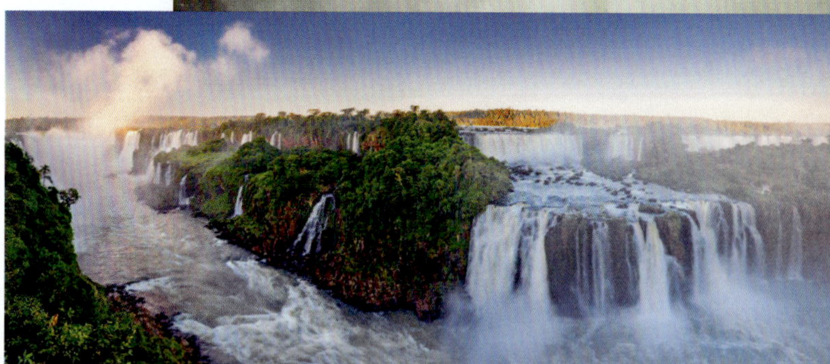

*The Iguaçu Falls, in the tri-border area where Brazil, Argentina and Paraguay join, are among the largest and most impressive in the world. On the Argentinian and Brazilian sides, the area is designated as a national park. Since the mid-1980s, this park and the waterfalls have been on the list of UNESCO Natural Heritage sites.*

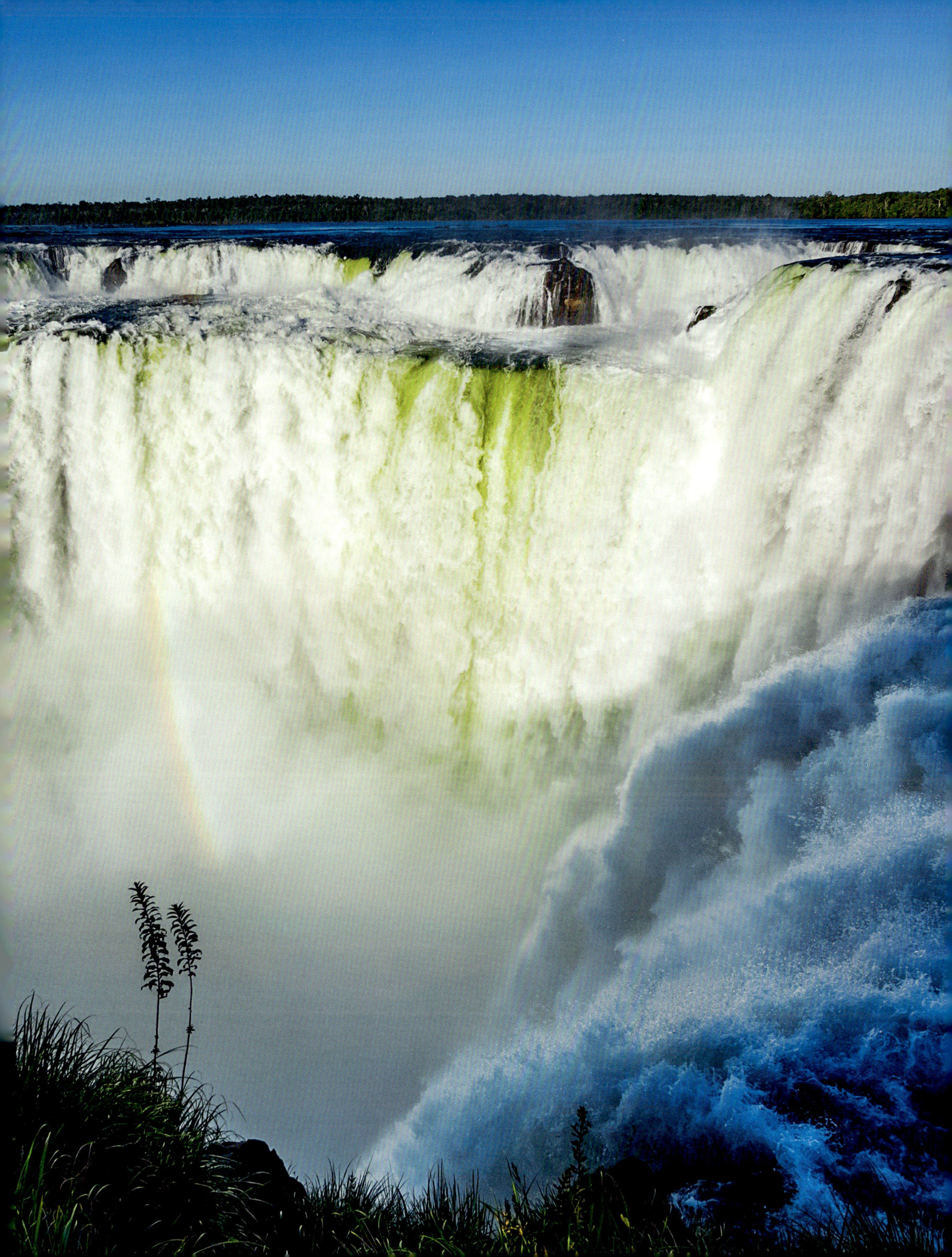

FOUR ELEMENTS

WATER

# SALTO ANGEL

The highest waterfall in the world

The Salto Angel – the highest free-falling waterfall in the world, which falls 979 m (3,212 ft) – lies in the second largest national park in Venezuela, Canaima National Park, in the southeast of the country, close to its borders with Guyana and Brazil. It is named after the American bush pilot and gold prospector, Jimmie Angel (1899 - 1956), who discovered it in 1933 – or, more precisely, rediscovered it. A Venezuelan had already sighted the waterfall in 1910, but it was Angel who first made it famous. The largest single step of this mighty natural spectacle measures 805 m (2,641 ft), with a second step measuring 200 m (656 ft) adjoining it. The water plunges down from a mountain ledge of the Auyan-Tepui, a mesa which measures 700 km$^2$ (270 sq mi), and then forms the tributary of the Rio Carrao. The Salto Angel is fed by the rainfall on the plateau, which often falls during violent thunderstorms.

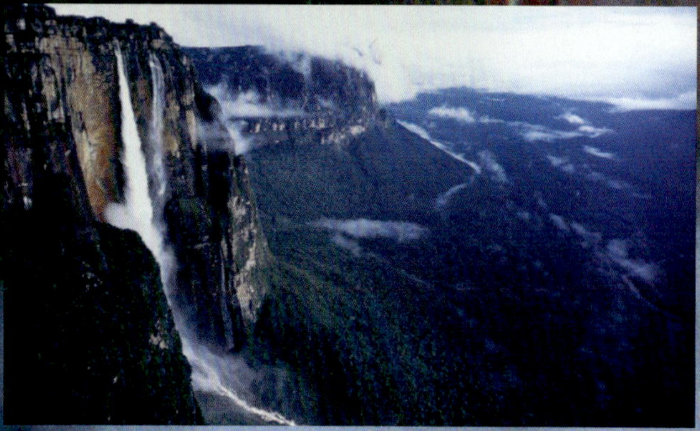

*The massive height and strong upwinds mean that, in the dry season when water levels are lower, a large part of the waters of the Salto Angel is dispersed in small water droplets.*

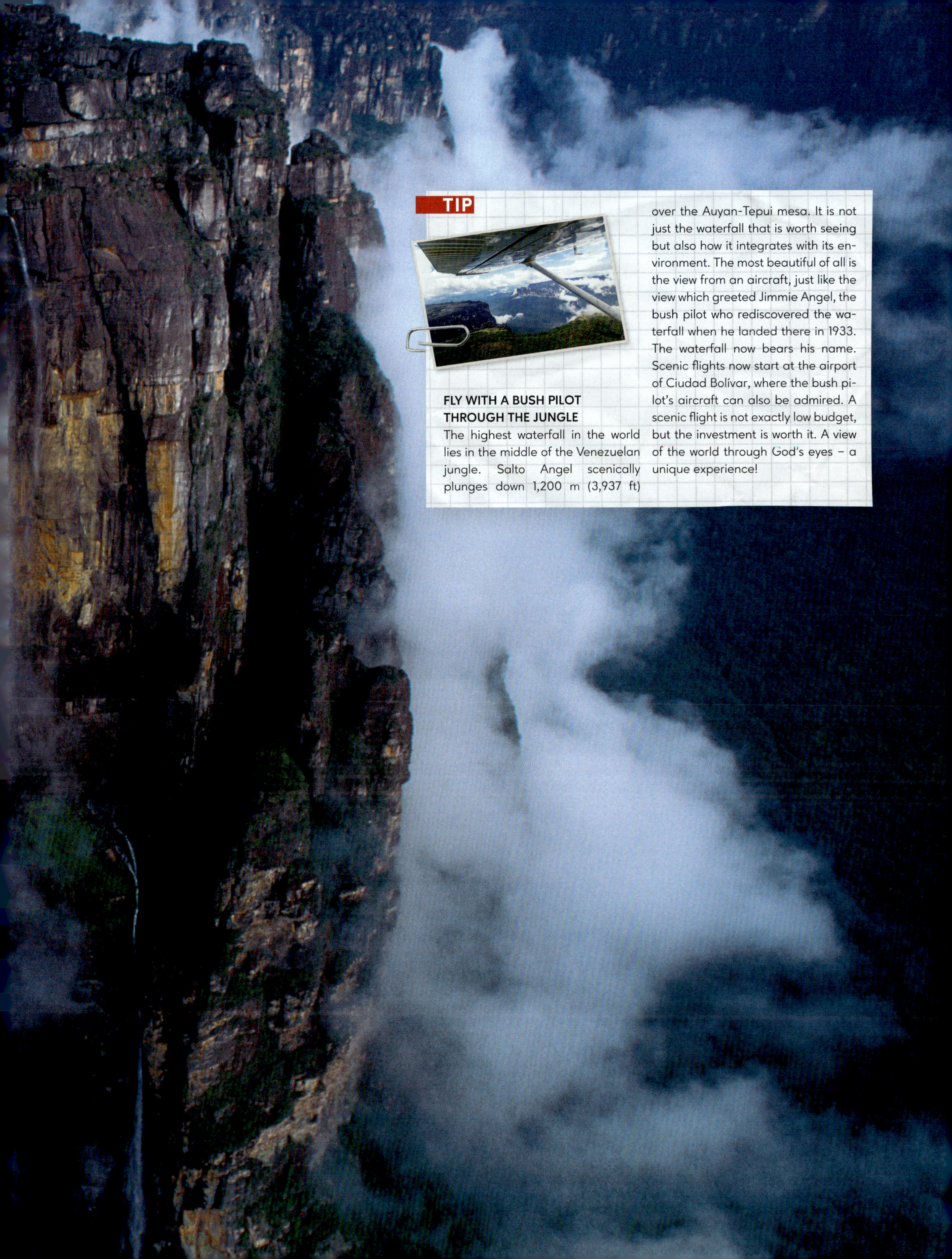

## TIP

**FLY WITH A BUSH PILOT THROUGH THE JUNGLE**

The highest waterfall in the world lies in the middle of the Venezuelan jungle. Salto Angel scenically plunges down 1,200 m (3,937 ft) over the Auyan-Tepui mesa. It is not just the waterfall that is worth seeing but also how it integrates with its environment. The most beautiful of all is the view from an aircraft, just like the view which greeted Jimmie Angel, the bush pilot who rediscovered the waterfall when he landed there in 1933. The waterfall now bears his name. Scenic flights now start at the airport of Ciudad Bolívar, where the bush pilot's aircraft can also be admired. A scenic flight is not exactly low budget, but the investment is worth it. A view of the world through God's eyes – a unique experience!

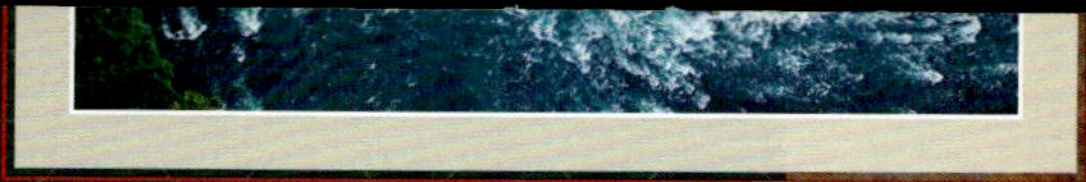

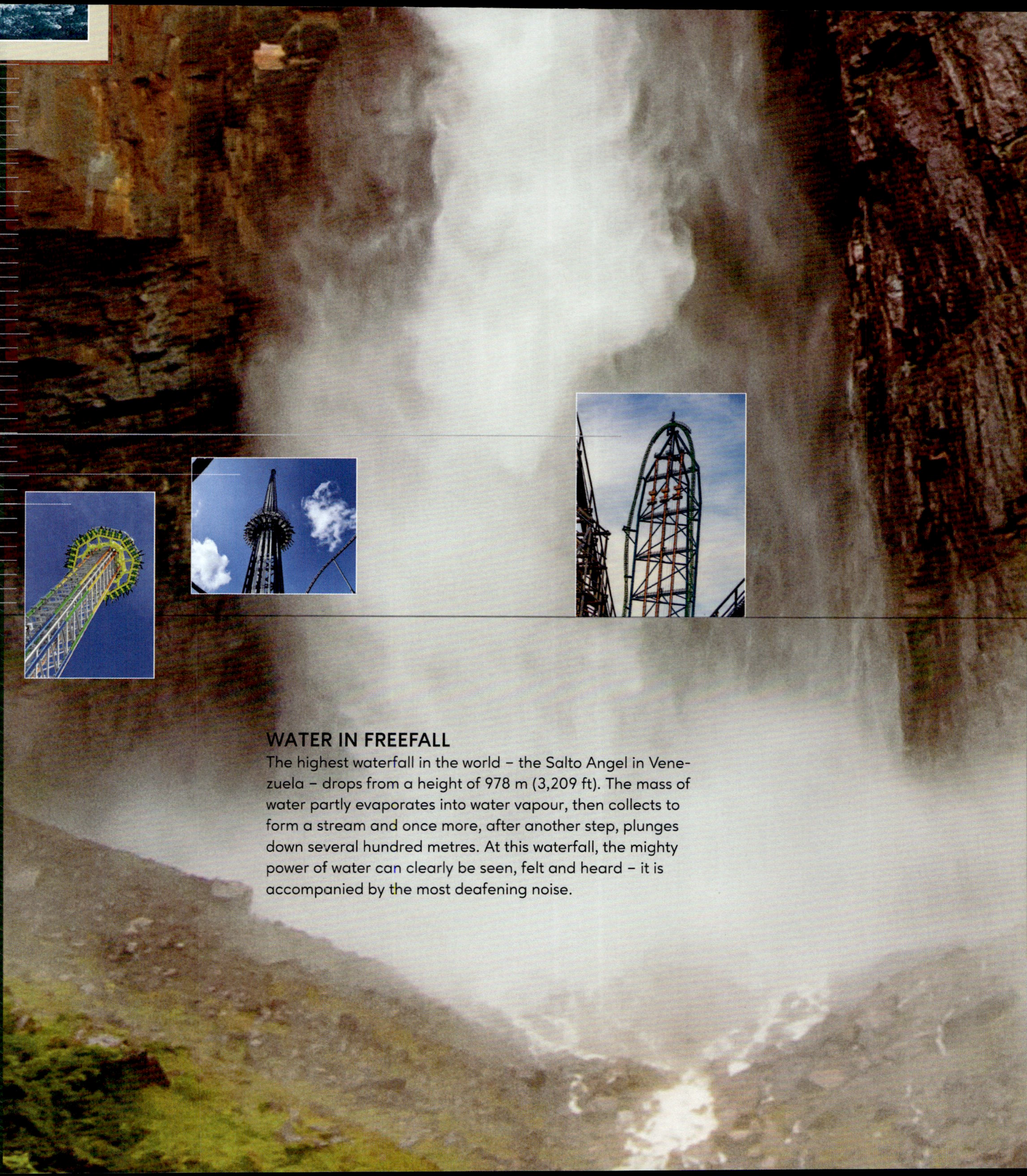

ZUMANJARO DROP OF DOOM, NEW JERSEY

Drop: 126 m (413 ft)

Rate of fall: 145 km/h (90 mi/h)

HIGHLANDER, HANSA-PARK, LÜBECK

Drop: 103 m (338 ft)

Rate of fall: 120 km/h (75 mi/h)

SKYFALL, OKTOBERFEST, MUNICH

Drop: 80 m (262 ft)

Rate of fall: 96 km/h (60 mi/h)

300 m (984 ft)

200 m (656 ft)

100 m (328 ft)

0 m

### WATER IN FREEFALL

The highest waterfall in the world – the Salto Angel in Venezuela – drops from a height of 978 m (3,209 ft). The mass of water partly evaporates into water vapour, then collects to form a stream and once more, after another step, plunges down several hundred metres. At this waterfall, the mighty power of water can clearly be seen, felt and heard – it is accompanied by the most deafening noise.

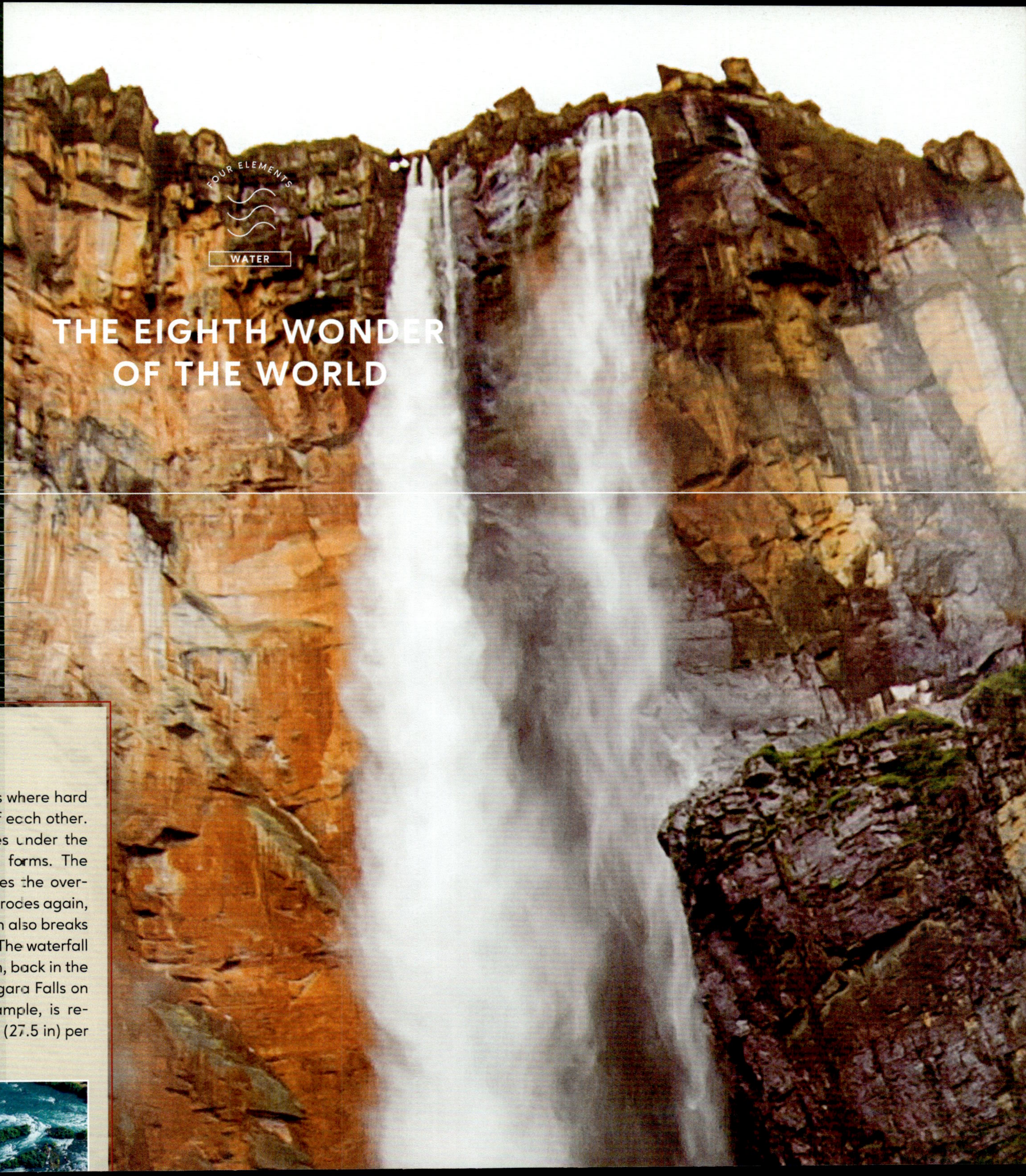


# THE EIGHTH WONDER OF THE WORLD

SALTO ÁNGEL, VENEZUELA

Drop: 979 m (3,212 ft)

Rate of fall: very fast!

978 m (3208,66 ft)

900 m (2,953 ft)

DID YOU KNOW?

### Migrating waterfalls

Waterfalls generally form in places where hard and soft layers of rock lie on top of each other. If the softer rock gradually erodes under the layer of hard rock, an overhang forms. The massive weight of the water causes the overhang to break off. The softer rock erodes again, and another overhang forms, which also breaks off due to the weight of the water. The waterfall thus measurably retreats upstream, back in the direction of the river's source. Niagara Falls on the US-Canadian border, for example, is retreating at a rate of around 70 cm (27.5 in) per year.

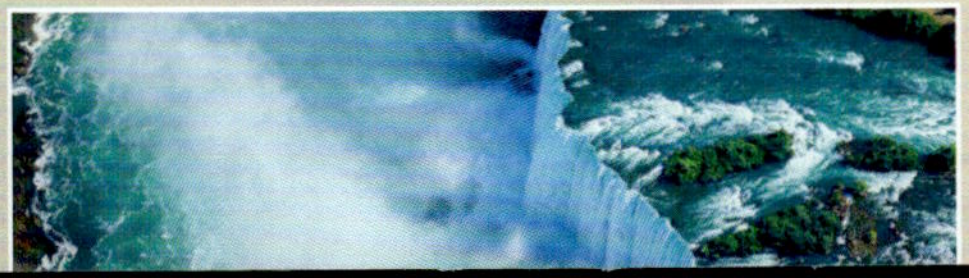

FOUR ELEMENTS

WATER

# SOUTH AMERICA

Waterfalls in South America: The highest flows and the highest waterfalls

The Salto Angel – the highest free-falling waterfall in the world with a drop of 979 m (3,212 ft) – is located in the second largest national park in Venezuela, Canaima National Park, in the southeast of the country, close to its borders with Guyana and Brazil. It is named after the American bush pilot and gold prospector, Jimmie Angel (1899 - 1956), who discovered it in 1933 – or, more precisely, rediscovered it. A Venezuelan had already sighted the waterfall in 1910, but it was Angel who first made it famous. The largest single step of this mighty natural spectacle measures 805 m (2,641 ft), with a second step measuring 200 m (656 ft) adjoining it. The water plunges down from a mountain ledge of the Auyan-Tepui, a mesa which spans 700 km$^2$ (270 sq mi), and then forms the tributary of the Rio Carrao. The Salto Angel is fed by the rainfall on the plateau, which often falls during violent thunderstorms.

## SALTO PARÁ

**Venezuela, 3,500 m$^3$ (123,601 cu ft) /s**

In terms of the volume of water, the Salto Pará is the most powerful waterfall in South America. In the middle of the rainforest, approximately halfway to the mouth of the Orinoco, the waters of the Caura River – a major tributary of the Orinoco – tumble from a height of 60 m (197 ft). The waterfall is in the shape of a half-moon and stretches in a funnel shape for 7 km (4.5 mi), until it reaches a large sandbank.

## CACHOEIRA DE PAULO AFONSO

**Brazil, 2,800 m$^3$ (98,881 cu ft) /s**

The waters of the Rio São Francisco fall over the Cachoeira de Paulo Afonso in two steps from a height of about 84 m (276 ft); the largest step here measures 59 m (194 ft). The waterfall is very narrow – around 18 m (59 ft) wide – but still carries large quantities of water. The waterfalls are located in the Brazilian state of Bahia where the Rio São Francisco breaks through the coastal mountains.

## KAIETEUR FALLS

**Guyana, 660 m$^3$ (23,308 cu ft) /s**

The Kaieteur Falls are located in the eponymous national park in central Guyana. The water descends 251 m (824 ft) over a width of around 113 m (371 ft); the waterfall is thus roughly five times as high as Niagara Falls.

## THE HIGHEST WATERFALLS

1. **SALTO ANGEL** Venezuela, 979 m (3,212 ft)
2. **CATARATAS LAS TRES HERMANAS** Peru, 914 m (2,999 ft)
3. **CATARATA YUMBILLA** Peru, 896 m (2,940 ft)
4. **CATARATA GOCTA** Peru, 771 m (2,530 ft)
5. **SALTO KUKENAAM** Venezuela, 674 m (2,211 ft)
6. **SALTO YUTAJÁ** Venezuela, 671 m (2,201 ft)
7. **SALTO RORAIMA** Venezuela, 610 m (2,001 ft)
8. **SALTO DEL IGUAPO** Venezuela, 600 m (1,969 ft)
9. **CATARATAS LA CHINATA** Peru, 580 m (1,903 ft)
10. **CACHOEIRA DO PILAO** Brazil, 524 m (1,719 ft)

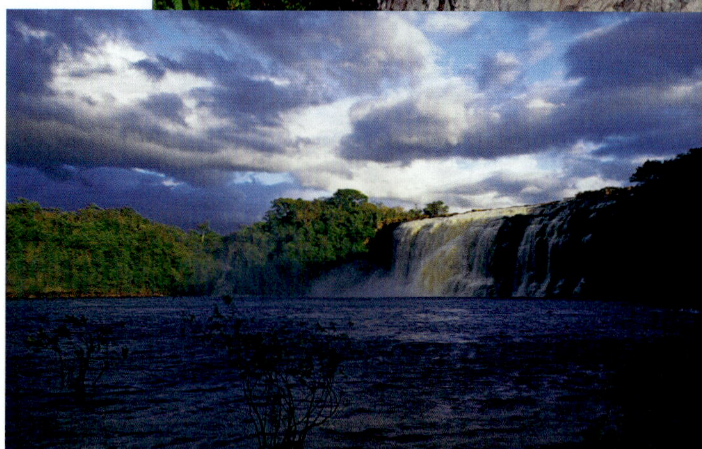

## SALTO HACHA/SALTO SAPO

**Venezuela, n/s**

Visitors do not come to the Canaima National Park just to see Salto Angel, the highest waterfall in the world. The Salto Hacha and the smaller Salto Sapo in the northwest of the park are just as popular. Both these falls are best reached by boat over the lagoon (picture, above), where the waters are tinged red due to the dissolved tannic acid content. The smaller Salto Sapo has a unique feature: visitors can walk under the waterfall between slippery rock walls – and, of course, get wet in the process. However, during the dry period, the Salto Sapo is just a trickle.

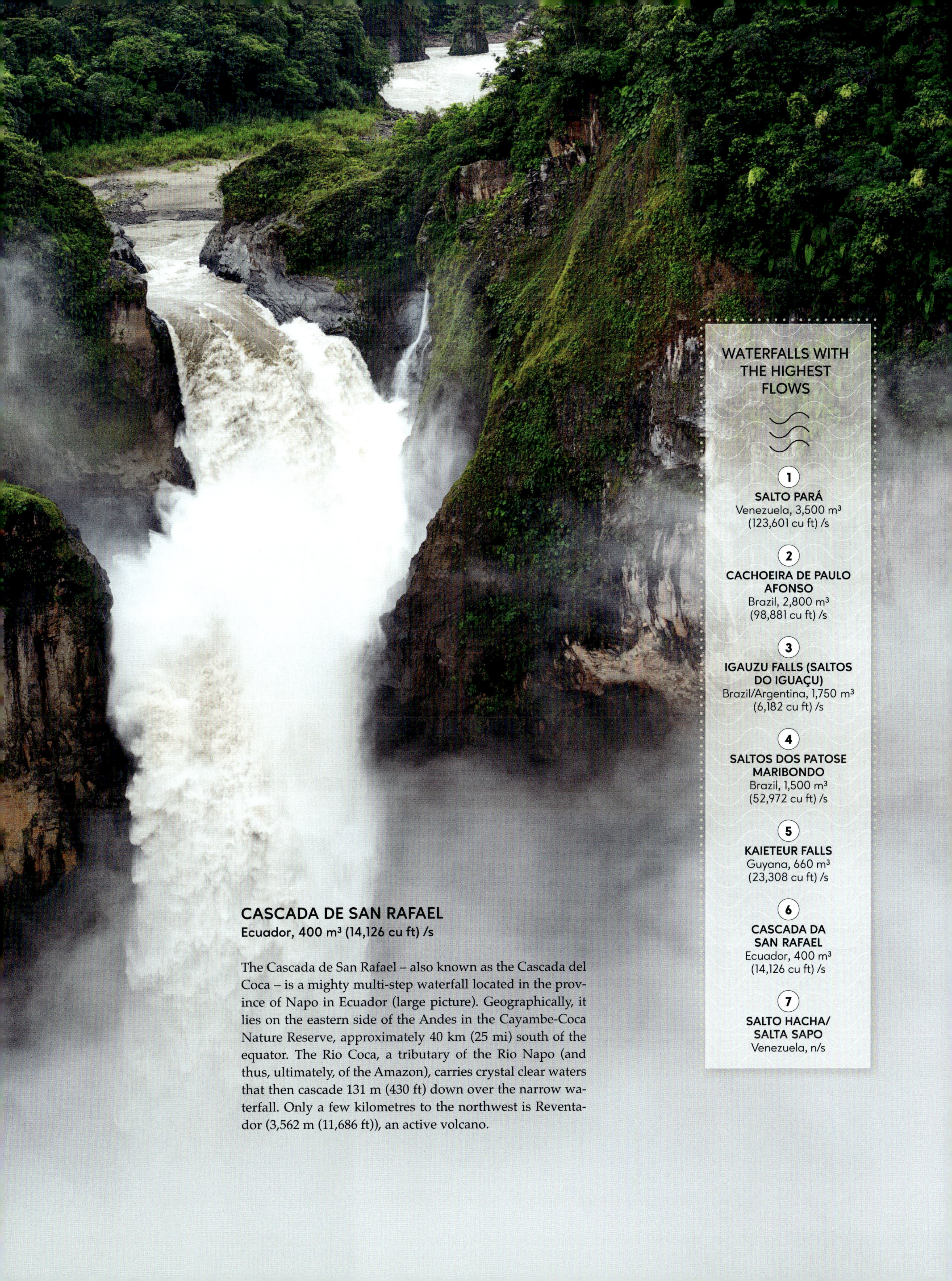

## CASCADA DE SAN RAFAEL

**Ecuador, 400 $m^3$ (14,126 cu ft) /s**

The Cascada de San Rafael – also known as the Cascada del Coca – is a mighty multi-step waterfall located in the province of Napo in Ecuador (large picture). Geographically, it lies on the eastern side of the Andes in the Cayambe-Coca Nature Reserve, approximately 40 km (25 mi) south of the equator. The Rio Coca, a tributary of the Rio Napo (and thus, ultimately, of the Amazon), carries crystal clear waters that then cascade 131 m (430 ft) down over the narrow waterfall. Only a few kilometres to the northwest is Reventador (3,562 m (11,686 ft)), an active volcano.

### WATERFALLS WITH THE HIGHEST FLOWS

1. **SALTO PARÁ**
Venezuela, 3,500 $m^3$ (123,601 cu ft) /s

2. **CACHOEIRA DE PAULO AFONSO**
Brazil, 2,800 $m^3$ (98,881 cu ft) /s

3. **IGAUZU FALLS (SALTOS DO IGUAÇU)**
Brazil/Argentina, 1,750 $m^3$ (6,182 cu ft) /s

4. **SALTOS DOS PATOSE MARIBONDO**
Brazil, 1,500 $m^3$ (52,972 cu ft) /s

5. **KAIETEUR FALLS**
Guyana, 660 $m^3$ (23,308 cu ft) /s

6. **CASCADA DA SAN RAFAEL**
Ecuador, 400 $m^3$ (14,126 cu ft) /s

7. **SALTO HACHA/ SALTA SAPO**
Venezuela, n/s

FOUR ELEMENTS

WATER

# IGAUZU FALLS (SALTOS DO IGUAÇU)

The most powerful waterfall in South America

These only rank as the third largest waterfalls in the continent of South America, but they are certainly the most beautiful, which has prompted UNESCO to place them on the list of natural heritage sites worthy of protection.

In the tri-border area of Brazil, Argentina and Paraguay, the Iguaçu River (which is called the Iguazú in Argentina) plunges down 62 m (203 ft) over a length of 2,700 m (8,859 ft). The waterfall is not a closed curtain of water like the Victoria Falls in Africa; more than 270 individual waterfalls can be distinguished along its entire width. The largest single waterfall is Garganta do Diabo (Garganta del Diablo in Spanish), meaning the Devil's Throat. It is a U-shaped canyon 150 m (492 ft) wide and 82 m (269 ft) high, over which an approximately 400-m (1,312-ft) wide wall of water plunges. The Devil's Throat is located on the border between Brazil and Argentina. With a flow rate of 1,750 $m^3$ (61,801 cu ft) per second, the Iguazu, or Iguaçu, Falls are the eighth largest waterfall in the world. Two-thirds of the waterfall lies in Argentina and one-third in Brazil. The surrounding area has been classified as a national park in both countries and, since 1984/86, has been designated a UNESCO Natural Heritage site. Visitors can access the falls from the neighbouring border towns of Foz do Iguaçu in Brazil and Puerto Iguazú in Argentina. There are also connections from the town of Ciudad del Este in Paraguay. On the Brazilian side, there is a walkway with an extension that leads to the Devil's Throat. From Argentina, visitors are transported on a narrow-gauge railway to the various walkways.

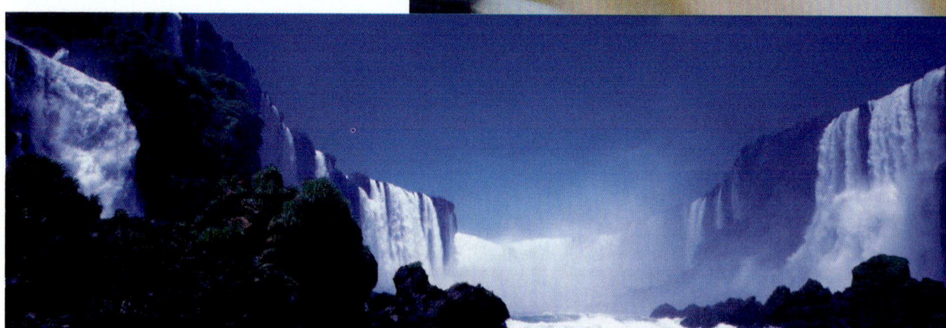

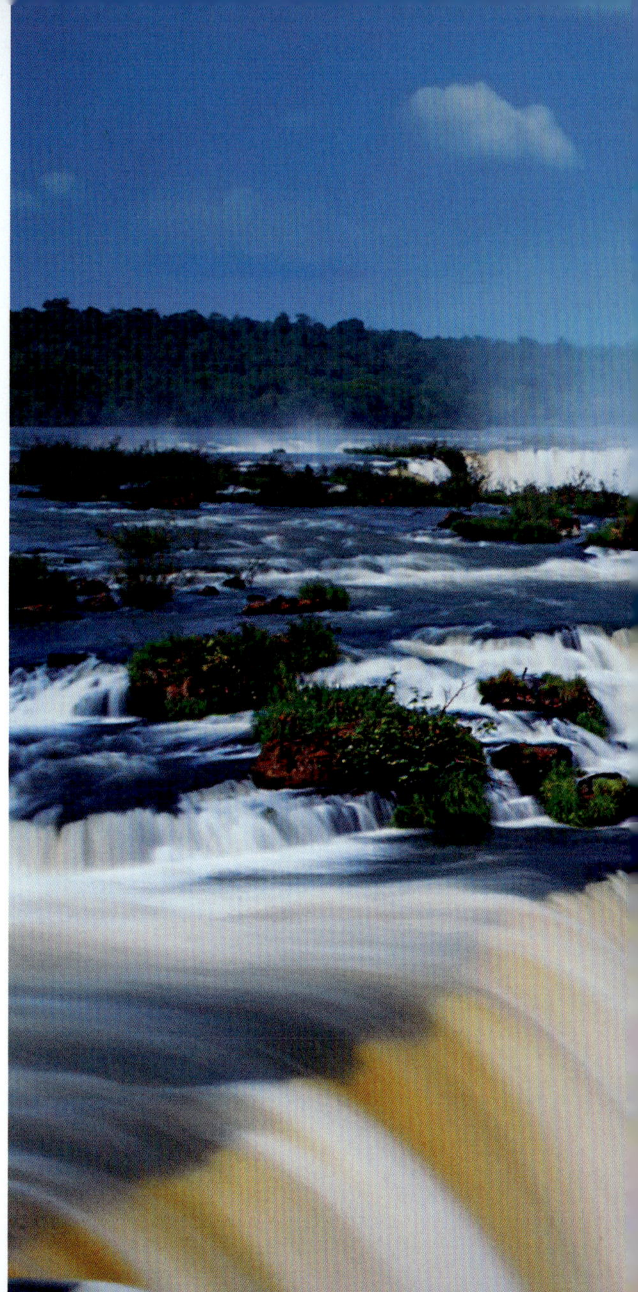

Above: The Iguaçu Falls, seen from the Brazilian side – in the haze beyond lies the Devil's Throat.

At the Iguaçu Falls, the sunlight is refracted in the spray of water droplets, displaying all the colours of the rainbow. The Iguaçu has numerous small islands which lie almost at the edge of the waterfall (large picture).

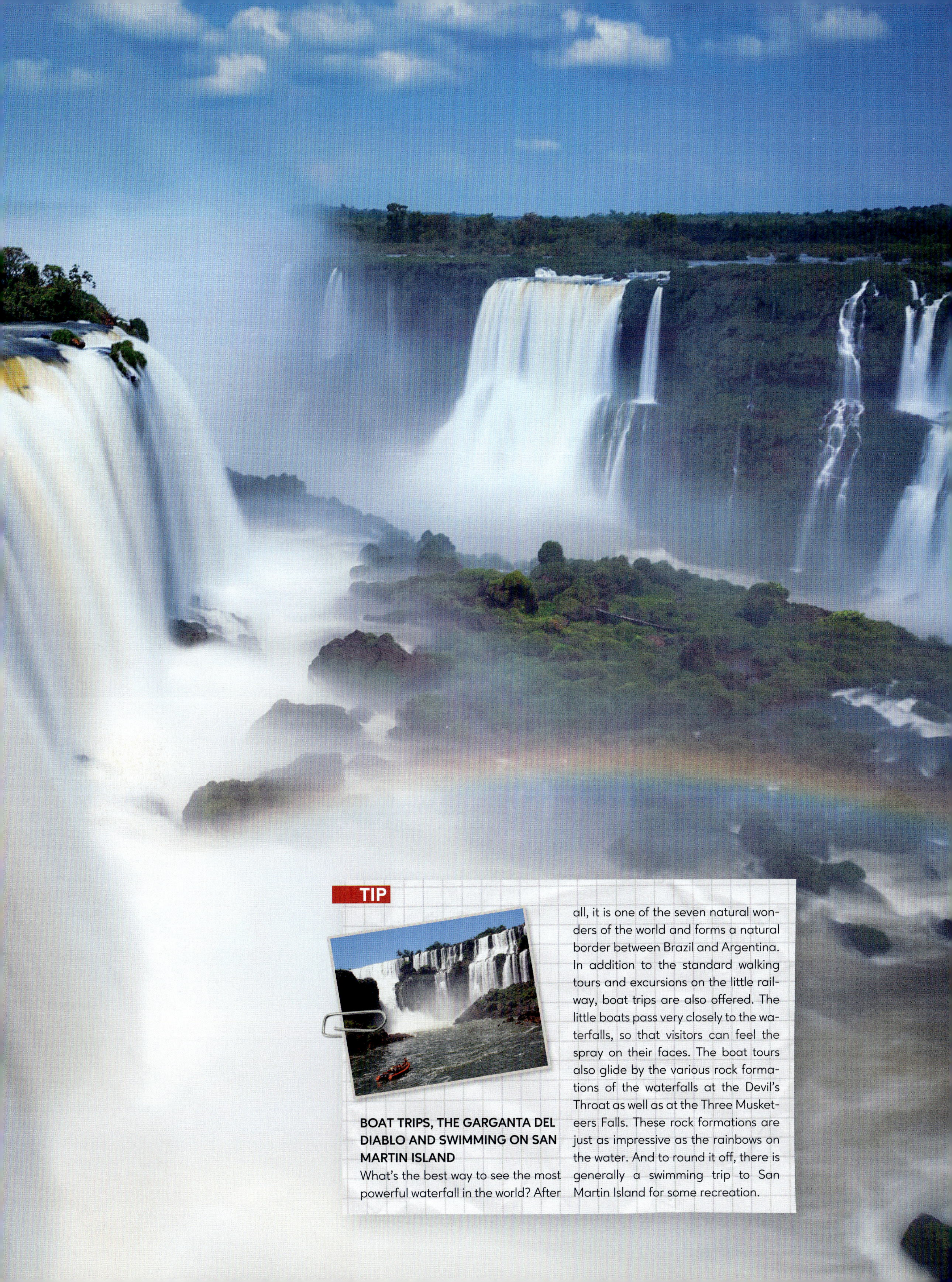

TIP

**BOAT TRIPS, THE GARGANTA DEL DIABLO AND SWIMMING ON SAN MARTIN ISLAND**

What's the best way to see the most powerful waterfall in the world? After all, it is one of the seven natural wonders of the world and forms a natural border between Brazil and Argentina. In addition to the standard walking tours and excursions on the little railway, boat trips are also offered. The little boats pass very closely to the waterfalls, so that visitors can feel the spray on their faces. The boat tours also glide by the various rock formations of the waterfalls at the Devil's Throat as well as at the Three Musketeers Falls. These rock formations are just as impressive as the rainbows on the water. And to round it off, there is generally a swimming trip to San Martin Island for some recreation.

FOUR ELEMENTS

WATER

# AFRICA

## Waterfalls of Africa: The highest flow and the highest waterfalls

According to the rankings of the highest waterfalls in the world, Tugela Falls in South Africa, with a drop of 948 m (3,110 ft), take second place after Salto Angel in Venezuela. Here the waters fall down in five steps. Tugela Falls are located in the northern Drakensberg mountains, a range that runs parallel to the southeast coast of South Africa and has elevations which can reach heights up to and over 3,000 m (9,843 ft). The area is part of Royal Natal National Park. The waterfall is fed by the Tugela River which rises just a few kilometres away in the Mont-Aux-Sources mountain. Tugela means 'sudden' in the language of the Zulu people.

### The highest waterfalls

1. **Tugela Falls**
   South Africa, 948 m (3,110 ft)
2. **Mutarazi Falls**
   South Africa, 762 m (2,500 ft)
3. **Cascades de Trou de Fer**
   Réunion, 725 m (2,379 ft)
4. **Cascade Blanche**
   Réunion, 640 m (2,099 ft)
5. **Ndedema Falls**
   South Africa, 460 m (1,509 ft)
6. **Vivienne Falls**
   Kenya, 460 m (1,509 ft)
7. **Gura Falls**
   Kenya, 305 m (1,001 ft)
8. **Cascade de Fleurs Jaunes**
   Réunion, 300 m (984 ft)
9. **Rianbavy**
   Madagascar, 250 m (820 ft)
10. **Chutes de Kambadaga**
    Guinea, 249 m (817 ft)
11. **Pungwe Falls**
    Zimbabwe, 243 m (797 ft)
12. **Elands River Falls**
    South Africa, 228 m

### Waterfalls with the highest flow

1. **Chutes Inga**
   Dem. Rep. Congo, 42,500 m³ (1,500,873 cu ft) /s
2. **Chutes de Livingstone**
   Dem. Rep. Congo, 35,000 m³ (1,126,013 cu ft) /s
3. **Boyoma Falls**
   Dem. Rep. Congo, 17,000 m³ (600,349 cu ft) /s
4. **Victoria Falls**
   Zimbabwe/Zambia, 1,100 m³ (38,846 cu ft) /s
5. **Kongou Falls**
   Gabon, 900 m³/s
6. **Epupa Falls**
   Namibia, 500 m³ (17,657 cu ft) /s
7. **Augrabies Falls**
   South Africa, 310 m³ (10,948 cu ft) /s
8. **Murchison Falls**
   Uganda, 300 m³ (10,594 cu ft) /s
9. **Ruacana Falls**
   Namibia/Angola, 280 m³ (9,888 cu ft) /s

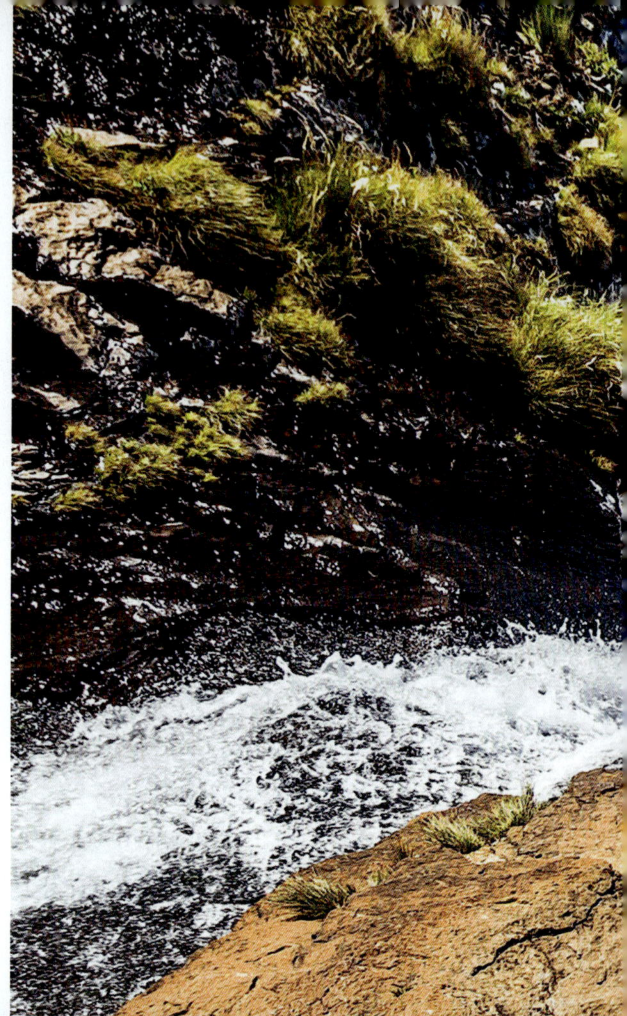

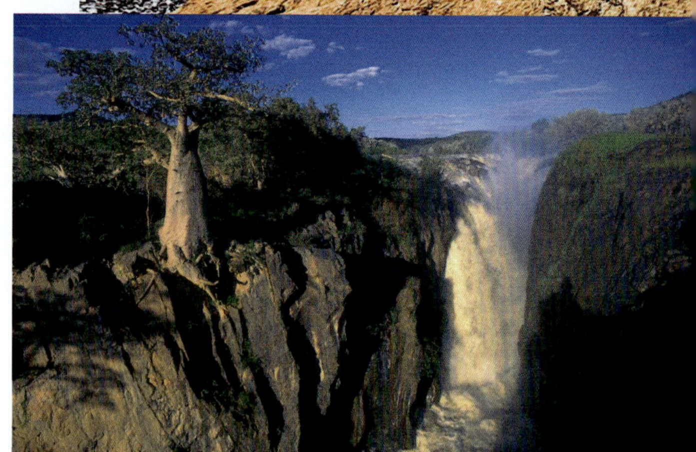

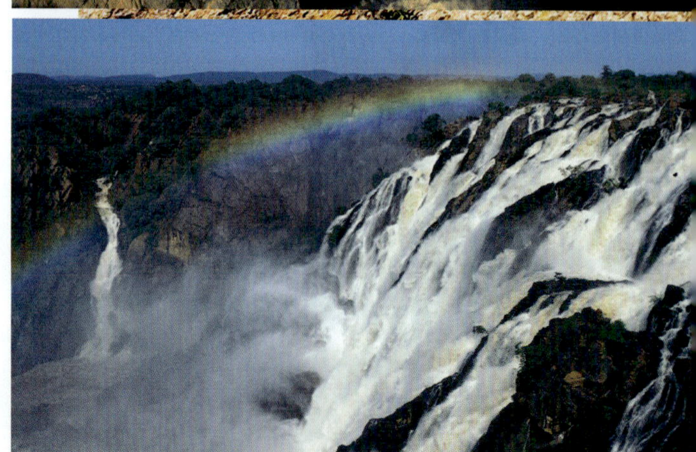

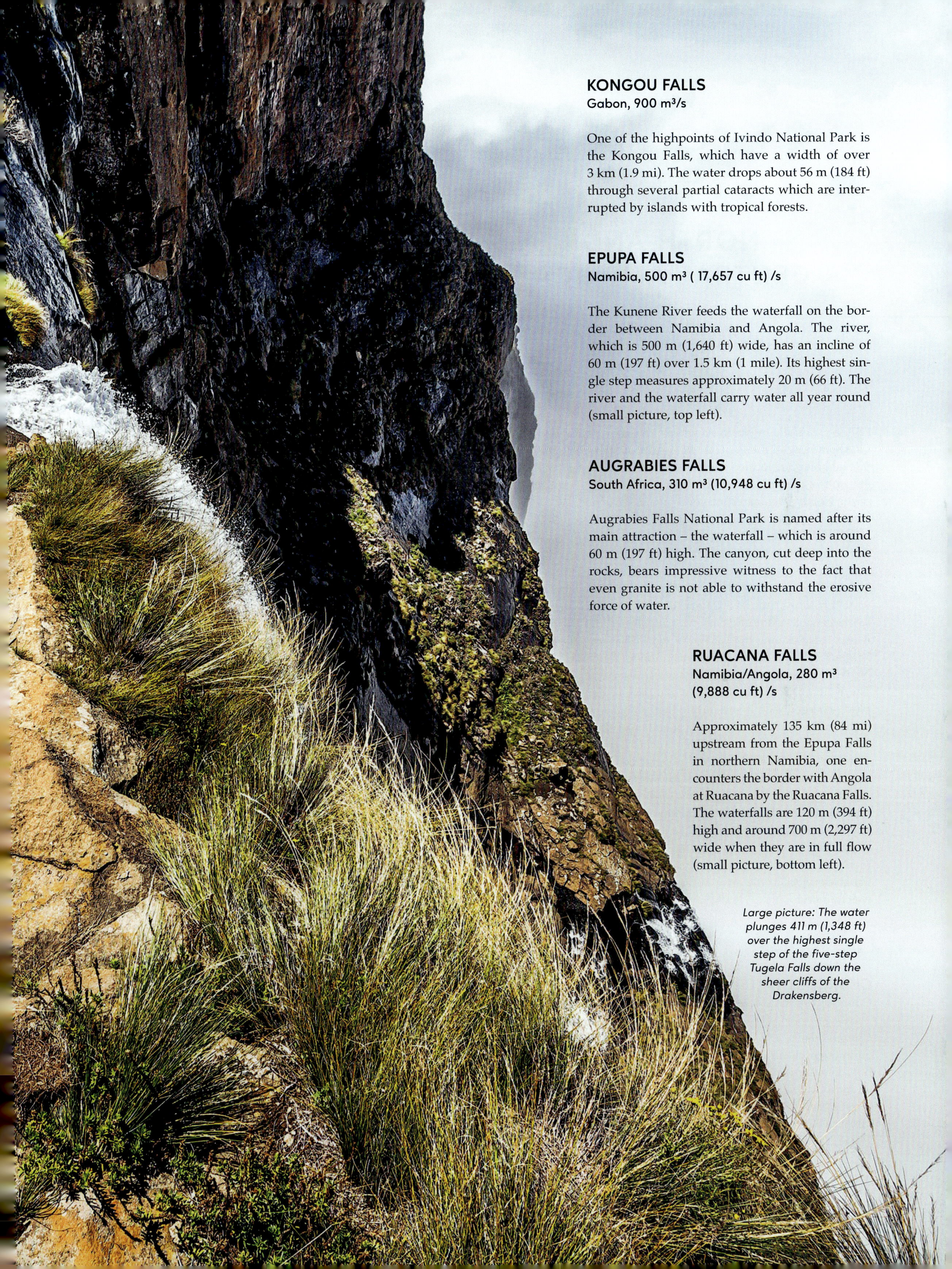

## KONGOU FALLS

**Gabon, 900 m³/s**

One of the highpoints of Ivindo National Park is the Kongou Falls, which have a width of over 3 km (1.9 mi). The water drops about 56 m (184 ft) through several partial cataracts which are interrupted by islands with tropical forests.

## EPUPA FALLS

**Namibia, 500 m³ ( 17,657 cu ft) /s**

The Kunene River feeds the waterfall on the border between Namibia and Angola. The river, which is 500 m (1,640 ft) wide, has an incline of 60 m (197 ft) over 1.5 km (1 mile). Its highest single step measures approximately 20 m (66 ft). The river and the waterfall carry water all year round (small picture, top left).

## AUGRABIES FALLS

**South Africa, 310 m³ (10,948 cu ft) /s**

Augrabies Falls National Park is named after its main attraction – the waterfall – which is around 60 m (197 ft) high. The canyon, cut deep into the rocks, bears impressive witness to the fact that even granite is not able to withstand the erosive force of water.

## RUACANA FALLS

**Namibia/Angola, 280 m³ (9,888 cu ft) /s**

Approximately 135 km (84 mi) upstream from the Epupa Falls in northern Namibia, one encounters the border with Angola at Ruacana by the Ruacana Falls. The waterfalls are 120 m (394 ft) high and around 700 m (2,297 ft) wide when they are in full flow (small picture, bottom left).

*Large picture: The water plunges 411 m (1,348 ft) over the highest single step of the five-step Tugela Falls down the sheer cliffs of the Drakensberg.*

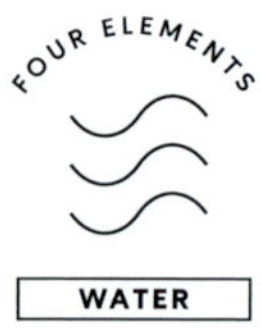


# NORTH AND CENTRAL AMERICA

Waterfalls with the highest flow and the highest waterfalls

The waterfalls in North America carry more water and are higher than those in the Central American countries. The highest waterfall is, however, located on the volcanic island of Hawaii, the fiftieth state of the USA, which, geologically speaking, is not actually part of the North American continent.

## VIRGINIA FALLS

**Canada, 1,000 m³ (35,315 cu ft) /s**

Nahanni National Park lies in the Mackenzie Mountains in the Northwest Territories, around 500 km (310 mi) west of Yellowknife. Still fairly inaccessible today, the park was established in 1972. Six years later, it was listed among the UNESCO World Natural Heritage sites. The Virginia Falls are at the southern summit of the park. The wild South Nahanni River drops 96 m (315 ft) across a width of 250 m (820 ft) here. The waterfall is named after the daughter of Fenley Hunter, the explorer charged with investigating the geological features of the South Nahanni in 1928. Hunter was a member of the Royal Geographical Society of London and of the New Yorker Explorers Club. He was an expert in cartography and very experienced in using the sextant for onsite surveying. He discovered the Virginia Falls and surveyed it on 21 August 1928, his daughter's birthday. When he submitted the maps and drawings to the Canadian authorities (the Geological Survey of Canada), he requested that the waterfall be named after his daughter. His wish was granted on account of his many services. The then prime minister, Pierre Trudeau (1919 - 2000), also rendered particular services to the area; because of his actions, the national park was founded.

## CHURCHILL FALLS

**Canada, 990 m³ (34,961 cu ft) /s**

Churchill Falls, also known as Grand Falls or Hamilton Falls, is located in the McLean Canyon in Newfoundland (formerly known as the Bowdoin Canyon). Ever since a large part of the water has been diverted for a hydroelectric power plant, the river no longer looks quite as impressive, except when it floods. The waters of the Hamilton River fall in two drops totalling 92 m (302 ft); the larger drop alone measures 75 m (246 ft). The waterfall has an average width of 46 m (151 ft).

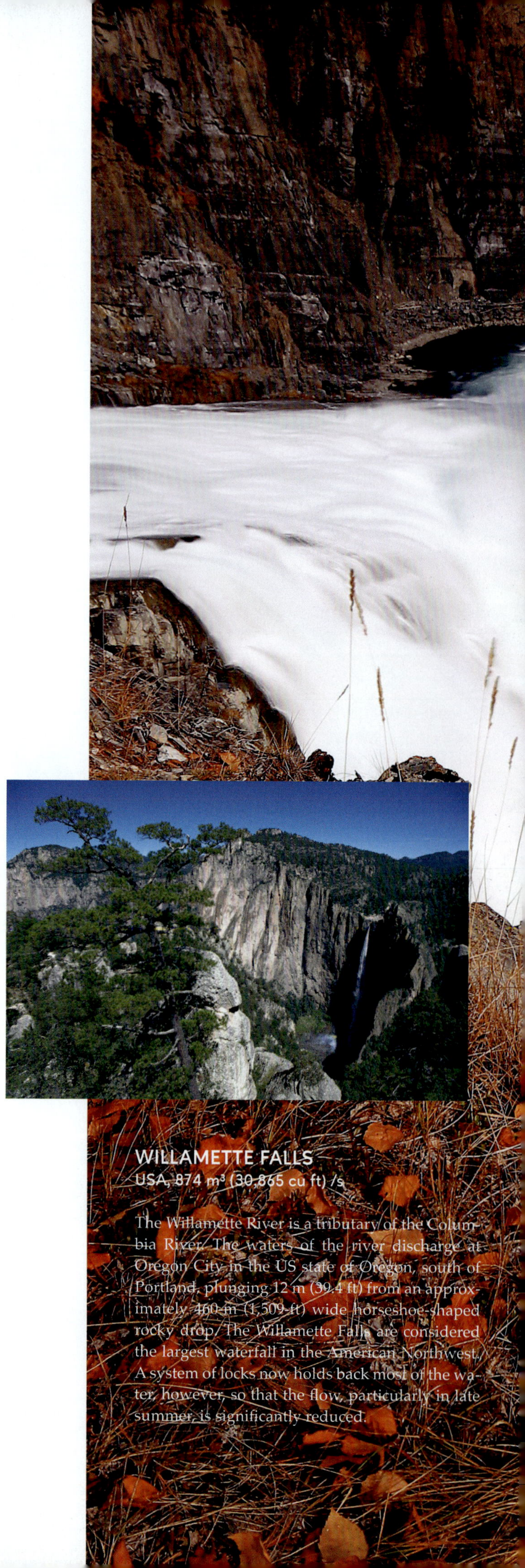

## WILLAMETTE FALLS

**USA, 874 m³ (30,865 cu ft) /s**

The Willamette River is a tributary of the Columbia River. The waters of the river discharge at Oregon City in the US state of Oregon, south of Portland, plunging 12 m (39.4 ft) from an approximately 460-m (1,509-ft) wide horseshoe-shaped rocky drop. The Willamette Falls are considered the largest waterfall in the American Northwest. A system of locks now holds back most of the water, however, so that the flow, particularly in late summer, is significantly reduced.

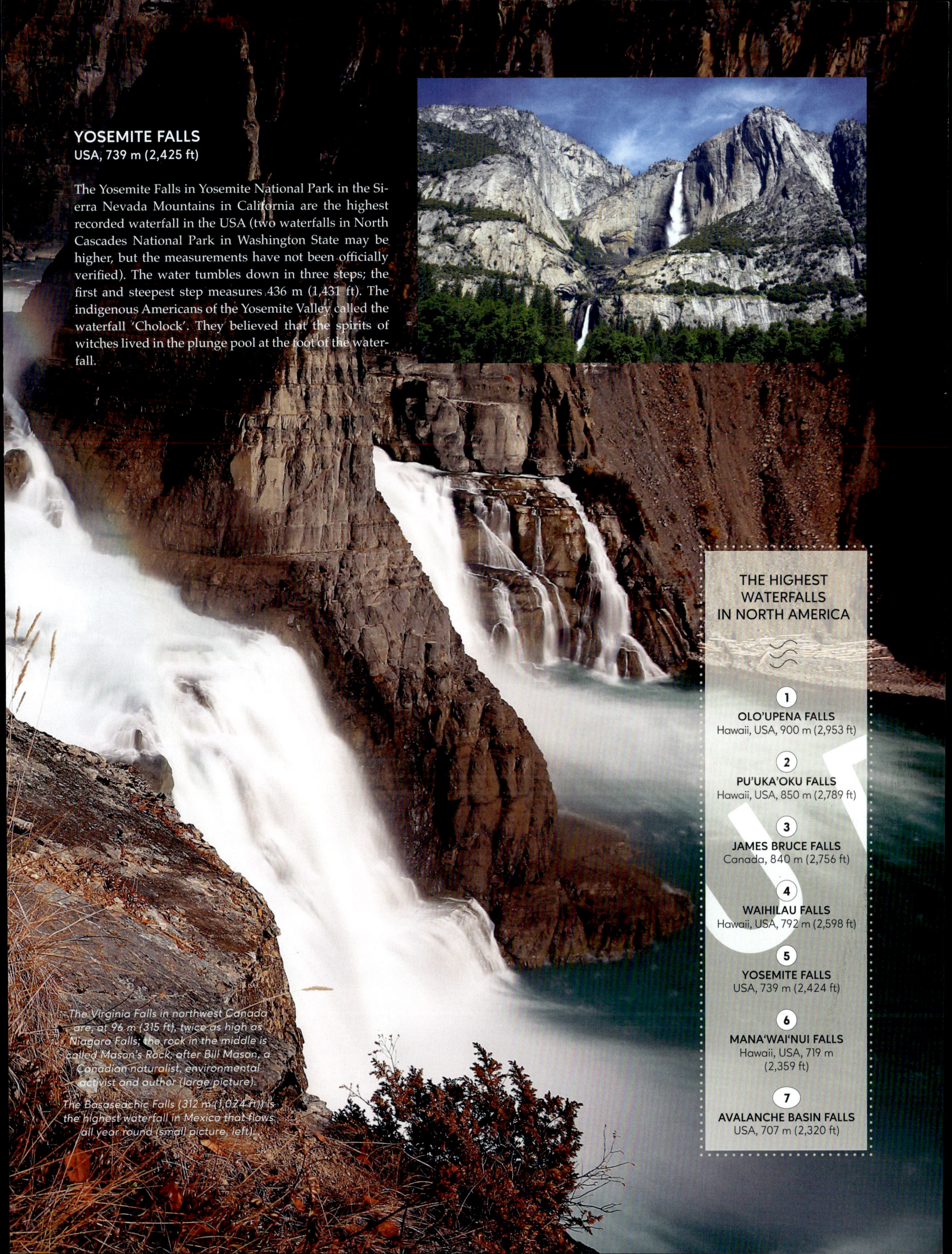

## YOSEMITE FALLS
**USA, 739 m (2,425 ft)**

The Yosemite Falls in Yosemite National Park in the Sierra Nevada Mountains in California are the highest recorded waterfall in the USA (two waterfalls in North Cascades National Park in Washington State may be higher, but the measurements have not been officially verified). The water tumbles down in three steps; the first and steepest step measures 436 m (1,431 ft). The indigenous Americans of the Yosemite Valley called the waterfall 'Cholock'. They believed that the spirits of witches lived in the plunge pool at the foot of the waterfall.

### THE HIGHEST WATERFALLS IN NORTH AMERICA

1. **OLO'UPENA FALLS** Hawaii, USA, 900 m (2,953 ft)
2. **PU'UKA'OKU FALLS** Hawaii, USA, 850 m (2,789 ft)
3. **JAMES BRUCE FALLS** Canada, 840 m (2,756 ft)
4. **WAIHILAU FALLS** Hawaii, USA, 792 m (2,598 ft)
5. **YOSEMITE FALLS** USA, 739 m (2,424 ft)
6. **MANA'WAI'NUI FALLS** Hawaii, USA, 719 m (2,359 ft)
7. **AVALANCHE BASIN FALLS** USA, 707 m (2,320 ft)

*The Virginia Falls in northwest Canada are, at 96 m (315 ft), twice as high as Niagara Falls; the rock in the middle is called Mason's Rock, after Bill Mason, a Canadian naturalist, environmental activist and author (large picture).*

*The Basaseachic Falls (312 m (1,024 ft)) is the highest waterfall in Mexico that flows all year round (small picture, left).*

FOUR ELEMENTS

WATER

# NIAGARA FALLS

The mightiest
waterfall in North America

The mighty Niagara Falls can be explored from the American or the Canadian side. Clad in waterproof oilskins, visitors can enter the Cave of the Winds – the area directly behind the waterfall.

Niagara Falls is one of the most famous waterfalls in the world; one-third of the Falls lies in the US state of New York and two-thirds in the Canadian province of Ontario. Approximately 2,500 $m^3$ (88,287 cu ft) of water per second flows over the steep cliffs here, confirming Niagara Falls' place as seventh on the list of the largest waterfalls in the world. The Niagara River, which joins Lake Ontario to Lake Erie, has a vertical drop of 52 m (171 ft) across 792 m (2,599 ft) on the Canadian side (the Horseshoe Falls); in the US section, however, it measures 363 m (1,191 ft), and the drop is only 21 m (69 ft) due to a rockfall. Between these lies Goat Island, an uninhabited island belonging to the state of New York. Niagara Falls have been well-developed in terms of tourism. In the past, newlyweds often spent their honeymoon here. In the low season and at night, the flow rate of the water is reduced; up to 90 per cent of the waters are diverted via a system of weirs and used by several power stations to generate electricity.

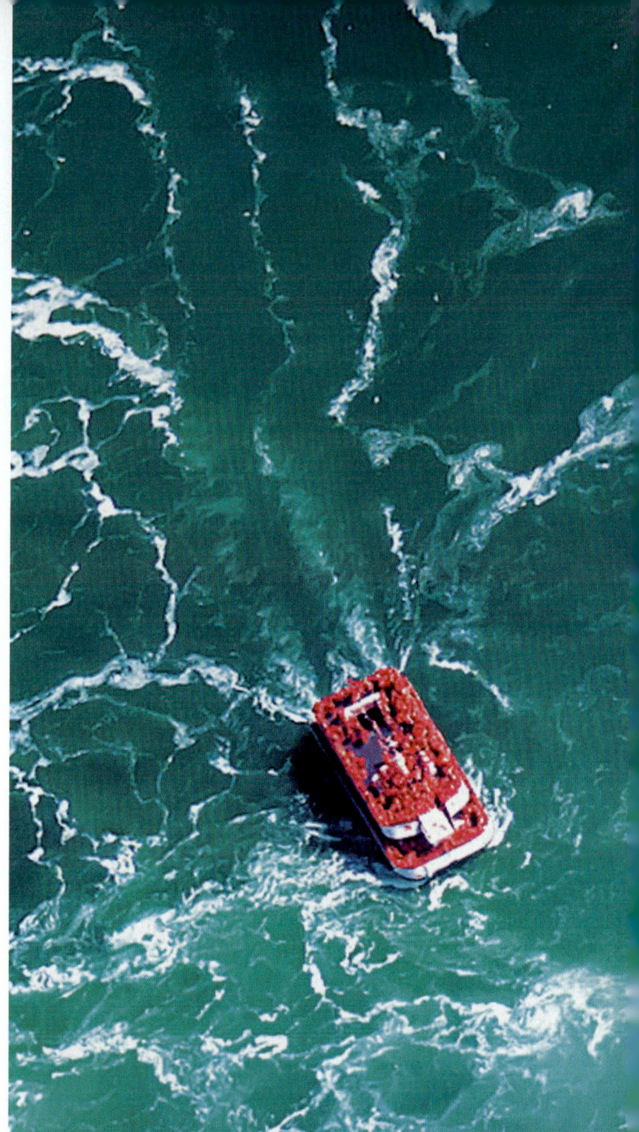

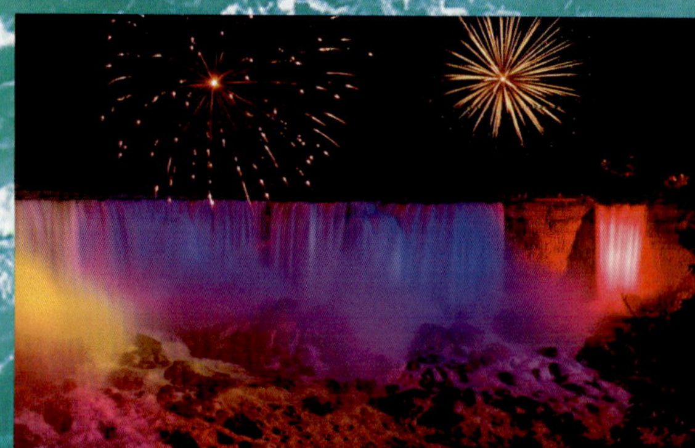

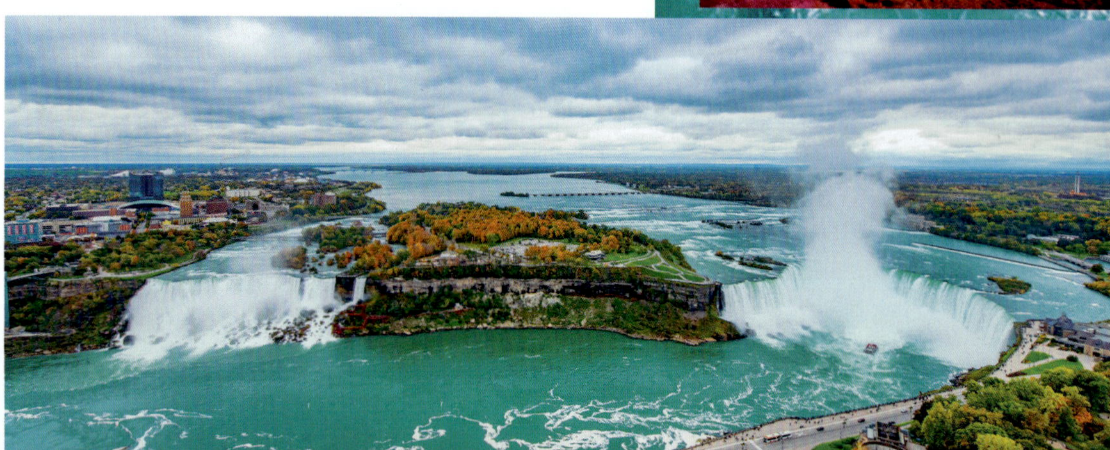

TIP

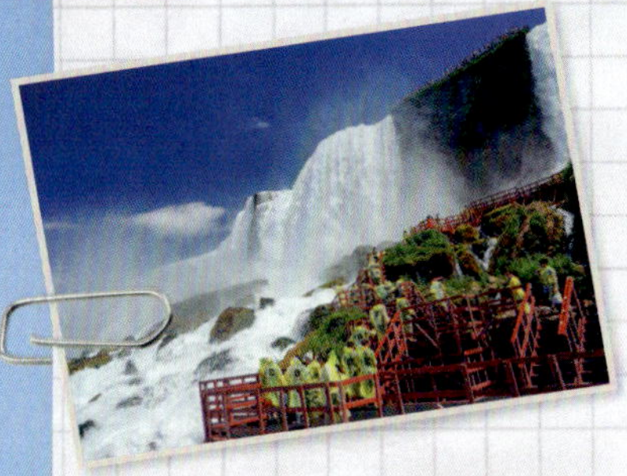

**EXCURSIONS TO THE CAVE OF THE WINDS – IN RAIN GEAR BEHIND THE WATERFALL**

Do you want to take a look behind the waterfall and peek through the veil of the most famous waterfall in the world? Well, who wouldn't? That is why tourists are offered the chance to visit the Cave of the Winds. An elevator takes them to a site where they can do exactly that. However, before setting off, they need to don rain ponchos and waterproof shoes. The elevator takes visitors 53 m (174 ft) into the canyon, where they can walk behind the waterfall on a wooden platform. The attraction is open from spring to fall; in winter, it is dismantled to prevent damage from the weather. The highpoint is the Hurricane Deck - you really cannot get any closer to the waterfall than that!

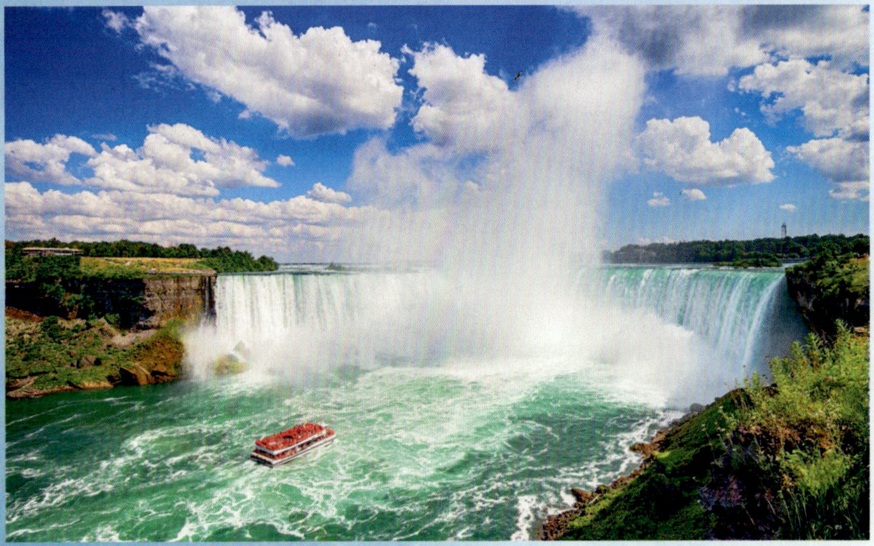

FOUR ELEMENTS

WATER

# EUROPE

The highest flows and the highest waterfalls

In contrast to the giant waterfalls in other parts of the world, the waterfalls in Europe tend to be on a more modest scale. The Rheinfall (Rhine Falls), however, is twenty-third in the world ranking.

## RHINE FALLS (RHEINFALL)

**Switzerland, 370 m³ (13,066 cu ft) /s**

Approximately 4 km (2.5 mi) west of Schaffhausen, the waters of the Rhine plunge down 23 m (75 ft) over rocky ground. On average, they flow at a rate of 370 m³ (13,066 cu ft) per second, across a width of 150 m (492 ft); the Rhine Falls therefore rank as the largest waterfall in Europe. In the summer months when rain is plentiful, the water flow averages 700 m³ (24,721 cu ft) per second. Since the last glacial period (the Würm or Weichsel glaciation), the Rhine has flowed in its current bed of hard Malm limestone. The Rhine developed approximately 14,000 to 17,000 years ago in a transition area between hard limestone and scree gullies which dates back to the earlier Riss Glacial period, 200,000 years ago. The Rhine Falls cliffs are the remains of the limestone flank, which was originally much steeper – visitors can climb the large cliffs. The site is a tourist magnet. It can be accessed on both sides via well-developed trails.

## DETTIFOSS

**Iceland, 200 m³ (7,063 cu ft) /s**

In northeastern Iceland, in the south of the Jökulsárgljúfur National Park, the grey-brown waters of the Jökulsá á Fjöllum, a glacial river, plunge 45 m (148 ft) across a width of 100 m (328 ft) at the Dettifoss waterfall. Approximately 30 km (19 mi) further north, the river discharges into the Norwegian Sea. With a flow rate averaging 200 m³ (7,063 cu ft) per second, it is, due to its double drop, more dynamic than the Rhine Falls. The water level varies considerably, however, depending on the season.

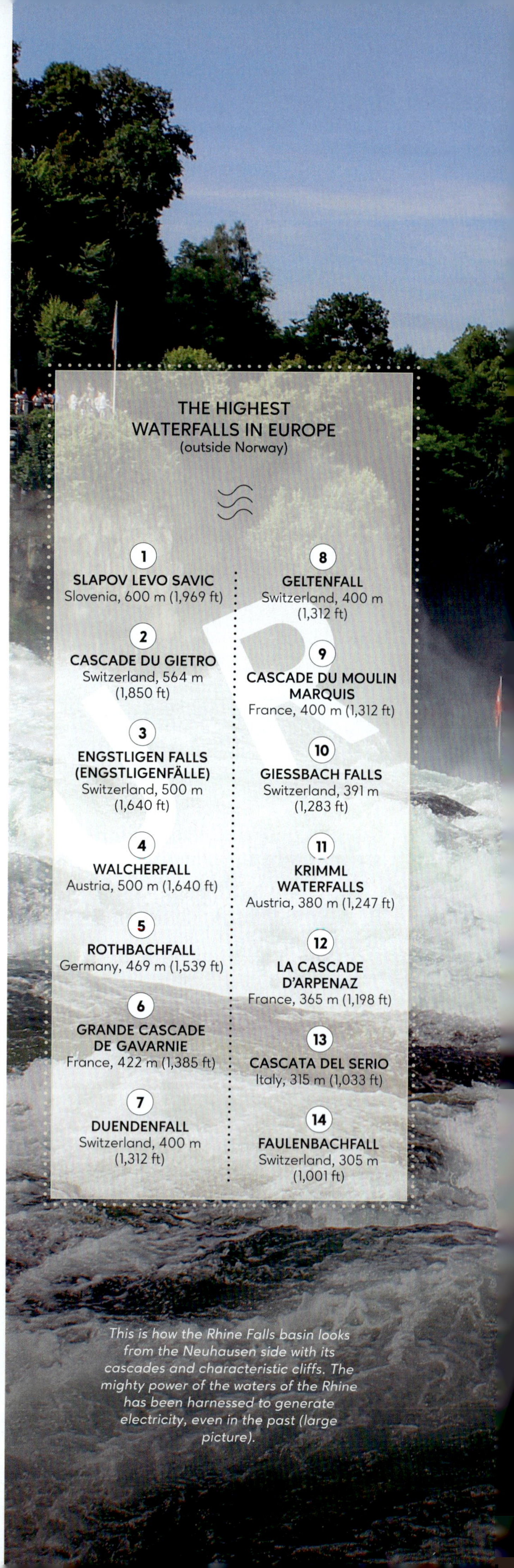

### THE HIGHEST WATERFALLS IN EUROPE
(outside Norway)

1. **SLAPOV LEVO SAVIC** Slovenia, 600 m (1,969 ft)
2. **CASCADE DU GIETRO** Switzerland, 564 m (1,850 ft)
3. **ENGSTLIGEN FALLS (ENGSTLIGENFÄLLE)** Switzerland, 500 m (1,640 ft)
4. **WALCHERFALL** Austria, 500 m (1,640 ft)
5. **ROTHBACHFALL** Germany, 469 m (1,539 ft)
6. **GRANDE CASCADE DE GAVARNIE** France, 422 m (1,385 ft)
7. **DUENDENFALL** Switzerland, 400 m (1,312 ft)
8. **GELTENFALL** Switzerland, 400 m (1,312 ft)
9. **CASCADE DU MOULIN MARQUIS** France, 400 m (1,312 ft)
10. **GIESSBACH FALLS** Switzerland, 391 m (1,283 ft)
11. **KRIMML WATERFALLS** Austria, 380 m (1,247 ft)
12. **LA CASCADE D'ARPENAZ** France, 365 m (1,198 ft)
13. **CASCATA DEL SERIO** Italy, 315 m (1,033 ft)
14. **FAULENBACHFALL** Switzerland, 305 m (1,001 ft)

*This is how the Rhine Falls basin looks from the Neuhausen side with its cascades and characteristic cliffs. The mighty power of the waters of the Rhine has been harnessed to generate electricity, even in the past (large picture).*

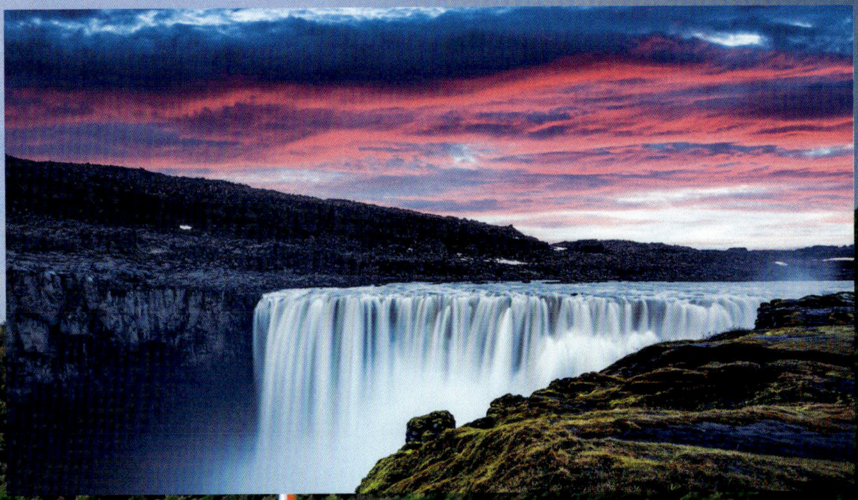

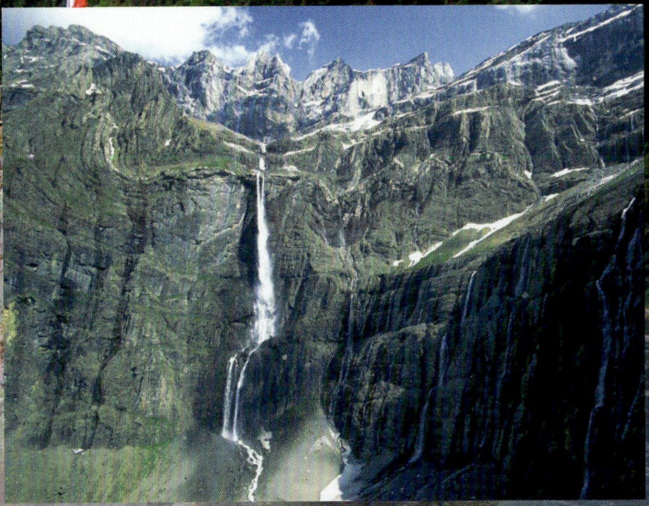

*The Dettifoss provides a fantastic backdrop – and not just at sunset. This is why it has already written film history, in the film Prometheus, with top director Ridley Scott (top).*

*The Grande Cascade de Gavarnie in the French section of the Pyrenees falls in three steps to the valley floor of the Cirque de Gavarnie (bottom).*

## HAFRAGILSFOSS

Iceland, 200 $m^3$ (7,063 cu ft) /s

Only 2 km (1.2 mi) north of the Dettifoss, the Jökulsá á Fjöllum surmounts a further step that is 91 m (299 ft) wide and 27 m (89 ft) deep – the Hafragilsfoss. The Hafragilsfoss lies on the southeastern border of the Jökulsárgljúfur National Park. It is the second highest waterfall on the same river course, after the Dettifoss. A challenging circular route, which in some places only consists of narrow, precipitous ledges secured with rope, leads through the gorge past the Hafragilsfoss. In winter, the roads to the park are virtually unpassable.

## STORFORSEN

Sweden, 186 $m^3$ (6,569 cu ft) /s

Storforsen refers to a mighty cataract in the province of Noorbotten in the north of Sweden, roughly 38 km (24 mi) northwest of Älvsbyn. The rapids stretch over a distance of 5 km (3 mi) in which they drop 80 m (262 ft); the biggest waterfall, with a drop of 60 m (197 ft), occurs in the last 2 km (1.2 mi). The small tributary of the Piteälven (Pite River Basin) carries the most water in June when it flows at a rate of 870 $m^3$ (30,724 cu ft) per second. The area around the rapids is now a nature reserve, which visitors can access via trails and boardwalks.

## SELFOSS

Iceland, 200 $m^3$ (7,063 cu ft) /s

The Selfoss lies a few hundred metres south of the Dettifoss. Here, the Jökulsá á Fjöllum, which means 'glacial river in the mountains', narrows. Across a width of 183 m (600 ft), the grey-green waters plunge down 11 m (36 ft) before reaching the Dettifoss shortly afterwards and then the Hafragilsfoss. The Jökulsá á Fjöllum, at 206 km (128 mi) long, is the second longest river on the island. It rises at Vatnahökull, the largest glacier in Iceland, which forms a 1,000-m (3,280-ft) thick ice cap over the southeast of the volcanic island.

# ASIA

The highest flows and the highest waterfalls

Asia has a massive river system that contains many powerful waterfalls and rapids. The Chutes de Khone on the Mekong River heads the list of waterfalls by a large margin.

The highest waterfall in Asia is in Japan. It is the Hannoki-no-taki near the city of Tateyama. But this waterfall, 500 m (1,640 ft) high, only flows when the snow melt occurs between April and July. It forms a double waterfall with the Shomyo-daki, which is around 350 m (1,148 ft) high and flows all year round. India has a whole series of high waterfalls. The Kunchikal Falls at Agumbe are considered the second highest in Asia. The waterfalls in the south of Thailand also have impressive drops; they receive their water from the Nakhon Si Thammarat mountains. The Allwagwag Falls in the Philippines are located on the Island of Mindanao at Davao. The Baizhang Pu waterfall receives its water from the Huangshan Mountain in China.

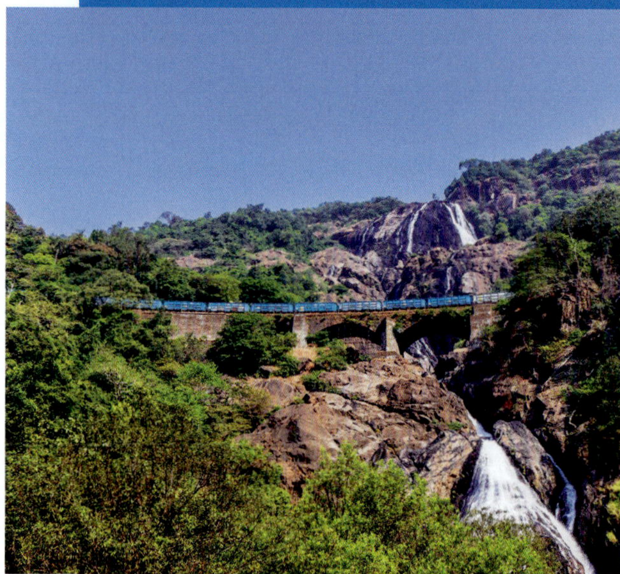

## CHUTES DE KHONE

**Laos/Mekong, 11,600 m³ (409,650 cu ft) /s**

The Chutes de Khone is a series of cascades on the Mekong River in the Laos province of Champasak on the border with Cambodia. The Chutes forms a cascade waterfall south of the inland delta of Si Phan Don, which means 'four thousand islands'. The Chutes are almost 11 km (6.8 mi) wide, and the water falls more than 21 m (69 ft) over a stretch measuring 10 km (6.2 mi). The Chutes de Khone is the widest waterfall in the world, the largest waterfall in Asia and the fourth largest in the world. Its average flow rate of 11,600 $m^3$ (409,650 cu ft) per second can rise to almost 50,000 $m^3$ in the rainy season. Due to the Chutes de Khone, the route to China on the Mekong is unnavigable. In the 19th century, French colonists tried in vain to get through with their ships. The railway line, which circumnavigates the most powerful rapids, first made it possible to transport ships to the other side of the Chutes.

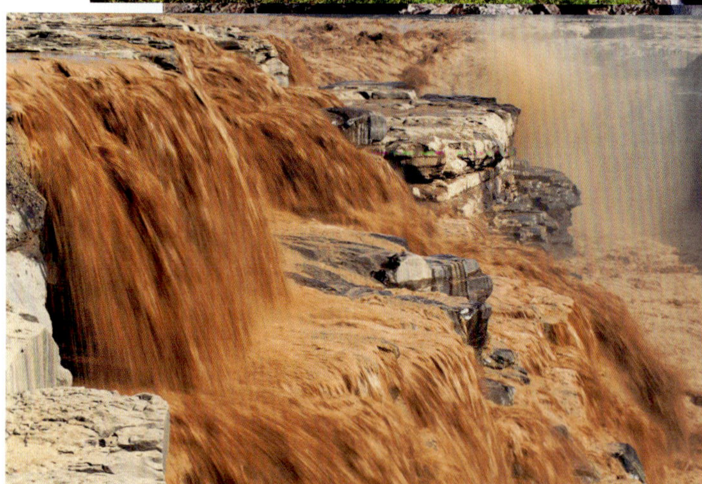

## HUKOU FALLS

**China/Huang He, 1,000 m³ (35,315 cu ft) /s**

The Hukou Falls, also known as Hukou Pubu, lie in the middle reaches of the Yellow River (Huang He) between the provinces of Shaanx and Shanxi. The waterfall is 30 m (98 ft) to 50 m (164 ft) wide, depending on the season; the drop is 30 m. When it floods, 8,000 $m^3$ (282,517 cu ft) of water per second can fall. The waterfall gives Huang He Hukou Waterfall National Park its name.

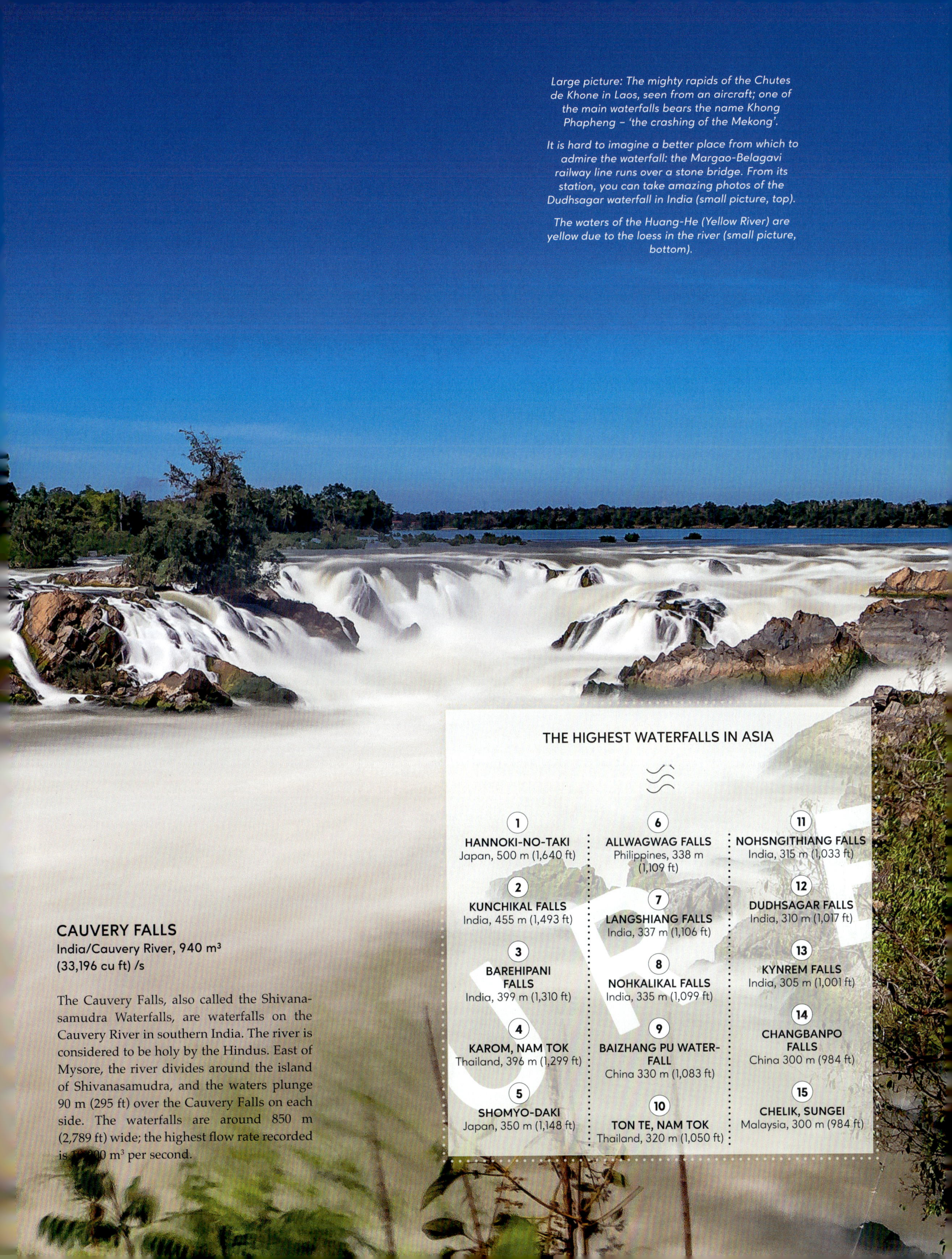

*Large picture: The mighty rapids of the Chutes de Khone in Laos, seen from an aircraft; one of the main waterfalls bears the name Khong Phapheng – 'the crashing of the Mekong'.*

*It is hard to imagine a better place from which to admire the waterfall: the Margao-Belagavi railway line runs over a stone bridge. From its station, you can take amazing photos of the Dudhsagar waterfall in India (small picture, top).*

*The waters of the Huang-He (Yellow River) are yellow due to the loess in the river (small picture, bottom).*

## CAUVERY FALLS

**India/Cauvery River, 940 m³ (33,196 cu ft) /s**

The Cauvery Falls, also called the Shivanasamudra Waterfalls, are waterfalls on the Cauvery River in southern India. The river is considered to be holy by the Hindus. East of Mysore, the river divides around the island of Shivanasamudra, and the waters plunge 90 m (295 ft) over the Cauvery Falls on each side. The waterfalls are around 850 m (2,789 ft) wide; the highest flow rate recorded is [illegible]0 m³ per second.

### THE HIGHEST WATERFALLS IN ASIA

1. **HANNOKI-NO-TAKI** – Japan, 500 m (1,640 ft)
2. **KUNCHIKAL FALLS** – India, 455 m (1,493 ft)
3. **BAREHIPANI FALLS** – India, 399 m (1,310 ft)
4. **KAROM, NAM TOK** – Thailand, 396 m (1,299 ft)
5. **SHOMYO-DAKI** – Japan, 350 m (1,148 ft)
6. **ALLWAGWAG FALLS** – Philippines, 338 m (1,109 ft)
7. **LANGSHIANG FALLS** – India, 337 m (1,106 ft)
8. **NOHKALIKAL FALLS** – India, 335 m (1,099 ft)
9. **BAIZHANG PU WATERFALL** – China 330 m (1,083 ft)
10. **TON TE, NAM TOK** – Thailand, 320 m (1,050 ft)
11. **NOHSNGITHIANG FALLS** – India, 315 m (1,033 ft)
12. **DUDHSAGAR FALLS** – India, 310 m (1,017 ft)
13. **KYNREM FALLS** – India, 305 m (1,001 ft)
14. **CHANGBANPO FALLS** – China 300 m (984 ft)
15. **CHELIK, SUNGEI** – Malaysia, 300 m (984 ft)

FOUR ELEMENTS
WATER

# AUSTRALIA/OCEANIA

The highest flows and the highest waterfalls

Although Australia is 29 times as big as New Zealand, the small island state clearly dominates the list of the highest waterfalls in this region of the world. A total of 16 waterfalls have a drop of 300 m (984 ft) or more. New Zealand has ideal conditions in terms of its geography. The island state is mountainous, traversed by multiple rivers and streams and has a steep fjord coastline in the southwest. In terms of climate, most of its territory is in the temperate, rainy zone. In contrast, large areas of the interior of Australia are desert, with many of the rivers here only flowing intermittently. Only the coastal areas in the south and southwest are classed as temperate zones in terms of their climate. One of the waterfalls from the many islands in Oceania first appears in thirteenth place: Vaipo, a waterfall located on Nuku Hiva with a 350 m (1,148 ft) drop. Nuku Hiva is the largest island of the Marquesas Islands group belonging to French Polynesia.

## Browne Falls
**836 m (2,742 ft) high**

The waters of Browne Falls cascade over the cliffs of Fiordland National Park on South Island in New Zealand, discharging into the Doubtful Sound on the Tasmanian Sea.

## Bluff Falls
**600 m (1,969 ft) high**

Bluff Falls are also located in Fiordland National Park. This is the largest park in New Zealand and is located in the southwest. Since 1990, it has been administrated as a UNESCO World Natural Heritage site.

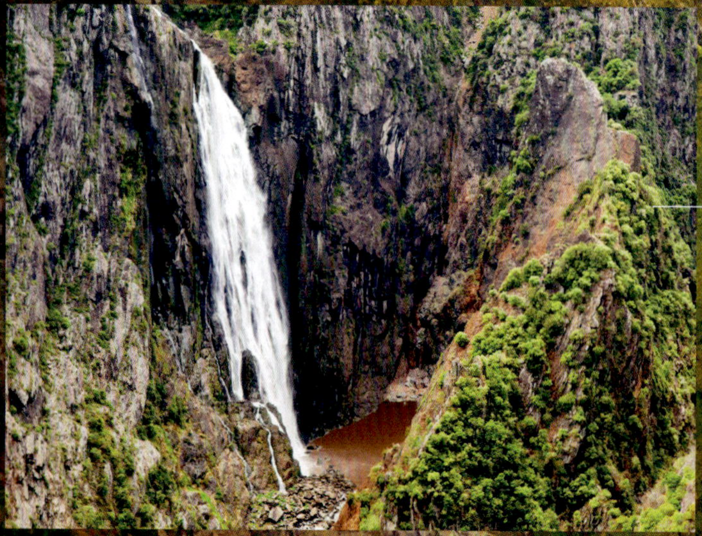

## Wollomombi Falls
**424 m (1,391 ft) high**

The highest waterfall in Australia only carries water intermittently. It lies in the Oxley Wild Rivers National Park in New South Wales, approximately 450 km (280 mi) north of Sydney.

*The Sutherland Falls are in Fiordland National Park on South Island in New Zealand (large picture).*

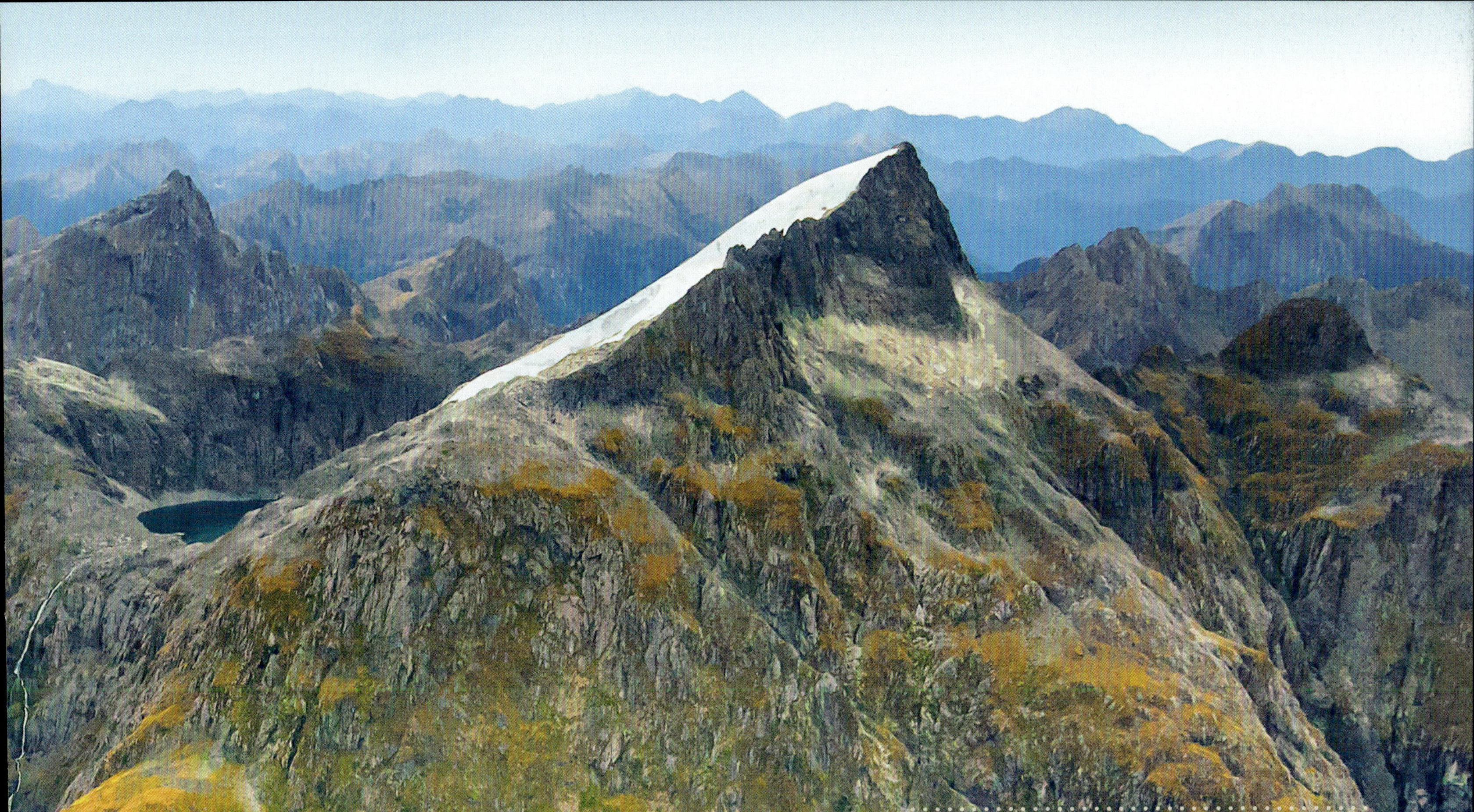

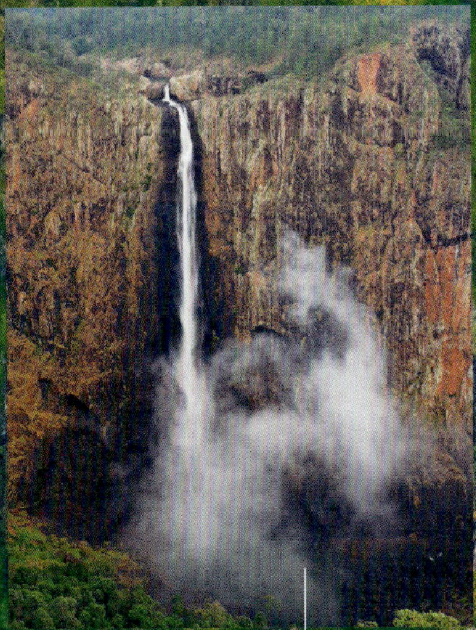

## Wallaman Falls
268 m (879 ft) high

The highest single-step waterfall in Australia is an impressive sight in the humid tropics of Queensland, which have been a UNESCO World Natural Heritage site since 1988.

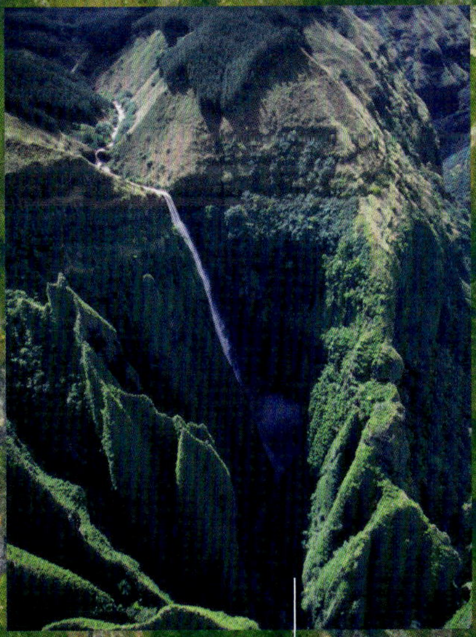

## Vaipo Falls
350 m (1,148 ft) high

On the volcanic island of Nuku Hiva in French Polynesia, you can discover the Vaipo Falls (also known as the Ahuii Falls) 15 km (9 mi) southwest of the town of Taiohae.

## THE HIGHEST WATERFALLS IN AUSTRALIA/OCEANIA

1. **BROWNE FALLS**
   New Zealand, 836 m (2,743 ft)
2. **LAKE CHAMBERLAIN FALLS**
   New Zealand, 700 m (2,297 ft)
3. **LAKE UNKNOWN FALLS**
   New Zealand, 680 m (2,231 ft)
4. **HIDDEN FALLS**
   New Zealand, 660 m (2,165 ft)
5. **BLUFF FALLS**
   New Zealand, 600 m (1,968 ft)
6. **WISHBONE FALLS**
   New Zealand, 600 m (1,968 ft)
7. **SUTHERLAND FALLS**
   New Zealand, 580 m (1,903 ft)
8. **DOUGLAS FALLS**
   New Zealand, 540 m (1,772 ft)
9. **WOLLOMOMBI FALLS**
   Australia, 424 m (1,391 ft)
10. **TIN MINE FALLS**
    Australia, 421 m (1,381 ft)
11. **GERARD FALLS**
    New Zealand, 420 m (1,378 ft)
12. **HIRERE FALLS**
    New Zealand, 420 m (1,378 ft)
13. **VAIPO (AHUII FALLS)**
    French Polynesia, 350 m (1,148 ft)

FOUR ELEMENTS

WATER

# ICEFIELDS AND GLACIERS

During the last glacial period – the Würm or Weichsel Glaciation – large parts of Northern Europe were covered in Fennoscandian ice sheets, and a vast ice cap lay over the Alps. But for the last 10,000 years, the icefields, or ice sheets, have been retreating. In contrast to glaciers, where the direction of the flow is determined by the surrounding tectonics, icefields reshape the underlying relief. The largest island in the world (Greenland) and an entire continent (Antarctica) lie under the two thickest ice caps. Icefields are distinguished from ice sheets, when they are larger than 50,000 km² (19,305 sq mi), and the smaller forms are ice caps.

### The longest glaciers in the world

1 **Lambert Glacier**
Antarctica, 400 km (248.5 mi)

2 **Beardmore Glacier**
Antarctica, 160 km (99 mi)

3 **Lillie Glacier**
Antarctica, 160 km (99 mi)

4 **Byrd Glacier**
Antarctica, 135 km (84 mi)

5 **Nimrod Glacier**
Antarctica, 135 km (84 mi)

6 **Hubbard Glacier**
Alaska/Canada, 122 km (76 mi)

7 **Vatnajökull**
Iceland, 100 km (62 mi)

8 **Nabesna Glacier**
Alaska, 80 km (50 mi)

9 **Bering Glacier**
Alaska, 76 km (47 mi)

### The largest icefields in the world

1 **Antarctic Ice Sheet**
approx. 12 million km²
(4,633,226 sq mi)

2 **Greenland Ice Sheet**
approx. 1.7 million km²
(656,374 sq mi)

3 **Kluane Icefield**
18,000 km² (6,950 sq mi)

4 **South Patagonian Icefield**
16,800 km² (6,486 sq mi)

5 **Devon Ice Cap**
12,000 km² (4,633 sq mi)

6 **Barnes Ice Cap**
6,000 km² (2,317 mi)

7 **Penny Ice Cap**
4,400 km² (1,989 sq mi)

8 **North Patagonian Icefield**
4,100 km² (1,583 sq mi)

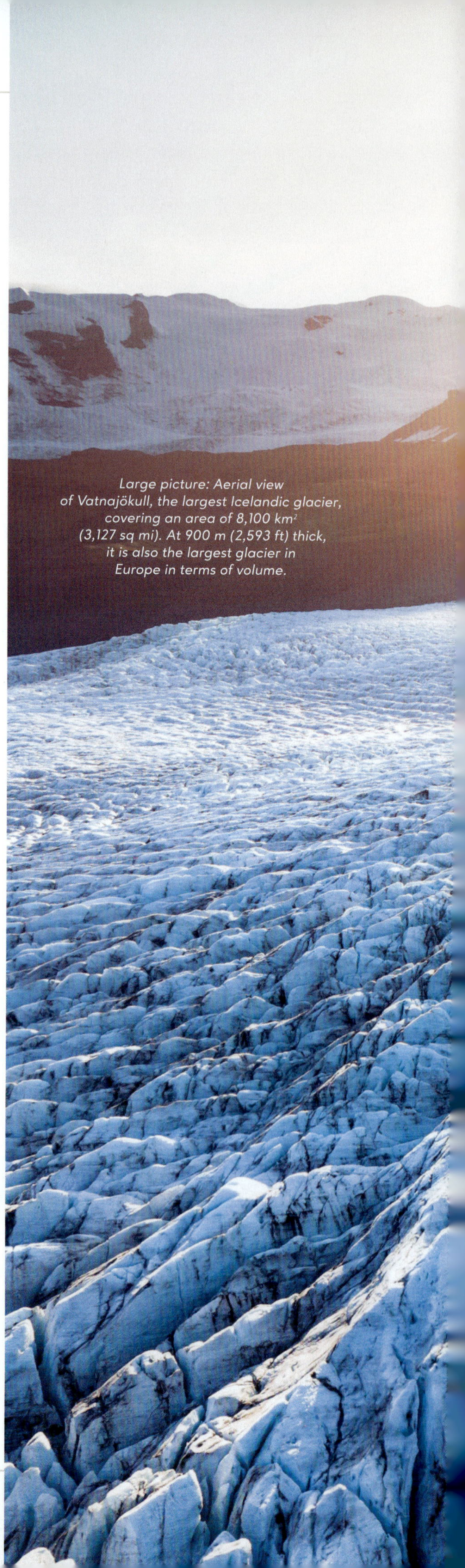

*Large picture: Aerial view of Vatnajökull, the largest Icelandic glacier, covering an area of 8,100 km² (3,127 sq mi). At 900 m (2,593 ft) thick, it is also the largest glacier in Europe in terms of volume.*

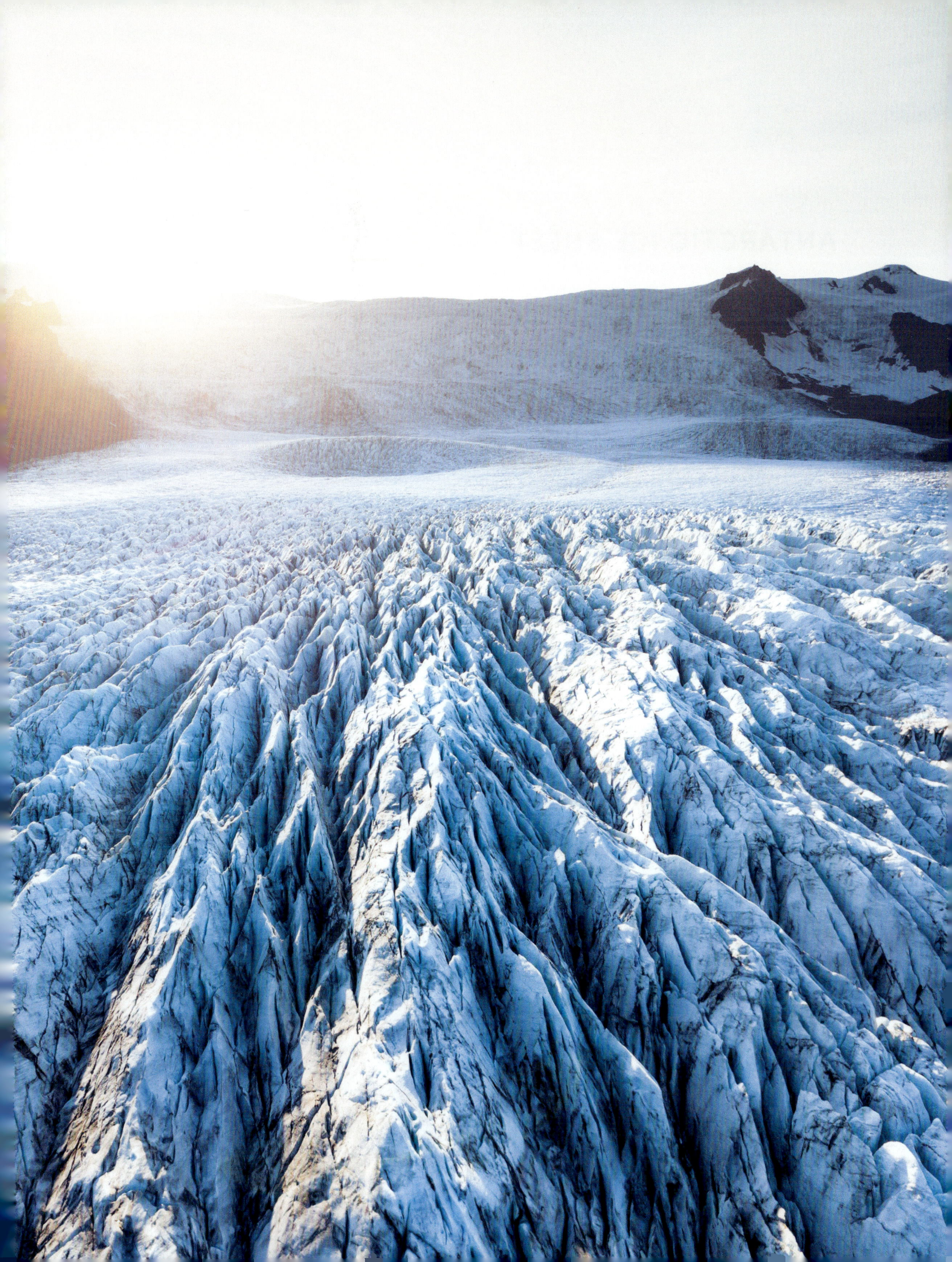

FOUR ELEMENTS

WATER

# ANTARCTIC ICE SHEET

The largest continuous icefield in the world

The Antarctic Ice Sheet, with a surface area of more than 12 million $km^2$ (4,633,226 sq mi) and a volume of 26 million $km^3$ (5,719,203 billion gal), is the largest continuous icefield on earth. If this massive body of ice were to melt, sea levels around the world would rise by more than 61 m (200 ft). In total, this ice mass contains around 90 per cent of all the freshwater on the earth's surface. The Antarctic Ice Sheet is in turn surrounded by huge ice shelves – floating ice sheets which are connected to the mainland by glaciers. The largest among these are the Ross, Filchner-Ronne and Amery ice shelves.

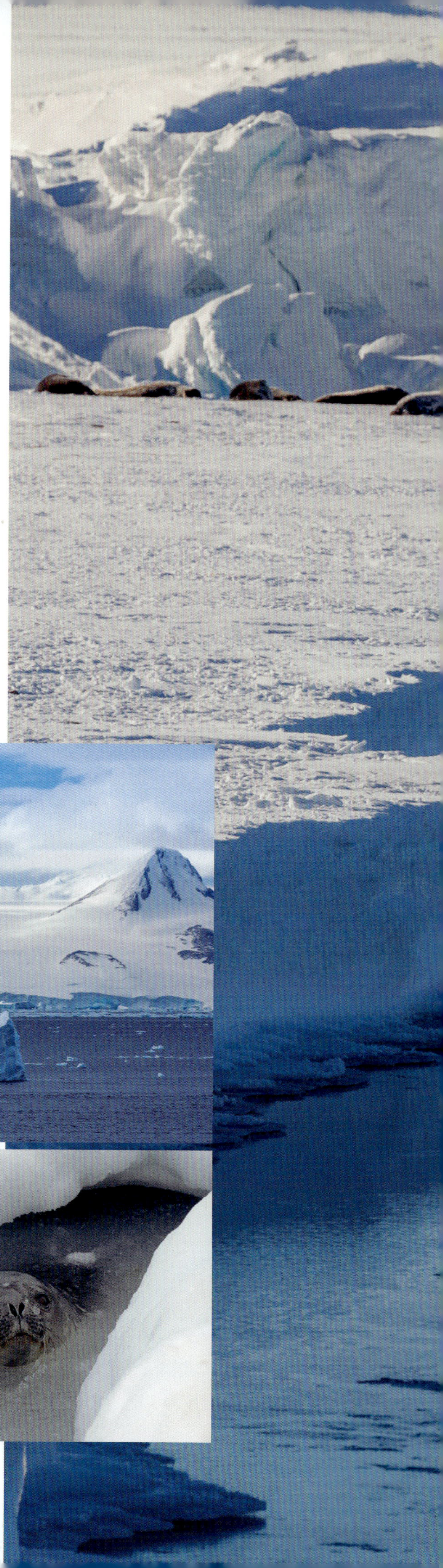

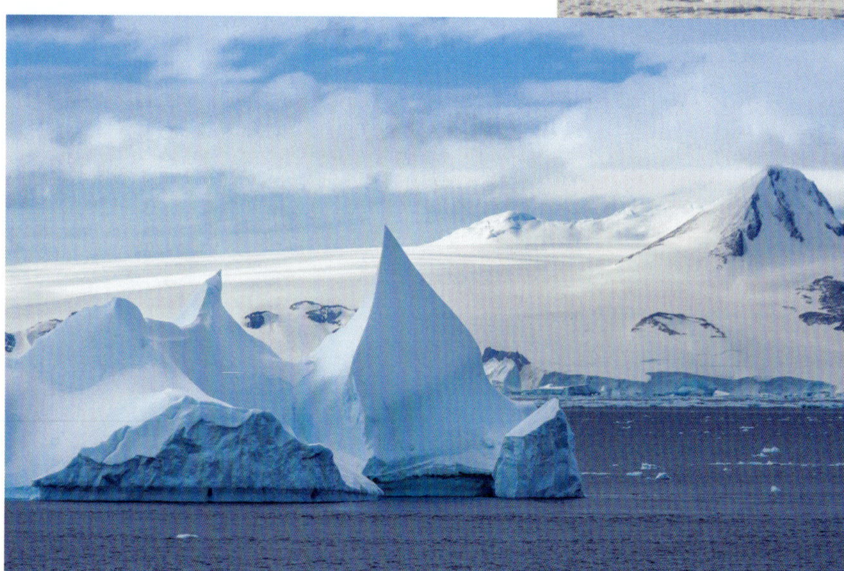

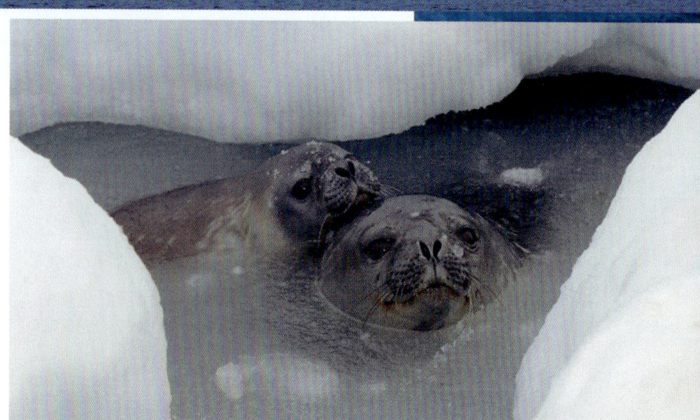

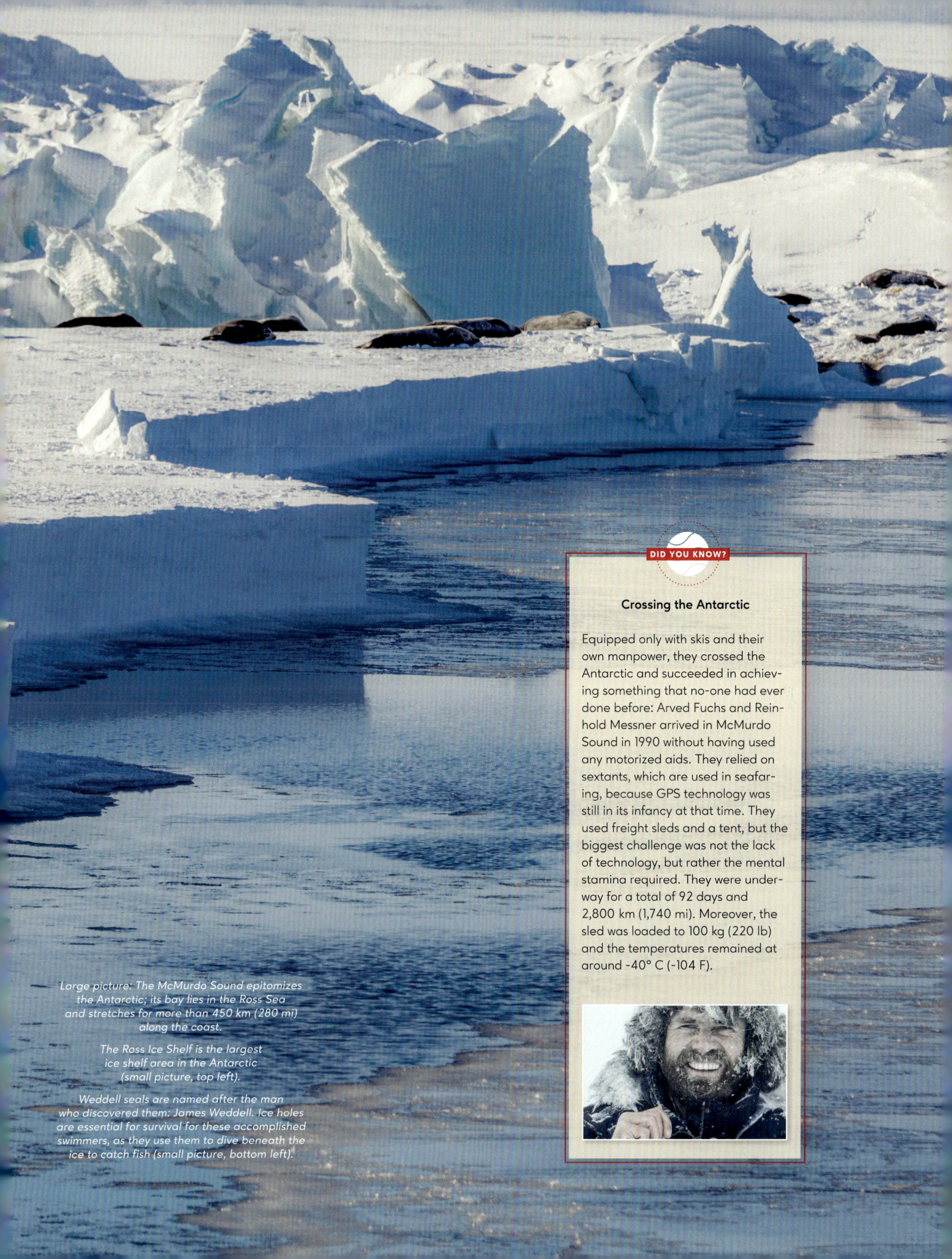

DID YOU KNOW?

### Crossing the Antarctic

Equipped only with skis and their own manpower, they crossed the Antarctic and succeeded in achieving something that no-one had ever done before: Arved Fuchs and Reinhold Messner arrived in McMurdo Sound in 1990 without having used any motorized aids. They relied on sextants, which are used in seafaring, because GPS technology was still in its infancy at that time. They used freight sleds and a tent, but the biggest challenge was not the lack of technology, but rather the mental stamina required. They were underway for a total of 92 days and 2,800 km (1,740 mi). Moreover, the sled was loaded to 100 kg (220 lb) and the temperatures remained at around -40° C (-104 F).

*Large picture: The McMurdo Sound epitomizes the Antarctic; its bay lies in the Ross Sea and stretches for more than 450 km (280 mi) along the coast.*

*The Ross Ice Shelf is the largest ice shelf area in the Antarctic (small picture, top left).*

*Weddell seals are named after the man who discovered them: James Weddell. Ice holes are essential for survival for these accomplished swimmers, as they use them to dive beneath the ice to catch fish (small picture, bottom left).*

FOUR ELEMENTS

WATER

# LAMBERT GLACIER

The longest and largest glacier in the world

The Lambert Glacier forms the main inflow of the Amery Ice Shelf on the Ingrid Christensen Coast. The ice sheet glides through glacial troughs in various glacier streams over the continental margins of the eastern Antarctic. Moving at a speed of around 2.5 m (8 ft) a day, the ice makes its way towards the shelf ice plateaus. The incoming sunlight reveals even the tiniest grooves and fissures. Some of the structures are pushed back by strong winds. Individual ridges called 'nunataks' (a term taken from the Inuit language), rise out of the ice and determine the direction of the ice flow.

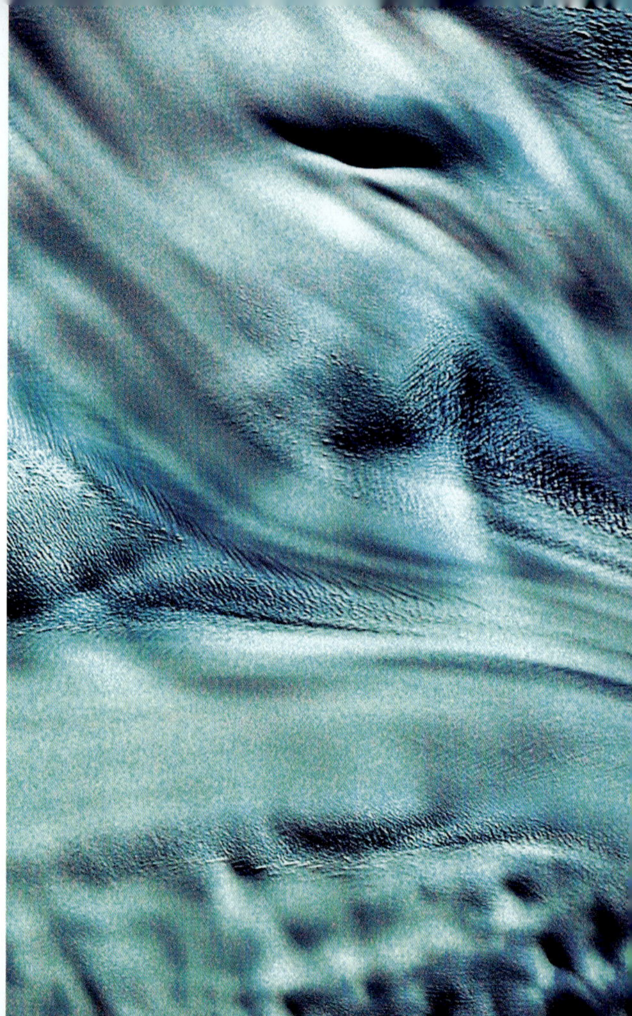

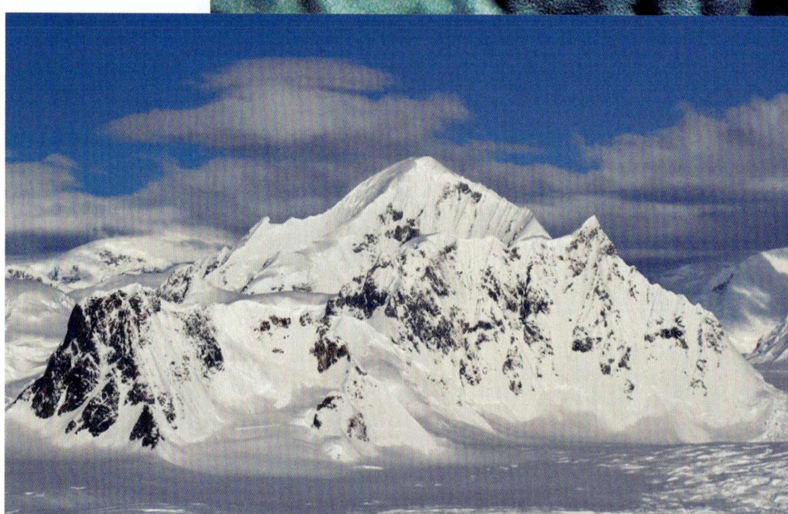

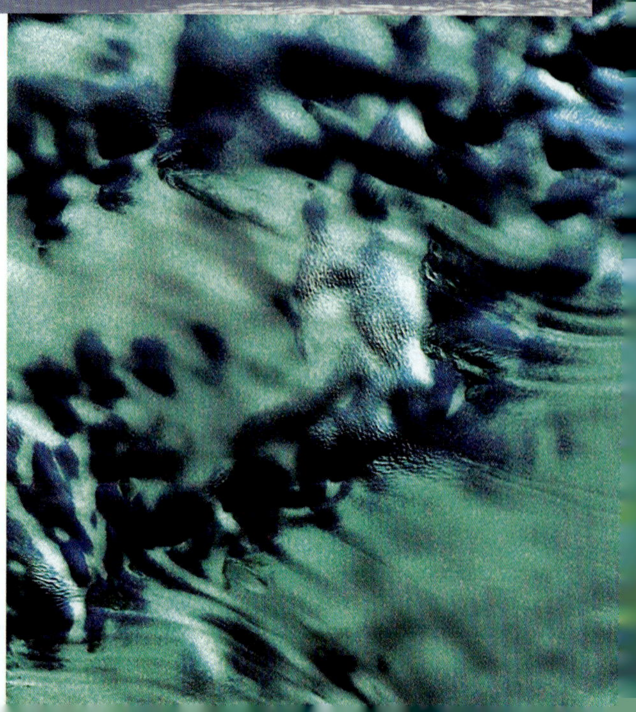

## THE LONGEST GLACIERS IN ANTARCTICA

1. **LAMBERT GLACIER** 400 km (248.5 mi)
2. **BEARDMORE GLACIER** 160 km (99 mi)
3. **LILLIE GLACIER** 160 km (99 mi)
4. **BYRD GLACIER** 135 km (84 mi)
5. **NIMROD GLACIER** 135 km (84 mi)
6. **SHACKLETON GLACIER** 100 km (62 mi)
7. **LEVERETT GLACIER** 80 km (50 mi)
8. **MINNESOTA GLACIER** 65 km (60 mi)

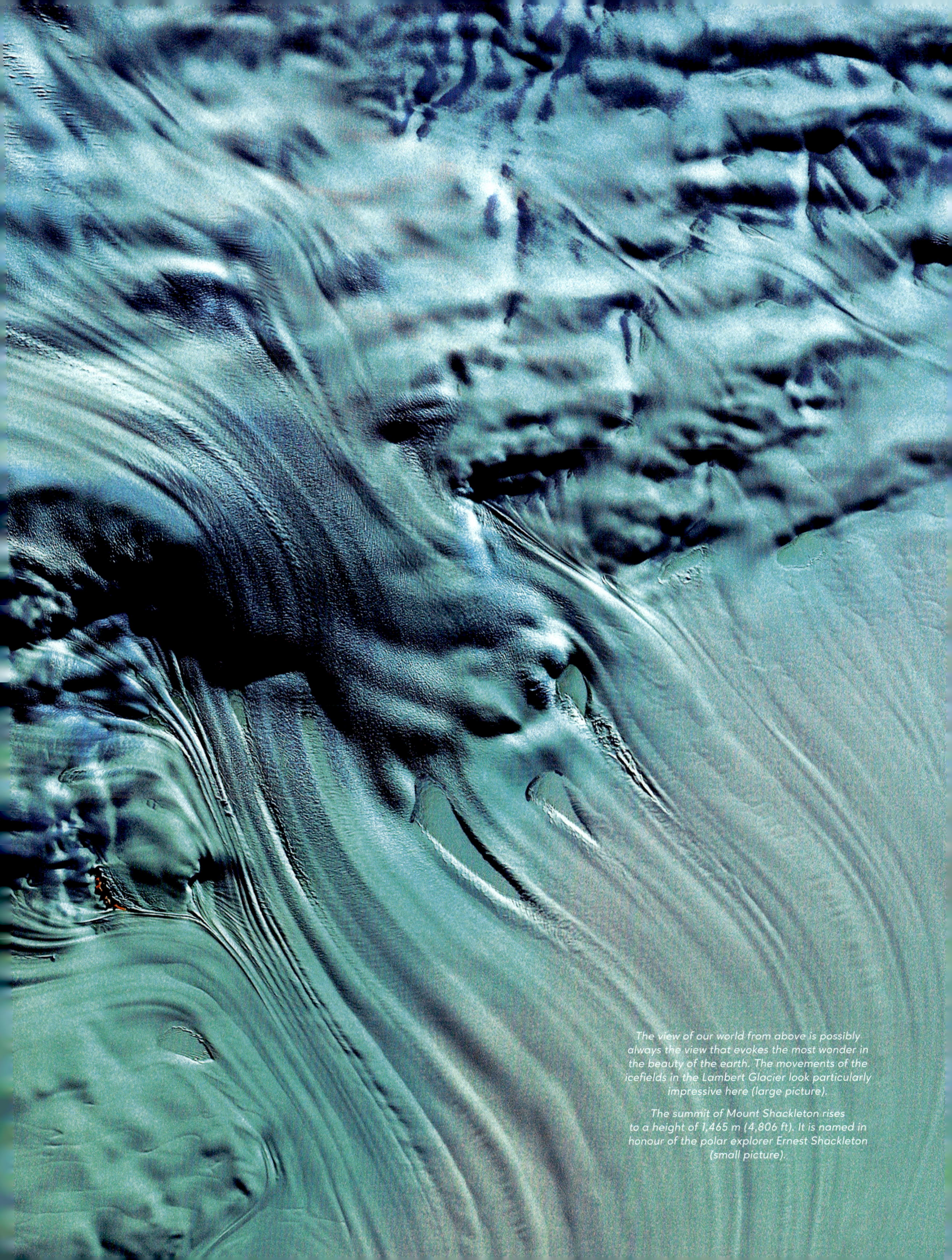

The view of our world from above is possibly always the view that evokes the most wonder in the beauty of the earth. The movements of the icefields in the Lambert Glacier look particularly impressive here (large picture).

The summit of Mount Shackleton rises to a height of 1,465 m (4,806 ft). It is named in honour of the polar explorer Ernest Shackleton (small picture).

FOUR ELEMENTS

WATER

# ICEBREAKERS

The largest and strongest icebreakers in the world

Ships that can frequently break through ice several metres thick are used for a variety of different purposes. In the polar regions, they keep navigation channels open for merchant shipping. In harsh winters, when even normally ice-free waters freeze over, they cut channels through the ice, thus opening up the routes for shipping. Some operate as research vessels; others are converted to cruise ships providing adventure trips to the Arctic or Antarctic. The *Arctic Sunrise* icebreaker is used by Greenpeace for protest actions. In order to be able to fulfil their remit, icebreakers are fitted with ultra high-performance engines and a shell that is much more resilient than that of standard ships. The bow is specially shaped to push through the ice and crush it underneath the ship. Modern icebreakers have equipment that enhances this capability and are able to pulverize even thick pack ice. Some of the ships can be made to sway by allowing large quantities of water to move back and forth between the bow and the stern, thus increasing the pressure on the ice. Several Russian nuclear-powered icebreakers have equipment which enables them to produce steam, which is used to 'thaw out' the ice in front of the bow. While these methods may be effective, they are far from environmentally friendly.

## 50 LET POBEDY – THE MOST POWERFUL ICEBREAKER IN THE WORLD

The Russian nuclear-powered Arktika giants are the only icebreakers that can pulverize ice several metres thick. The largest ship in this class, with a displacement of 25,840 tonnes (28,483 tons), is the *50 Let Pobedy*, which was commissioned on 7 April 2007. It operates in the Arctic Ocean and also provides cruises in the area. Fitted with two nuclear reactors and two steam turbines, this colossus achieves speeds of up to 21.4 knots per hour. Like all icebreakers, its construction is disproportionately wide.

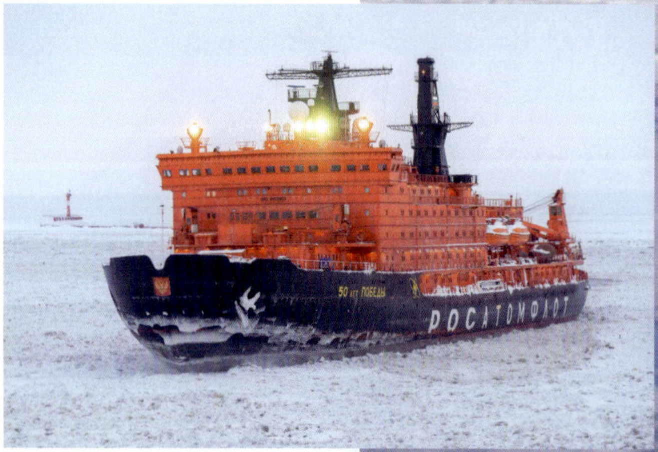

### THE LARGEST ICEBREAKERS IN THE WORLD
(Displacement capacity in tonnes (t))

1. **ROSSIYA**
Russia, 23,000 t (25,353 ton)

2. **SOVETSKY SOYUZ**
Russia, 22,120 t (24,383 ton)

3. **VIKTOR CHERNOMYRDIN**
Russia, 22,000 t (24,250 ton)

4. **WAIGATSCH**
Russia, 21,100 t (23,259 ton)

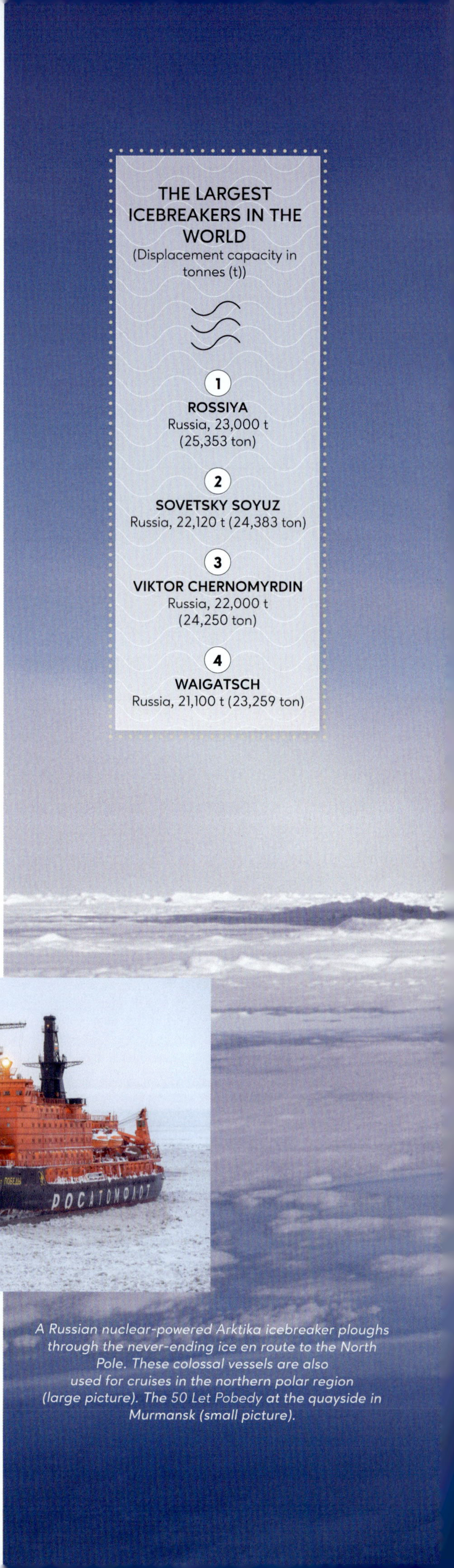

*A Russian nuclear-powered Arktika icebreaker ploughs through the never-ending ice en route to the North Pole. These colossal vessels are also used for cruises in the northern polar region (large picture). The 50 Let Pobedy at the quayside in Murmansk (small picture).*

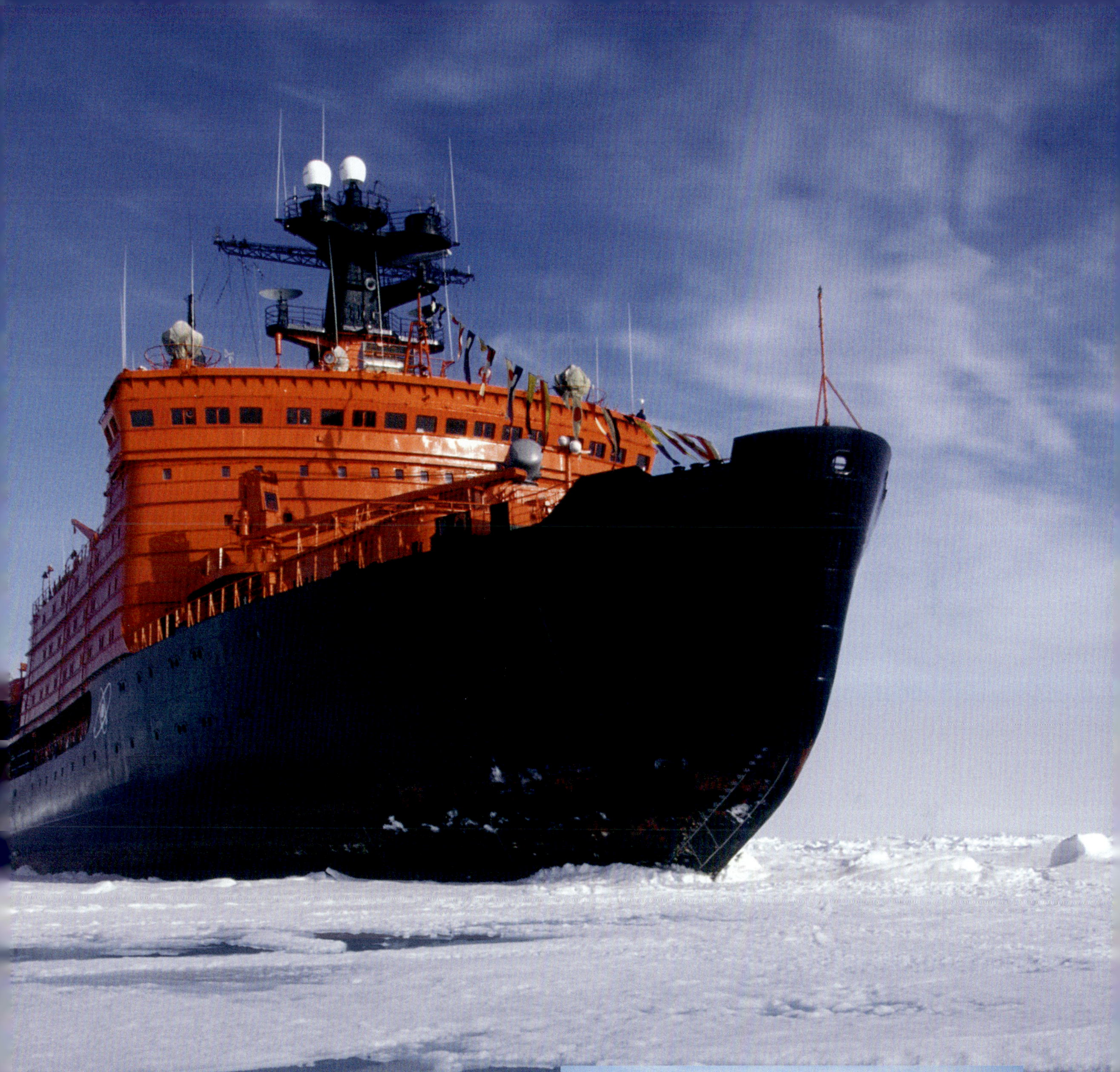

## SS MANHATTAN – OIL TANKER AND ICEBREAKER

At 307 m (1,007 ft) long and 40 m (131 ft) wide, the SS *Manhattan* is currently the largest icebreaker ever built. The oil tanker, built in the USA in 1962, was refitted some years later and equipped with a spoon-shaped icebreaker bow. In 1969, the ship was the first to undertake a run through the Northwest Passage, as those responsible wanted to test if it was possible to transport oil from northern Alaska by sea. Although the SS *Manhattan* overcame the Passage on its first trip, it failed on its second run in April 1970. The project was literally put on ice. The SS *Manhattan* continued to operate as an oil tanker until 1987, when it was scrapped.

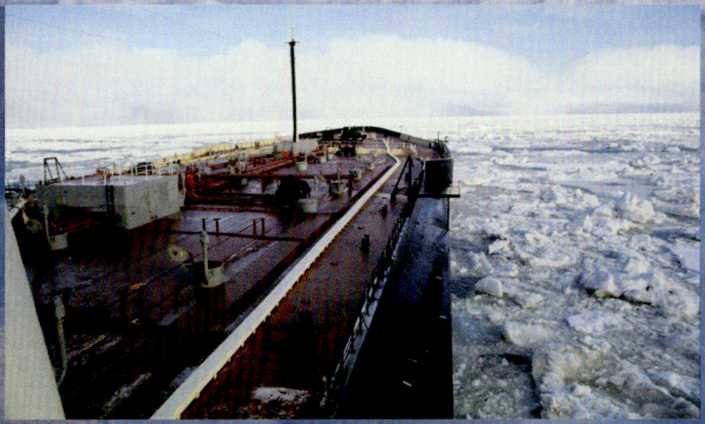

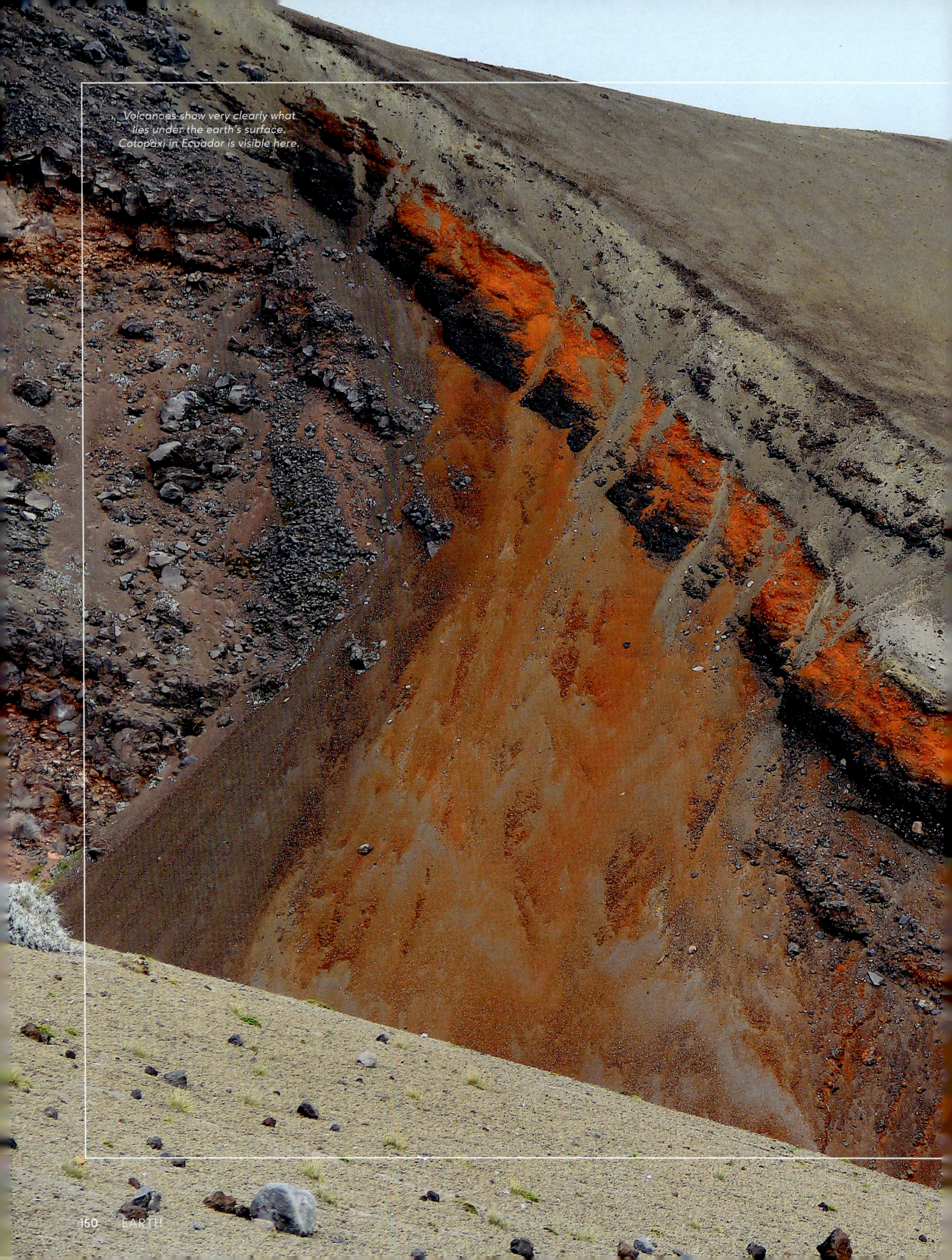

Volcanoes show very clearly what lies under the earth's surface. Cotopaxi in Ecuador is visible here.

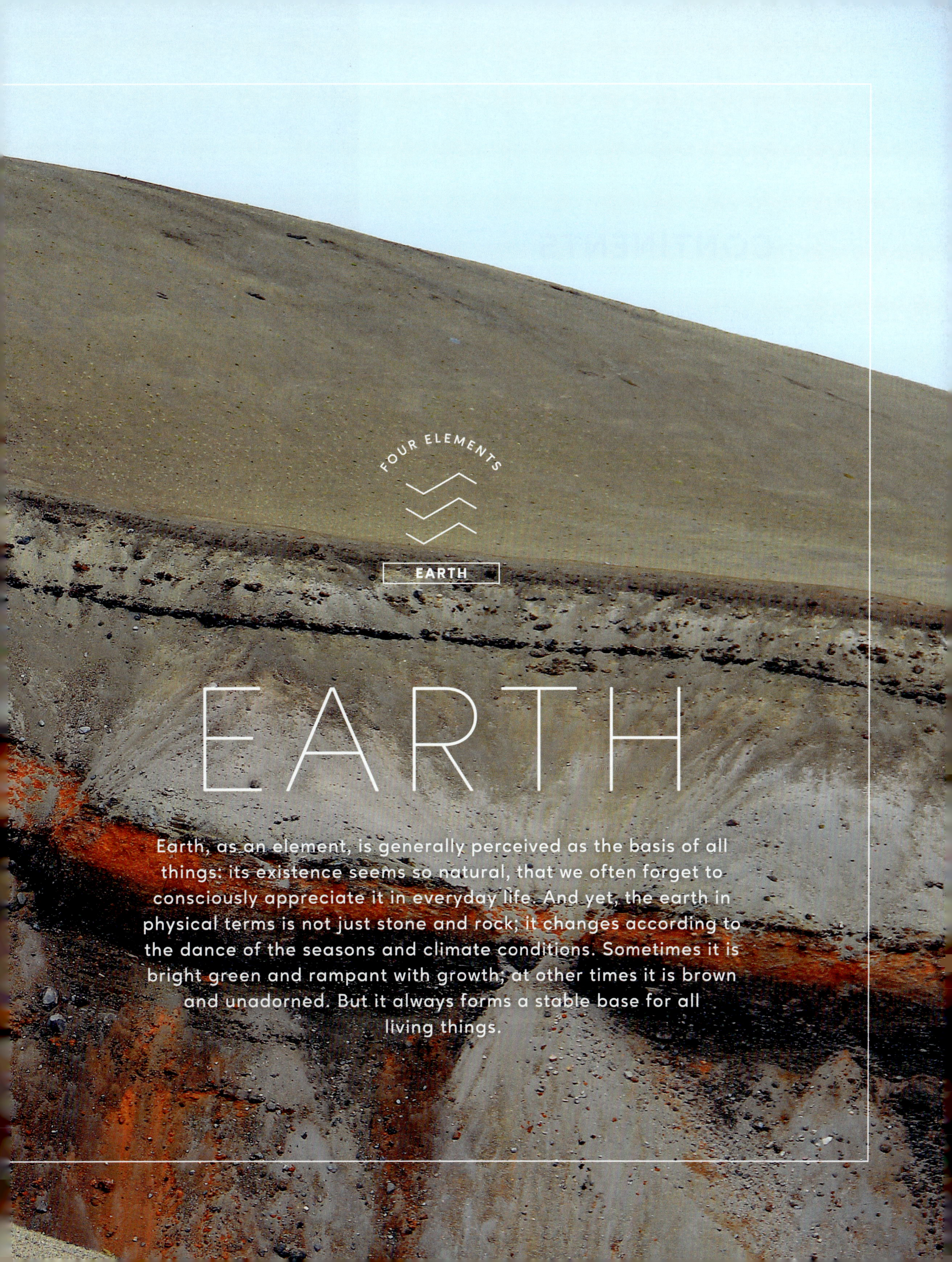

FOUR ELEMENTS

EARTH

# EARTH

Earth, as an element, is generally perceived as the basis of all things: its existence seems so natural, that we often forget to consciously appreciate it in everyday life. And yet, the earth in physical terms is not just stone and rock; it changes according to the dance of the seasons and climate conditions. Sometimes it is bright green and rampant with growth; at other times it is brown and unadorned. But it always forms a stable base for all living things.

FOUR ELEMENTS

EARTH

# CONTINENTS

Five, six, or seven? The question of how many continents there are is not as straightforward as it seems

The earth's landmasses – which make up 30 per cent of the earth's surface – are divided into continents which are more or less completely separate from each other. In principle, there is no difference between an island and a continent: both are defined as landmasses that rise up out of the water. Nevertheless, geoscientists have agreed to use the term 'continent' for landmasses above a certain size. However, there is no agreement on the question of how many continents there are. Those who view Europe and Asia, where the partition has purely historical and cultural reasons, and North and South America each as one geographical unit arrive at a total of five continents: Eurasia, Africa, America, Australia and Antarctica. Other people talk about six or seven different continents: Europe, Asia, Africa, North America, South America, Australia and Antarctica. The term 'continent' means 'connected land' (Latin: *terra continens*). Today, the division of the oceans and landmasses seems to be relatively stable, but, in terms of the earth's history, they represent just a moment in time, as the continental plates are constantly shifting.

## Surface area of the continents

1 **Asia**
44.4 million km² (17,142,936 sq mi)

2 **Africa**
30.3 million km² (11,698,895 sq mi)

3 **North America**
24.9 million km² (9,613,944 sq mi)

4 **South America**
17.8 million km² (6,872,618 sq mi)

5 **Antarctica**
13.2 million km² (5,06,548 sq mi)

6 **Europe**
10.5 million km² (4,054,073 sq mi)

7 **Australia/Oceania**
8.5 million km² (3,281,868 sq mi)

## Length of the continents' coastlines

1 **North America**
75,600 km (46,976 mi)

2 **Asia**
69,000 km (42,875 mi)

3 **Europe**
37,900 km (23,550 mi)

4 **Africa**
30,500 km (18,952 mi)

5 **South America**
28,700 km (17,833 mi)

6 **Australia/Oceania**
25,000 km (15,534 mi)

7 **Antarctica**
18,000 km (11,185 mi)

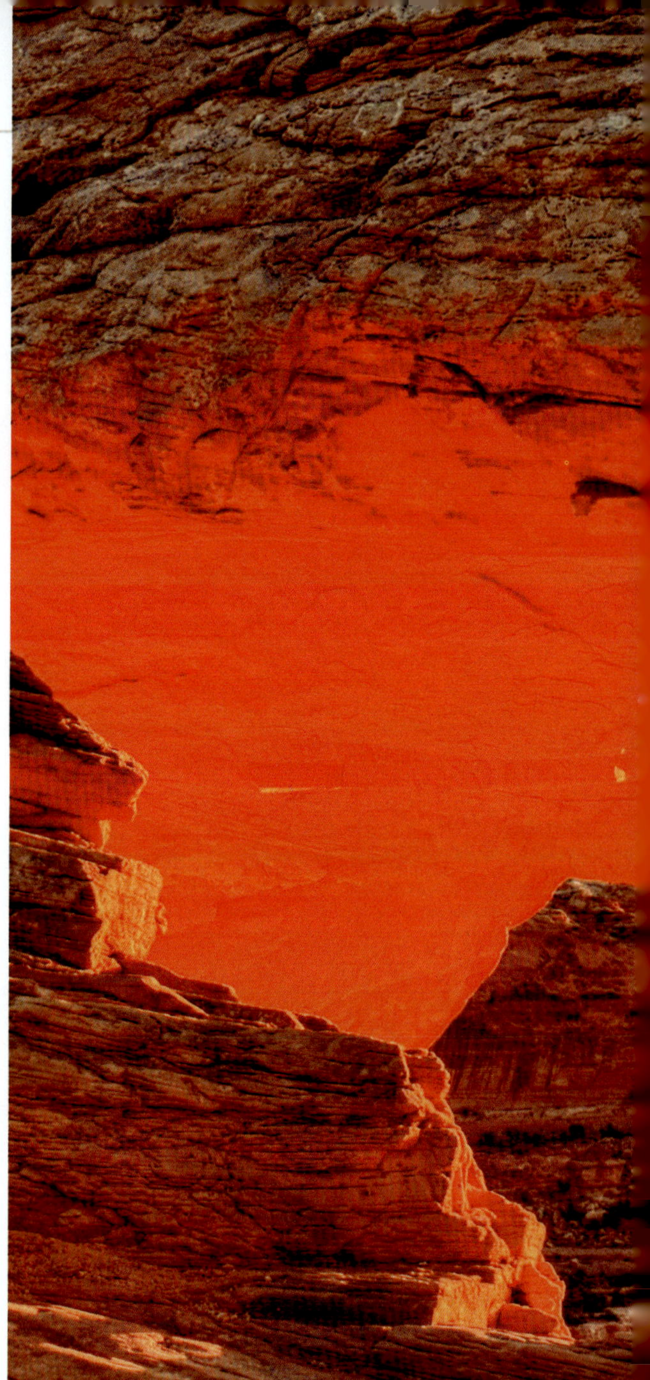

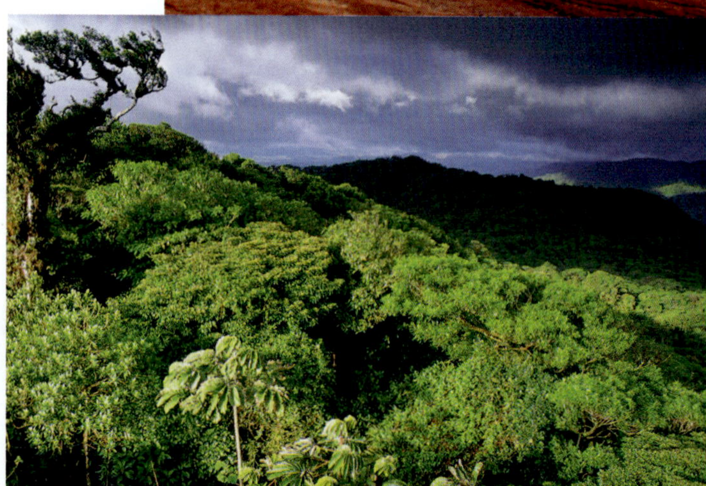

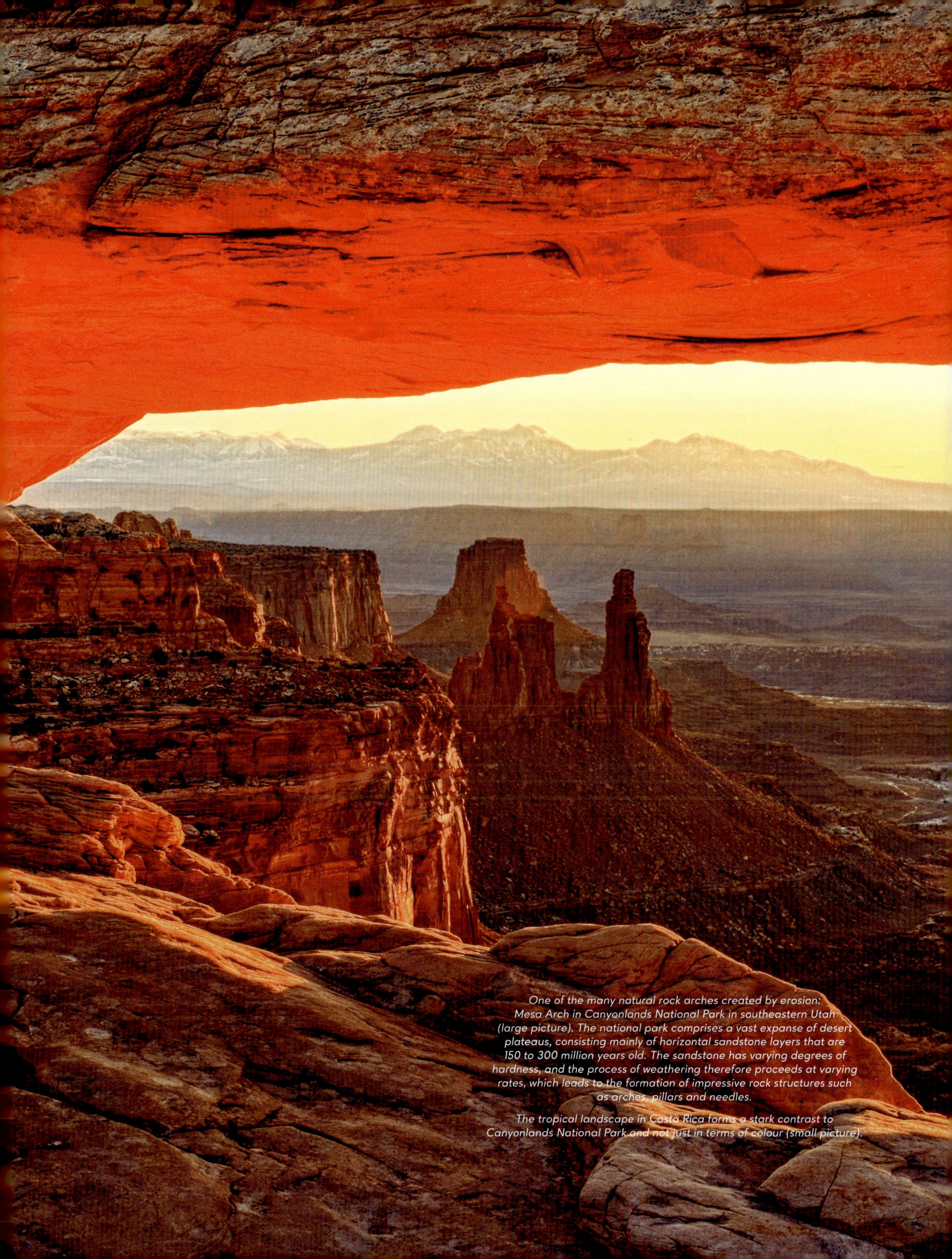
One of the many natural rock arches created by erosion: Mesa Arch in Canyonlands National Park in southeastern Utah (large picture). The national park comprises a vast expanse of desert plateaus, consisting mainly of horizontal sandstone layers that are 150 to 300 million years old. The sandstone has varying degrees of hardness, and the process of weathering therefore proceeds at varying rates, which leads to the formation of impressive rock structures such as arches, pillars and needles.

The tropical landscape in Costa Rica forms a stark contrast to Canyonlands National Park and not just in terms of colour (small picture).

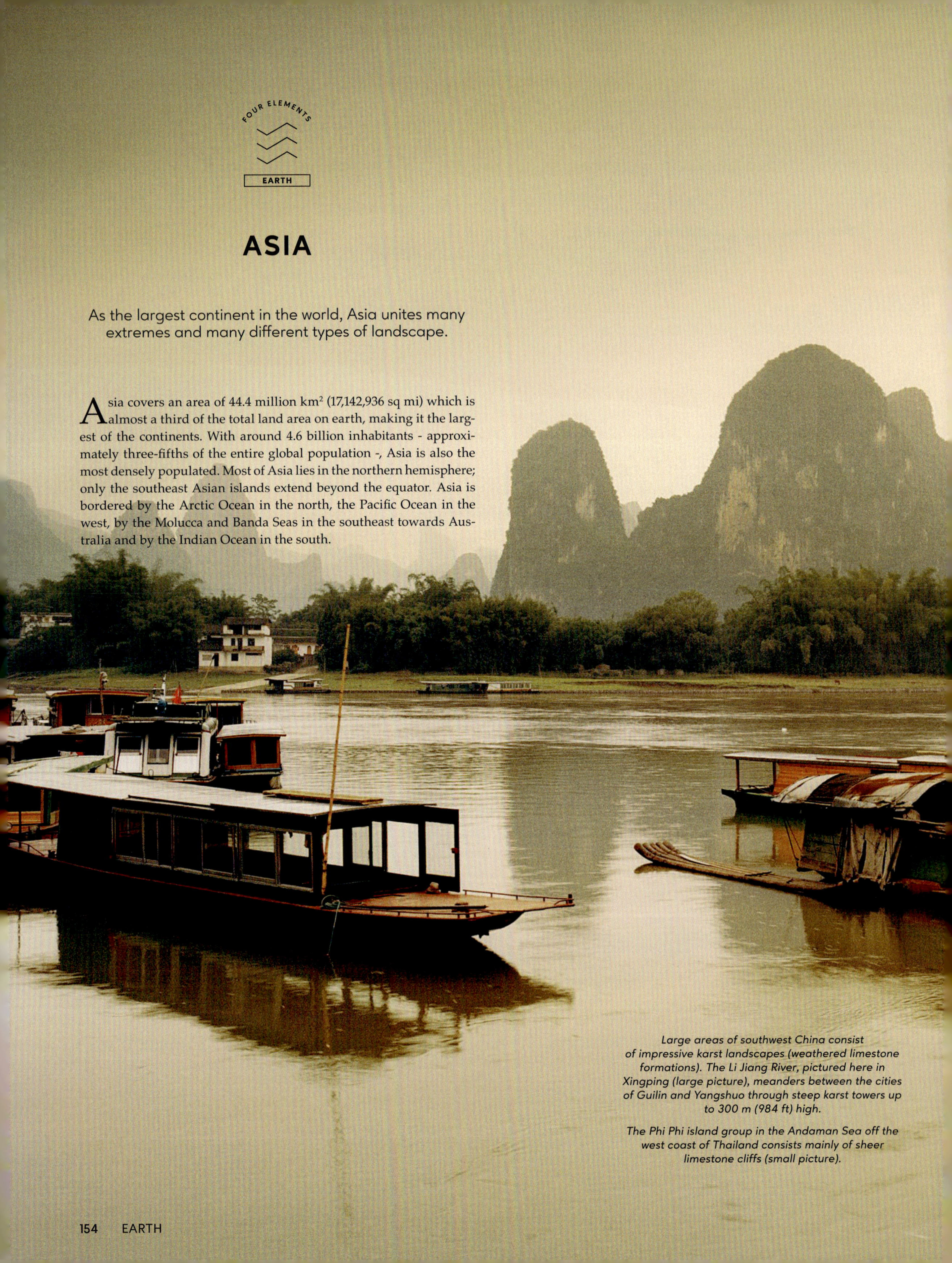

FOUR ELEMENTS

EARTH

# ASIA

As the largest continent in the world, Asia unites many extremes and many different types of landscape.

Asia covers an area of 44.4 million km² (17,142,936 sq mi) which is almost a third of the total land area on earth, making it the largest of the continents. With around 4.6 billion inhabitants - approximately three-fifths of the entire global population -, Asia is also the most densely populated. Most of Asia lies in the northern hemisphere; only the southeast Asian islands extend beyond the equator. Asia is bordered by the Arctic Ocean in the north, the Pacific Ocean in the west, by the Molucca and Banda Seas in the southeast towards Australia and by the Indian Ocean in the south.

*Large areas of southwest China consist of impressive karst landscapes (weathered limestone formations). The Li Jiang River, pictured here in Xingping (large picture), meanders between the cities of Guilin and Yangshuo through steep karst towers up to 300 m (984 ft) high.*

*The Phi Phi island group in the Andaman Sea off the west coast of Thailand consists mainly of sheer limestone cliffs (small picture).*

Asia is separated from America by the shallow Bering Strait, which connects the Arctic and Pacific Oceans; the Isthmus of Suez links it with Africa.

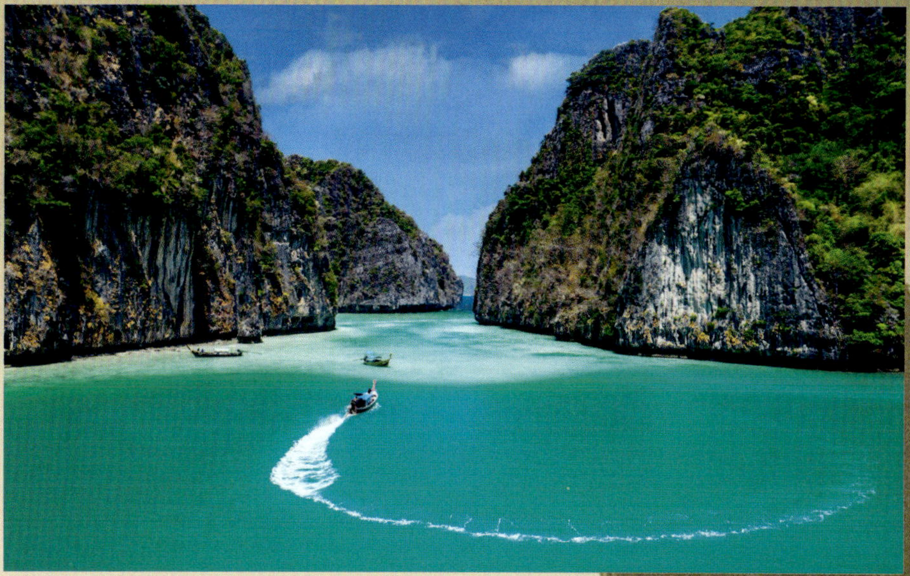

Asia is a continent of superlatives and contrasts, which explains how one finds the highest mountain and the deepest basin in the world – Mount Everest and the Dead Sea - in Asia. Because the continent stretches from the polar regions in the north to the inner tropics in the south, the major geographic regions, as well as the characteristic vegetation and climate zones, reflect an analogous diversity. The north is characterized, above all, by the lowlands of western and northern Siberia as well as the Central Siberian Plateau. In the southwest, a series of fold mountains dominate, stretching from the islands in the Mediterranean up to the Himalayas, with the eight-thousanders as the highest summits. South of this, highlands and lowlands (the Deccan and Tibetan Plateaus and the Turan Lowland) extend, alternating with basins (the Tarim Basin) and wide plains. From north to south, tundra and taiga (the largest forest area in the world), coniferous boreal forest, forest steppes, steppes, deserts (Gobi, Rub' al-Khali), arid savanna, tropical rainforest and monsoon forests can all be found.

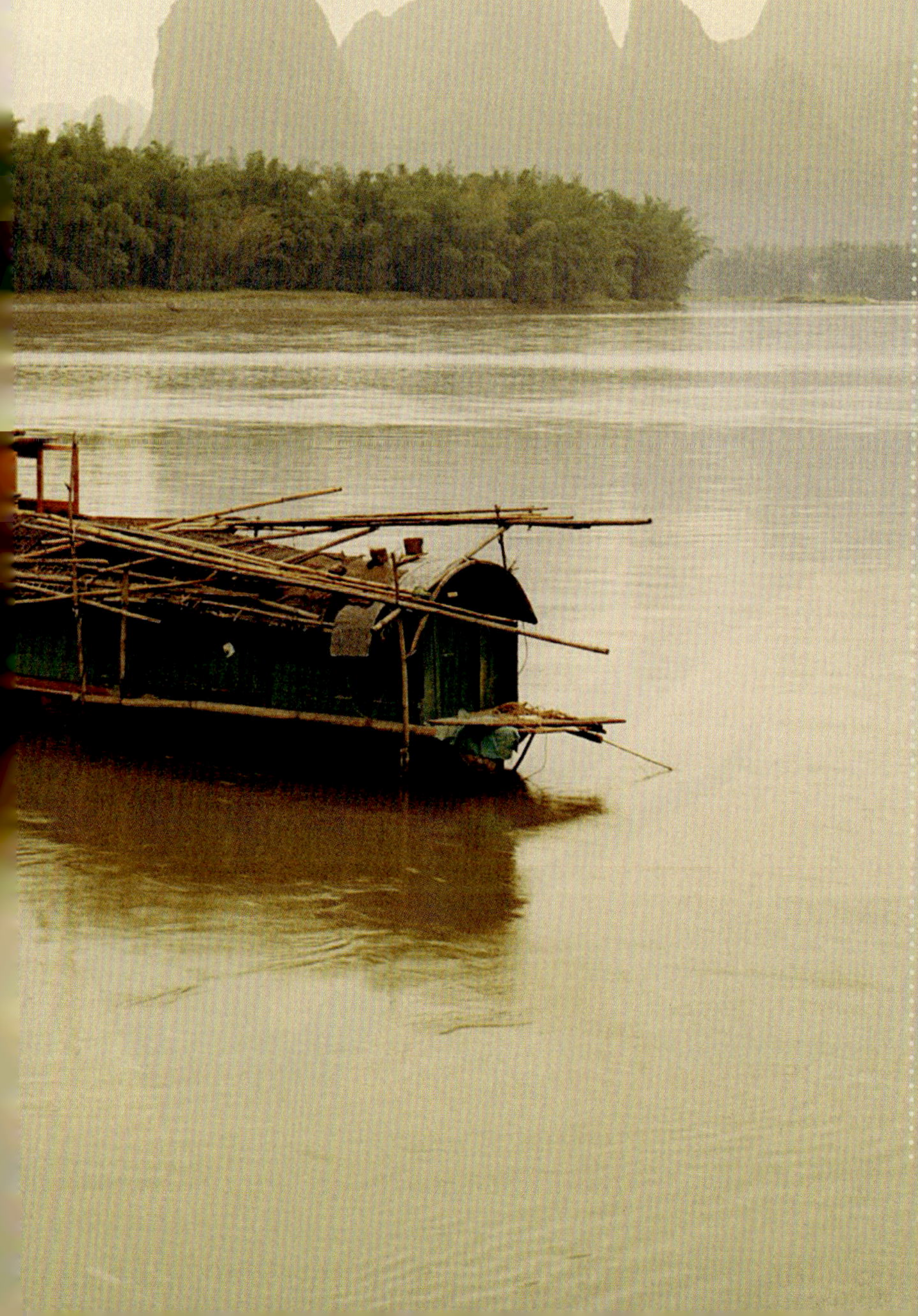

## ASIA – THE GEOGRAPHICAL SUPERLATIVES

1 **AREA**
North-south 8,500 km (5,282 mi)
West-east 11,000 km (6,835 mi)

2 **NORTHERNMOST POINT**
Cape Chelyuskin, Russia, 77°43' N

3 **SOUTHERNMOST POINT**
Southwest Pacific Malaysia, 1°16' N

4 **WESTERNMOST POINT**
Cape Baba, Turkey
Longitude 26°03' E

5 **EASTERNMOST POINT**
Cape Dezhnev, Russia
Longitude 169°4' W

6 **HIGHEST MOUNTAIN**
Himalayas
Mount Everest, 8,850 m (29,035 ft)

7 **LOWEST POINT**
Dead Sea, -400 m (-1,312 ft)

8 **LONGEST RIVER**
Yangtze Kiang, 6,380 km (3,964 mi)

9 **LARGEST LAKE**
Caspian Sea, 371,000 km²

10 **DEEPEST LAKE**
Lake Baikal, Russia, 1,637 m (5,371 ft)

11 **LARGEST ISLAND**
Borneo, 748,168 km² (288,869 sq mi)

# AFRICA

The second largest continent is often referred to as the 'cradle of humankind'.

The landmass of the African continent lies between the Indian and Atlantic Oceans. With an area of 30.3 million km$^2$ (11,698,895 sq mi) - 20 per cent of the earth's surface -, it is the second largest continent in the world in terms of its size as well as its population (around 2.3 billion, or 30.6 per cent of the global population).

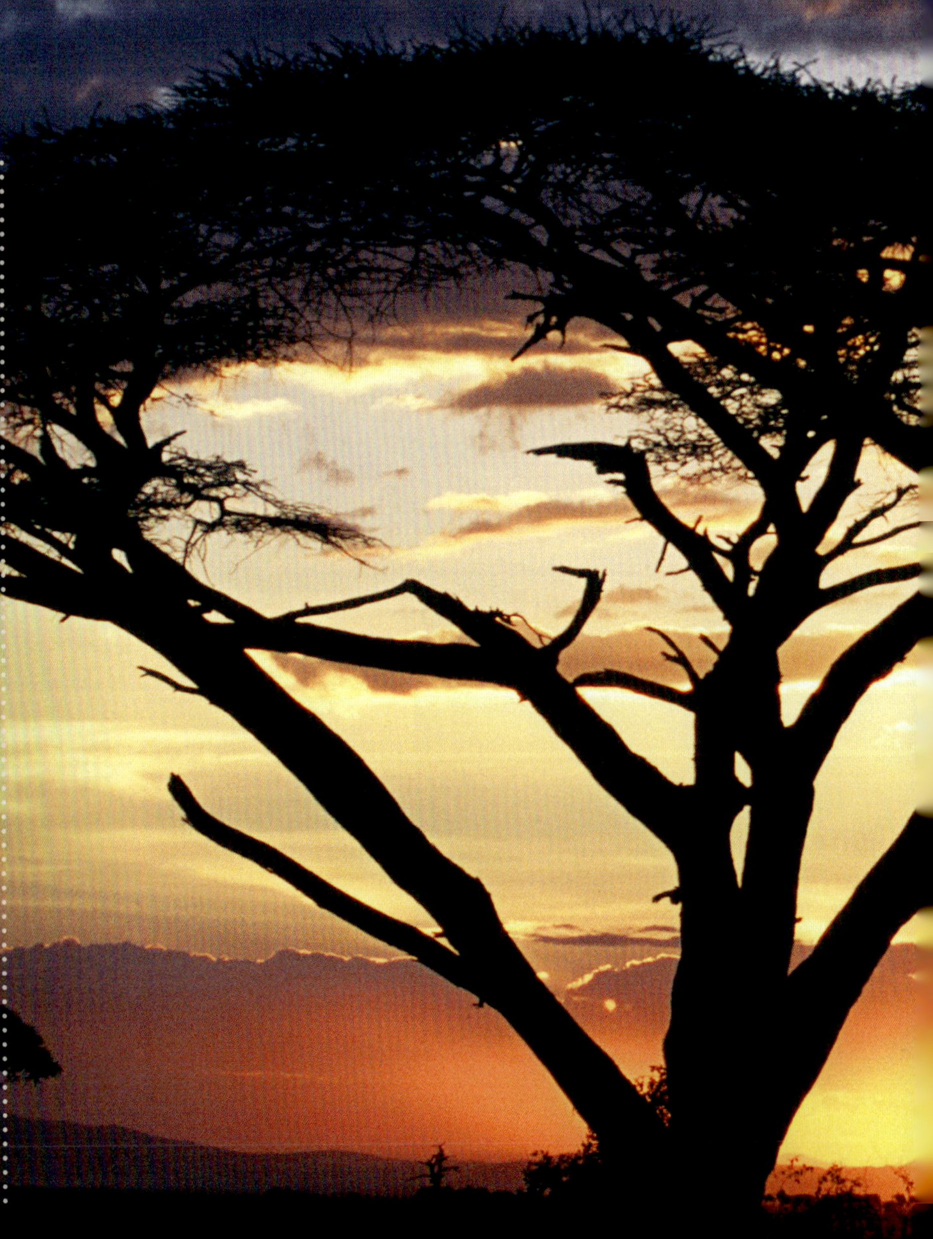

## AFRICA – THE GEOGRAPHICAL SUPERLATIVES

1. **AREA**
North-south 7,500 km (4,660 mi)
West-east 6,000 km (3,728 mi)

2. **NORTHERNMOST POINT**
Bizerta, Tunisia
37°16' S, 9°52' E

3. **SOUTHERNMOST POINT**
Cape Agulhas, South Africa
34°49' S, 20°0' E

4. **WESTERNMOST POINT**
Cape Verde Peninsula, Senegal
14°44' N, 17°31' W

5. **EASTERNMOST POINT**
Ras Hafun, Somalia
10°25' N, 51°16' E

6. **HIGHEST MOUNTAIN**
Kilimanjaro, Tanzania
Kibo, 5,895 m (19,340 ft)

7. **LOWEST POINT**
Lake Assal, Djibouti,
-153 m (-502 ft)

8. **LONGEST RIVER**
Nile, 6,671 km (4,145 mi)

9. **LARGEST LAKE**
Lake Victoria, East Africa,
68,870 km$^2$ (26,69 sq mi)

10. **DEEPEST LAKE**
Lake Tanganyika, East Africa, 1,470 m (4,823 ft)

11. **LARGEST ISLAND**
Madagascar,
587,042 km$^2$
(226,658 sq mi)

Africa stretches along both sides of the equator, and reaches its southernmost end point at the Cape of Good Hope. The Mediterranean Sea in Europe forms its northern border. At the Strait of Gibraltar, the continent approaches to within 14 km (8.7 mi) of Europe. The Red Sea forms a natural boundary with Asia, and, since 1869, the Suez Canal has formed an artificial one.

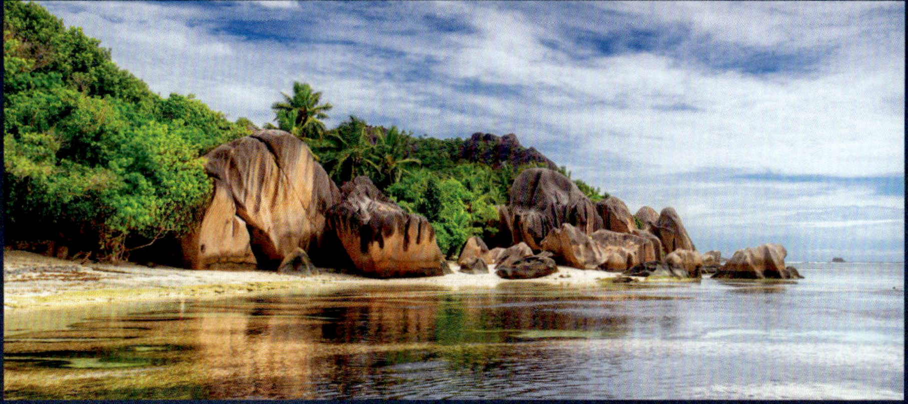

Its geographical location, positioned across the equator, also defines Africa's climatic conditions. Seventy-five per cent of its surface lies within the tropics and thus forms the largest continuous tropical landmass on earth. To the north and south, the subtropical savanna and semi-desert zones (Sahel regions) adjoin. While the Namib Desert in the south is one of the smaller examples, the Sahara Desert, the largest desert in the world, defines almost the entire northern third of the continent.

Mountain ridges and vast basins are the dominant features in Africa's relief. The Maghreb is shaped by the Sahara Basin and the massifs of the Al Hajar and Atlas Mountains. The Niger River and Lake Chad form vast basins in the north, while, in central and southern Africa, the Congo River and the Okavango River do the same.

Apart from Madagascar in the east, there are only a few offshore islands in the Gulf of Guinea – otherwise the coast of the African continent is mainly uninterrupted.

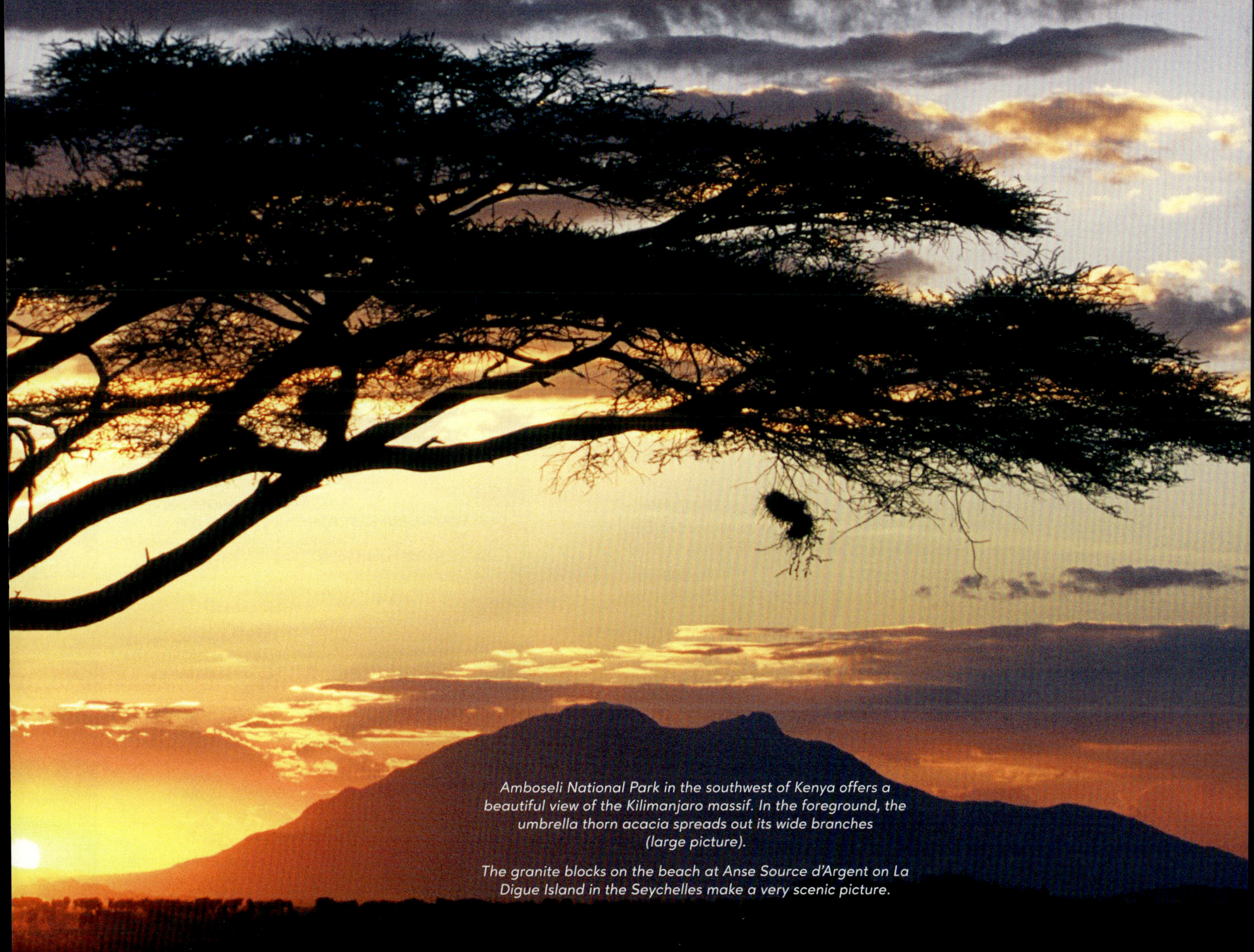

*Amboseli National Park in the southwest of Kenya offers a beautiful view of the Kilimanjaro massif. In the foreground, the umbrella thorn acacia spreads out its wide branches (large picture).*

*The granite blocks on the beach at Anse Source d'Argent on La Digue Island in the Seychelles make a very scenic picture.*

FOUR ELEMENTS

EARTH

# NORTH AND CENTRAL AMERICA

From Canada to Panama – North and Central America stretch from the cold north to the Caribbean

North America is the third largest continent and the largest landmass in the western hemisphere. It stretches from the glacial world of Alaska down to the Caribbean and encompasses almost 25 million $km^2$ (9,652,554 sq mi), including Greenland, Mexico, the countries of the Central American isthmus and the islands of the Caribbean – thereby occupying over 16 per cent of all the land on earth. More than three-quarters of this landmass is accounted for by two countries: the USA and Canada. North America is bounded in the north by the Arctic Ocean, in the east by the Atlantic Ocean and in the west by the Pacific.

North America stretches from the Arctic Ocean to the Caribbean Sea. It is connected to South America by the Central American isthmus, which is only 50 km (31 mi) wide at its narrowest point, whereby the latter is as much part of the North American landmass, physically and geographically, as the islands of the American Mediterranean (the Caribbean Sea and the Gulf of Mexico') and Greenland are. The continent extends 6,000 km (3,728 mi) from west to east, that is, from Alaska to Nova Scotia.

The American landscape is immensely varied. Three major geographic regions running from north to south can be distinguished: firstly, in the west, the Cordilleras include several mountain ranges and with a length of 15,000 km (9,320 mi), are the longest fold mountains in the world. The geologically young Rocky Mountains are the main mountain range of the North American Cordilleras. The second major geographic region rises in the shape of the Appalachian Mountains, a forested low mountain range in the east, surrounded by wide coastal lowlands. The Appalachians are approximately 400 million years old and are therefore among the oldest mountains on earth. Thirdly, the inland plains situated between the mountain ranges, feature vast, open landscapes (the Great Plains) and the five Great Lakes, which, with a total area of 244,000 $km^2$ (94,209 sq mi), form the largest inland freshwater area in the world.

Deep inlets are a characteristic feature of the North American coastline. The continent is tectonically very active as it is the meeting place of the North American and Pacific plates; volcanic eruptions and earthquakes are frequent.

The average population density is low compared to the industrial countries of Europe; its only densely populated urban centres can be found on the east coast of the USA.

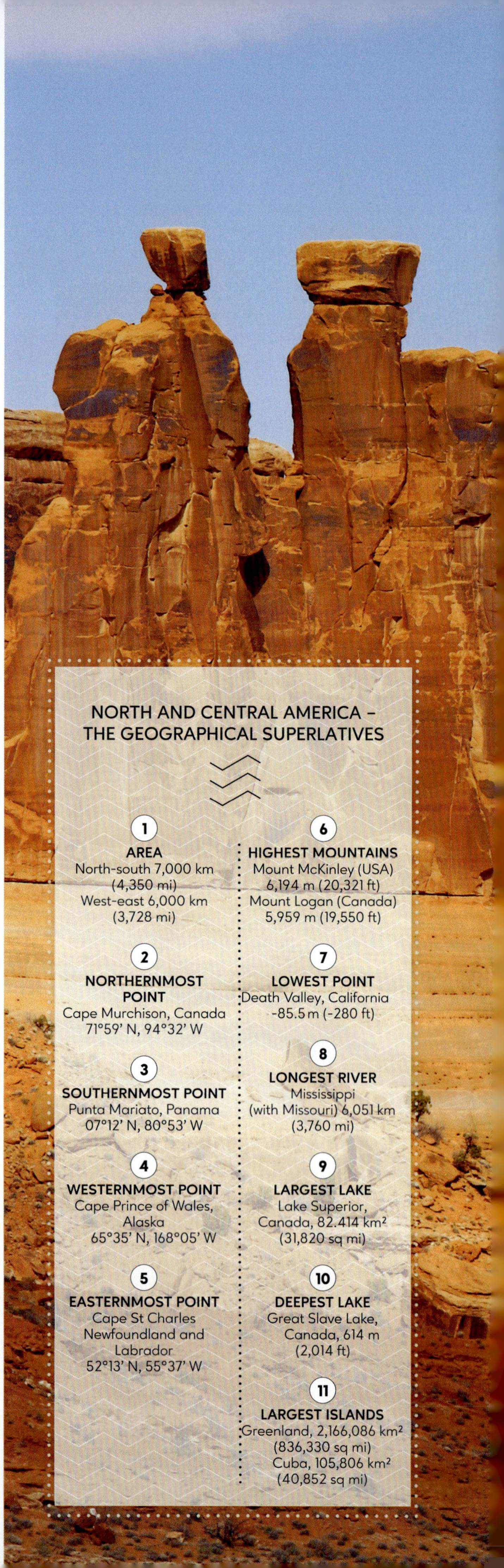

## NORTH AND CENTRAL AMERICA – THE GEOGRAPHICAL SUPERLATIVES

1 **AREA**
North-south 7,000 km (4,350 mi)
West-east 6,000 km (3,728 mi)

2 **NORTHERNMOST POINT**
Cape Murchison, Canada
71°59' N, 94°32' W

3 **SOUTHERNMOST POINT**
Punta Mariato, Panama
07°12' N, 80°53' W

4 **WESTERNMOST POINT**
Cape Prince of Wales, Alaska
65°35' N, 168°05' W

5 **EASTERNMOST POINT**
Cape St Charles Newfoundland and Labrador
52°13' N, 55°37' W

6 **HIGHEST MOUNTAINS**
Mount McKinley (USA) 6,194 m (20,321 ft)
Mount Logan (Canada) 5,959 m (19,550 ft)

7 **LOWEST POINT**
Death Valley, California
-85.5 m (-280 ft)

8 **LONGEST RIVER**
Mississippi (with Missouri) 6,051 km (3,760 mi)

9 **LARGEST LAKE**
Lake Superior, Canada, 82.414 $km^2$ (31,820 sq mi)

10 **DEEPEST LAKE**
Great Slave Lake, Canada, 614 m (2,014 ft)

11 **LARGEST ISLANDS**
Greenland, 2,166,086 $km^2$ (836,330 sq mi)
Cuba, 105,806 $km^2$ (40,852 sq mi)

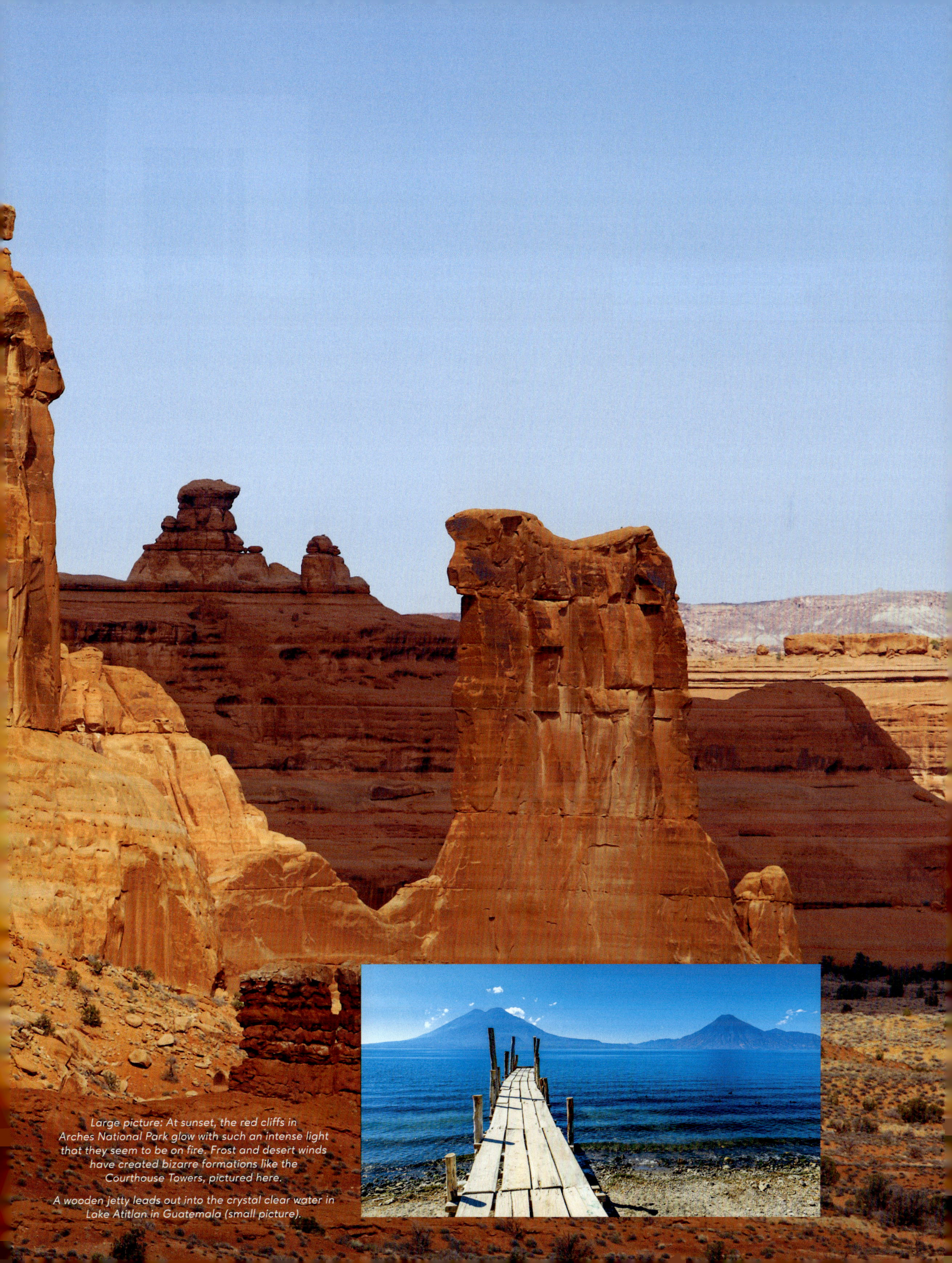

*Large picture: At sunset, the red cliffs in Arches National Park glow with such an intense light that they seem to be on fire. Frost and desert winds have created bizarre formations like the Courthouse Towers, pictured here.*

*A wooden jetty leads out into the crystal clear water in Lake Atitlan in Guatemala (small picture).*

# WHO CONSTRUCTS THE LOVELIEST ARCHES – HUMANS OR NATURE?

FOUR ELEMENTS

EARTH

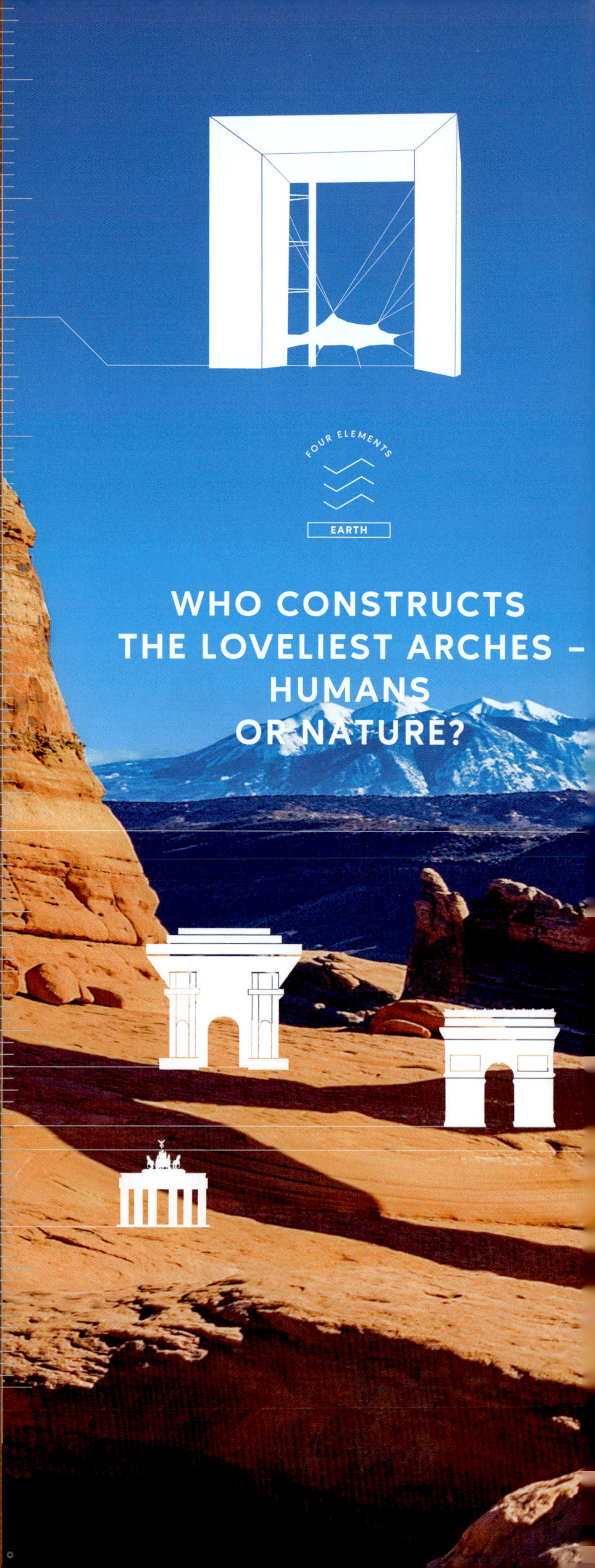


**GRANDE ARCHE**
Height of arch: approx. 88 m (289 ft)
Construction time: approx. 5 yrs
Visitors per yr: approx. 1.8 million

## GRAND ARCHE

The Grande Arche was constructed in the 1980s as a modern version of a classical triumphal arch in the high-rise district of La Défense west of Paris. The almost cubiform structure houses 35 floors of offices and conference rooms in the main building. The white sail stretched beneath the arch lends the austere construction a sense of lightness and transparency.

**KAESŎNMUN (PJONGYANG)**
Height of arch: approx. 27 m (89 ft)
Construction time: approx. 1 yr
Visitors per yr: approx. 240,000

**ARC DE TRIOMPHE**
Height of arch: approx. 29 m (95 ft)
Construction time: approx. 30 yrs
Visitors per yr: approx. 1.8 million

**DELICATE ARCH**
Height of arch: approx. 20 m (66 ft)
Period of creation: approx. 100 million yrs
Visitors per yr: approx. 1.5 million

**BRANDENBURGER TOR**
Height of arch: approx. 13.5 m (44 ft)
Construction period: approx. 3 yrs
Visitors per yr: approx. 6 million

## DELICATE ARCH

Arches National Park in the southwest of the USA is well-known for its natural sandstone arches, numbering more than 2,000. The most famous of these and one of the most-photographed, particularly at sunset, is the Delicate Arch.

The extreme desert climate in northern Utah, the deep porous salt deposits and harder upper sandstone layers have created this unique natural architectural feature. Wind, heat, frost and subsequent thaws over the course of millions of years resulted in the creation of fissures in the rock, the gradual leaching of the salt deposits, and, finally, the creation of the arches.
A hike along the Delicate Arch Trail to this fascinating free-standing rock arch is one of the main attractions on a visit to Arches National Park. The route passes over fairly smooth rocks to a height of around 150 m (492 ft) – this can be very challenging when the temperature is above 30° C (86 F), and it can also be slippery when it rains. The hike takes two to three hours, but with the fantastic view at the top, every struggle or strain is forgotten.

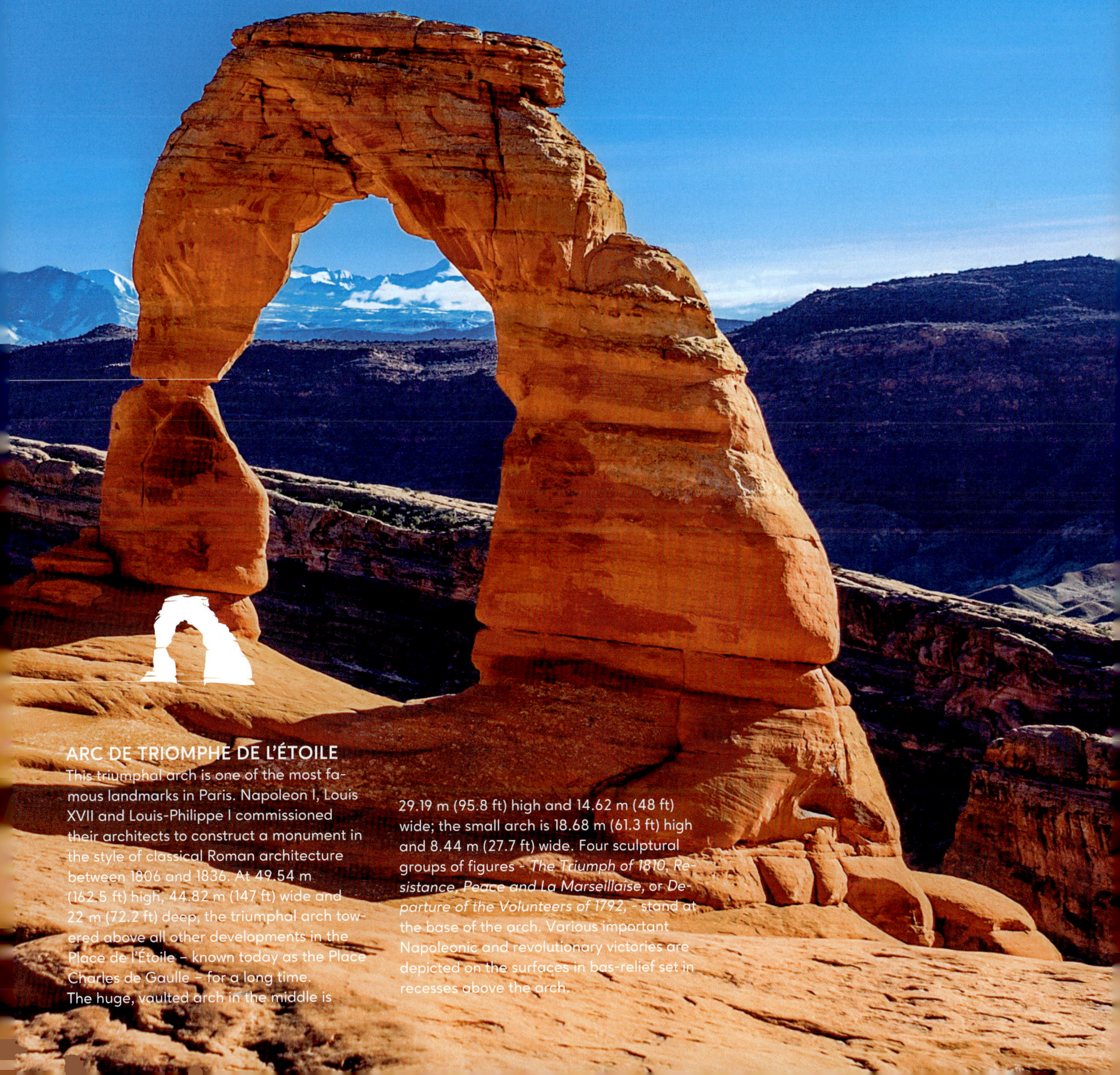

## ARC DE TRIOMPHE DE L'ÉTOILE

This triumphal arch is one of the most famous landmarks in Paris. Napoleon I, Louis XVII and Louis-Philippe I commissioned their architects to construct a monument in the style of classical Roman architecture between 1806 and 1836. At 49.54 m (162.5 ft) high, 44.82 m (147 ft) wide and 22 m (72.2 ft) deep, the triumphal arch towered above all other developments in the Place de l'Étoile – known today as the Place Charles de Gaulle – for a long time.
The huge, vaulted arch in the middle is 29.19 m (95.8 ft) high and 14.62 m (48 ft) wide; the small arch is 18.68 m (61.3 ft) high and 8.44 m (27.7 ft) wide. Four sculptural groups of figures - *The Triumph of 1810, Resistance, Peace and La Marseillaise*, or *Departure of the Volunteers of 1792*, - stand at the base of the arch. Various important Napoleonic and revolutionary victories are depicted on the surfaces in bas-relief set in recesses above the arch.

FOUR ELEMENTS

EARTH

# SOUTH AMERICA

The population density in South America is comparatively low, but, instead, it does have a great number of different climates

South America is the fourth largest continent in the world, after Asia, Africa and North America. With an overall area of 17.8 million km² (6,872,618 sq mi), it accounts for around 12 per cent of the earth's land surface. In the north, South America is linked to North America by the tongue of land in Central America; in the south, it reaches closer to Antarctica than any other continent.

In contrast to other continents, South America seems a relatively solid landmass; the coast has few bays and the relief is fairly uniform, except for the mountainous region of the Andes. Three major areas define the surface of this continent: the high mountain range of the Andes in the west, three river plains and three mountainous areas east of the Andes.

The towering cordillera also continues in the southern part of the Americas along the Pacific coast. It reaches its widest point with the Andes, the second highest mountain range in the world after the Himalayas – approximately 700 km (435 mi) west to east on a level with the Tropic of Capricorn. The Andes run north to south down the entire continent of South America and are the longest mountain chain in the world. Between the mountain ridges of the cordillera, the upland plains spread out. To the east, it connects to the lowlands. Between the mountain region of Guiana in the north and the Brazilian mountain region, the Amazon Basin stretches out, with the Amazon running through it. This lowland covers 4 million km² (1,544,409 sq mi) and is the largest tropical rainforest in the world. The alluvial plains of the Gran Chaco and the Pampas connect with it; these lead on to Patagonia, the third mountain region. The lowlands of the Orinoco lie in the far north of the continent, and the river plain of the Rio Paraguay and the Rio Paraná lies in the south of the continent and merges into a subtropical alluvial landscape.

The vast north-south range of the continent gives rise to extreme variations in climate. So in the south, the humid heat of the Amazon Rainforest is followed by the cool, temperate region of Patagonia, and the Tierra del Fuego latitudes are influenced by the polar regions. The twelve countries of South America, the Galápagos Islands and the English and French overseas territories of the Falkland Islands and French Guiana are home to more than 370 million people, amounting to five per cent of the global population.

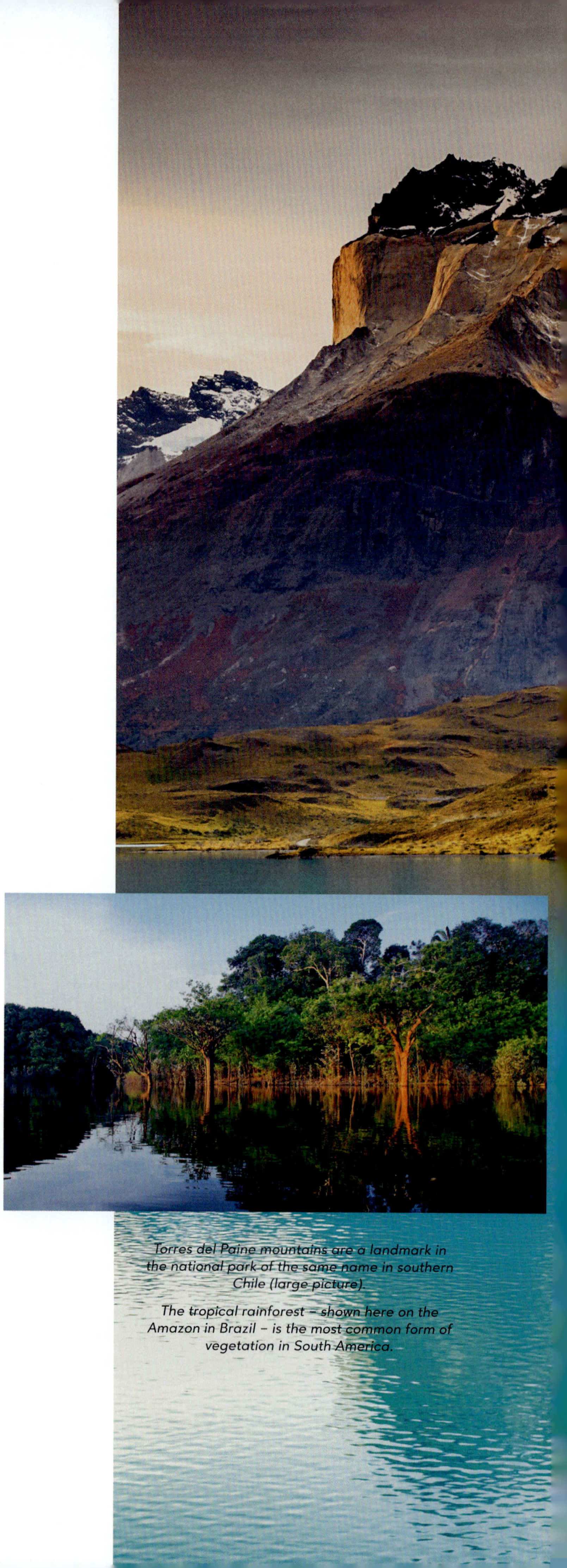

*Torres del Paine mountains are a landmark in the national park of the same name in southern Chile (large picture).*

*The tropical rainforest – shown here on the Amazon in Brazil – is the most common form of vegetation in South America.*

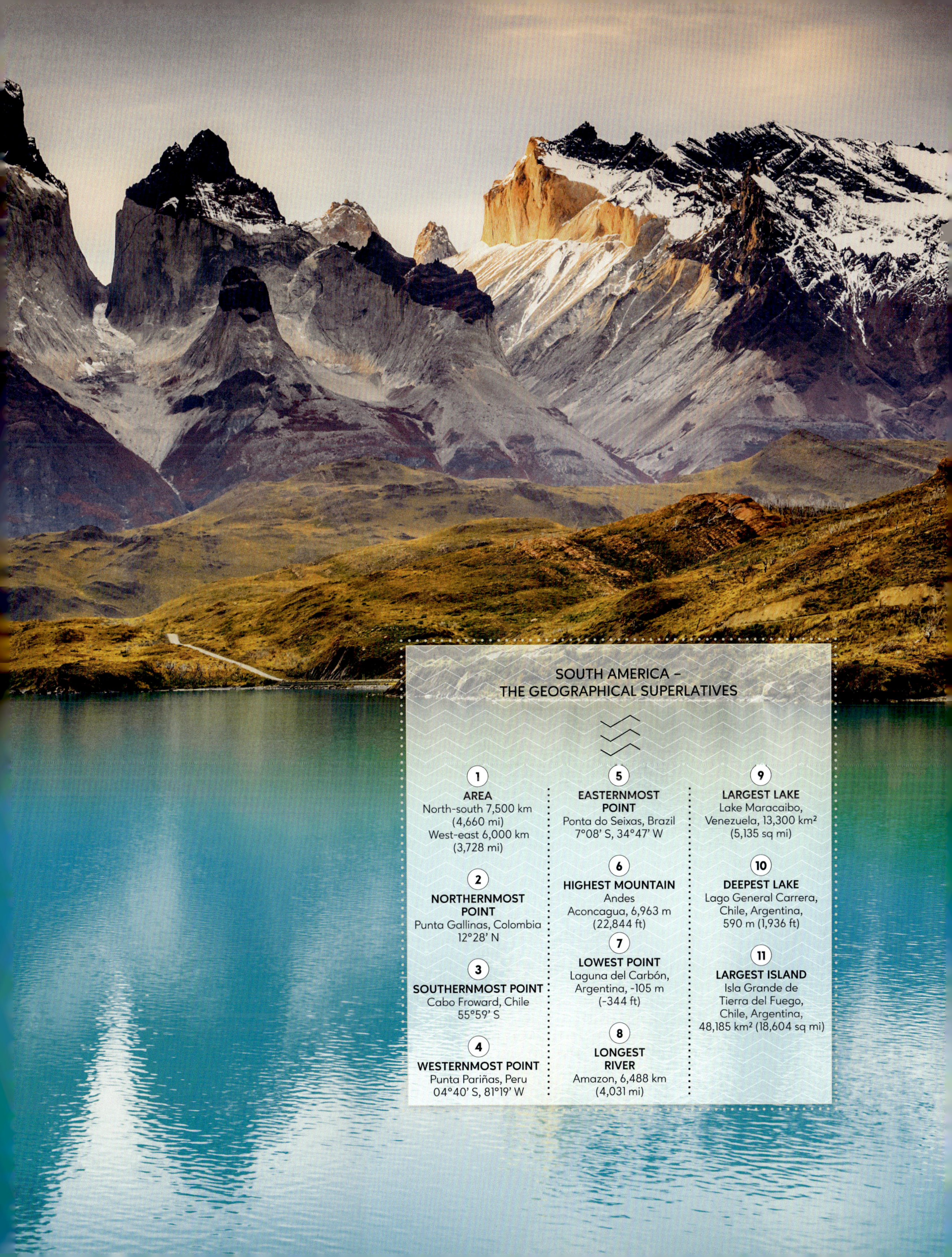

SOUTH AMERICA –
THE GEOGRAPHICAL SUPERLATIVES

1
**AREA**
North-south 7,500 km (4,660 mi)
West-east 6,000 km (3,728 mi)

2
**NORTHERNMOST POINT**
Punta Gallinas, Colombia
12°28' N

3
**SOUTHERNMOST POINT**
Cabo Froward, Chile
55°59' S

4
**WESTERNMOST POINT**
Punta Pariñas, Peru
04°40' S, 81°19' W

5
**EASTERNMOST POINT**
Ponta do Seixas, Brazil
7°08' S, 34°47' W

6
**HIGHEST MOUNTAIN**
Andes
Aconcagua, 6,963 m (22,844 ft)

7
**LOWEST POINT**
Laguna del Carbón, Argentina, -105 m (-344 ft)

8
**LONGEST RIVER**
Amazon, 6,488 km (4,031 mi)

9
**LARGEST LAKE**
Lake Maracaibo, Venezuela, 13,300 km² (5,135 sq mi)

10
**DEEPEST LAKE**
Lago General Carrera, Chile, Argentina, 590 m (1,936 ft)

11
**LARGEST ISLAND**
Isla Grande de Tierra del Fuego, Chile, Argentina, 48,185 km² (18,604 sq mi)

FOUR ELEMENTS

EARTH

# ANTARCTICA

It is fortunate that the peaceful use of Antarctica is regulated by a treaty

The continent of Antarctica covers an area of 12.4 million $km^2$ (4,787,666 sq mi); only 200,000 $km^2$ (77,220 sq mi) of this is ice-free rock. This landmass, which is the size of Europe, is covered by a layer of ice approximately 2 km (1.25 mi) thick. Towards the middle of the continent, this layer can be over 4.5 km (2.80 mi) thick, making Antarctica also the highest continent.

The Antarctic region extends to the Antarctic Circle on a latitude of 66°33′ south. It is bounded by the Antarctic Convergence or Antarctic Polar Front at a latitude of 50° south. The size of the continent varies depending on the season. In the winter half-year, it grows to around double its size due to the increase in the ice mass.

The Antarctic, also called the southern polar region, is the southernmost part of the earth and extends almost completely beyond the other side of the Antarctic Circle. Establishing the borders of the Antarctic is not done according to the mainland coastline but according to the Antarctic Convergence, a clearly delineated zone on the southern margins of the Atlantic, Indian and Pacific Oceans. The Convergence lies between 48 and 60° latitude south, where colder, northward-flowing waters from the Antarctic meet warmer, southward-flowing waters. If, in calculating the area, the land domains of offshore shelf ice plateaus are also included, with the sea and the islands pertaining to them, the total area arrived at is 53 million $km^2$ (20,463,414 sq mi). Weddell Sea on the Atlantic side and Ross Sea on the Pacific side cut deep into the Antarctic mainland. These incisions have led to the division of the continental mass into East Antarctica, with an area of more than 10 million $km^2$ (3,861,021 sq mi), and West Antarctica, with 2.3 million $km^2$ (888,035 sq mi). On this, the coldest continent, more than 90 per cent of the total global ice mass is concentrated around the South Pole. The volume of this ice sheet is estimated to be 24 million $km^3$ (5,279,264.5 billion gal), making it the largest continuous ice mass in the world. The thick ice layer covering Antarctica moves to the margins as shelf ice or flows out as icebergs and ice floes ('calving').

The Transantarctic Mountain range (the fifth longest cordillera in the world), runs between East and West Antarctica over a length of 3,500 km (2,175 mi) and reaches heights of over 4,500 m (14,764 ft). The McMurdo Dry Valleys form part of the mountain range – an area of just under 5,000 $km^2$ (1,930 sq mi) in East Antarctica, which has been ice-free for millions of years and is dryer than the Sahara or Atacama deserts. However hostile to life the dry valleys may be, living things can still be found here, albeit only microscopically small ones.

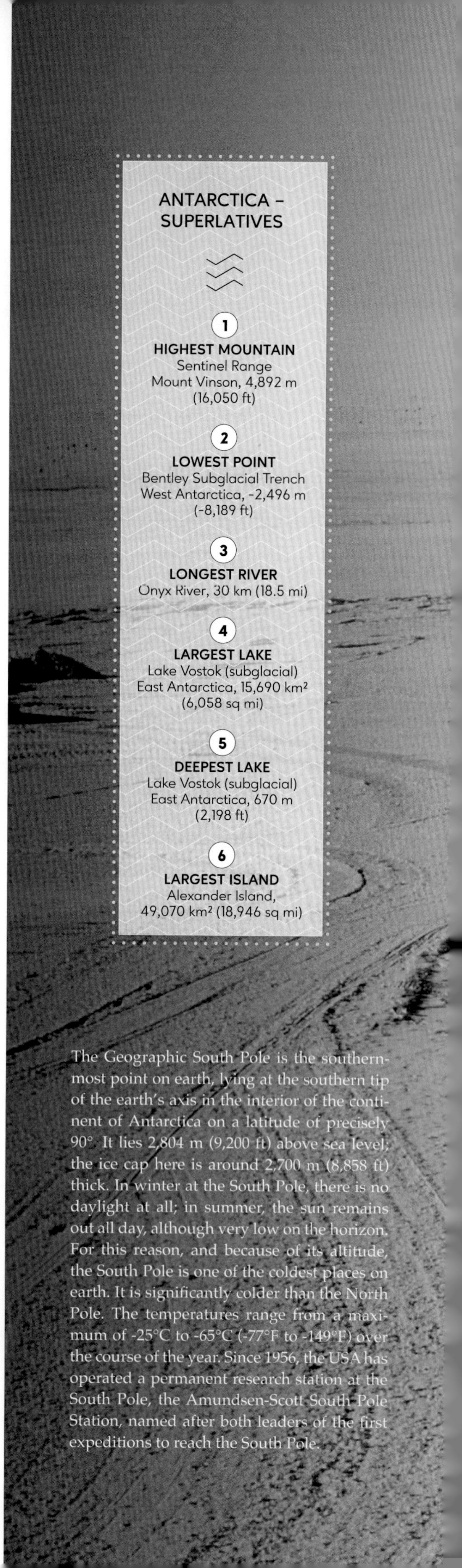

## ANTARCTICA – SUPERLATIVES

1

**HIGHEST MOUNTAIN**
Sentinel Range
Mount Vinson, 4,892 m
(16,050 ft)

2

**LOWEST POINT**
Bentley Subglacial Trench
West Antarctica, -2,496 m
(-8,189 ft)

3

**LONGEST RIVER**
Onyx River, 30 km (18.5 mi)

4

**LARGEST LAKE**
Lake Vostok (subglacial)
East Antarctica, 15,690 $km^2$
(6,058 sq mi)

5

**DEEPEST LAKE**
Lake Vostok (subglacial)
East Antarctica, 670 m
(2,198 ft)

6

**LARGEST ISLAND**
Alexander Island,
49,070 $km^2$ (18,946 sq mi)

The Geographic South Pole is the southernmost point on earth, lying at the southern tip of the earth's axis in the interior of the continent of Antarctica on a latitude of precisely 90°. It lies 2,804 m (9,200 ft) above sea level; the ice cap here is around 2,700 m (8,858 ft) thick. In winter at the South Pole, there is no daylight at all; in summer, the sun remains out all day, although very low on the horizon. For this reason, and because of its altitude, the South Pole is one of the coldest places on earth. It is significantly colder than the North Pole. The temperatures range from a maximum of -25°C to -65°C (-77°F to -149°F) over the course of the year. Since 1956, the USA has operated a permanent research station at the South Pole, the Amundsen-Scott South Pole Station, named after both leaders of the first expeditions to reach the South Pole.

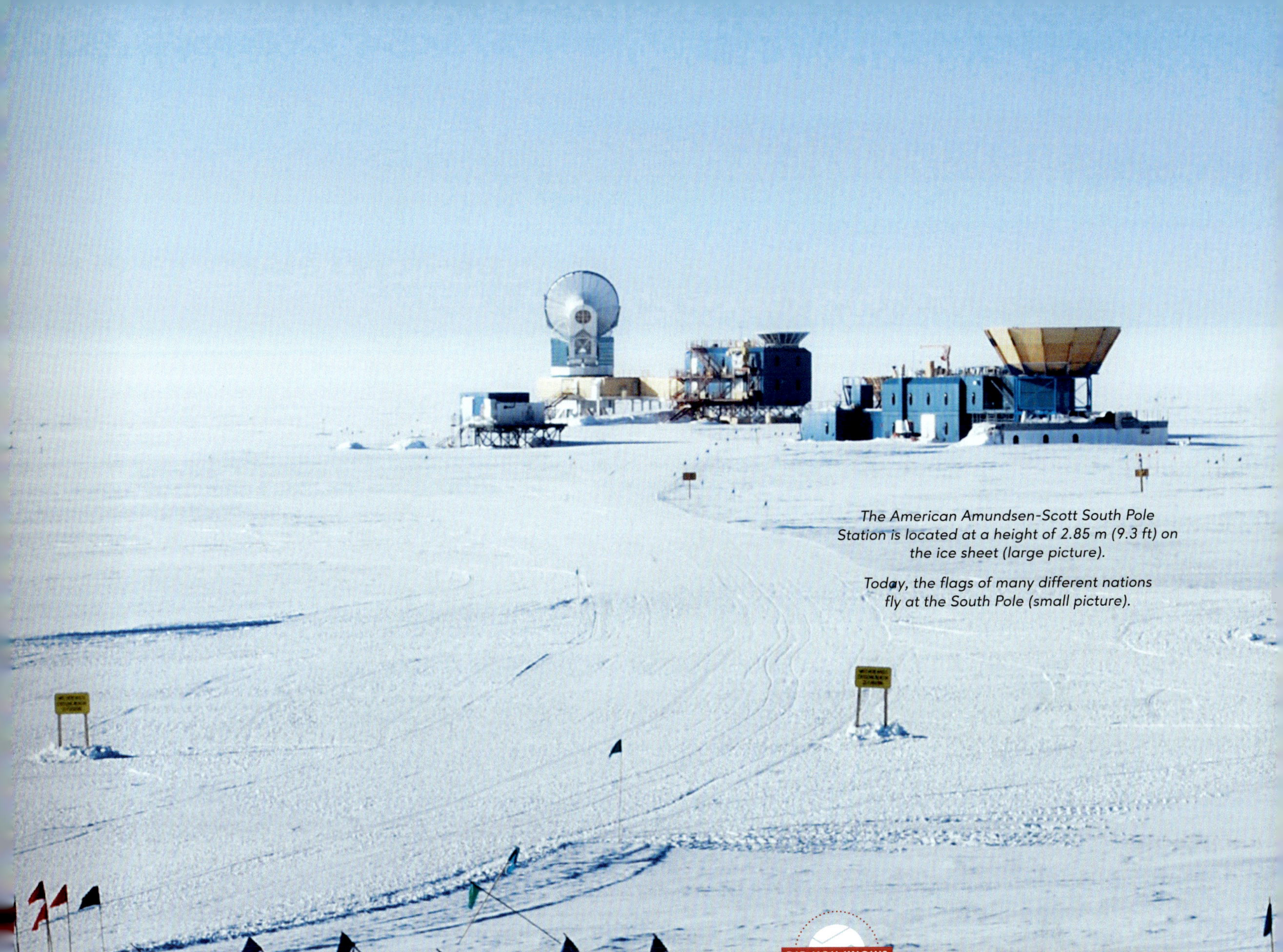

*The American Amundsen-Scott South Pole Station is located at a height of 2.85 m (9.3 ft) on the ice sheet (large picture).*

*Today, the flags of many different nations fly at the South Pole (small picture).*

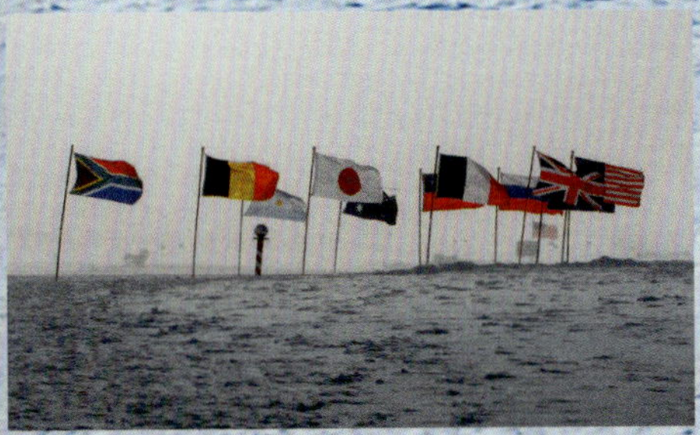

DID YOU KNOW?

**The race between Roald Amundsen and Robert Scott**

In 1911, the Englishman, Robert Falcon Scott (1868-1912), and the Norwegian, Roald Amundsen (1872-1928), together with their respective expedition teams, started a race to reach the South Pole. The winner of this exciting, but also tragic contest was Amundsen who, together with four of his team, reached the South Pole on 14 December 1911 and named his camp *Polheim* ('Home at the Pole'). When Scott, on 17 January 1912, also reached the Pole, he found a tent there with the Norwegian flag flying – he had arrived too late and had come in second. On the return journey, all the members of the British expedition, including Scott, died of malnutrition and severe frostbite. It was not until 45 years after Amundsen and Scott that the next person set foot on the South Pole. In 1956, Admiral George Dufek arrived by plane to make preparations for the construction of the American Amundsen-Scott Research Station.

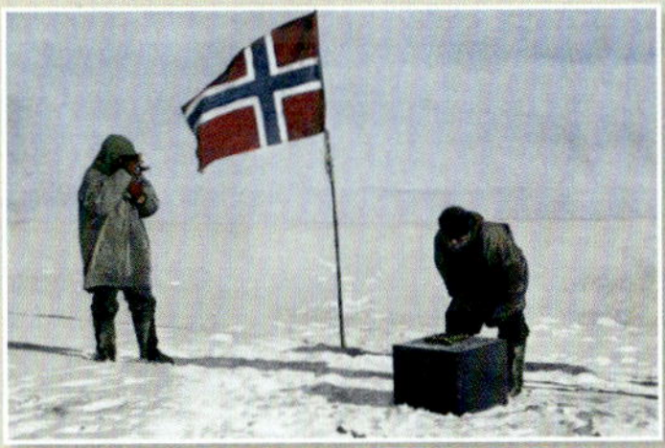

*14 December 1911*
*The goal has been reached.*

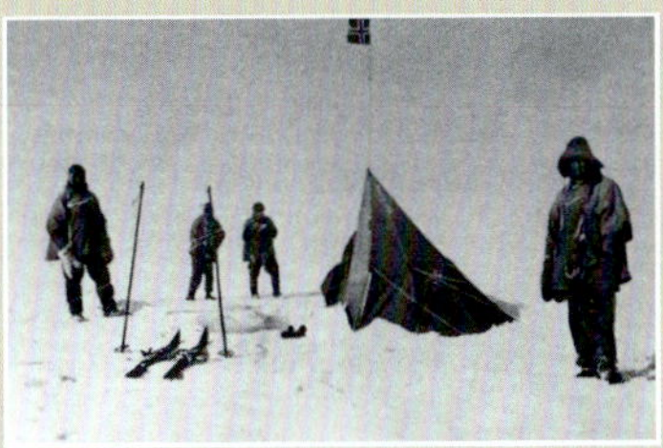

*Scott finds Amundsen's tent at the Pole.*

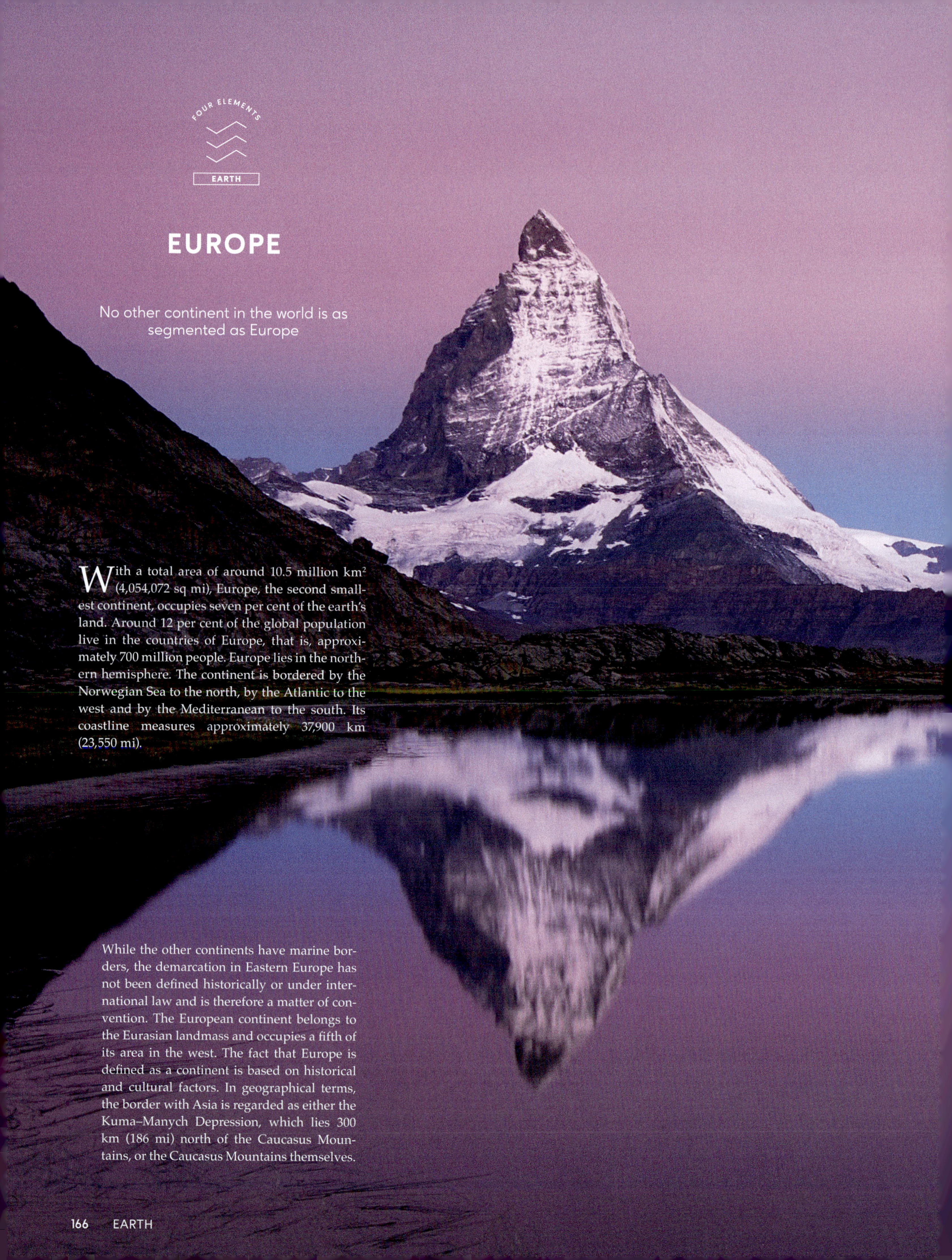

FOUR ELEMENTS

EARTH

# EUROPE

No other continent in the world is as segmented as Europe

With a total area of around 10.5 million km² (4,054,072 sq mi), Europe, the second smallest continent, occupies seven per cent of the earth's land. Around 12 per cent of the global population live in the countries of Europe, that is, approximately 700 million people. Europe lies in the northern hemisphere. The continent is bordered by the Norwegian Sea to the north, by the Atlantic to the west and by the Mediterranean to the south. Its coastline measures approximately 37,900 km (23,550 mi).

While the other continents have marine borders, the demarcation in Eastern Europe has not been defined historically or under international law and is therefore a matter of convention. The European continent belongs to the Eurasian landmass and occupies a fifth of its area in the west. The fact that Europe is defined as a continent is based on historical and cultural factors. In geographical terms, the border with Asia is regarded as either the Kuma–Manych Depression, which lies 300 km (186 mi) north of the Caucasus Mountains, or the Caucasus Mountains themselves.

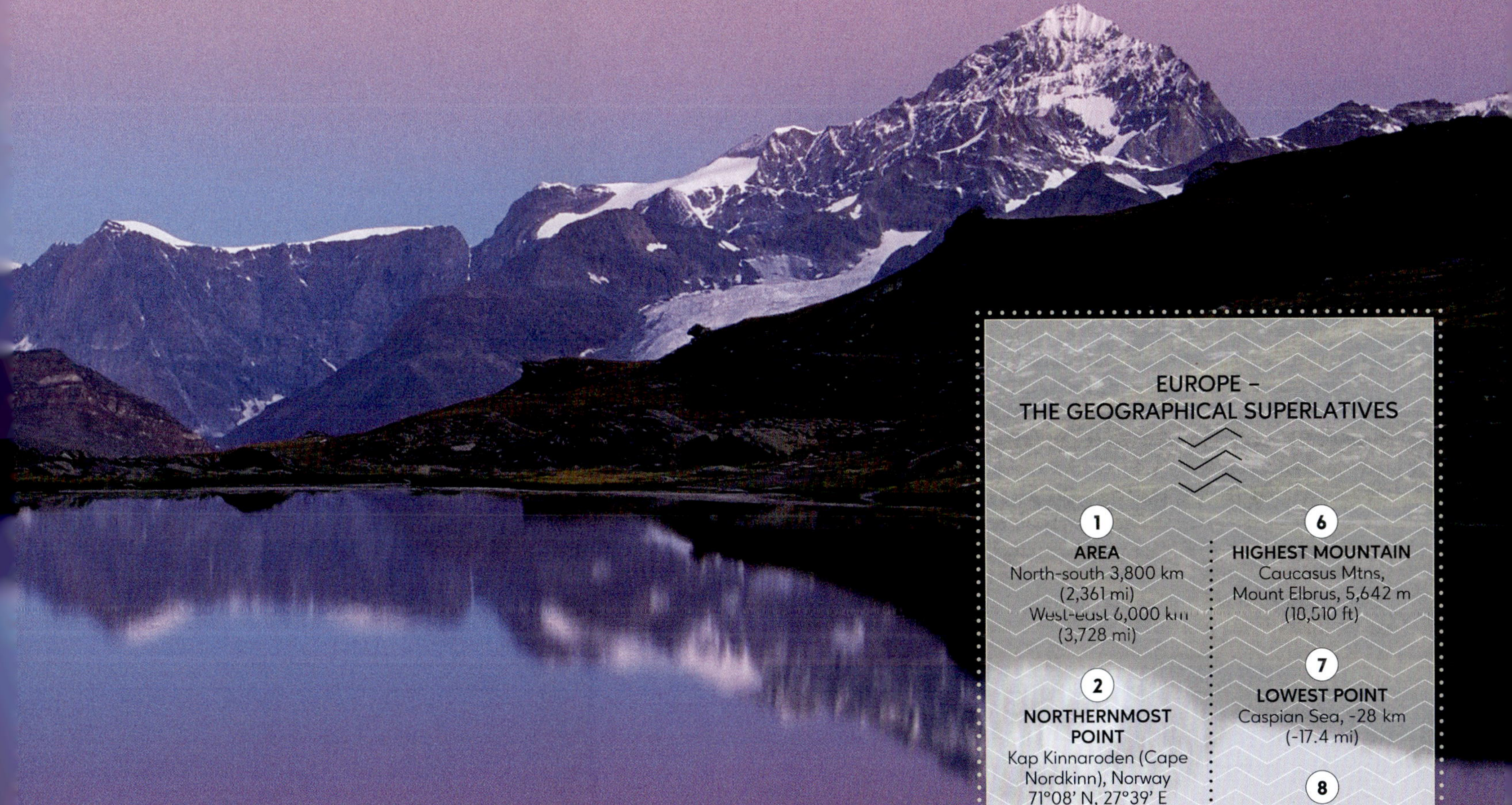

*Because of its striking shape, the Matterhorn – a Swiss landmark – is one of the most famous mountains in the world (large picture).*

*The Cliffs of Moher in Ireland are also well-known (small picture).*

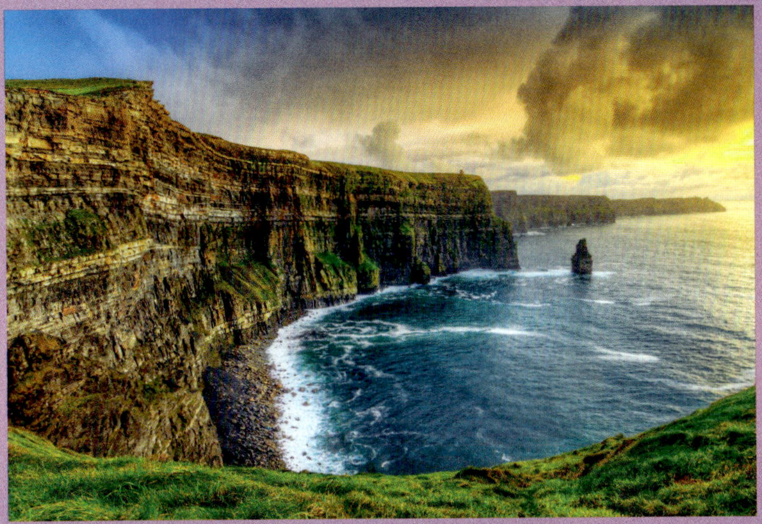

The relief of the continent of Europe – the most segmented of the continents – is so heterogenous that the comparatively small continent can be divided up into many different landscape types: The landscape alternates between mountains and basins, between lowlands and low mountain ranges and between jagged coastlines and numerous islands. The continent of Europe has the highest number of peninsulas.

The landscape of Northern Europe was mainly formed during the ice age and is defined by the rugged coasts of the northwest, barren plateaus, mountain ranges and flat areas with many lakes. The Central European or Central German Upland Range connects to it, with wooded low mountain ranges, hilly country and dells in western and central Europe. The high fold mountains of the Pyrenees, the Alps and the Carpathian Mountains continue in the mountain ranges of the South European peninsulas – the Peloponnese, Iberian and Apennine peninsulas. In addition, highlands and plateaus are found in the south. The landscape of Eastern Europe is defined by the East European plain which stretches to the Ural Mountains and the Ural River and to the Arctic Ocean. Europe is the continent with the largest variety of forms within the smallest space.

## EUROPE – THE GEOGRAPHICAL SUPERLATIVES

**1 AREA**
North-south 3,800 km (2,361 mi)
West-east 6,000 km (3,728 mi)

**2 NORTHERNMOST POINT**
Kap Kinnaroden (Cape Nordkinn), Norway
71°08' N, 27°39' E

**3 SOUTHERNMOST POINT**
Punta de Tarifa, Spain
36°0' N, 5°36' W

**4 WESTERNMOST POINT**
Cabo da Roca, Portugal
38°46' N, 9°30' W

**5 EASTERNMOST POINT**
Ural Mountains, Russia
Longitude 67° E

**6 HIGHEST MOUNTAIN**
Caucasus Mtns, Mount Elbrus, 5,642 m (10,510 ft)

**7 LOWEST POINT**
Caspian Sea, -28 km (-17.4 mi)

**8 LONGEST RIVER**
Volga, 3,534 km (2,196 mi)

**9 LARGEST LAKE**
Lake Ladoga, 17,703 km² (6,835 sq mi)

**10 DEEPEST LAKE**
Hornindalsvatnet, Norway, 514 m (1,686 ft)

**11 LARGEST ISLAND**
Great Britain, 229,883 km² (88,758 sq mi)

FOUR ELEMENTS

EARTH

# AUSTRALIA/OCEANIA

Red continent, green island – the colours of the countries of Australia and Oceania attract much attention

The smallest continent in the world, covering 8.5 million km² (3,281,868 sq mi) in total, encompasses, in addition to the main landmass of Australia, the island of Tasmania off the northern coast and several smaller islands as well as the Great Barrier Reef, the largest coral reef in the world. The name of the continent comes from the Latin (*terra australis*) and means 'southern land', referring to its position in the southern hemisphere on both sides of the Tropic of Capricorn and on the boundary between the Indian and Pacific Oceans.

The continent of Australia is bordered by the Timor and Arafura Seas and the Torres Strait in the north and by the Coral Sea and the Tasman Sea in the east. The Bass Strait lies to the south and the Indian Ocean to the west. In contrast to other continents, the coast of Australia has very few indentations, excepting the Gulf of Carpentaria in the north. The Australian continent can be divided into three major geographic regions: the tableland of western Australia, which features bizarre rock landscapes such as the Kimberley region and vast deserts such as the Great Sandy Desert. The Great Dividing Range runs along the entire eastern coast, and the Great Artesian Basin acts as a connecting link between western and eastern Australia and is also a lowland area known as 'the outback', with vast deserts, such as the Great Victoria Desert. The most striking elevations in the middle of the country are the Uluru and the Olgas.

New Zealand consists of two islands. South Island was formed approximately 100 million years ago when two continental plates collided. Hills, plains, lakes and fjords are the result of glacial processes in the ice age. On North Island, there are many volcanoes, which are located on the interface of the Australian and Pacific plates - this explains the numerous hot springs and geysers. New Zealand is often described as a 'green island' because of its sparse population and its unique and mostly untouched natural environment.

The designation 'Oceania', consolidates, culturally and economically, the island world of the Pacific with the islands of Melanesia, Micronesia and Polynesia as well as New Guinea and New Zealand into a semi-continent. In terms of the tectonics and geography, however, this is incorrect, as the regions lie on different plates (the Australian, Pacific and other smaller plates). Over 15 million people live on the islands, which number more than 7,000 and have a land area of around 1.3 million km² (501,933 sq mi).

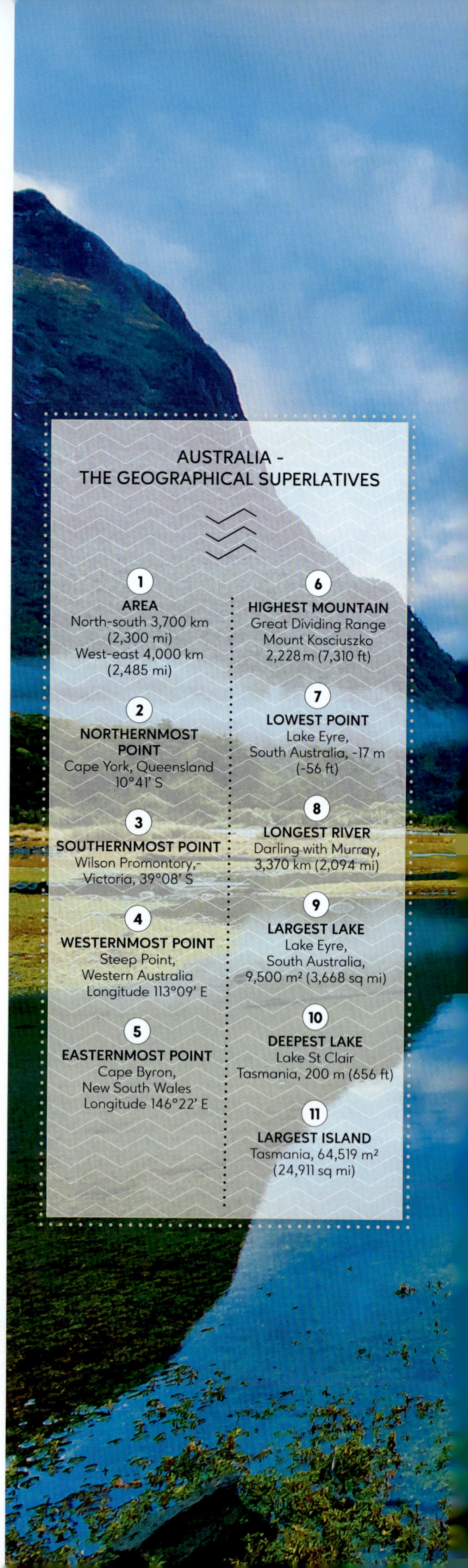

## AUSTRALIA - THE GEOGRAPHICAL SUPERLATIVES

1 **AREA**
North-south 3,700 km (2,300 mi)
West-east 4,000 km (2,485 mi)

2 **NORTHERNMOST POINT**
Cape York, Queensland
10°41' S

3 **SOUTHERNMOST POINT**
Wilson Promontory, Victoria, 39°08' S

4 **WESTERNMOST POINT**
Steep Point, Western Australia
Longitude 113°09' E

5 **EASTERNMOST POINT**
Cape Byron, New South Wales
Longitude 146°22' E

6 **HIGHEST MOUNTAIN**
Great Dividing Range
Mount Kosciuszko
2,228 m (7,310 ft)

7 **LOWEST POINT**
Lake Eyre, South Australia, -17 m (-56 ft)

8 **LONGEST RIVER**
Darling with Murray, 3,370 km (2,094 mi)

9 **LARGEST LAKE**
Lake Eyre, South Australia,
9,500 m² (3,668 sq mi)

10 **DEEPEST LAKE**
Lake St Clair
Tasmania, 200 m (656 ft)

11 **LARGEST ISLAND**
Tasmania, 64,519 m² (24,911 sq mi)

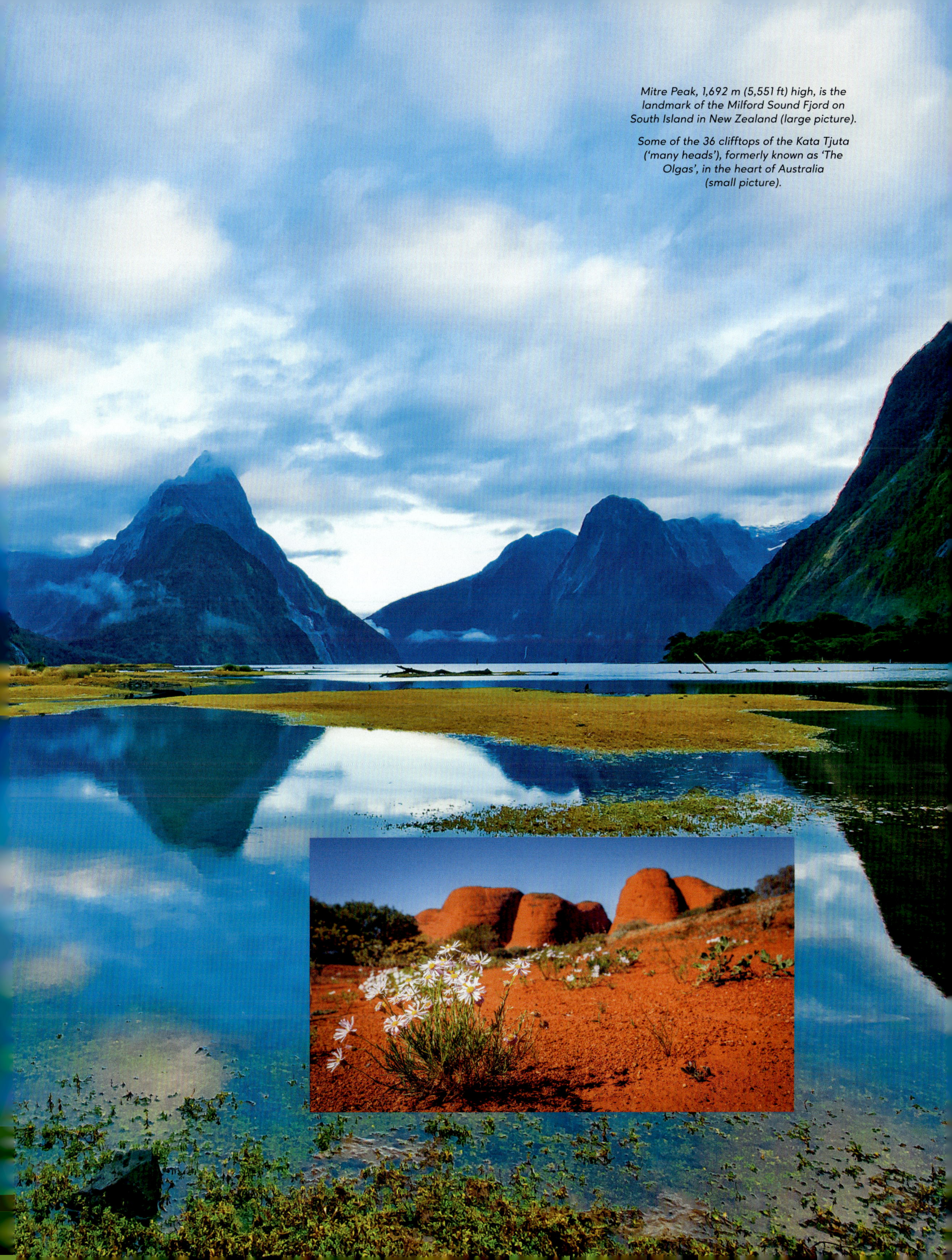

Mitre Peak, 1,692 m (5,551 ft) high, is the landmark of the Milford Sound Fjord on South Island in New Zealand (large picture).

Some of the 36 clifftops of the Kata Tjuta ('many heads'), formerly known as 'The Olgas', in the heart of Australia (small picture).

# EARTHQUAKES

The most powerful tremors on earth

Earthquakes are natural tremors in the earth's crust and mantel. Friction occurs, particularly at the edges of the continental plates where these are moving apart or towards each other; this friction is discharged by jerking movements of the earth's crust – which can lead to tectonic earthquakes. On the San Andreas Fault in California, for instance, where the Pacific plate drifts past the North American plate, earthquakes occur frequently. The strength of an earthquake is measured on the Richter Scale; in this process, the energy released by the earthquake is registered on a seismograph.

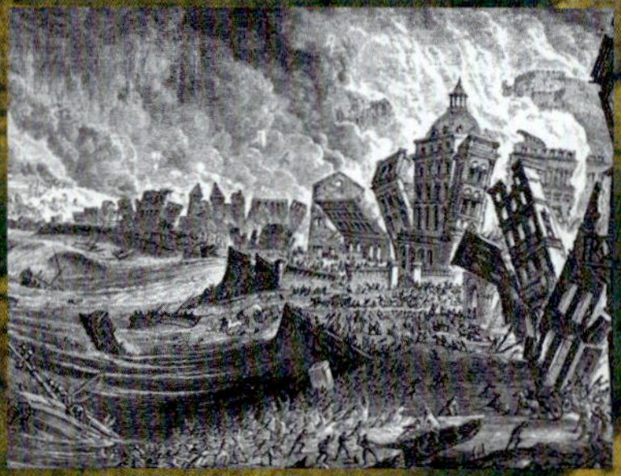

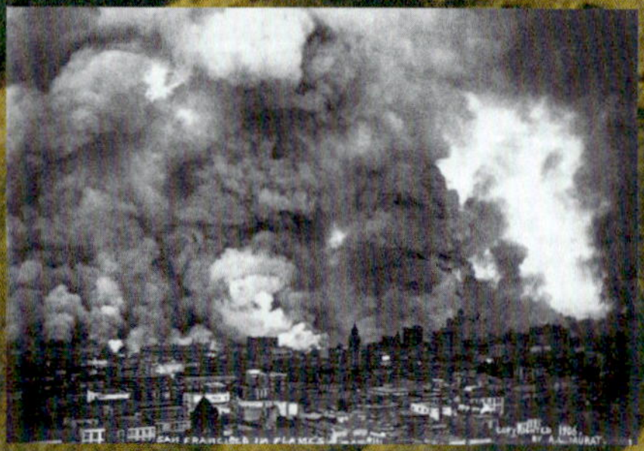

## EARTHQUAKES IN HISTORY

On 1 November 1755, an earthquake and the subsequent fires destroyed Lisbon, the capital of Portugal (above). The quake, which claimed between 30,000 and 100,000 lives, is thought to have been a magnitude of 8.5 to 9 on the Richter Scale. The catastrophe is regarded as having been the impetus for modern seismology.

One of the worst natural catastrophes in the USA was the San Francisco earthquake on 18 April 1906, which claimed 3,000 lives (right). It was a magnitude of 7.8 to 8.8 on the Richter Scale and shook the entire area along the San Andreas Fault for hundreds of miles.

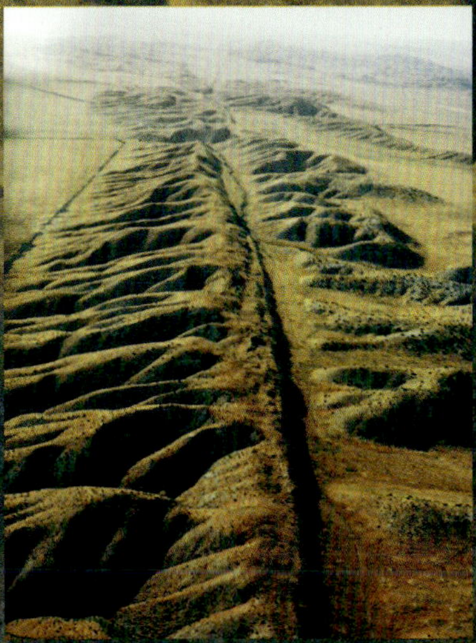

The San Andreas Fault stretches 1,100 km (684 mi) from Mexico to San Francisco (large picture and small picture, top).

## THE MOST POWERFUL EARTHQUAKES ON EARTH

Strength on the Richter Scale

1. **CHILE** (Valdivia), 1960, 9.5
2. **ALASKA** ('Good Friday Earthquake') 1964, 9.2
3. **SEAQUAKE OFF SUMATRA** 2004, 9.1
4. **RUSSIA (KAMCHATKA)** 1952, 9.0
5. **EARTHQUAKE OFF CHILE** 2010, 8.8
6. **EARTHQUAKE OFF ECUADOR** 1906, 8.8
7. **ALASKA (ANDREANOF ISLANDS)** 1957, 8.7
8. **ALASKA (RAT ISLAND)** 1965, 8.7
9. **EARTHQUAKE OFF NORTHERN SUMATRA** 2005, 8.6
10. **INDIA (ASSAM)** 1950, 8.6

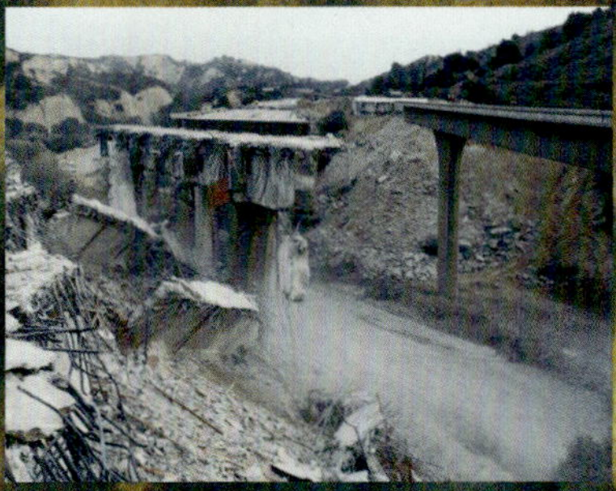

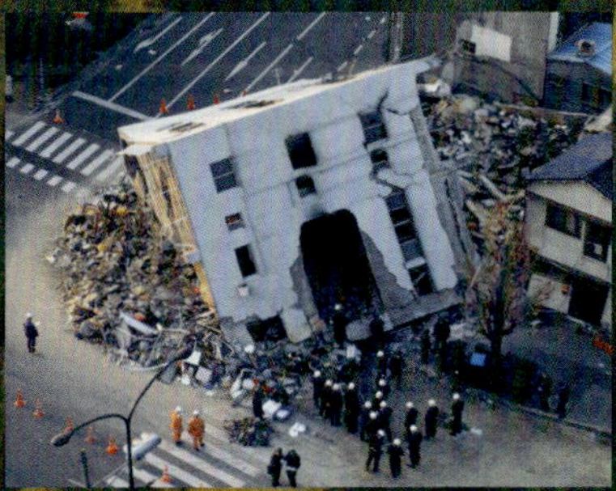

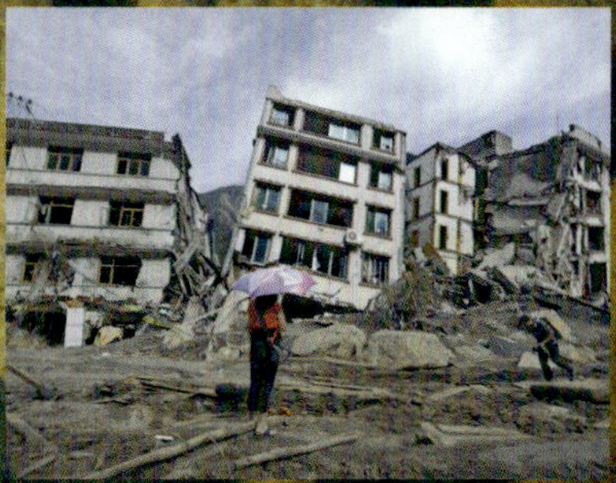

## EARTHQUAKES IN THE 20TH AND 21ST CENTURIES

Mexico City lies in a tectonically active region where earthquakes are a common occurrence. In 1985, this city of over one million inhabitants was devastated by an earthquake with a magnitude of 8.1. More than 10,000 people died. In 1994, an earthquake in California measuring 6.7 claimed 60 lives (top); in Kobe, Japan in 1995, there were tremors with a magnitude of 7.3 (bottom). One of the most catastrophic earthquakes was the seaquake in the Indian Ocean off the island of Sumatra on 26 December 2004, which led to the deaths of 230,000 people. On 12 May 2008, the Great Sichuan (or Wenchuan) Earthquake, with a magnitude of 8.0, shook the Chinese province of Sichuan and claimed 86,000 lives (right). Earthquakes often unleash natural catastrophes, such as in the Sulawesi earthquake in 2018, which together with the ensuing tsunami, claimed more than 4,000 lives.

FOUR ELEMENTS

EARTH

# ISLANDS

With over 230,000 islands, Norway is the country with the most islands, followed by Sweden and Canada

In area they outstrip the Benelux countries many times over and feature in every type of climate region. The largest islands in the world, such as Greenland, lie under a thick ice sheet or, like New Guinea, are covered in dense tropical rainforest. Between these two extremes, there are many hybrid forms, depending on the climate zone. The common feature they share is that they are all completely surrounded by water, and frequently, because of their size, separate oceans from marginal seas and other marine zones.

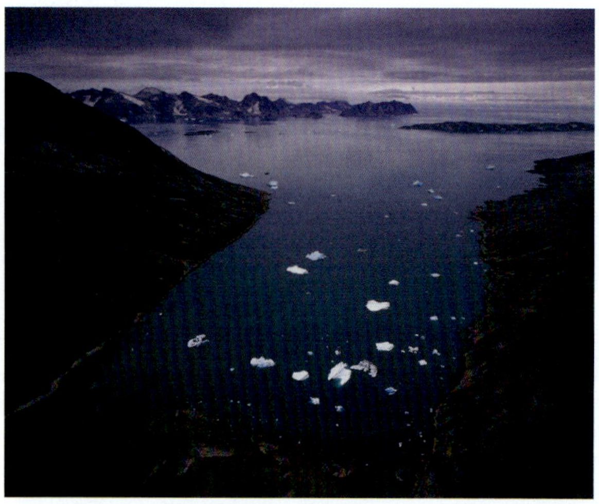

*Large picture: The Cape Reinga Lighthouse has been keeping watch over North Cape on North Island in New Zealand since 1941. For visitors, it is a fantastic lookout point.*

*Above: The Sermiligaq Fjord on Greenland.*

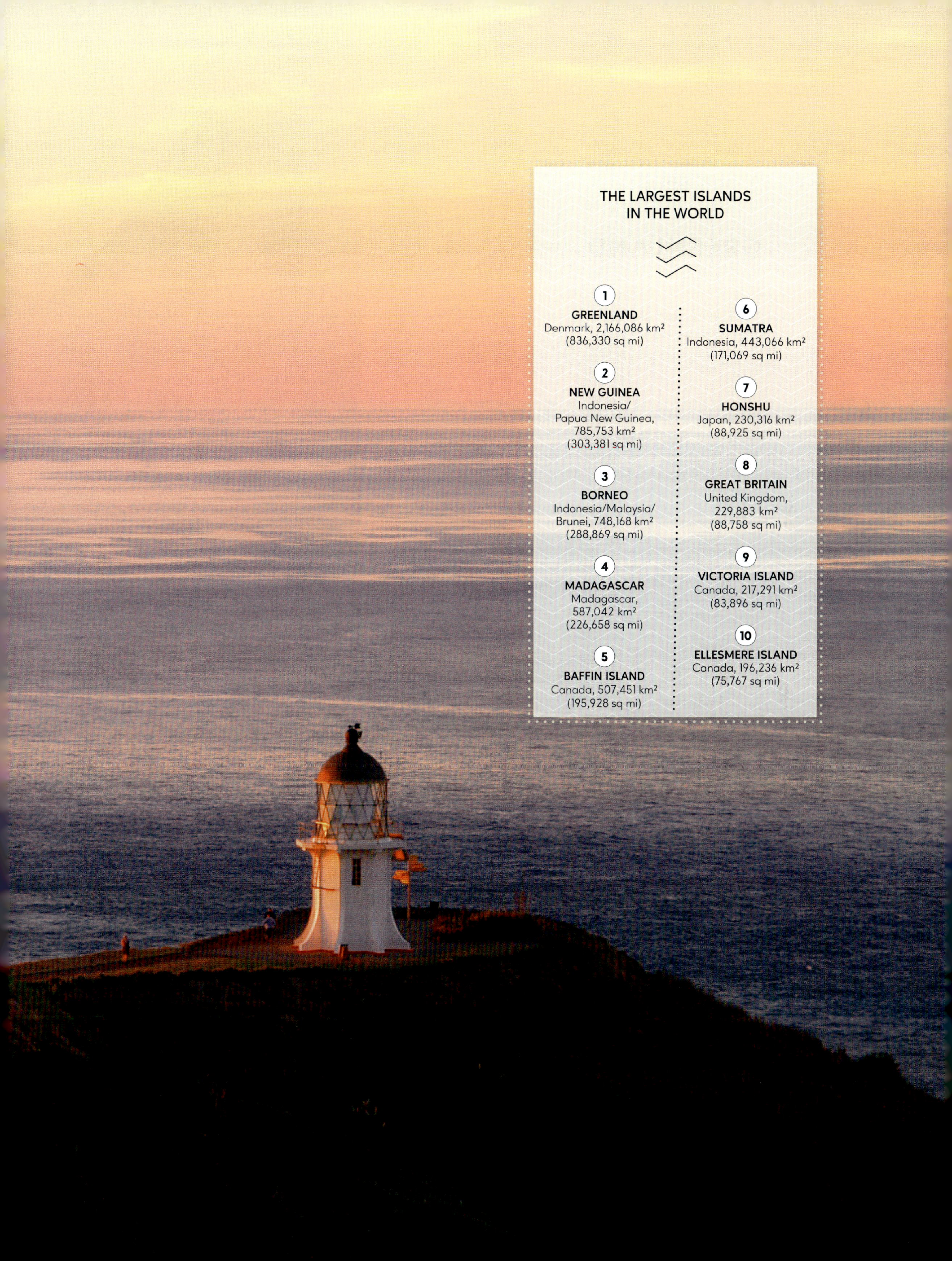

## THE LARGEST ISLANDS IN THE WORLD

1. **GREENLAND**
Denmark, 2,166,086 km²
(836,330 sq mi)

2. **NEW GUINEA**
Indonesia/
Papua New Guinea,
785,753 km²
(303,381 sq mi)

3. **BORNEO**
Indonesia/Malaysia/
Brunei, 748,168 km²
(288,869 sq mi)

4. **MADAGASCAR**
Madagascar,
587,042 km²
(226,658 sq mi)

5. **BAFFIN ISLAND**
Canada, 507,451 km²
(195,928 sq mi)

6. **SUMATRA**
Indonesia, 443,066 km²
(171,069 sq mi)

7. **HONSHU**
Japan, 230,316 km²
(88,925 sq mi)

8. **GREAT BRITAIN**
United Kingdom,
229,883 km²
(88,758 sq mi)

9. **VICTORIA ISLAND**
Canada, 217,291 km²
(83,896 sq mi)

10. **ELLESMERE ISLAND**
Canada, 196,236 km²
(75,767 sq mi)

FOUR ELEMENTS

EARTH

# GREENLAND

The largest island
in the world

Greenland is the largest island in the world. It is approximately 2,650 km (1,647 mi) long with a maximum width of roughly 1,000 km (621 mi) and an area of 2,166,086 km² (836,330 sq mi). The thick ice sheet can be up to 3,000 m (9,842 ft) thick. Covering about 80% of Greenland's total area, it is a relic of the last ice age. Large areas of the coast in the south and southwest are ice-free. Polar bears, reindeer, musk oxen and over 200 different bird species live in the predominantly subpolar to polar climate. Trees are only found in a very small marginal zone in the far south. Otherwise, tundra vegetation covers any coastal areas that are not frozen over.

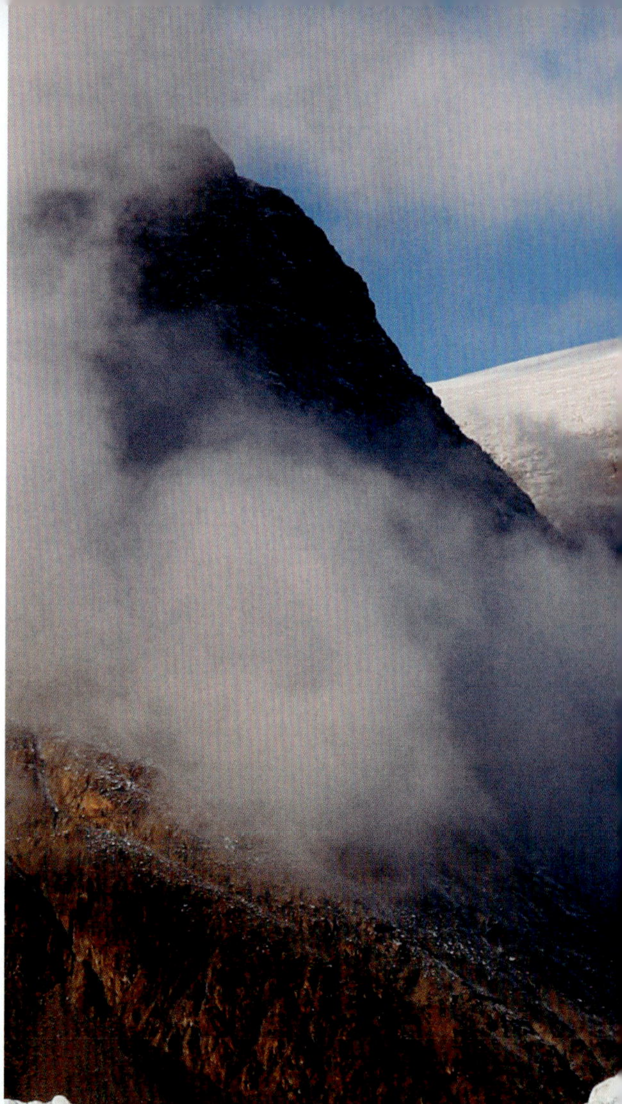

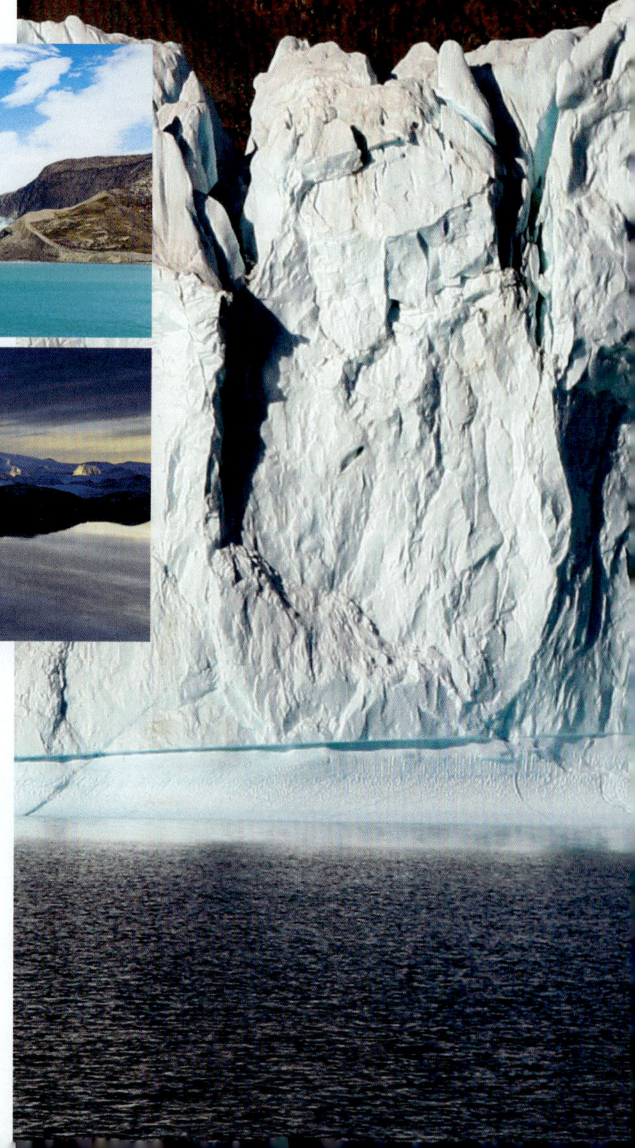

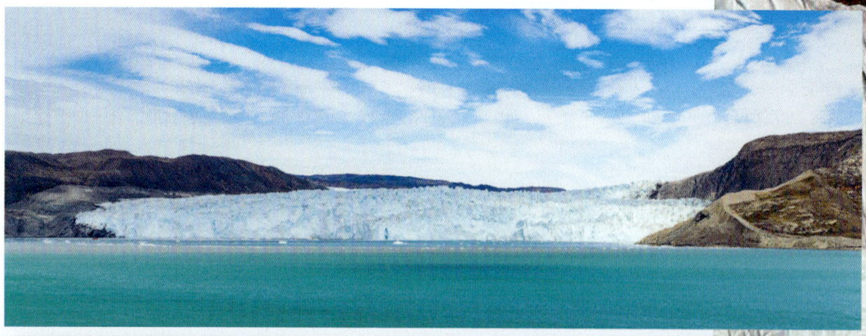

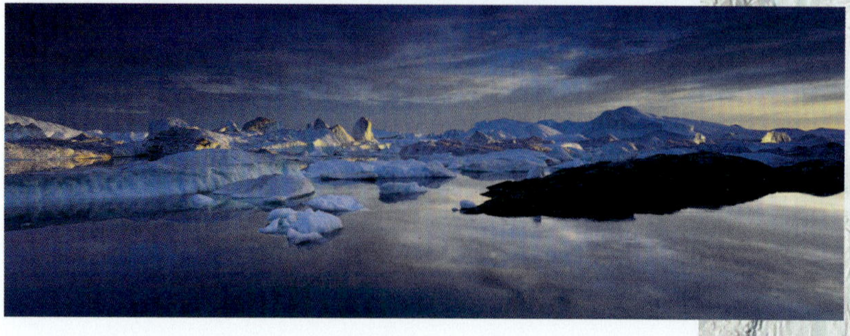

In many places in Greenland, mighty glaciers can be seen reaching to the ocean and calving into huge fragments on the coast. These chunks of glaciers drift through the bays and out to the open ocean. The two small pictures above show glaciers moving between the cliffs towards the coast. The small picture below shows a section of Disko Bay in western Greenland filled with icebergs.

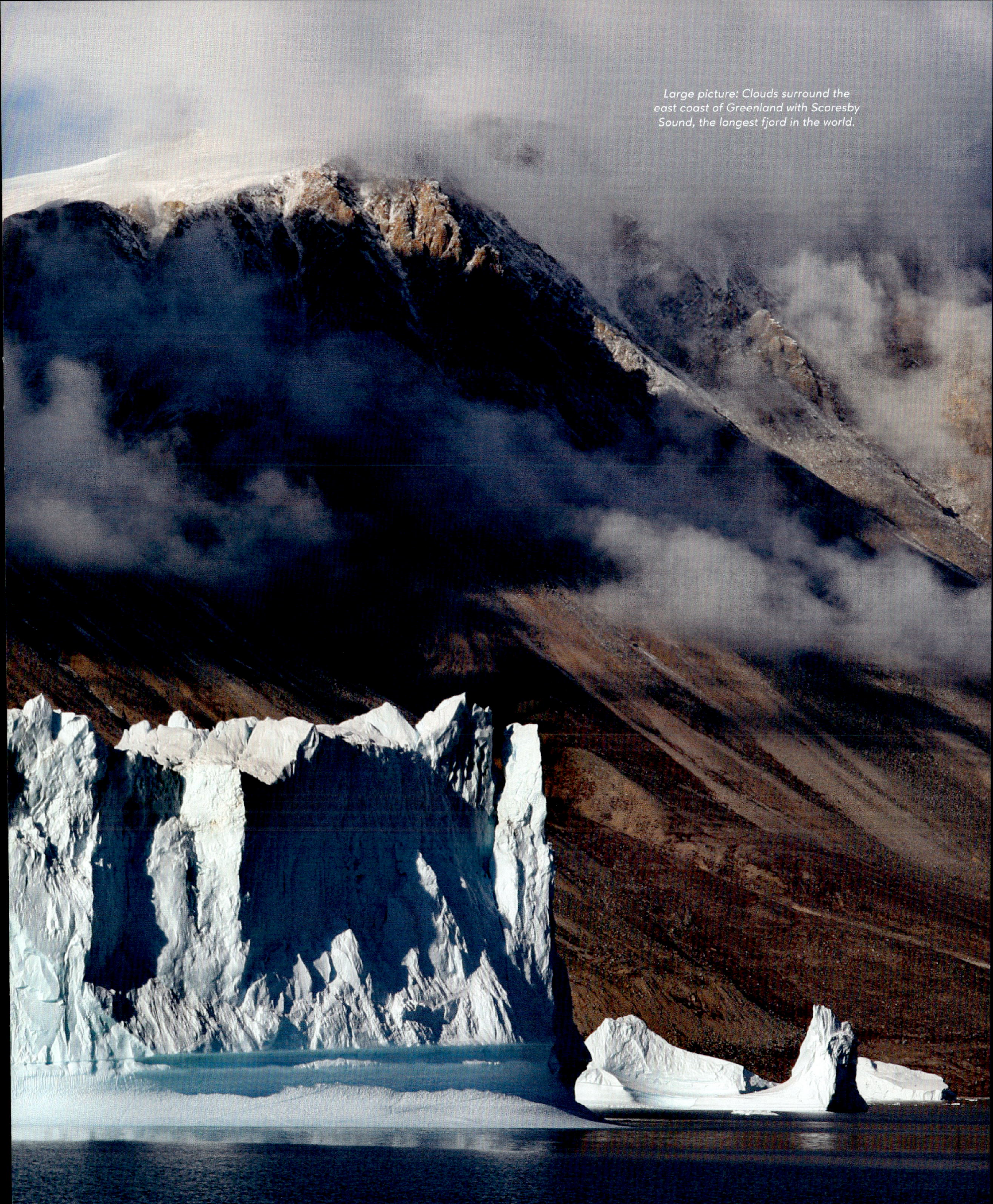

*Large picture: Clouds surround the east coast of Greenland with Scoresby Sound, the longest fjord in the world.*

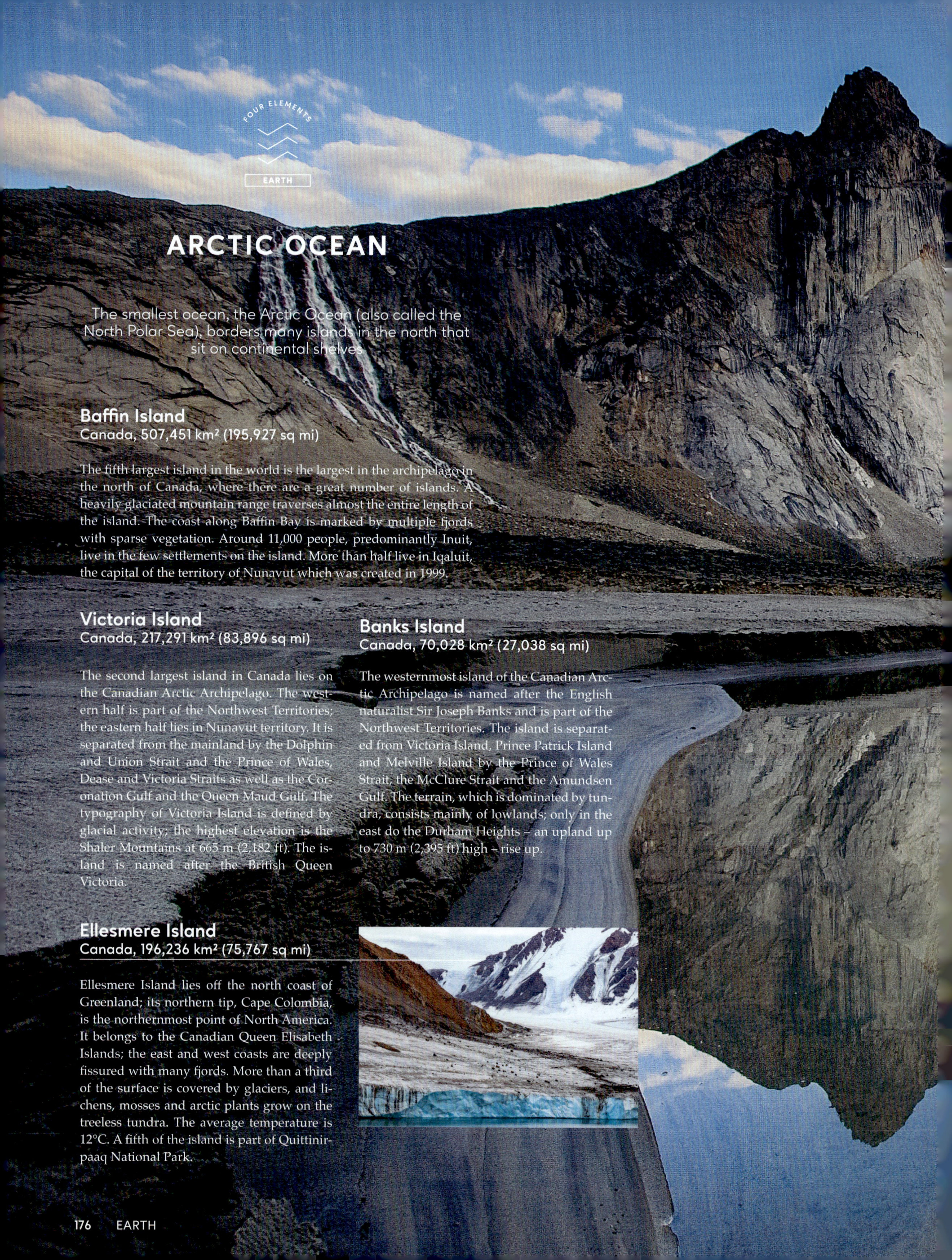

# ARCTIC OCEAN

The smallest ocean, the Arctic Ocean (also called the North Polar Sea), borders many islands in the north that sit on continental shelves

## Baffin Island

**Canada, 507,451 km² (195,927 sq mi)**

The fifth largest island in the world is the largest in the archipelago in the north of Canada, where there are a great number of islands. A heavily glaciated mountain range traverses almost the entire length of the island. The coast along Baffin Bay is marked by multiple fjords with sparse vegetation. Around 11,000 people, predominantly Inuit, live in the few settlements on the island. More than half live in Iqaluit, the capital of the territory of Nunavut which was created in 1999.

## Victoria Island

**Canada, 217,291 km² (83,896 sq mi)**

The second largest island in Canada lies on the Canadian Arctic Archipelago. The western half is part of the Northwest Territories; the eastern half lies in Nunavut territory. It is separated from the mainland by the Dolphin and Union Strait and the Prince of Wales, Dease and Victoria Straits as well as the Coronation Gulf and the Queen Maud Gulf. The typography of Victoria Island is defined by glacial activity; the highest elevation is the Shaler Mountains at 665 m (2,182 ft). The island is named after the British Queen Victoria.

## Banks Island

**Canada, 70,028 km² (27,038 sq mi)**

The westernmost island of the Canadian Arctic Archipelago is named after the English naturalist Sir Joseph Banks and is part of the Northwest Territories. The island is separated from Victoria Island, Prince Patrick Island and Melville Island by the Prince of Wales Strait, the McClure Strait and the Amundsen Gulf. The terrain, which is dominated by tundra, consists mainly of lowlands; only in the east do the Durham Heights – an upland up to 730 m (2,395 ft) high – rise up.

## Ellesmere Island

**Canada, 196,236 km² (75,767 sq mi)**

Ellesmere Island lies off the north coast of Greenland; its northern tip, Cape Colombia, is the northernmost point of North America. It belongs to the Canadian Queen Elisabeth Islands; the east and west coasts are deeply fissured with many fjords. More than a third of the surface is covered by glaciers, and lichens, mosses and arctic plants grow on the treeless tundra. The average temperature is 12°C. A fifth of the island is part of Quittinirpaaq National Park.

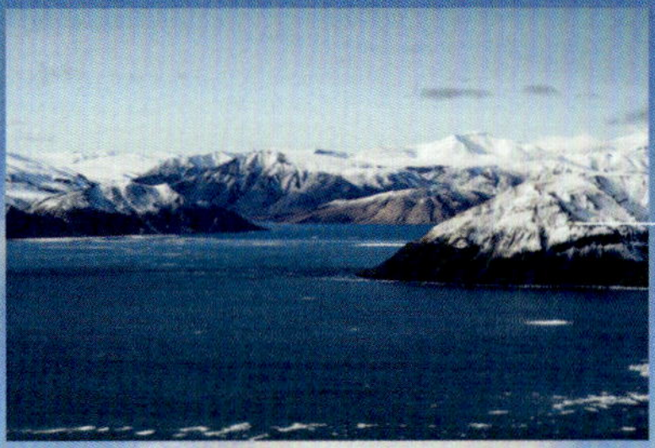

## Axel Heiberg Island

**Canada, 43,178 km² (16,671 sq mi)**

The island, part of the Queen Elizabeth Islands, is the seventh largest in Canada and is cleft by multiple fjords. A third of this uninhabited island is covered in glaciers and ice. In 1985, a petrified forest from the Eocene period was found in the east of the island, with fossils dating back 40 to 50 million years and Chinese swamp cypresses and redwoods up to 50 m (164 ft) tall and 1,000 years old.

## Melville Island

**Canada, 42,149 km² (16,274 sq mi)**

This hilly, uninhabited island is the fourth largest of the Queen Elizabeth Islands in Canada's Arctic Archipelago and rises up 776 m (2,546 ft) at its highest point. In the north, it is bordered by the Hazen Strait, in the northeast by the Byam Martin canal and in the south by the Viscount Melville Sound and the McClure Strait. The western half of Melville Island lies within the Northwest Territories, while the eastern half is in Nunavut territory. The central area has only sparse vegetation and is partially covered in ice.

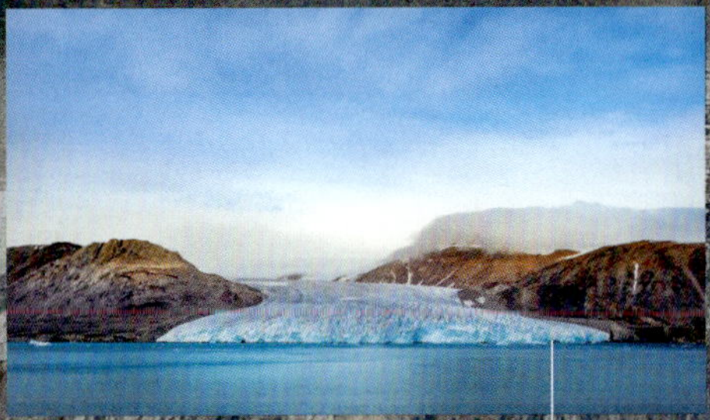

## Devon Island

**Canada, 55,247 km² (21,330 sq mi)**

With an area of 55,247 km² (21,330 sq mi), this is the largest uninhabited island in the world. A third of the surface area is covered in an ice cap – the 1,920-m (6,299-ft) high Devon Ice Cap in the Treuter Mountains. Devon Island is also known for the Haughton Crater, which has a diameter of approximately 20 km (12.4 mi).

## Southampton Island

**Canada, 41,214 km² (15,913 sq mi)**

Southampton Island lies at the northern edge of Hudson Bay, surrounded by the Roes Welcome Sound and the Frozen, Evans and Fisher Straits. The topography is mainly flat, except in the northeast where the Porsild Mountains rise up more than 600 m (1,968 ft), among which the highest summit is Mathiasen Mountain at 625 m (2,050 ft). Many mammal species live on the tundra vegetation: polar bears, caribou, lemmings and Arctic hares. The largest settlement is Coral Harbour with just under 800 inhabitants.

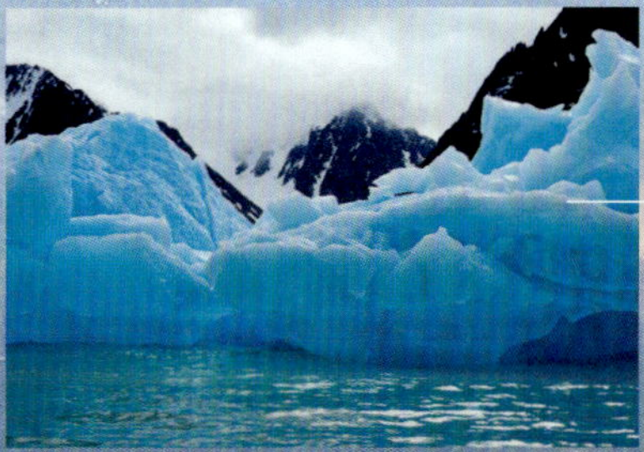

## Spitzbergen

**Norway, 37,673 km² (14,546 sq mi)**

Spitzbergen is the largest island of the Svalbard Archipelago and belongs to Norway. In German, all the islands of this archipelago are commonly referred to as Spitzbergen. The arctic flora consists almost entirely of mosses, lichens and ferns; large areas are without vegetation. In Ny-Ålesund on the west coast, an international base for Arctic research has been set up with the northernmost oceanic research laboratory and an observatory for studying the Northern Lights.

## Novaya Zemlya Northern Island

**Russia, 48,904 km² (18,882 sq mi)**

The northern island, Sewery Ostrow (Severny Island), is the fourth largest island in Europe. Together with the southern island and several smaller islands, it forms the Novaya Zemlya island group, which separates the Barents Sea from the Kara Sea. Tundra dominates the southern island, while the northern island is covered by glaciers. The island has several research stations for geophysics and meteorology.

FOUR ELEMENTS

EARTH

# ATLANTIC OCEAN

From the subpolar regions through the temperate latitudes to the Mediterranean and tropical regions, every climatic zone is represented here

## GREAT BRITAIN

**United Kingdom, 229,883 km² (88,758 sq mi)**

The largest island in Europe is the eighth largest island in the world. In spite of its size, the terrain is extremely varied. Geographically, Great Britain can be divided into two main zones – the uplands and the lowlands. Scotland and Wales, as well as the north, northwest and southwest of England lie in the upland zone. The most mountainous areas of the United Kingdom can be found in the Scottish Highlands, followed by the Scottish Lowlands and the Southern Uplands. On the rugged coasts, towering, majestic sheer cliffs alternate with sandy beaches.

## NEWFOUNDLAND

**Canada, 108,860 km² (42,031 sq mi)**

Newfoundland lies off the wide estuary of the Gulf of St Lawrence in eastern Canada. The rocky island with low mountain features is, in geological terms, a continuation of the Appalachian Mountains; it has a deeply indented fjord coastline and reaches a height of 814 m (2,670 ft). Forests and marshland define the interior of the island which, in its morphology, displays a low-hilled relief, sculpted in the last ice age.

## CUBA

**Cuba, 105,806 km² (40,852 sq mi)**

Cuba, in the Greater Antilles, is the largest island in the Caribbean and separates the Gulf of Mexico from the Caribbean Sea. In an area 1,200 km (746 mi) long and 120 km (74.5 mi) wide, there are three mountain formations in the western, central and eastern regions, interrupted by wide lowland plains. The highest point (1,974 m (6,476 ft)) lies in the Sierra Maestra.

## ICELAND

**Iceland, 103,000 km² (39,768 sq mi)**

The island of fire and ice lies on a hot spot on the Mid-Atlantic Ridge. There are around 130 active volcanoes here in addition to geysers, fumaroles and sulfur springs. Moreover, eleven per cent of the island is covered by glaciers. The main section, however, is occupied by a plateau 300 to 1,000 m (984 to 3,280 ft) high of rock and gravel desert lacking vegetation.

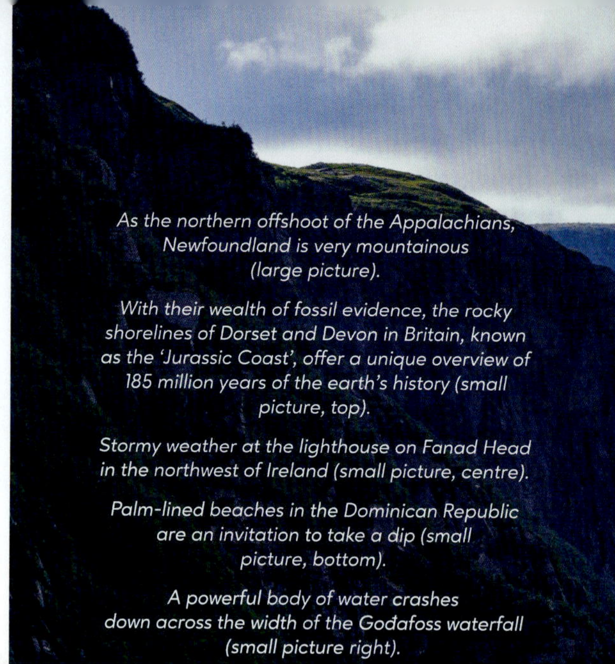

*As the northern offshoot of the Appalachians, Newfoundland is very mountainous (large picture).*

*With their wealth of fossil evidence, the rocky shorelines of Dorset and Devon in Britain, known as the 'Jurassic Coast', offer a unique overview of 185 million years of the earth's history (small picture, top).*

*Stormy weather at the lighthouse on Fanad Head in the northwest of Ireland (small picture, centre).*

*Palm-lined beaches in the Dominican Republic are an invitation to take a dip (small picture, bottom).*

*A powerful body of water crashes down across the width of the Godafoss waterfall (small picture right).*

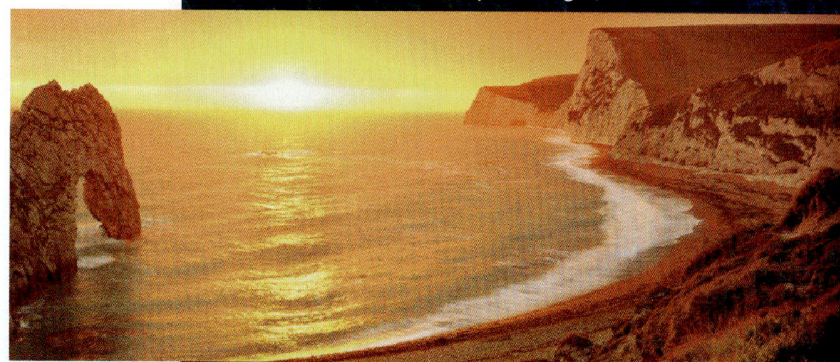

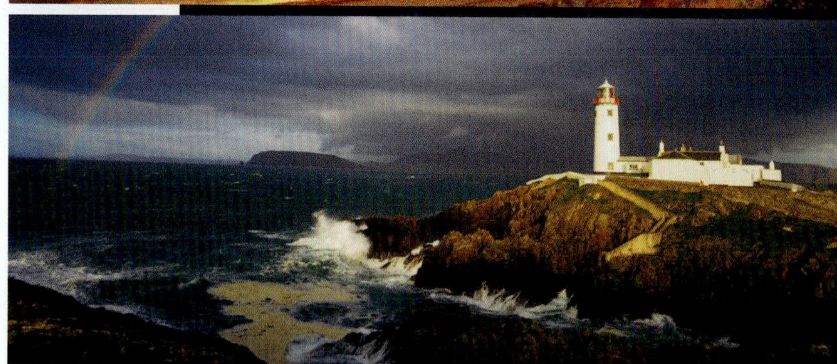

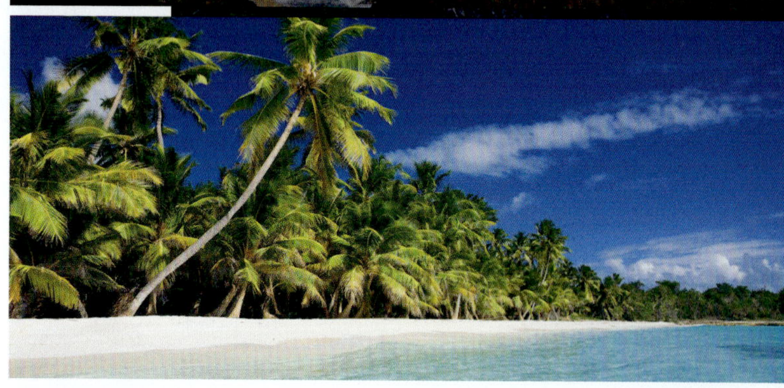

## IRELAND

**Ireland/United Kingdom, 81,638 km² (31,521 sq mi)**

Ireland, known as the Emerald Isle, is divided into the Republic of Ireland and Northern Ireland, which is part of Britain. The central section of the island is a lowland with abundant marshland and lakes, which, with its small groves of beech and oak, resembles a park. The lowland plain is surrounded by mountainous areas; the highest point is the Carrauntoohill at 1,038 m (3,405 ft). In contrast to the east coast, the west coast has numerous inlets and boasts several sheer cliffs.

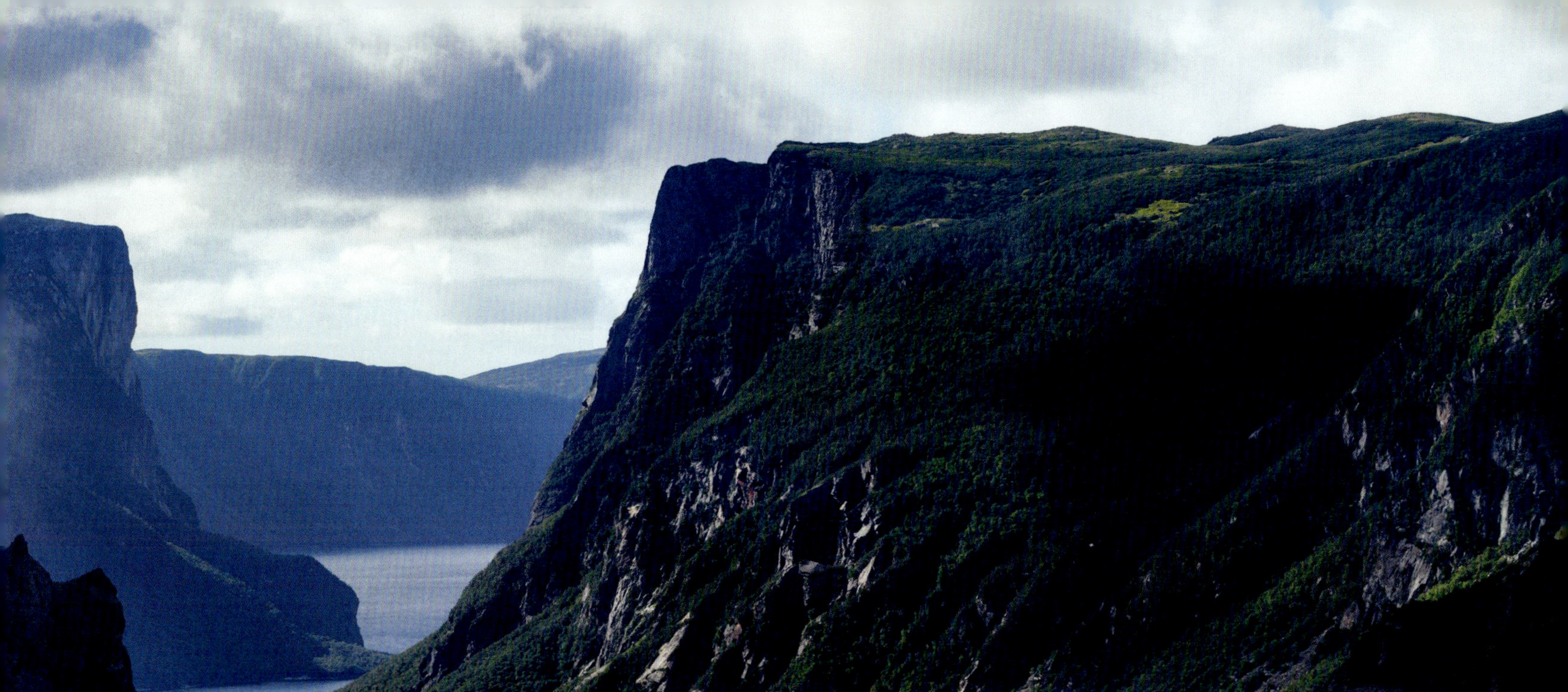

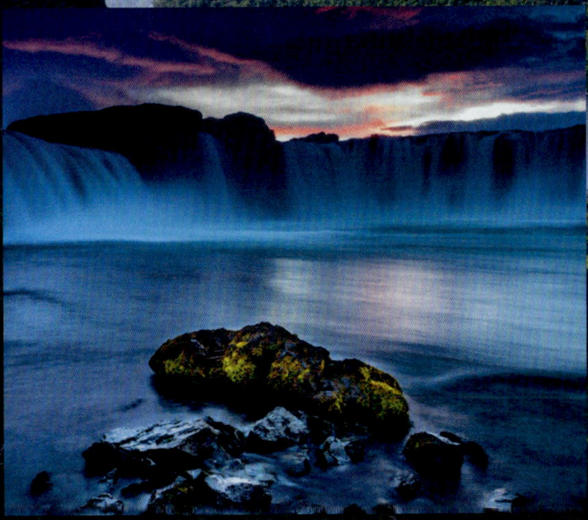

**TIP**

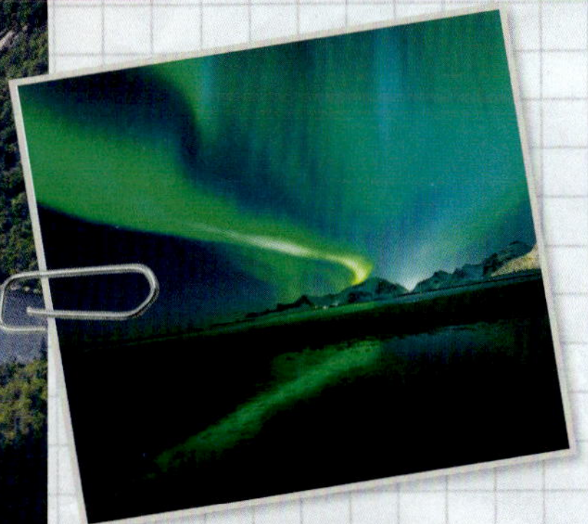

**BEST TIMES FOR A TRIP TO ICELAND FOR THE NORTHERN LIGHTS**

The timeframe is long: tour operators recommend September to April as the best time to see the Northern Lights, or polar lights, in Iceland. But this is only half the story, because to see this natural phenomenon, what you need above all is luck and the right weather conditions. It should be a starlit night; the moon should be faint rather than full, i.e. a new moon rather than a full moon, and there must be a lot of solar activity. There are now apps that predict when the polar lights will appear; even the meteorological office offers a forecast of prime times when activity might be seen. Then it is just a case of going outside, preferably far from towns or cities, so that light pollution levels are low.

## HISPANIOLA

**Dominican Rep./Haiti, 73,929 km$^2$ (28,544 sq mi)**

Hispaniola is the second largest island of the Greater Antilles. The Republic of Haiti is in the smaller, western part of the island, while the Dominican Republic is the larger, eastern area. Five mountain chains run across this mountainous island; they contain five of the highest mountains in the Caribbean, such as the Pico Duarte, at 3,175 m (10,417 ft) high. The climate is sub-tropical to tropical-humid. The natural vegetation forms rainforests and cloud forests.

## TIERRA DEL FUEGO

**Argentina/Chile, 47,404 m$^2$ (18,303 sq mi)**

Tierra del Fuego is separated from the South American mainland by the Magellan Straits. The Andes mountain chain ends here with the Darwin Cordillera. The climate in this region is subpolar – humid, maritime and often stormy. The vegetation on Tierra del Fuego transitions from forest in the interior of the island to tundra vegetation.

## SARDINIA

**Italy, 23,949 km$^2$ (9,247 sq mi)**

Sardinia is the second largest island in the Mediterranean, after Sicily. It has a very diverse relief. The central section is occupied by a mountainous area. The mountain region in the northeast features wild, jagged formations. The mountainous area in the former mining region of Iglesiente is separated from it by the Campidano Plain.

## JAMAICA

**Jamaica, 10,991 km$^2$ (4,244 sq mi)**

The third largest island in the Greater Antilles is predominantly mountainous – the Blue Mountains run through it from west to east, while there is extensive swampland in the coastal areas in the south. Sixty per cent of the island's surface is made up of limestone plateaus up to 900 m (2,953 ft) high. With the exception of a few areas with rainforest and cloud forest cover, the indigenous vegetation on the island has been largely destroyed by human activity.

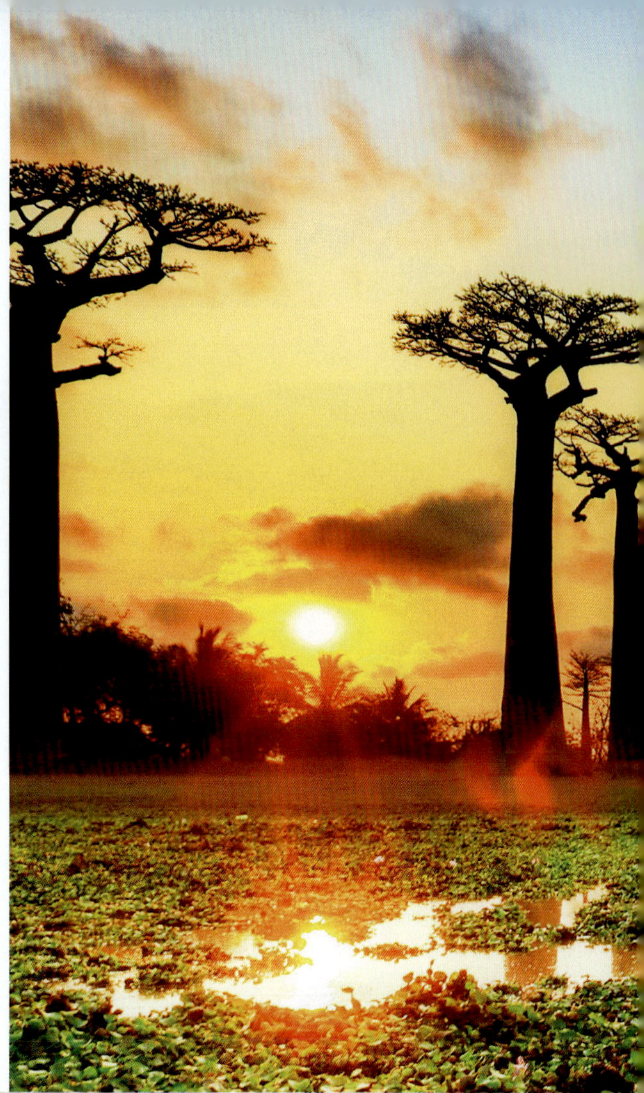

FOUR ELEMENTS

EARTH

# INDIAN OCEAN

At the edges of the five tectonic plates under the Indian Ocean, sections of the plates split off and then rise as islands, archipelagos and island arcs

## MADAGASCAR

**Madagascar, 587,042 km² (226,658 sq mi)**

The wide Mozambique Channel separates the fourth largest island in the world from the African continent. The east coast of the island, which is around 1,600 km (994 mi) long and up to 580 km (360 mi) wide, has very few inlets. The coast runs mainly in a straight line, almost as if drawn with a ruler. In contrast, the west coast is more deeply indented, with inlets and estuaries that reach into the alluvial plains. A 2,876-m (9,436-ft) high cordillera, which drops away steeply in scarps to the east, forms the backbone of the island. Madagascar's relative remoteness from the African continent has led to a high number of native species among its flora and fauna. Lemurs, a species of ape which are present in a large number of varieties and sizes, are the most well-known of these. The island's many other species have scarcely been explored. Their habitat is critically threatened by deforestation and conversion of the forests to agricultural land.

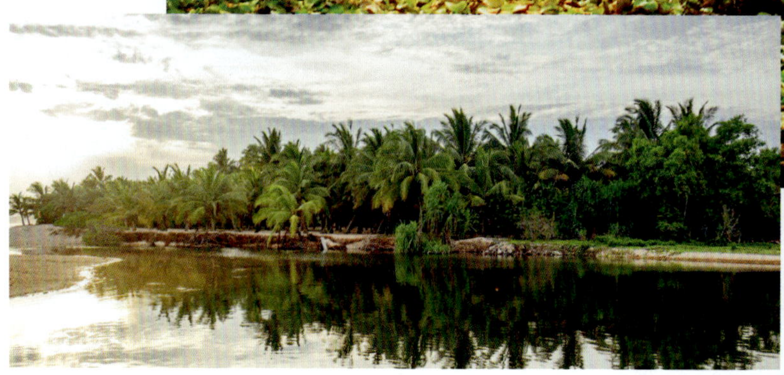

## SUMATRA

**Indonesia, 443,066 km² (171,069 sq mi)**

The sixth largest island in the world stretches over a length of 1,700 km (1,056 mi). The Barisan Mountains run parallel to the western shore; they consist of a chain of extinct and active volcanoes. The highest point is Kerinci at 3,805 m (12,484 ft). To the east, the mountain ridge transitions into a wide alluvial plain streaked with swamps. In the northwest lies Lake Toba which, with an area of 1,103 km² (426 sq mi), is the largest volcanic lake in the world. It was created 74,000 years ago by the eruption of a supervolcano.

## JAVA

**Indonesia, 138,794 km² (53,589 sq mi)**

Java, the most densely populated island in Indonesia, lies east of Sumatra. On this extremely fertile island, there are many active and extinct volcanoes, among them the Merapi, one of the most dangerous in the world. The Sunda Trench, the subduction zone of the Australian plate, lies south of the island. The plate pushes under the Eurasian Plate here, which is the reason for the volcanism in this area and the frequent earthquakes. The weather is determined to a large extent by the monsoons.

## SRI LANKA

**Sri Lanka, 65,268 km² (25,200 sq mi)**

This tropical island country lies only a few degrees north of the equator at the southern tip of India, with which it was probably connected by Adam's Bridge 6,000 years ago. The central highlands, with mountains up to 2,500 m (8,202 ft) high, are well-known for tea-growing, a relic from the British colonial period, and also divide the island climatically.

## TASMANIA

**Australia, 64,519 km² (24,911 sq mi)**

The Bass Strait separates Tasmania from the southern Australian mainland. Large areas in the west of the island are untouched wilderness, consisting of rainforest and moorland with a large number of endemic species. In terms of its fauna, there are also many endemic species, such as the famous Tasmanian devil.

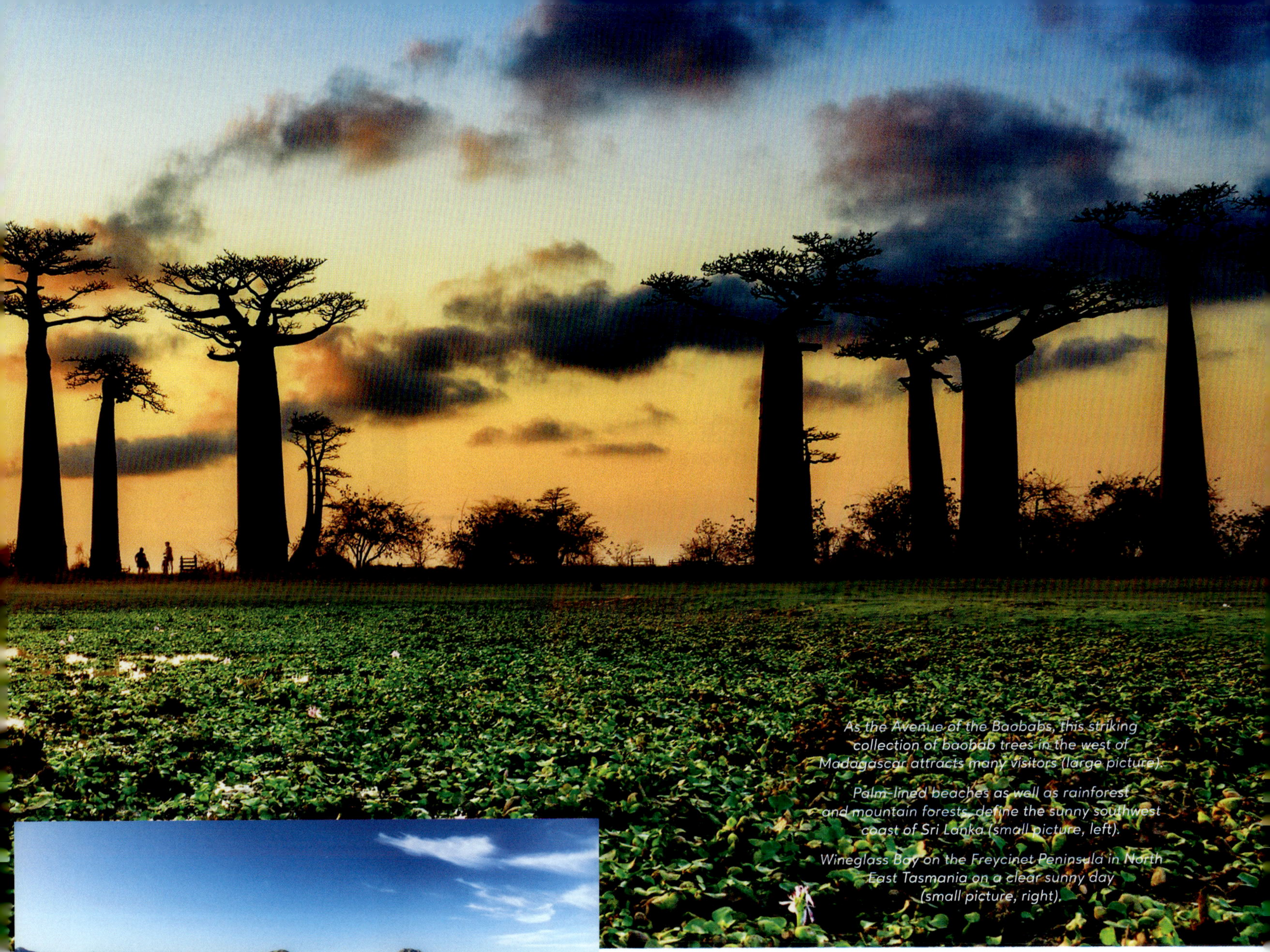

*As the Avenue of the Baobabs, this striking collection of baobab trees in the west of Madagascar attracts many visitors (large picture).*

*Palm-lined beaches as well as rainforest and mountain forests define the sunny southwest coast of Sri Lanka (small picture, left).*

*Wineglass Bay on the Freycinet Peninsula in North East Tasmania on a clear sunny day (small picture, right).*

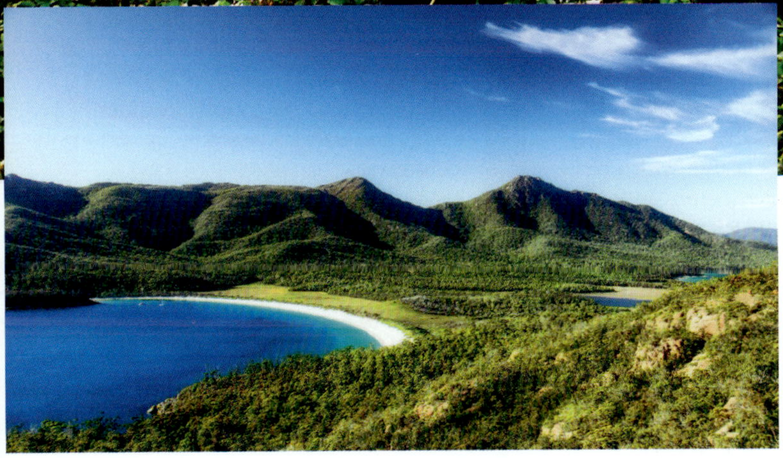

## FLORES

**Indonesia, 13,540 km² (5,228 sq mi)**

This island, around 400 km (248 mi) long and 70 km (43.5 mi) wide at its widest point, consists of a mountainous region with small lowland areas along the coast. A special feature is the Kelimutu volcano, which has three differently coloured crater lakes close together.

## TIMOR

**Indonesia / East Timor, 28,418 km² (10,972 sq mi)**

The island, approximately 500 km (310 mi) long and 80 km (50 mi) wide, is divided politically. The western half belongs to Indonesia, with the exception of a small enclave; the eastern part has been an independent state since 2002. The highest point is the extinct volcano, Tatamailau (2,963 m (9,721 ft)). Not much is left of the original vegetation of the tropical rainforest. Secondary forest and grassland predominate today, where the land is not being used for agriculture.

## SUMBAWA

**Indonesia, 14,386 km² (5,554 sq mi)**

Sumbawa lies between the islands of Lombok and Flores. The island is crossed by several mountain chains, on the slopes of which grow the remains of tropical rainforests. Mount Tambora, a volcano which devastated the land after a massive eruption in 1815, lies in the north of the island. The discharge of around 100 km³ of magma and 400 million tonnes (440,924,524 million tons) of sulfuric gas had a significant impact on the global climate at the time and led to a temporary cooling of the earth.

## SUMBA

**Indonesia, 10,711 km² (4,136 sq mi)**

The island of Sumba lies not far off from the Lesser Sunda Islands chain. The mountainous western area is damp and has sufficient rainfall for wet rice agriculture to be carried out. The east, however, is dry with tall grasses and remnants of a tree population and has more of a savanna character. Sumba has no active volcanoes. Its former abundance of sandalwood has been drastically reduced because of logging. Because of its remote location, traditional rituals and customs have remained alive here.

FOUR ELEMENTS

EARTH

# PACIFIC OCEAN

The islands of the Pacific Ocean lie mainly in the tropical and subtropical climatic zones. The South Pacific, or South Sea, represents an idealized island world

## NEW GUINEA

**Indonesia / Papua New Guinea, 785,753 km² (303,381 sq mi)**

The second largest island in the world lies north of Australia in close proximity to the equator. The island is divided politically: the west belongs to the Republic of Indonesia, and the east has belonged to the independent state of Papua New Guinea since 1975. Tropical rainforest dominates the island – around 75 per cent of the island is forested. With regards to flora, New Guinea is among the top ten leading regions for biodiversity. In terms of fauna, it also boasts a wide variety of indigenous species, among them the bird of paradise, the national symbol of Papua New Guinea.

## BORNEO

**Indonesia/Malaysia/Brunei, 748,168 km² (288,869 sq mi)**

The third largest island in the world straddles the equator in the permanently wet tropics. Rainforest covers the extensive lowlands and hilly regions. The island is relatively inaccessible and thus scarcely populated. The highest point is Mount Kinabalu in the north of the island at 4,095 m (13,435 ft). The biodiversity in flora and fauna species is very wide, but, due to deforestation, the home of the orangutan is under threat.

## HONSHU

**Japan, 230,316 km² (88,925 sq mi)**

Honshu, separated from Hokkaido by the Tsugaru Strait, is the largest island in Japan and forms part of the Circum-Pacific Belt, or Pacific Ring of Fire. There are numerous volcanoes here that are very active. They equalize the pressure which builds up due to the Pacific plate submerging beneath the Eurasian plate. Earthquakes, which are catastrophic for the densely populated island, are far from rare here. Mount Fuji, at 3,776 m (12,388 ft) high, is the highest volcano in Japan.

## SULAWESI

**Indonesia, 189,216 km² (73,056 sq mi)**

The irregular shape of this island, which has volcanic origins and which before Indonesian independence was still known as Celebes, makes this island particularly striking. The vegetation and climate – lush, thick rainforest and mountain forests with tropical temperatures – are similar to Borneo, its neighbour to the west, from which it is separated by the Makassar Strait. The island's relief is complex; the highest peak is Rantemario at 3,440 m (11,286 ft).

## SOUTH ISLAND, NEW ZEALAND

**New Zealand, 145,836 km² (56,308 sq mi)**

The Southern Alps and their foothills traverse the west of this island. To the east and south, they transition to mountainous areas and lowland plains. The highest point is Aoraki (Mount Cook) in the middle of the island. In the south, many glacial lakes and the characteristic fjord coast point to the landscape's formation during the ice age. The cool, temperate climate is accompanied by high rainfall.

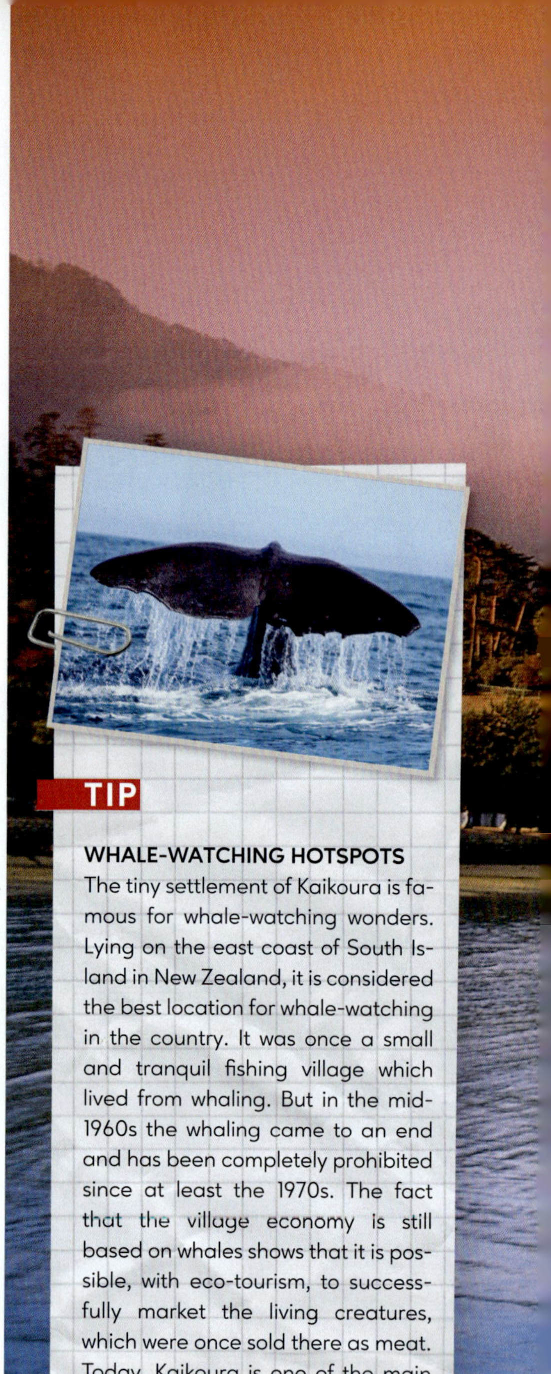

**TIP**

**WHALE-WATCHING HOTSPOTS**

The tiny settlement of Kaikoura is famous for whale-watching wonders. Lying on the east coast of South Island in New Zealand, it is considered the best location for whale-watching in the country. It was once a small and tranquil fishing village which lived from whaling. But in the mid-1960s the whaling came to an end and has been completely prohibited since at least the 1970s. The fact that the village economy is still based on whales shows that it is possible, with eco-tourism, to successfully market the living creatures, which were once sold there as meat. Today, Kaikoura is one of the main places to see whales, fur seals and elephant seals and to swim with dolphins.

## NORTH ISLAND, NEW ZEALAND

**New Zealand, 111,583 km² (43,082 sq mi)**

The 35-km (22-mi) wide Cook Strait separates North Island from South Island. Volcanic uplands lie at the centre of the subtropical North Island around Lake Taupo, at the edges of which thermal springs, hot springs, fumaroles, mud ponds and geysers rise to the surface from the earth's interior. Low mountain ranges run along the east coast of the island. The elongated northern section of the island has numerous bays with a wealth of islands to the east; in the west, by contrast, there are sheer, seemingly endless expanses of sandy beaches with no inlets.

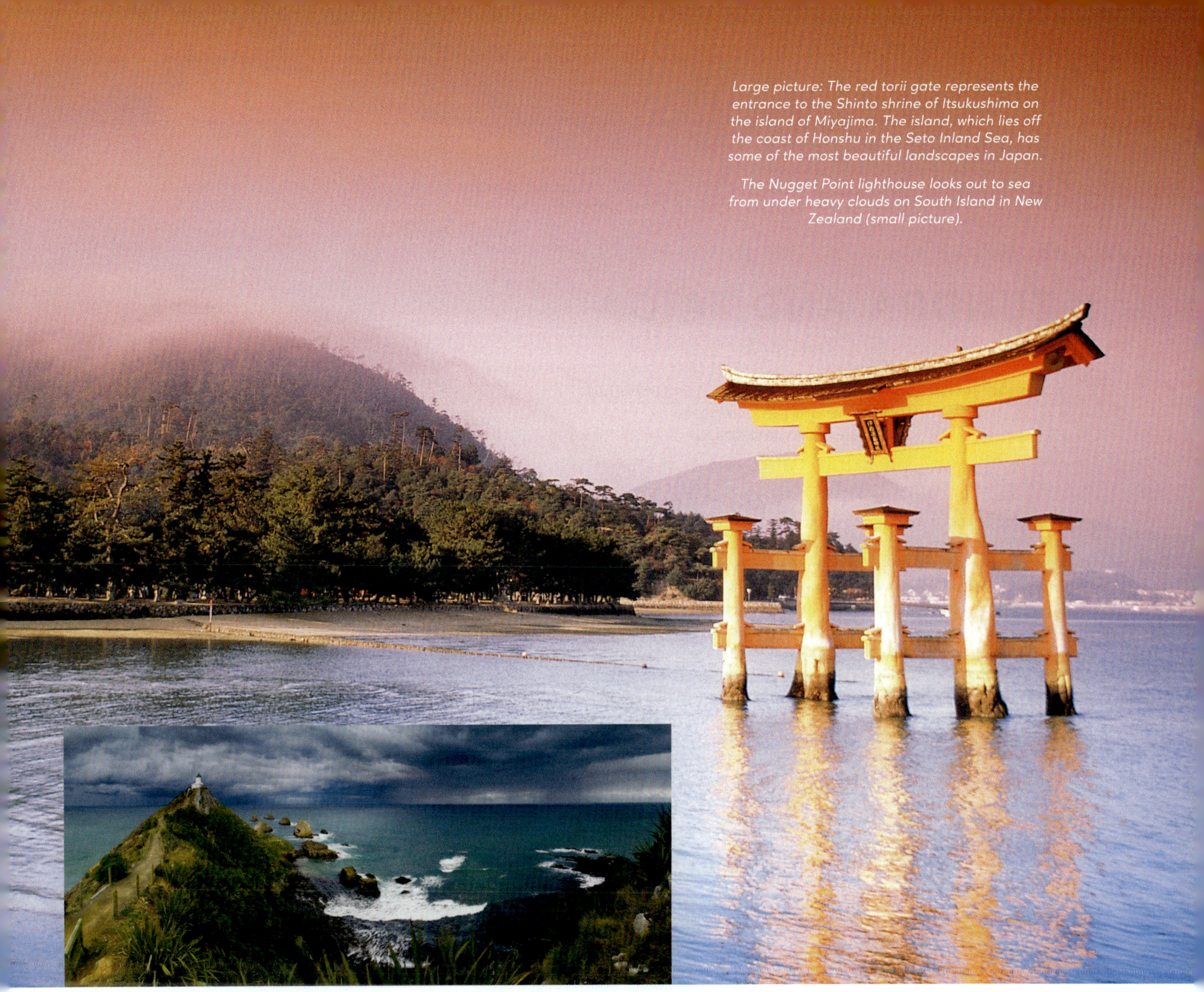

*Large picture: The red torii gate represents the entrance to the Shinto shrine of Itsukushima on the island of Miyajima. The island, which lies off the coast of Honshu in the Seto Inland Sea, has some of the most beautiful landscapes in Japan.*

*The Nugget Point lighthouse looks out to sea from under heavy clouds on South Island in New Zealand (small picture).*

## LUZON

**Philippines, 109,965 km² (42,458 sq mi)**

The largest island in the Philippines is one of the most densely populated in the world. Luzon is very mountainous; cordilleras run from north to south and form a boundary to wide lowland plains in the interior. There are many active volcanos, including Pinatubo and Mayon – the latter is known for its ideal typical shape as a stratovolcano. With Laguna de Bay, it has the second largest freshwater lake in southeast Asia.

## MINDANAO

**Philippines, 97,530 km² (37,657 sq mi)**

Among the many other mountains on Mindanao is Mount Apo, an extinct volcano 2,954 m (9,692 ft) high, which is also the highest summit in the Philippines. The island's tropical rainforest is home to a large number of native species as well as rare animals such as the Philippine eagle, with a wing span of up to 2.20 m (7.2 ft).

## HOKKAIDO

**Japan, 78,719 km² (30,394 sq mi)**

Cool summers and snowy winters characterize the climate in the northernmost and second largest island in Japan. The plateaus in the centre of the island, created as a result of volcanic activity, are encircled by lowland plains extending as far as the coast. There is frequent seismic activity on the island. In addition to several still active volcanoes, there are frequent and violent earth tremors and quakes on Hokkaido.

## SACHALIN

**Russia, 72,493 km² (27,990 sq mi)**

This elongated island in the east is Russia's largest island. It measures almost 1,000 km (621 mi) in length but is only between 24 and 160 km (15 and 100 mi) wide. Low, narrow mountain chains run across the length of the island. For the most part, the vegetation is pristine taiga.

FOUR ELEMENTS

EARTH

# TUNDRA AND TAIGA

Over the years, plants, animals and also humans have adapted to the unique climate found in the tundra and taiga

The steppes north of the coniferous boreal forest zones of Eurasia and North America, characterized by long, cold winters and short summers with low rainfall, are referred to as tundra. Due to the permafrost, which never completely thaws even in summer, only those plants and animals that can cope with the barren land and the low temperatures have established themselves in these regions. In summer, the steppe is populated by many different species of birds. The largest continuous and, in part, still wholly unexploited coniferous zone in the world, which lies south of the tundra and is only occasionally interrupted by vast marshlands, is the taiga. This type of landscape is also characterized by long, cold winters and short, cool summers with permafrost, making the taiga only habitable by animals and plants resistant to the cold.

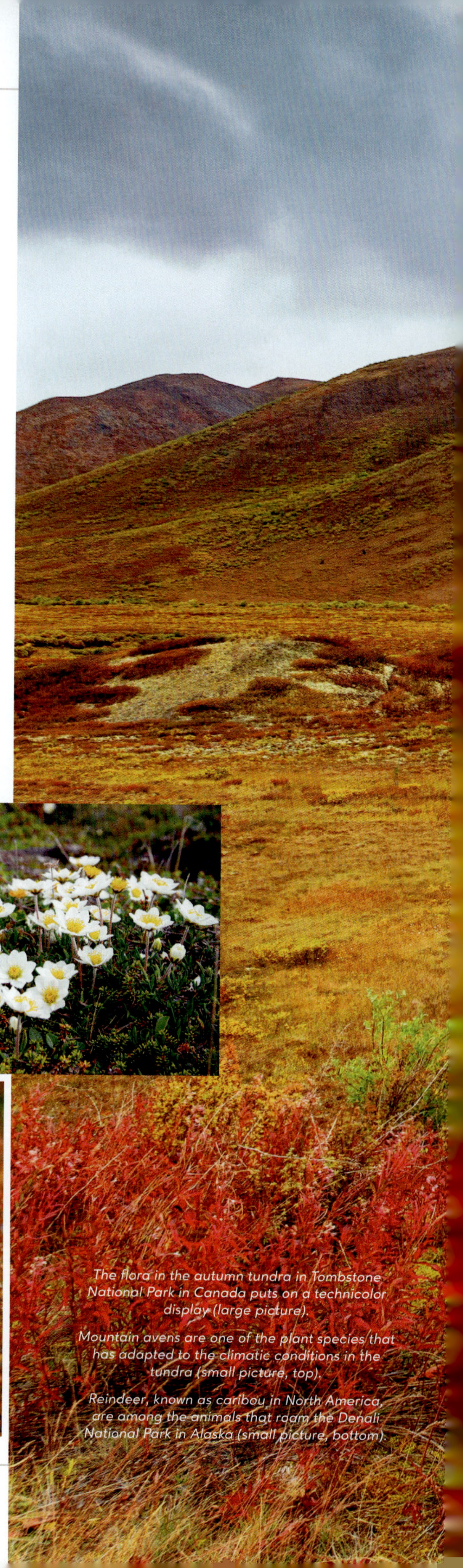

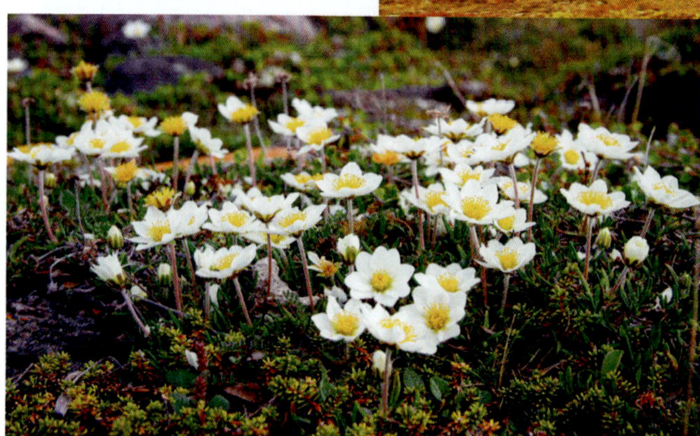

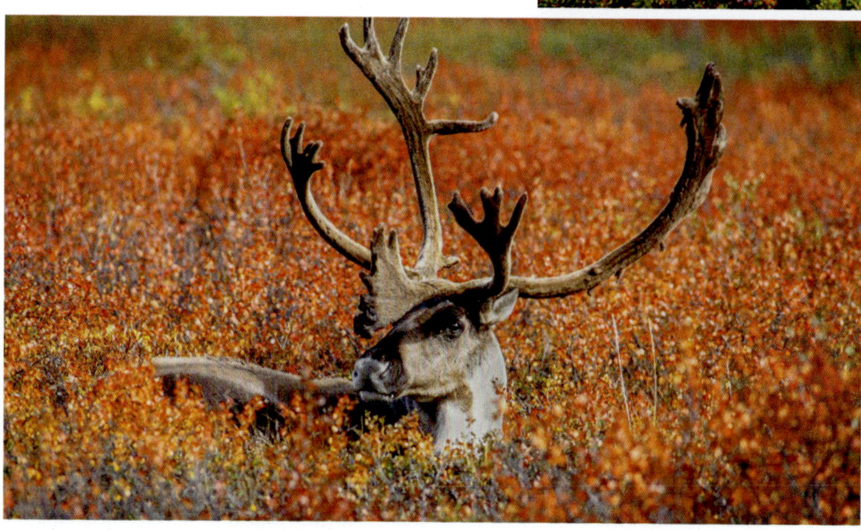

The flora in the autumn tundra in Tombstone National Park in Canada puts on a technicolor display (large picture).

Mountain avens are one of the plant species that has adapted to the climatic conditions in the tundra (small picture, top).

Reindeer, known as caribou in North America, are among the animals that roam the Denali National Park in Alaska (small picture, bottom).

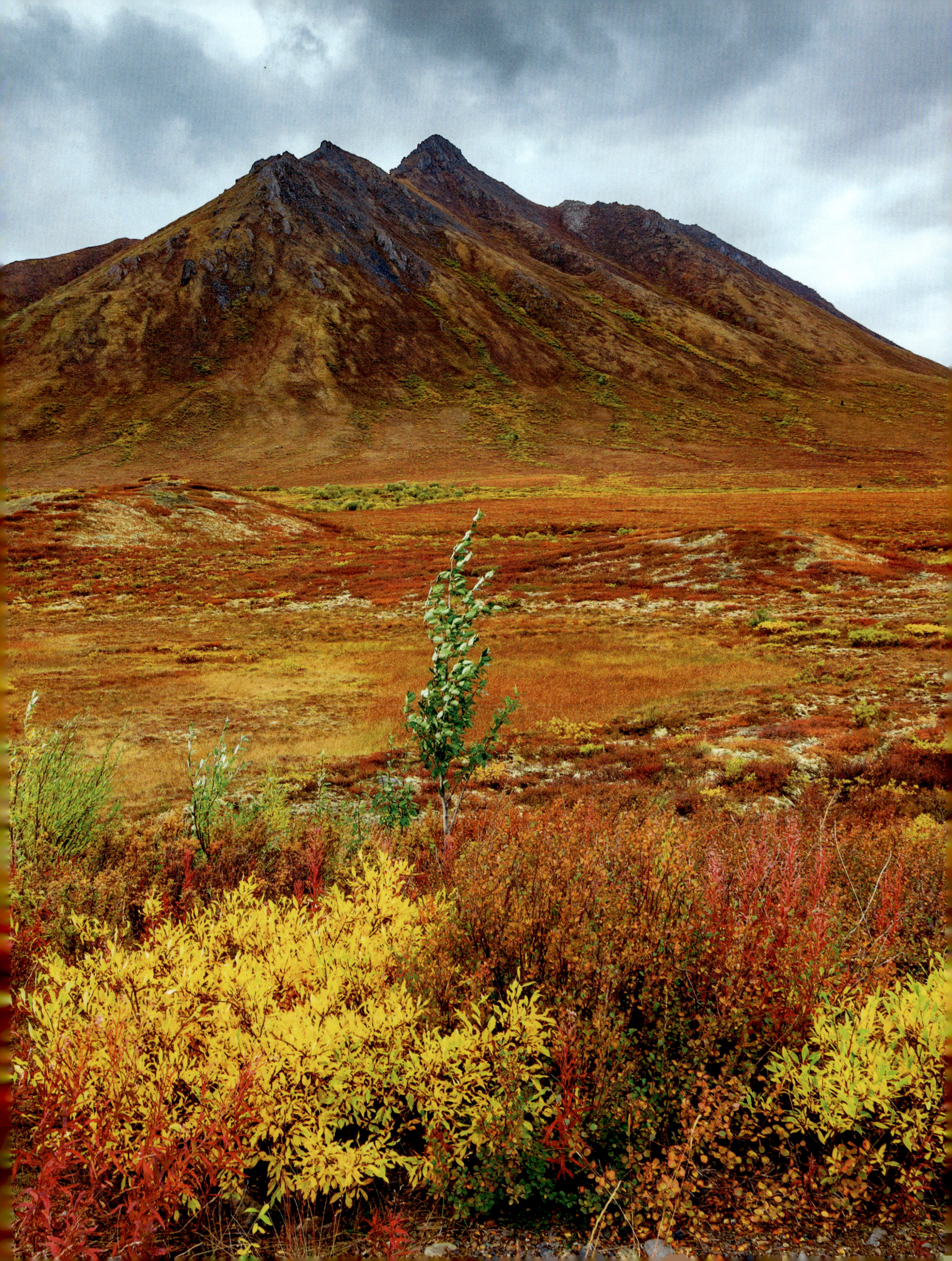

FOUR ELEMENTS

EARTH

# NORTH ASIA

The entirety of North Asia is characterized by a polar climate and a landscape which is predominantly tundra

Beyond the Arctic Circle, North Asia (or Siberia) stretches from roughly the island of Novaya Zemlya over the Byrranga Mountains in northern Siberia to the Anadyr Mountains in the extreme east of Asia. The tundra runs along the entire Siberian coast on the Arctic Ocean in a more or less narrow strip. It is an almost treeless, steppe-like landscape which transforms to a forest tundra where only a few small trees grow in the south and then transitions to the coniferous boreal forest of the taiga. The tundra climate (long, extremely cold winters and short, cool summers, when the sun never completely sets and only the top layer of the permafrost thaws) means that food tends to be scarce for the animals that live here – grass, berry bushes, mosses and lichen grow here in the summer months. But the musk oxen, reindeer, red deer, Arctic hares and lemmings seem to manage. In the winter, the reindeer scrape at the frozen ground with their hooves and antlers in order to find lichen. Among the feathered inhabitants of the tundra are snowy owls, red-breasted geese and common guillemots (murre).

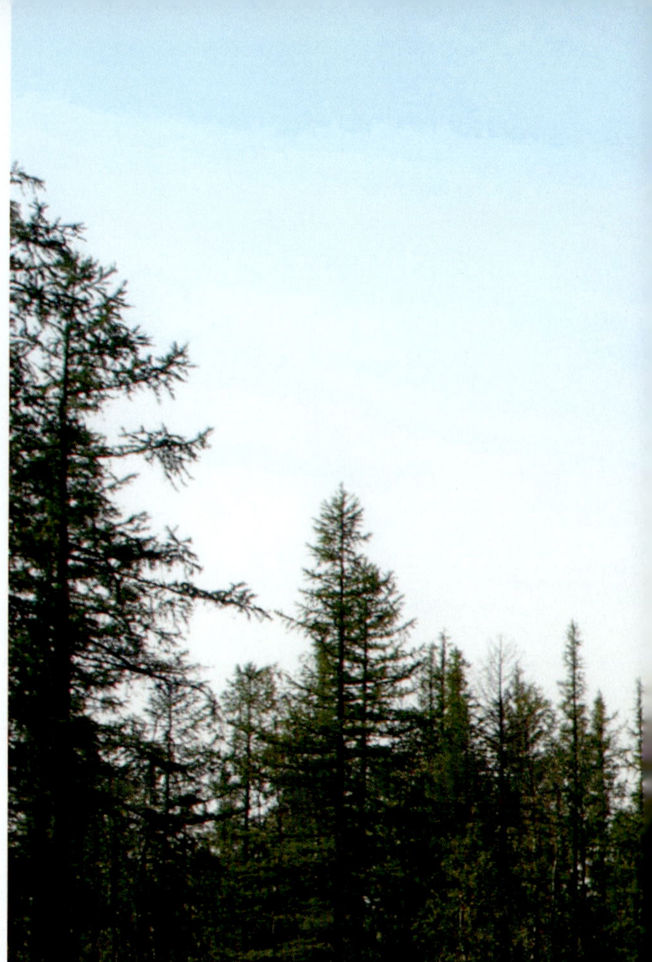

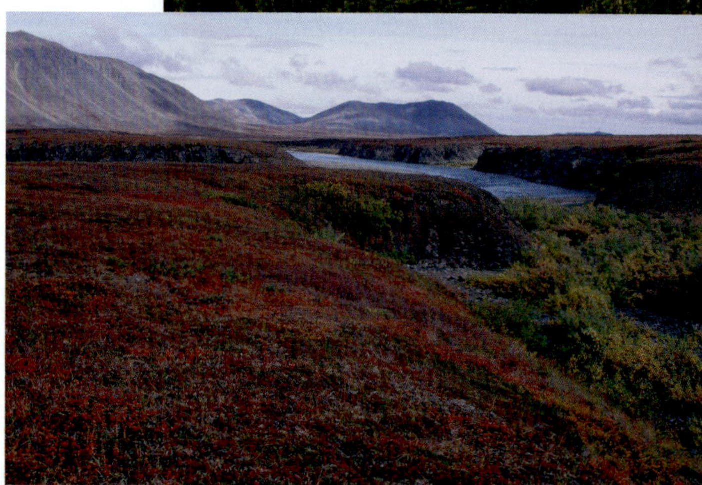

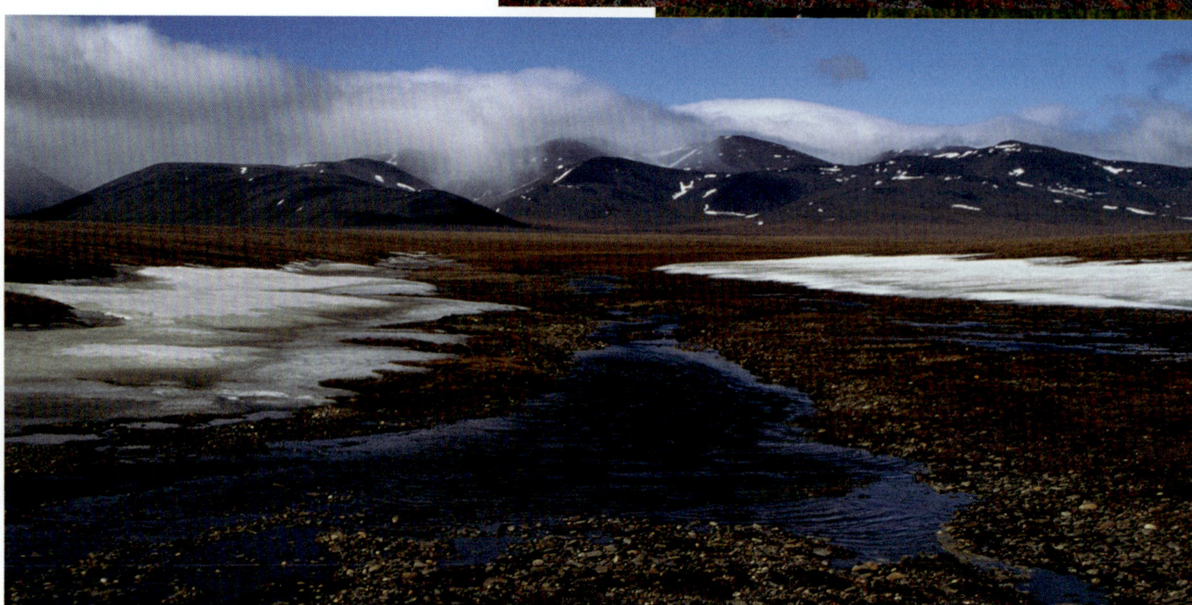

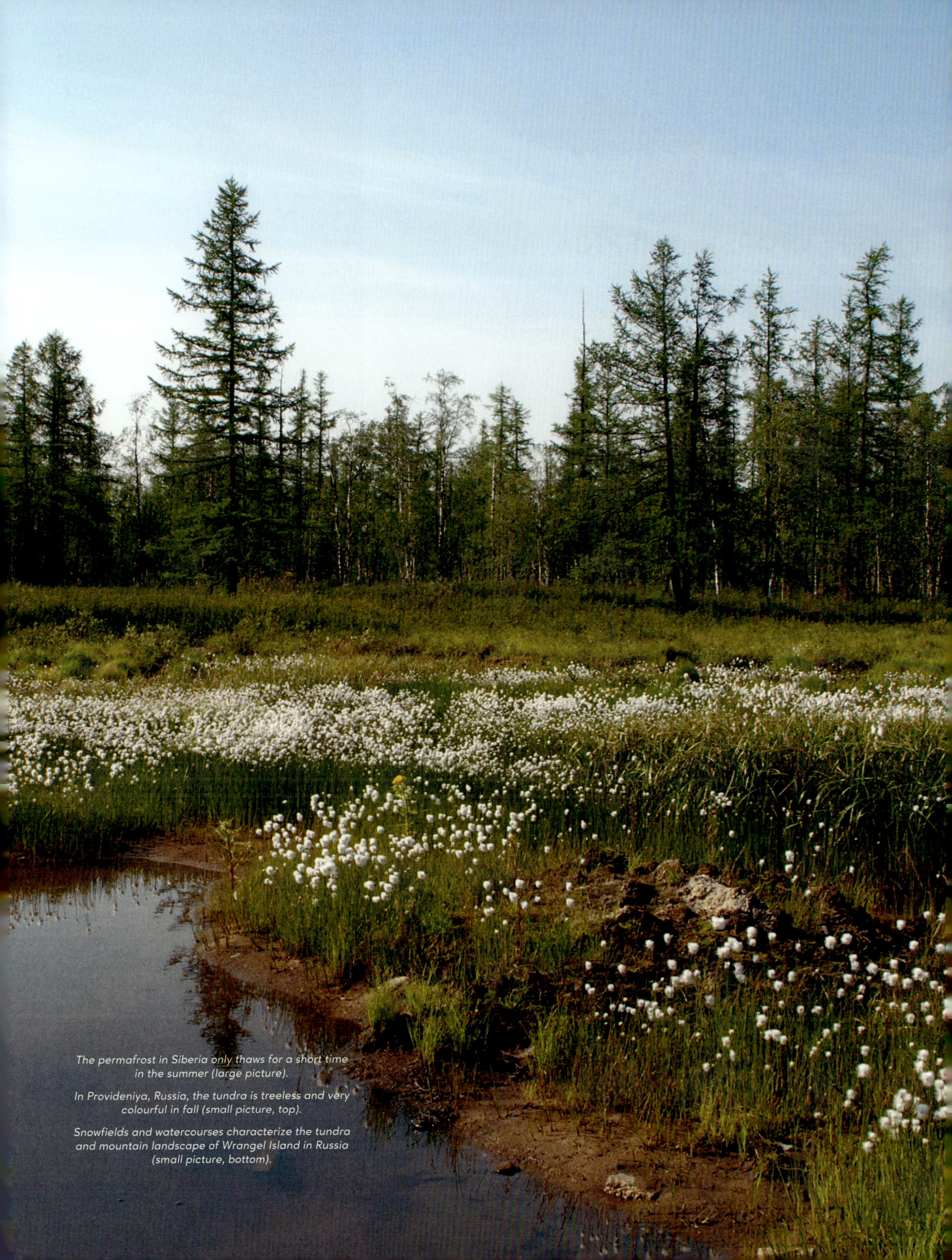

The permafrost in Siberia only thaws for a short time in the summer (large picture).

In Provideniya, Russia, the tundra is treeless and very colourful in fall (small picture, top).

Snowfields and watercourses characterize the tundra and mountain landscape of Wrangel Island in Russia (small picture, bottom).

# NORTH AMERICA

Beyond the Arctic tree line, the landscape of North America is defined by vast areas of tundra where living conditions are extreme.

The tundra in the North American subpolar regions is also characterized by permafrost and sparse vegetation. Plants survive the Arctic winter because of the protection provided by the snow cover. But, in May, the snowmelt begins along the Arctic coast. As soon as the temperature reaches freezing point, the vegetation starts a phenomenal growth spurt because the plants have continued growing in their snow tunnels and have already developed buds and leaves. In damp areas, bog moss plays a major part; in dry locations, lichens and bushes that will bear berries in the autumn thrive. On the stony southern slopes warmed by the low-lying sun, true flower gardens spread out. More than one hundred bird species breed in the Arctic circle in the summer, such as brent geese, long-tailed ducks and Arctic terns. Starting in April, herds of caribou migrate from the forests of Canada to the tundra. The populations of smaller mammals, such as lemmings, voles and mountain hares, increase rapidly and attract skuas and Arctic foxes. In fall, most of the animals once again move southwards.

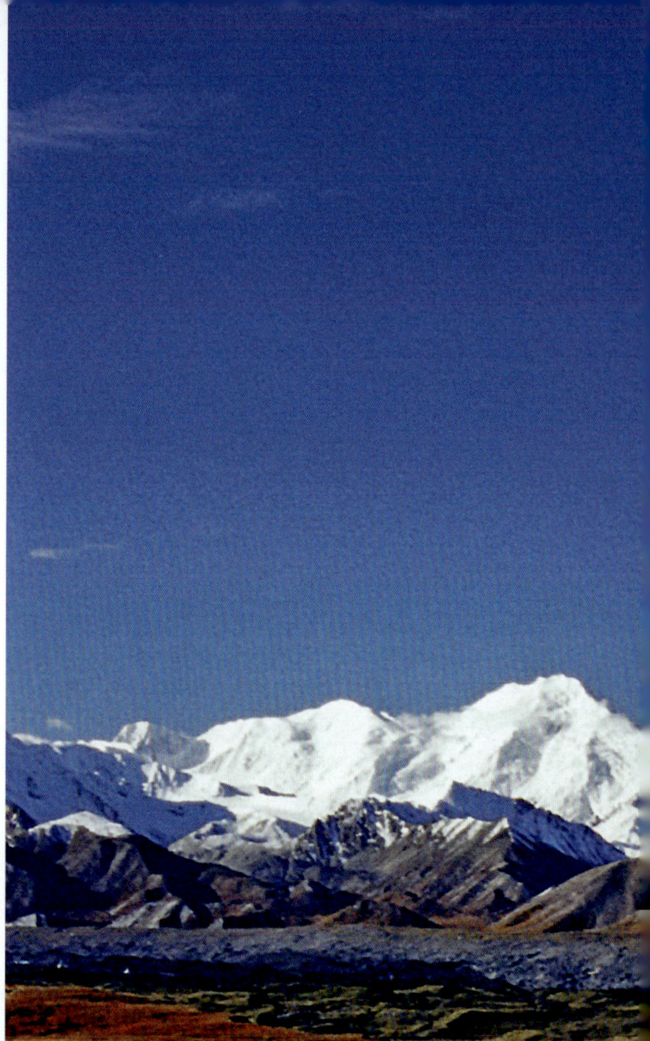

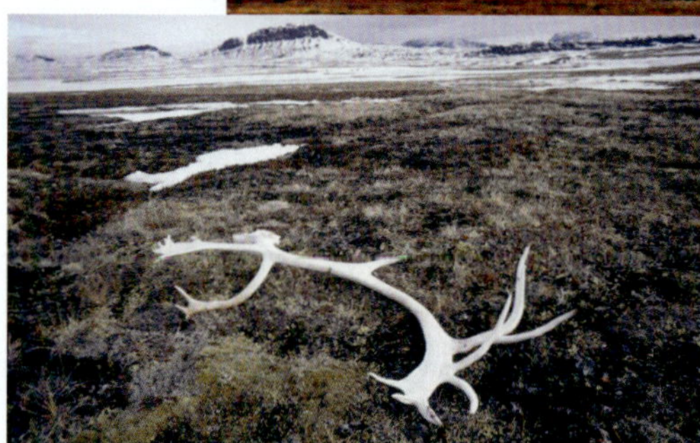

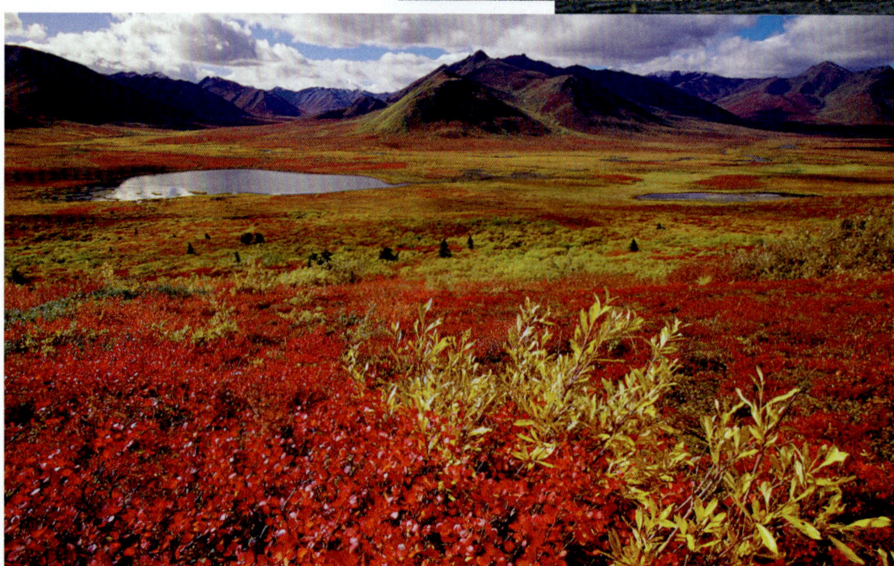

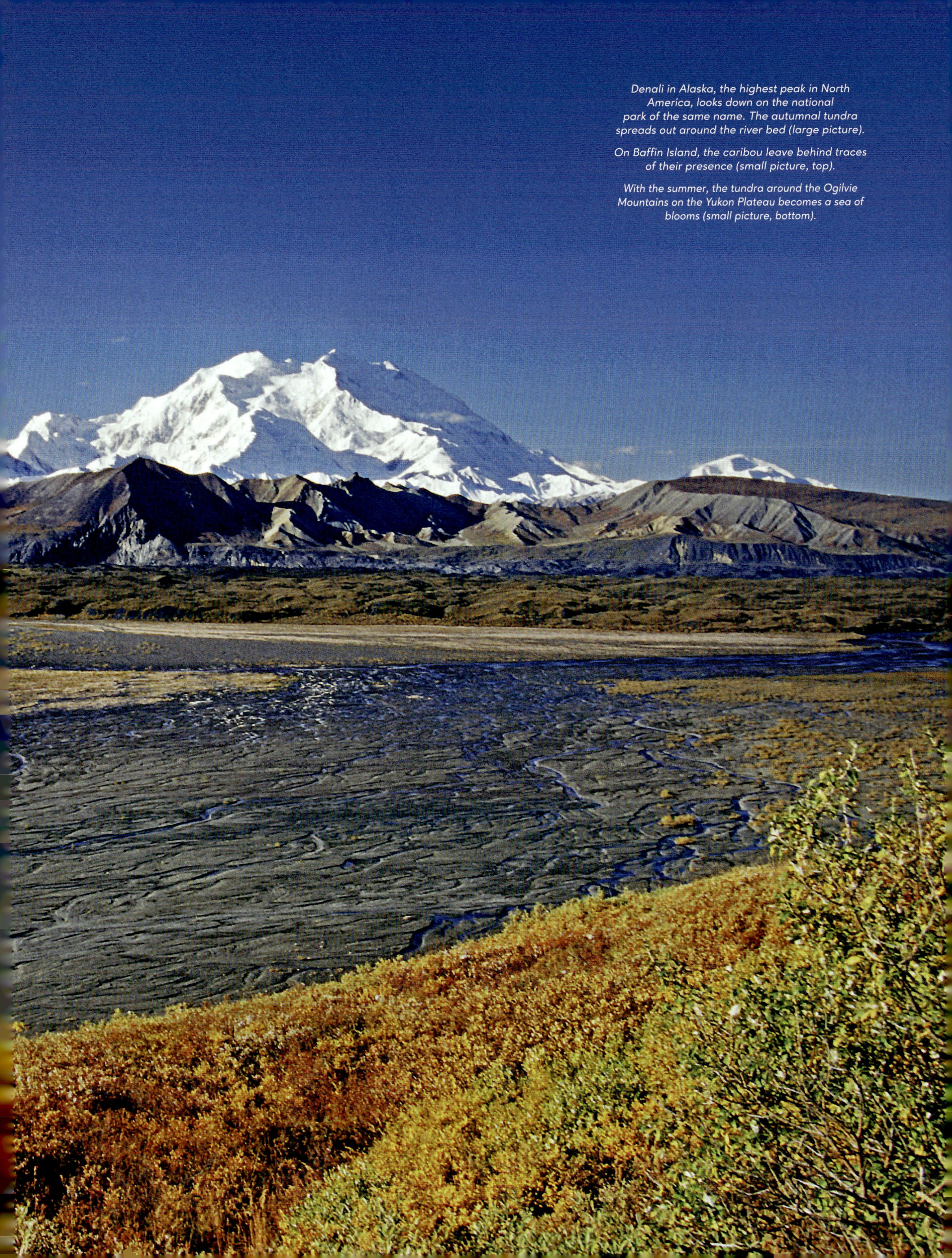

Denali in Alaska, the highest peak in North America, looks down on the national park of the same name. The autumnal tundra spreads out around the river bed (large picture).

On Baffin Island, the caribou leave behind traces of their presence (small picture, top).

With the summer, the tundra around the Ogilvie Mountains on the Yukon Plateau becomes a sea of blooms (small picture, bottom).

## ANIMAL MIGRATIONS

Many creatures living on the earth, in the air or underwater are constantly on the move. They travel in order to find food, water, a mate for reproduction or good conditions in which to rear their young.

A team working with the American zoologist, Kyle Joly, has recently carried out an investigation into which land mammals undertake the longest journeys. For this, existing data on this remarkable natural phenomenon was evaluated and supplemented with their own survey data. The five terrestrial mammals who showed the greatest wanderlust:

The Tibetan antelope (chiru, *Pantholops hodgsonii*) occupies fifth place with a distance of 700 km (435 mi). Each year between May and June, after mating, the females travel from the nature conservation areas in Tibet to the provinces of Xinjiang and Qinghai in China. Their destination is the area around the Zhuonai Lake, where they will bear their young.
The mule deer *(Odocoileus hemionus)*, with a distance of 772 km (480 mi), comes in at fourth place. These animals are spread over an area stretching from the mountainous regions of the Rocky Mountains to the northwest of Mexico. A herd of them migrates every summer from Wyoming to Idaho. The animals, otherwise quite wary of people, are not daunted by the feat of crossing two highways on their journey.

A pack of wolves from the Northwest Territories in Canada – the only predators on this list – is in third position. They cover, on average, over 1,016 km (630 mi) every year in their search for prey. While undertaking their analysis, the researchers also came across one lone wolf from Mongolia who was reported to have travelled 7,247 km (4,503 mi) in one year.

Reindeer *(Rangifer tarandus)* from the Russian Taymyr Peninsula – the northernmost part of any continental mainland in the world – achieved second place. Every year, in the Arctic spring, they travel 1,200 km (745 mi) with their newborn calves from the forests of the taiga north into the open tundra. An arduous journey, especially for their young. Prior to notable global climate changes, the river on this route used to freeze over, making it possible for the reindeer to walk across the ice. But now they often have to swim across the river. Many of the calves are too weak to do this, so they do not survive the journey. In 2000, there were still 500,000 wild reindeer on the Taymyr Peninsula – now there are just 380,000. This sharp drop in the reindeer population, however, is primarily due to unauthorized hunters who hunt the reindeer for their antlers.

First place among the wandering mammals goes to the caribou. These North American representatives of the reindeer family are somewhat larger and have longer legs than their relatives. With their annual journey of 1,350 km (839 mi), the caribou in the Northwest Territories and in Alaska are definitely the top long distance runners.

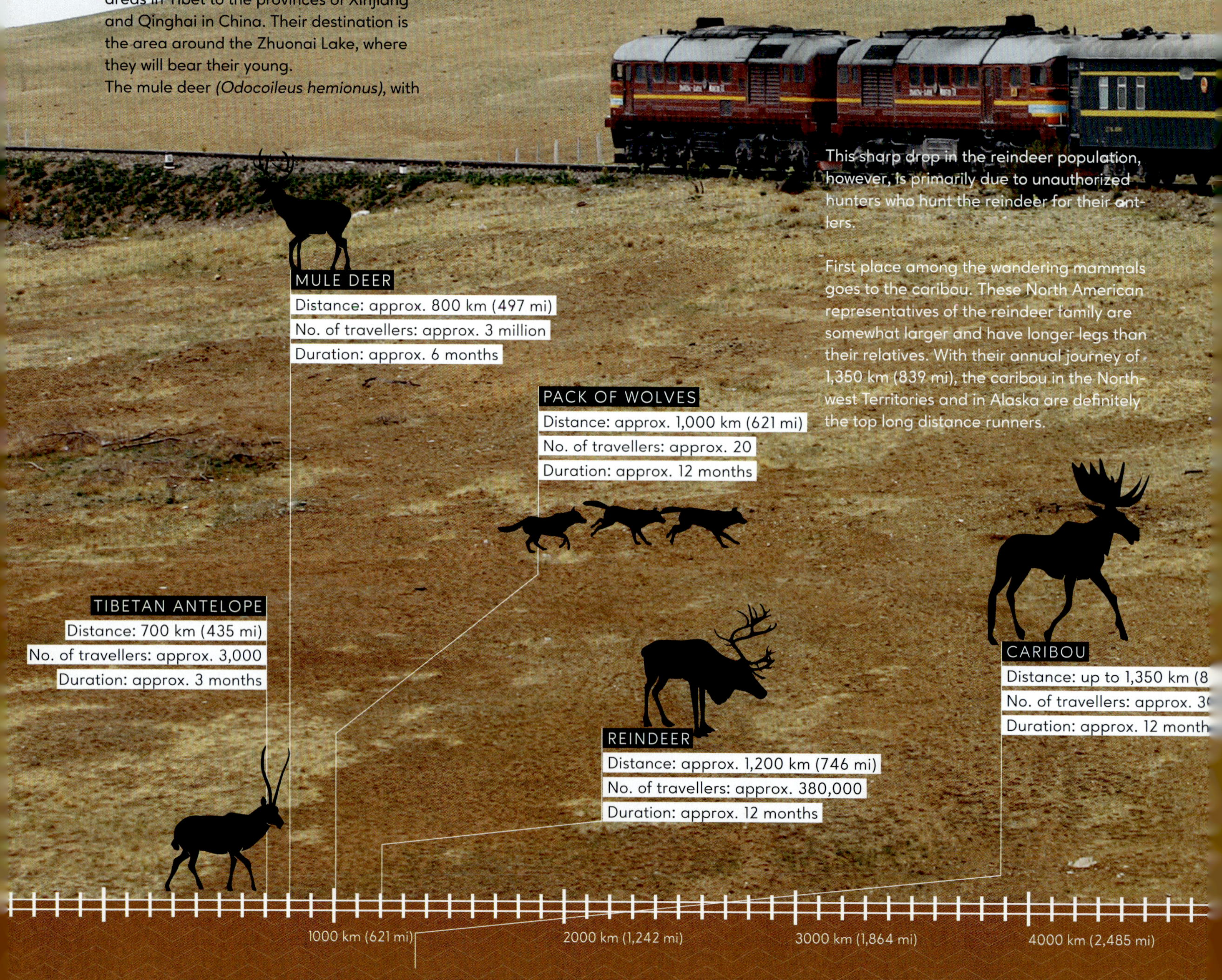

## THE LONGEST RAIL LINE IN THE WORLD

The Trans-Siberian Railway, or Transsib for short, is the longest continuous rail line in the world – at exactly 9,288.2 km (5,771 mi). It connects Moscow in Europe with Vladivostok in Asia.

Ever since construction started in 1891, Transsib has been a true legend. For 25 years, labourers, sometimes more than 90,000, toiled in freezing temperatures in winter and in scorching heat in the summer, surrounded by plagues of mosquitos. Thousands of the workers did not survive the construction work. With the inauguration at the bridge over the Amur River near Khabarovsk, the stretch was finally completed in 1916.

With the Trans-Siberian Railway, it was possible to open up the regions of Russia in the far east and mine the valuable raw materials in Siberia. Many new settlements were founded along the line. Novosibirsk in West Siberia was settled in 1893 during the construction of a rail bridge over the River Ob; today, it is the third largest city in Russia with a population of around 1.5 million.

Since 2002, the Trans-Siberian Railway has been electrified along its entire length. The line, which like almost all Russian rail lines, has a track gauge of 1,520 mm, has 396 stations and 89 cities, for instance, Yekaterinburg, Novosibirsk, Krasnoyarsk and Irkutsk on Lake Baikal.

On the nearly 7-day journey with the Trans-Siberian Railway from Moscow, travelling east to the Pacific, the train also passes through seven different time zones.

FOUR ELEMENTS

EARTH

## AND THEN EVERYTHING RETURNS!

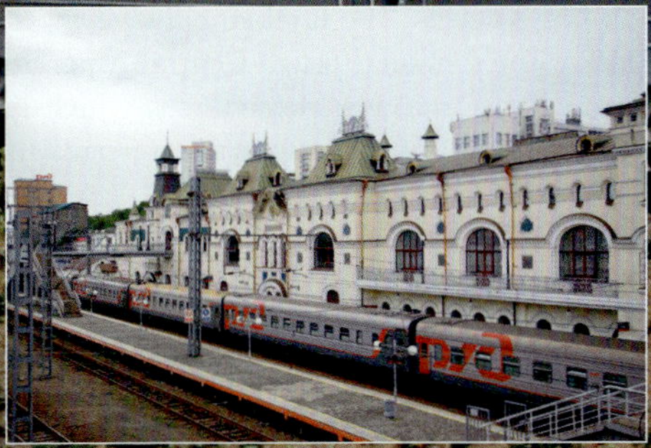

### FINAL STOP OF THE TRANSSIB

The port of Vladivostok is one of the major commercial centres in the East Asian region. It lies on the Sea of Japan, close to the borders with China and North Korea. In the 20th century, trade flourished following the completion of the final station on the Trans-Siberian Railway. The magnificent houses of the businessmen of that time can still be seen today.

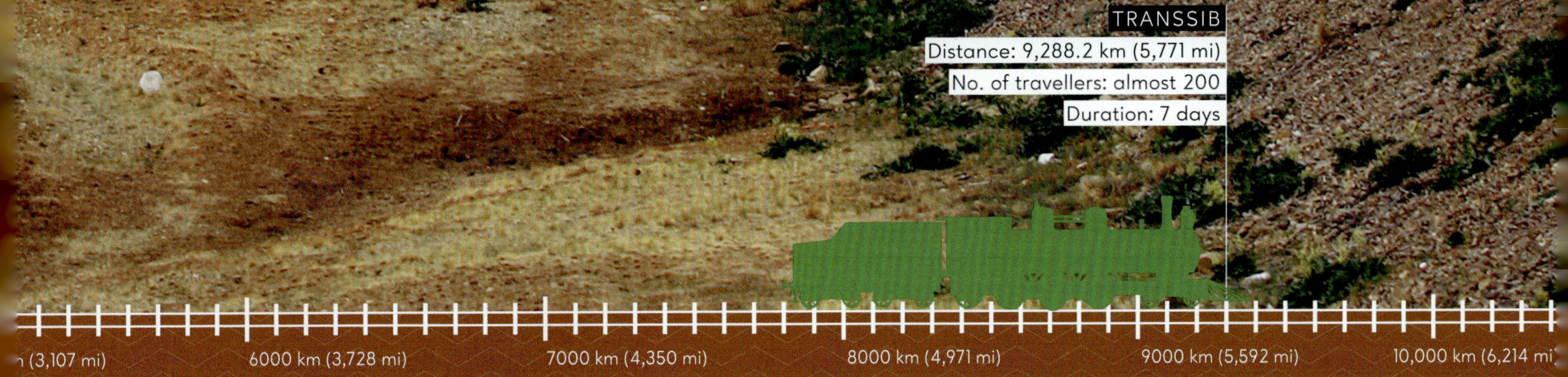

FOUR ELEMENTS

EARTH

# THE PAN-AMERICAN HIGHWAY

The Pan-American Highway stretches from Alaska to Tierra del Fuego and comprises a multitude of highways linked together – with a gap in Panama

The highway system covers almost 48,000 km (29,825 mi) overall, with the longest north-south section measuring 27,750 km (17,243 mi). Historically, the Pan-American Highway originated at an international conference in Buenos Aires. Officially, it begins in Laredo, Texas, on the border between the USA and Mexico, and runs from there in the direction of Mexico City. Canada and the USA are not, in fact, part of the Pan-American highway network. But for reasons of tourism, various routes in the USA and Canada were included in the system. There is a route that travels via Minneapolis or Denver as well as along the west coast via Los Angeles. The final stop, however, is always Prudhoe Bay in Alaska. The section through Central America from Laredo, Texas to the Panama Canal via Mexico City, which is particularly popular with North American tourists, is also referred to as the Inter-American Highway.

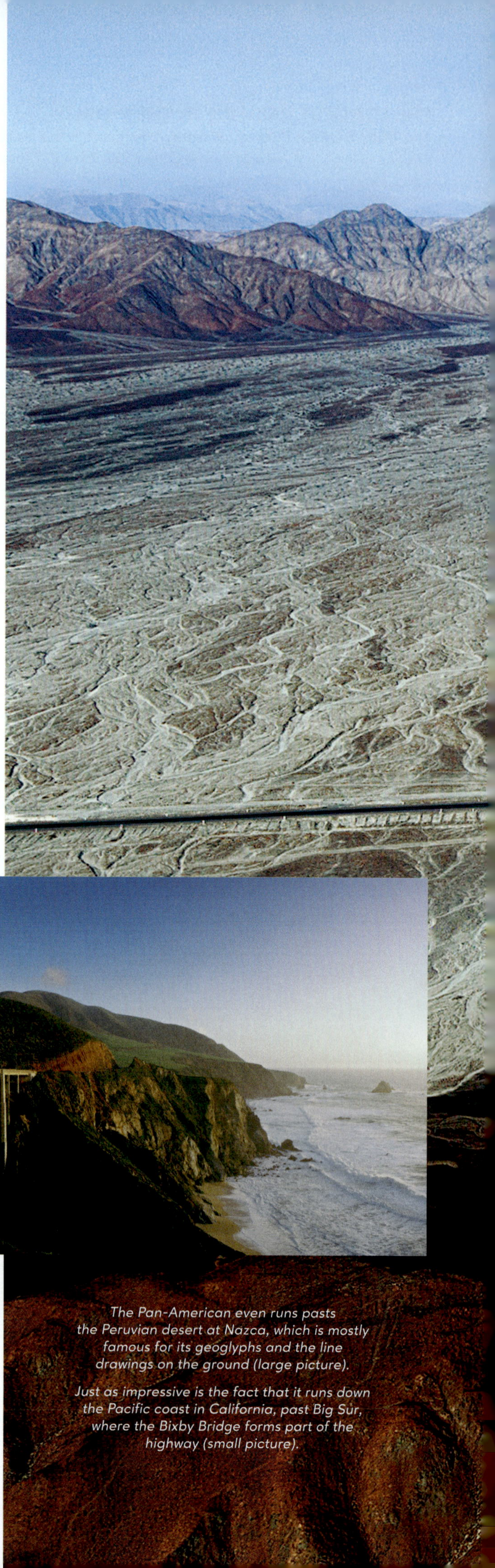

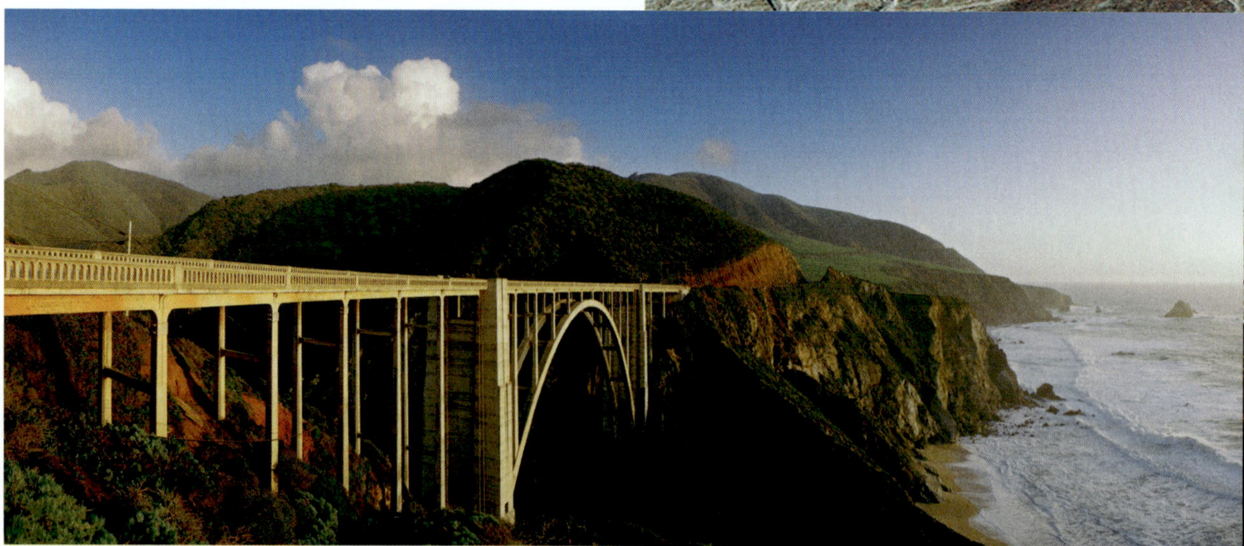

There is a gap of around 100 km (62 mi) in the Pan-American Highway network (the Darien Gap) between Yaviza in southeastern Panama and northwest Colombia This area contains an inaccessible mountainous swamp and jungle area where roads have not yet been built. In South America, the Pan American Highway fans out once again into various different routes. The southernmost route runs via Buenos Aires to Ushuaia in Tierra del Fuego in the far south of the continent.

*The Pan-American even runs pasts the Peruvian desert at Nazca, which is mostly famous for its geoglyphs and the line drawings on the ground (large picture).*

*Just as impressive is the fact that it runs down the Pacific coast in California, past Big Sur, where the Bixby Bridge forms part of the highway (small picture).*

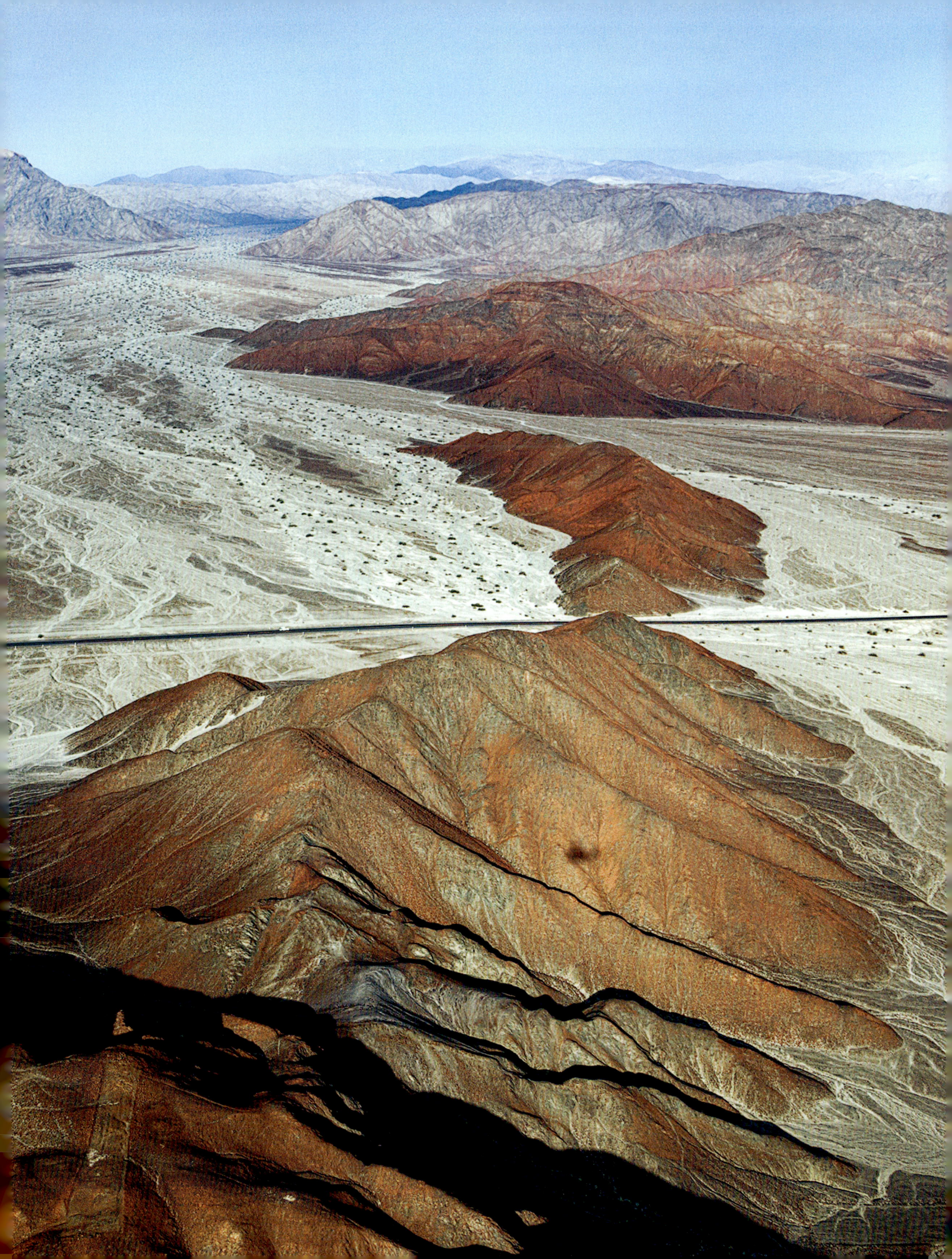

# SIBERIAN TAIGA

The largest forest region in the world

In northern Russia, the largest unified belt of conifer forest in the world stands with a width of almost 1,000 km (621 mi) and a length of 4,800 km (2,982 mi). A characteristic feature of this taiga region is its marshy or permanently frozen ground, which is often only covered by a sparse layer of herbage, low shrubs and mosses. The taiga landscape generally lies far from the ocean and is therefore influenced by a continental climate. Harsh winters with frequent snow are typical, with dry, warm summers, which may even be very hot in parts of Siberia. The temperature difference between summer and winter can be as much as 100°C (212°F). In the southern taiga, the growing season for vegetation, when temperatures are above 10°C (50°F), is only two to three months long; in more northern latitudes, it is even shorter. There are very few different species of fauna and flora, but the individual animals and plants occur in large numbers. The tree population comprises spruce, pine, fir trees and larch, among others. The taiga is generally inhabited by mammal species that are resistant to the cold. Many bird species overwinter in warmer climes; other animals hibernate or enter a period of torpidity. Large parts of the taiga are in the permafrost zone: this zone encompasses roughly one-fifth of the earth's surface, and the ground there is frozen all year round to a depth of up to 1,000 m (3,280 ft). The top metre of the ground may thaw in the summer months, but the resulting meltwater cannot drain away. This results in vast expanses of marshland.

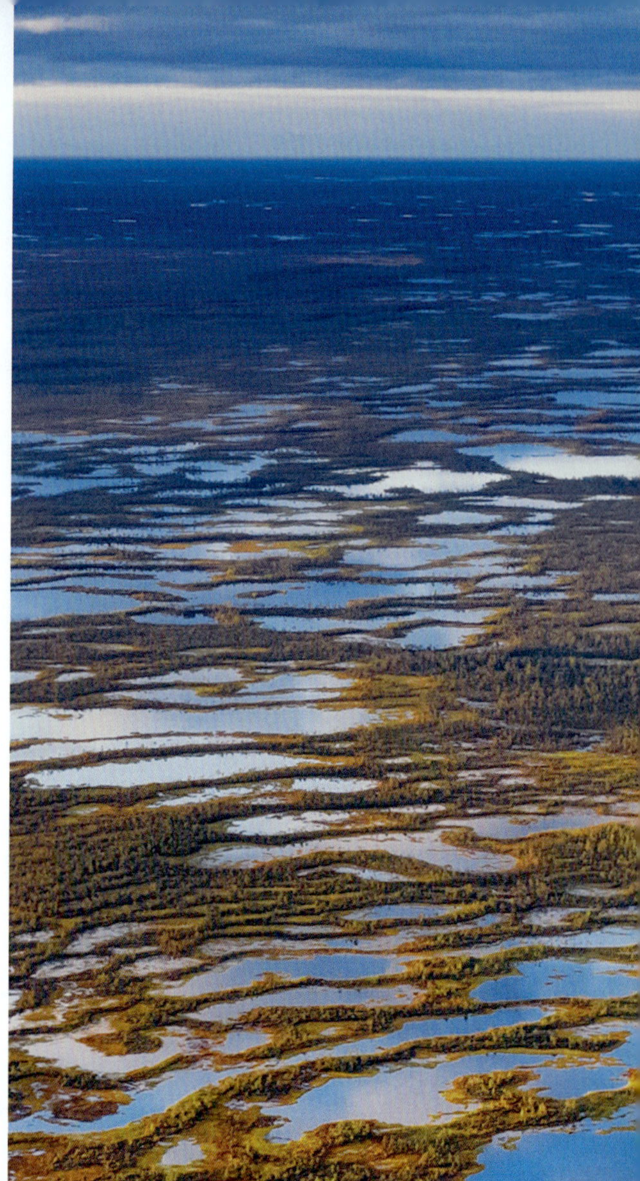

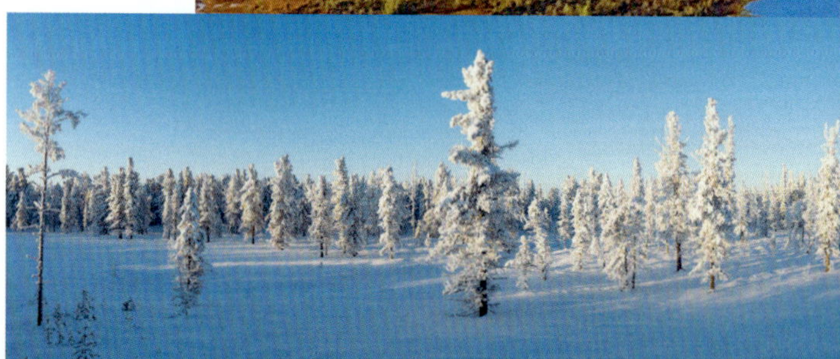

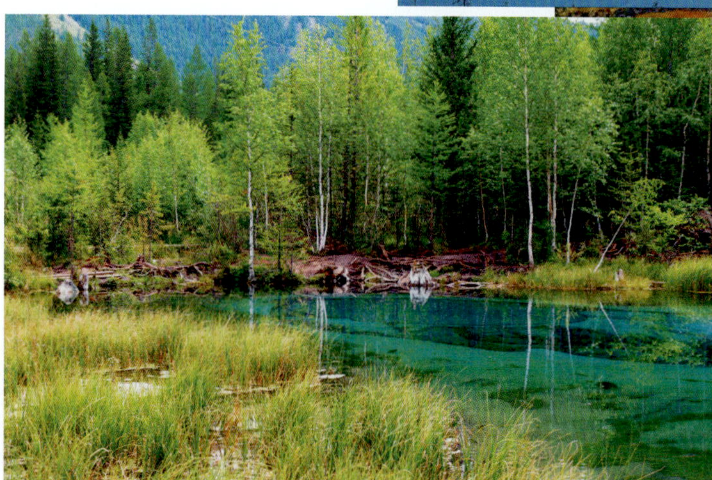

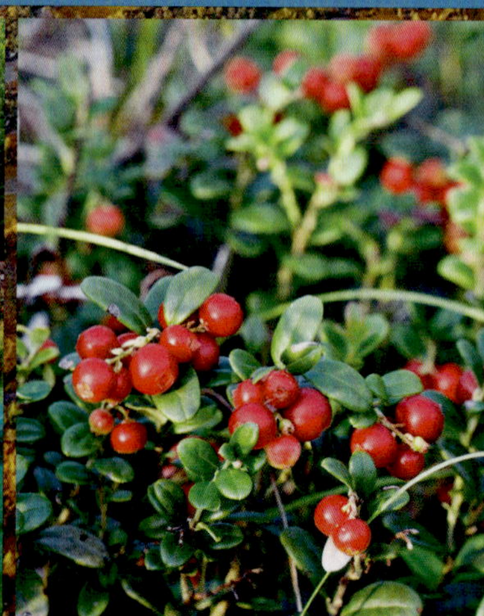

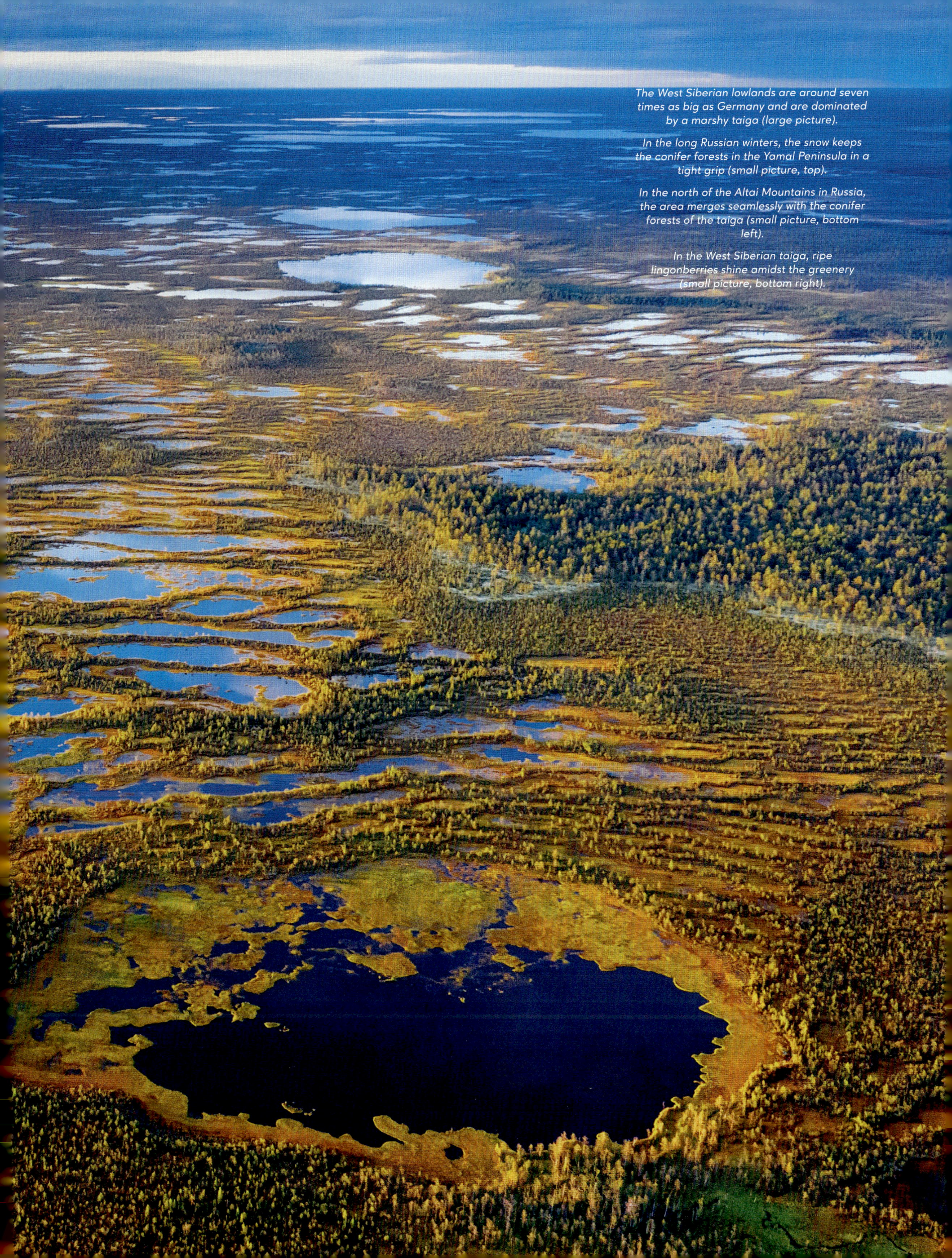

*The West Siberian lowlands are around seven times as big as Germany and are dominated by a marshy taiga (large picture).*

*In the long Russian winters, the snow keeps the conifer forests in the Yamal Peninsula in a tight grip (small picture, top).*

*In the north of the Altai Mountains in Russia, the area merges seamlessly with the conifer forests of the taiga (small picture, bottom left).*

*In the West Siberian taiga, ripe lingonberries shine amidst the greenery (small picture, bottom right).*

FOUR ELEMENTS

EARTH

# TIGERS, ELKS, BEARS

The largest animals
in the taiga

## SIBERIAN TIGERS – THE LARGEST BIG CATS IN THE WORLD

Around 450 Siberian or Amur tigers still live in the wild in the forests around the Amur River and its tributaries in the far east of Russia and the adjoining areas in China and North Korea. Although the tigers are a protected species, they are critically threatened with extinction by hunters. The main reason for this is that the Siberian Tiger is very sought after for East Asian medicine. It represents a very lucrative quarry for the hunters because almost all parts of the dead animal are used for medicinal purposes. The skin, bones, teeth, claws, blood – even the gall stones and genitalia are considered valuable for treatments for medical conditions such as rheumatism, epilepsy or skin or eye defects. Particular substances are supposed to increase intelligence and improve general well-being. These tiger medicines are sold for high prices not only in China, but also in the USA and in Europe. Furthermore, expensive coats can be made from tiger skin and exported for large profits. The tiger, which is up to 3 m (10 ft) long and weighs 300 kg (661 lb), is a solitary, nocturnal animal, ruling over a huge territory. Here, it hunts wild animals such as gorals, wild boars and deer, covering 15 to 20 km (9 to 12 mi) a day. To survive in the cold climate, it must consume around 10 kg (22 lb) of meat per day.

## ELKS – THE LARGEST SPECIES OF DEER

The natural habitat of the elk is the boreal conifer forests of Europe, Asia and North America. The largest member of the deer family is the North American elk. It can be found in Alaska, but also in Canada and the area around the Great Lakes in North America. The male, called a bull elk, can be up to 2.3 m (7.5 ft) at the shoulder and weigh up to 500 kg (1,102 lb). Its mighty antlers are very striking: each is 1 to 1.6 m (3.3 to 5.25 ft) wide with several tines and are shed in winter. The female elk, the cow, is smaller and without antlers.

In the rutting season, the elks gather in specific places. After mating, the females go their own way once more. When all the females have finished mating, the males also leave the rutting site. After an eight-month gestation period, the females give birth to one or two calves, who stay with the mother until she is once again due to calve, at which point she drives them away. The animals are solitary and diurnal. The maximum lifespan of elks is around 27 years although, in the wild, they seldom live past 15 years.

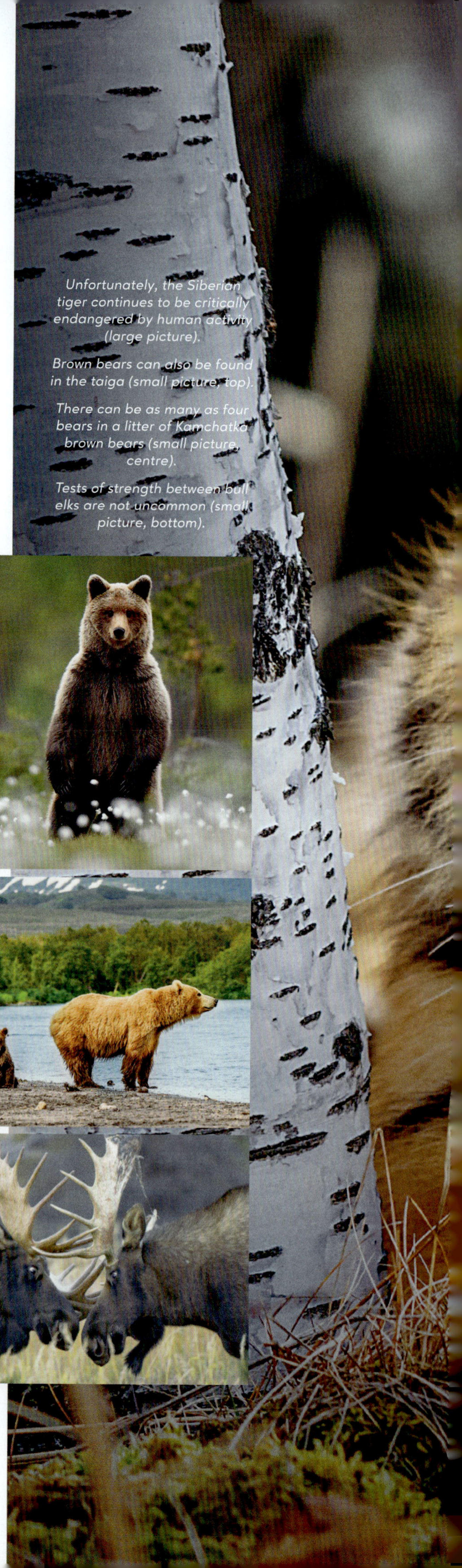

Unfortunately, the Siberian tiger continues to be critically endangered by human activity (large picture).

Brown bears can also be found in the taiga (small picture, top).

There can be as many as four bears in a litter of Kamchatka brown bears (small picture, centre).

Tests of strength between bull elks are not uncommon (small picture, bottom).

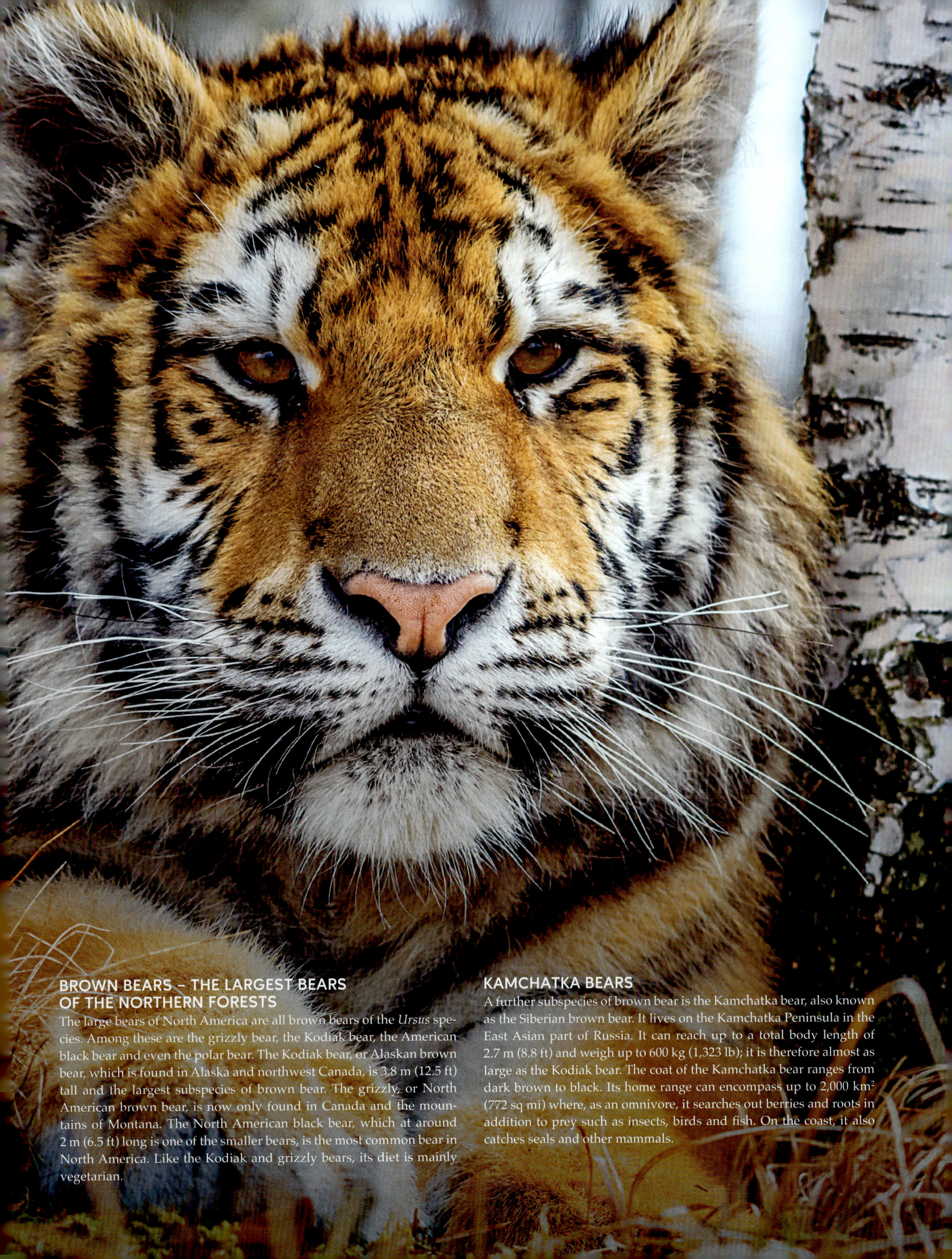

## BROWN BEARS – THE LARGEST BEARS OF THE NORTHERN FORESTS

The large bears of North America are all brown bears of the *Ursus* species. Among these are the grizzly bear, the Kodiak bear, the American black bear and even the polar bear. The Kodiak bear, or Alaskan brown bear, which is found in Alaska and northwest Canada, is 3.8 m (12.5 ft) tall and the largest subspecies of brown bear. The grizzly, or North American brown bear, is now only found in Canada and the mountains of Montana. The North American black bear, which at around 2 m (6.5 ft) long is one of the smaller bears, is the most common bear in North America. Like the Kodiak and grizzly bears, its diet is mainly vegetarian.

## KAMCHATKA BEARS

A further subspecies of brown bear is the Kamchatka bear, also known as the Siberian brown bear. It lives on the Kamchatka Peninsula in the East Asian part of Russia. It can reach up to a total body length of 2.7 m (8.8 ft) and weigh up to 600 kg (1,323 lb); it is therefore almost as large as the Kodiak bear. The coat of the Kamchatka bear ranges from dark brown to black. Its home range can encompass up to 2,000 $km^2$ (772 sq mi) where, as an omnivore, it searches out berries and roots in addition to prey such as insects, birds and fish. On the coast, it also catches seals and other mammals.

FOUR ELEMENTS

EARTH

# DECIDUOUS AND MIXED WOODLAND ZONES

Around 6,000 animal species make their homes in temperate deciduous and mixed woodland zones

The type of vegetation known as deciduous and mixed woodland is almost exclusively limited to the northern hemisphere: in Europe, mixed woodlands stretch from Great Britain across France and Central and Eastern Europe as far as the Urals. In East Asia, they can be found in northeast China, Korea and Japan. In North America, this form of vegetation occurs south of the Great Lakes all the way to the Atlantic, and as far as the Gulf of Mexico. A typical feature of deciduous and mixed woodland is the autumnal leaf fall, which serves primarily to protect the broadleaved trees from drying out in the cold season. Due to the last ice age, there are fewer deciduous and mixed woodland species than those of the tropical rainforests.

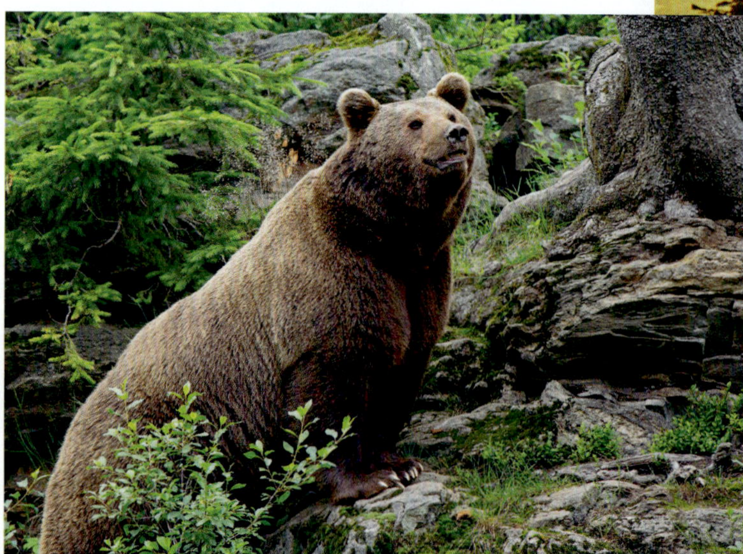

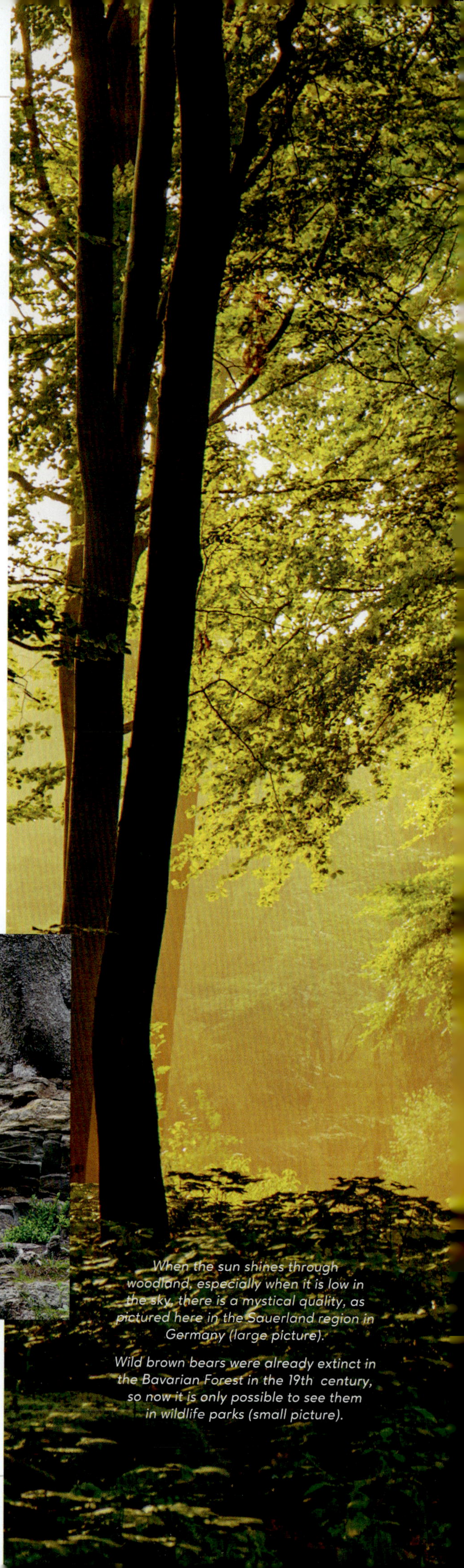

*When the sun shines through woodland, especially when it is low in the sky, there is a mystical quality, as pictured here in the Sauerland region in Germany (large picture).*

*Wild brown bears were already extinct in the Bavarian Forest in the 19th century, so now it is only possible to see them in wildlife parks (small picture).*

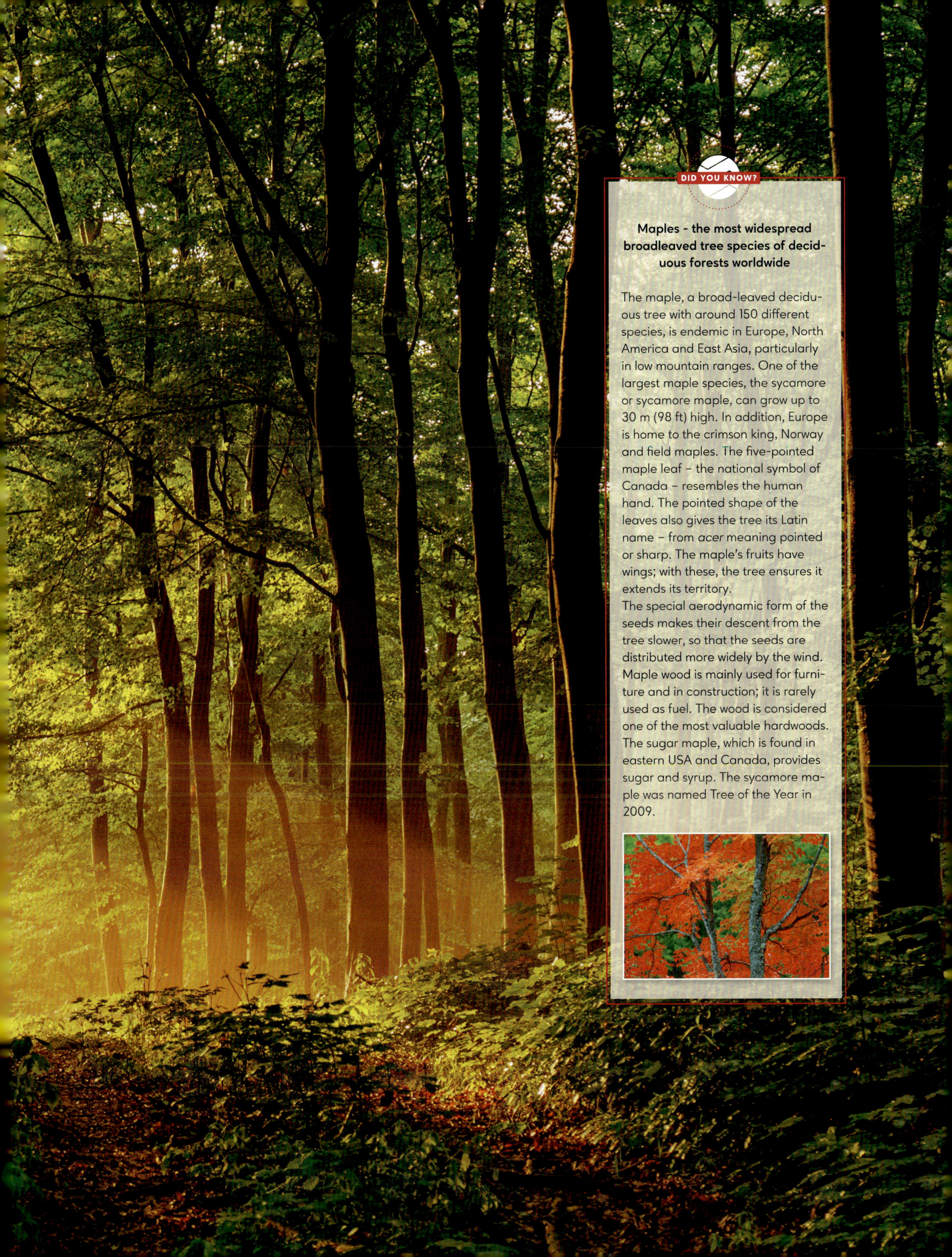

DID YOU KNOW?

**Maples - the most widespread broadleaved tree species of deciduous forests worldwide**

The maple, a broad-leaved deciduous tree with around 150 different species, is endemic in Europe, North America and East Asia, particularly in low mountain ranges. One of the largest maple species, the sycamore or sycamore maple, can grow up to 30 m (98 ft) high. In addition, Europe is home to the crimson king, Norway and field maples. The five-pointed maple leaf – the national symbol of Canada – resembles the human hand. The pointed shape of the leaves also gives the tree its Latin name – from *acer* meaning pointed or sharp. The maple's fruits have wings; with these, the tree ensures it extends its territory.
The special aerodynamic form of the seeds makes their descent from the tree slower, so that the seeds are distributed more widely by the wind. Maple wood is mainly used for furniture and in construction; it is rarely used as fuel. The wood is considered one of the most valuable hardwoods. The sugar maple, which is found in eastern USA and Canada, provides sugar and syrup. The sycamore maple was named Tree of the Year in 2009.

# WISENTS, LYNX

The largest forest animals
in Europe

## WISENTS – THE LARGEST FOREST ANIMALS IN EUROPE

Alongside the aurochs, the wisents, or European bison, were the wild cattle that lived in deciduous and mixed woodlands. As far back as the early Middle Ages, there were two different types of buffalo in the forests of Central Europe. The European bison is closely related to the American bison which lives on the steppes, but it differs in that its front is less prominent and is not a different colour to the rest of the coat. The long, woolly hair on its head and on the front of its body protect it from the cold. Wisents feed on grass, lichen, mosses, leaves and tree bark. By the beginning of the 20 th century, the wisent was almost extinct. Now 300 wisents live in the wild once again in the Białowieża National Park.

## LYNX – THE LARGEST FELINE PREDATOR IN EUROPE

The Eurasian lynx, with a body length between 80 cm and 1.2 m (2.6 to 4 ft) and a shoulder height of 50 to 70 cm (1.6 to 2.3 ft), is the largest feline predator in Europe. After bears and wolves, it is the third largest predator indigenous to Europe. After the lynx was deemed to have become extinct in western Europe, it was resettled in the 1950s. Now the animal can be found in low mountain ranges in Europe and in the Alps. Typical features of lynxes are tufted ears, a very short tail and prominent whiskers. They live as solitary animals and only converge during the mating season between February and April. After a gestation period of 70 days, they bear between two and five young, which are reared by the mother until the following spring.

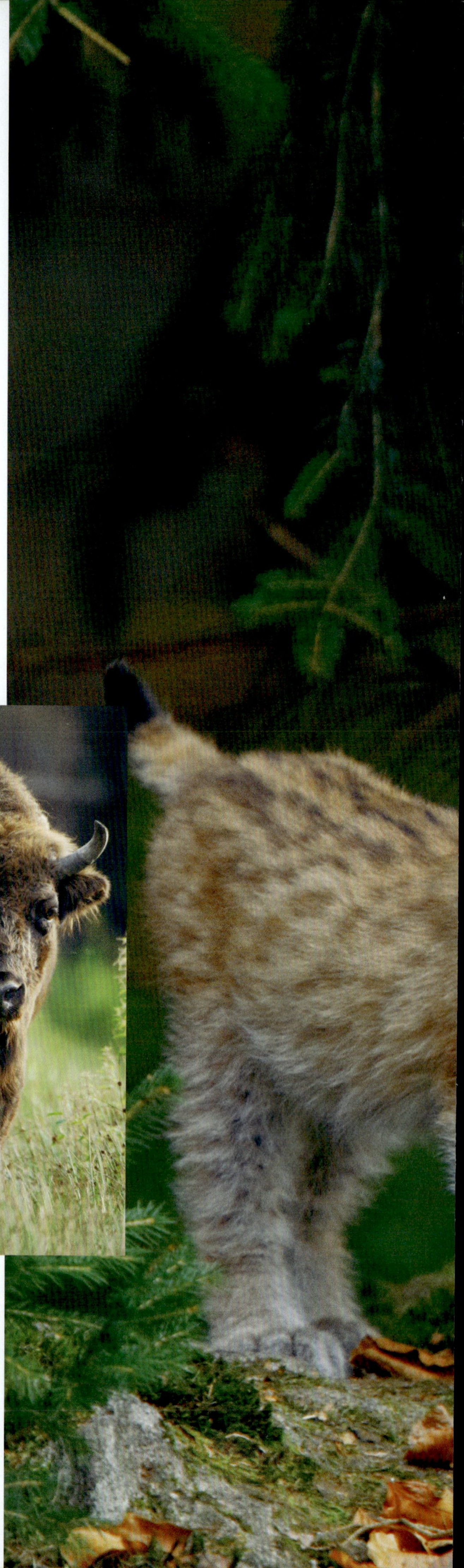

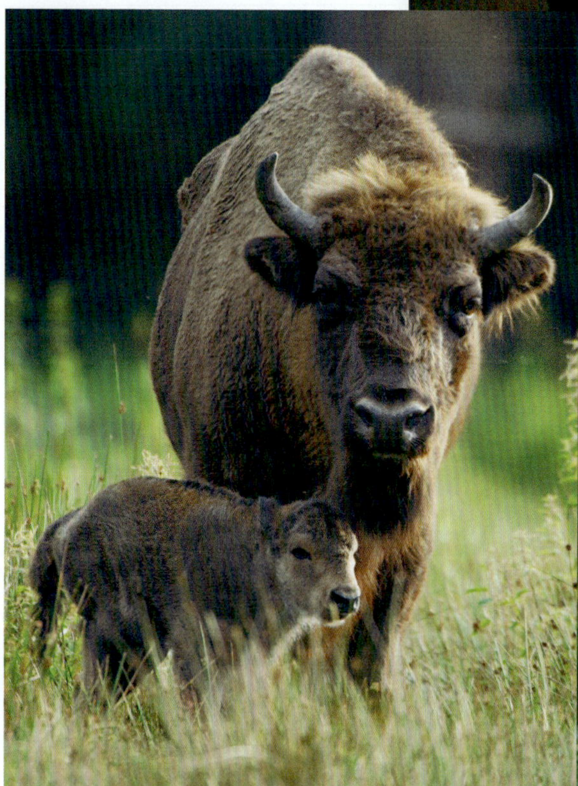

*The wisent reaches a head-body length of up to 3.3 m (10.8 ft), stands up to 2 m (6.5 ft) at the shoulder and can weigh up to a tonne (1.1 ton). This makes it the largest and heaviest terrestrial mammal in Europe (small picture).*

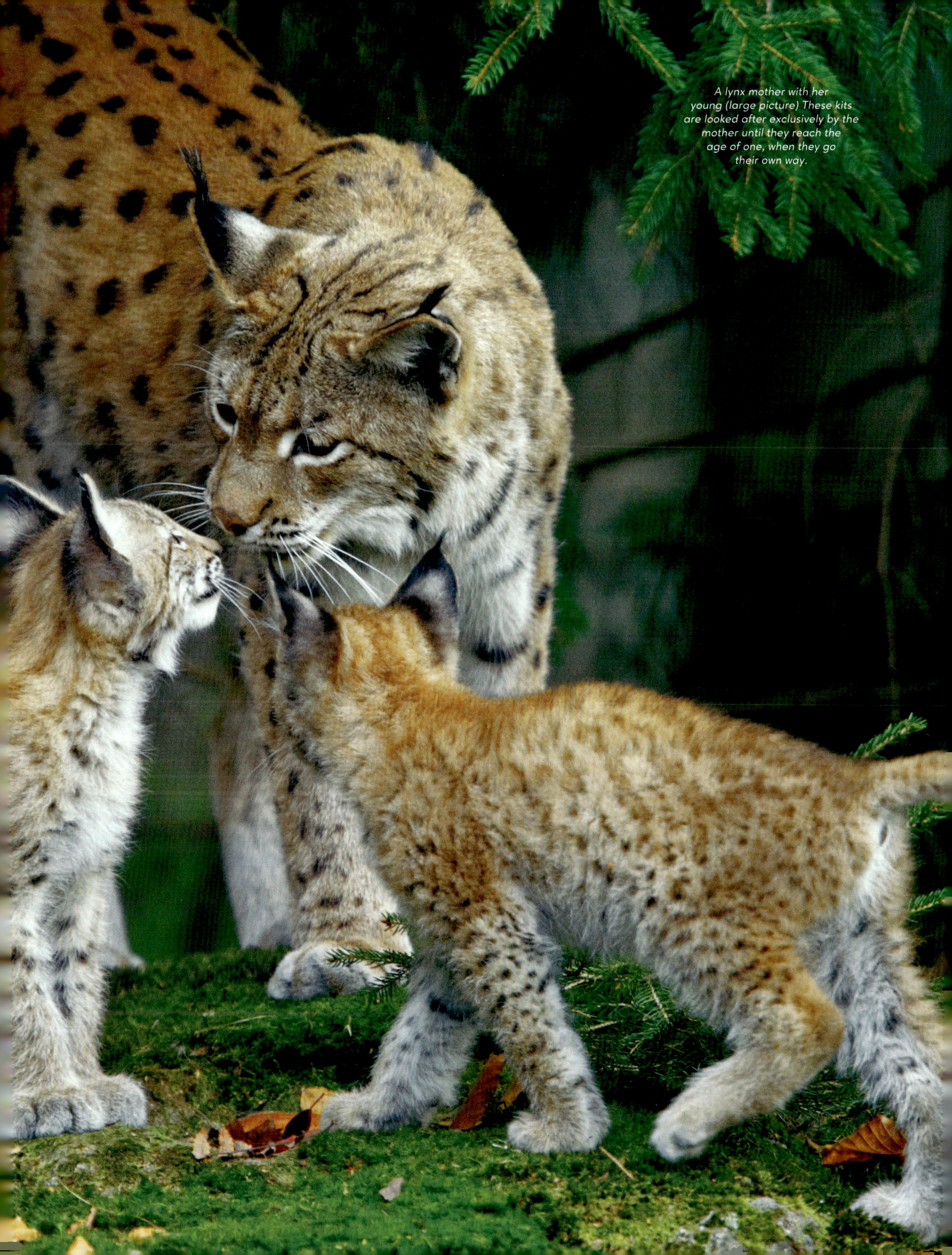

*A lynx mother with her young (large picture) These kits are looked after exclusively by the mother until they reach the age of one, when they go their own way.*

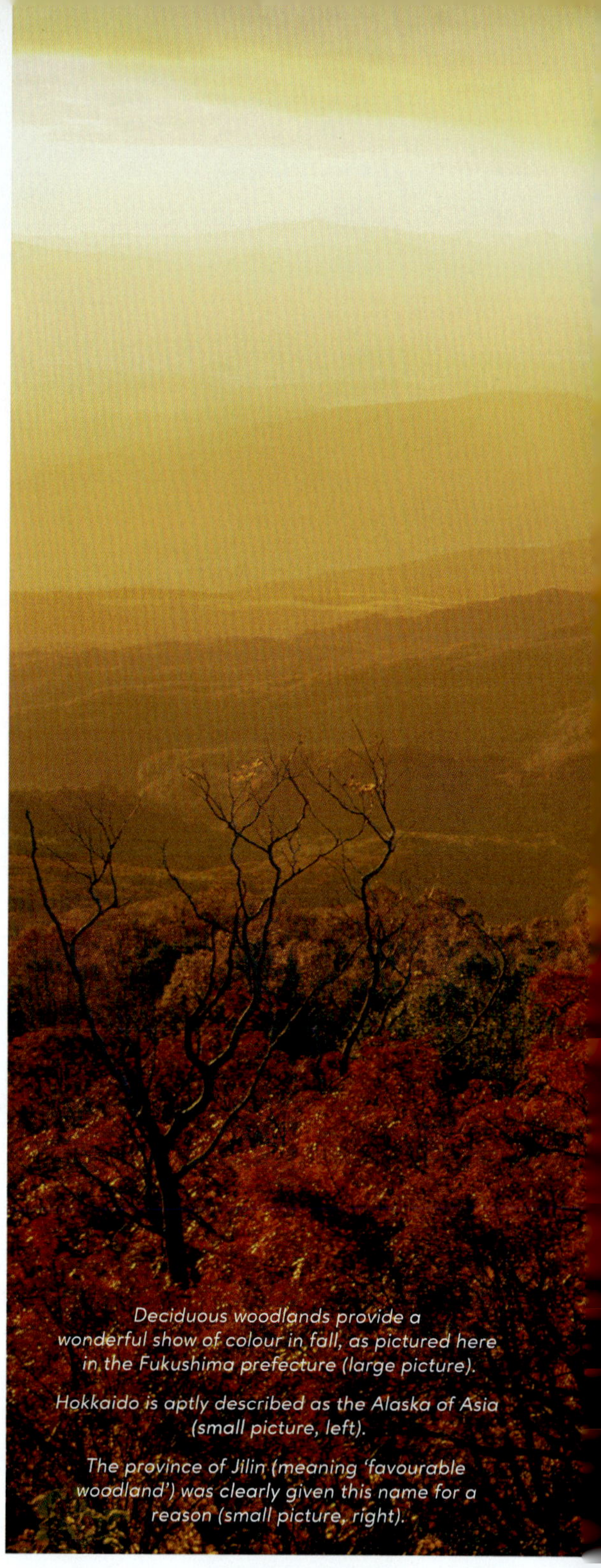

Deciduous woodlands provide a wonderful show of colour in fall, as pictured here in the Fukushima prefecture (large picture).

Hokkaido is aptly described as the Alaska of Asia (small picture, left).

The province of Jilin (meaning 'favourable woodland') was clearly given this name for a reason (small picture, right).

FOUR ELEMENTS

EARTH

# DECIDUOUS WOODLAND IN ASIA

In East Asia, the predominant form of vegetation is that of the forest steppes, i.e. the transition zone between mixed woodland zones and treeless steppes.

## EAST CHINA

In East China, there are large areas of mixed woodland with numerous endemic species such as the Siberian tiger (the largest feline predator in the world) and the Chinese pangolin. Many of the animals now living there first arrived in the region towards the end of the last ice age. East China also has extensive woodland areas, such as, for example, the Meili Snow Mountains on the border with Tibet. Rare birds and other animals not found anywhere else live in these areas of mixed woodland, and, in spring and summer, countless flowers bloom in the glades. Miyaluo in the Aba prefecture is famous for its show of colour in October, when the broadleaved trees glow with every shade from golden yellow to scarlet. The rich variety of fauna in these woodlands includes, for example, the rare common flying dragon. One of the most well-known mountains in China is Laoshan (Mount Lao)which borders the Yellow Sea in the east. In summer, the floodwaters make the waterfalls and streams sparkle; in fall, the leaves of the broadleaved trees and shrubs turn red and brown.

## JAPAN – HOKKAIDO

Anyone who visits Japan because of its centres of industry and big cities may not be aware that two-thirds of the country is covered in mountains, hills and woodlands that exhibit a rich and diverse range of flora and fauna. Hokkaido, Japan's second largest island, is sparsely populated and is often compared with Alaska because of its vast areas of natural landscape – with untouched woodlands, magnificent snow-covered summits, numerous mountain ranges and volcanic massifs, clear lakes and hot springs as well as a multitude of, in some cases very rare, animal and plant species. In the woodland areas of Hokkaido, for example, there are brown bears, red foxes, raccoon dogs and sika deer; in the marshland areas, the graceful red-crowned cranes can often be sighted. The winters on Hokkaido are long (from the beginning of December to the end of March) and severe (temperatures of -40°C (-40°F) are not infrequent); this is followed by a very brief spring when cherry trees and plum trees and lilacs all bloom at the same time. The summers are likewise short and warm (around 30°C (80 °F) on average, with 60 to 70 per cent humidity); in fall, which lasts around one and a half months, the mixed woodlands are bathed in a sea of colour. The cool climate, without the absence of Japan's otherwise typical rainy season, is one of the reasons why Hokkaido is also called Hanataikriku or 'a continent of flowers'. Many nature lovers visit during the blossoming period to marvel at the wondrous display of colour.

## JAPAN – HONSHU

Honshu, the largest island in Japan and the seventh largest island in the world, is divided into several different vegetation zones. In the north, rice-growing areas, woodlands and mountains predominate; in central Honshu, the highest mountains in the land rise up, such as Mount Fuji, Japan's national landmark. The densely populated coastal regions and the metropolitan area of Tokyo are characterized by industry interspersed with agricultural land. In the island's interior, there are numerous picturesque lakes as well as deciduous and mixed woodlands containing pines, oaks, chestnut oaks, willows, birches, beech, maples and Japanese cypresses – the tree vegetation corresponds with that of a temperate zone. Mulberry trees are grown on a large scale for the breeding of silkworms, and the Japanese lacquer tree, or sumac, provides the raw materials for Japanese and Chinese lacquerwork.

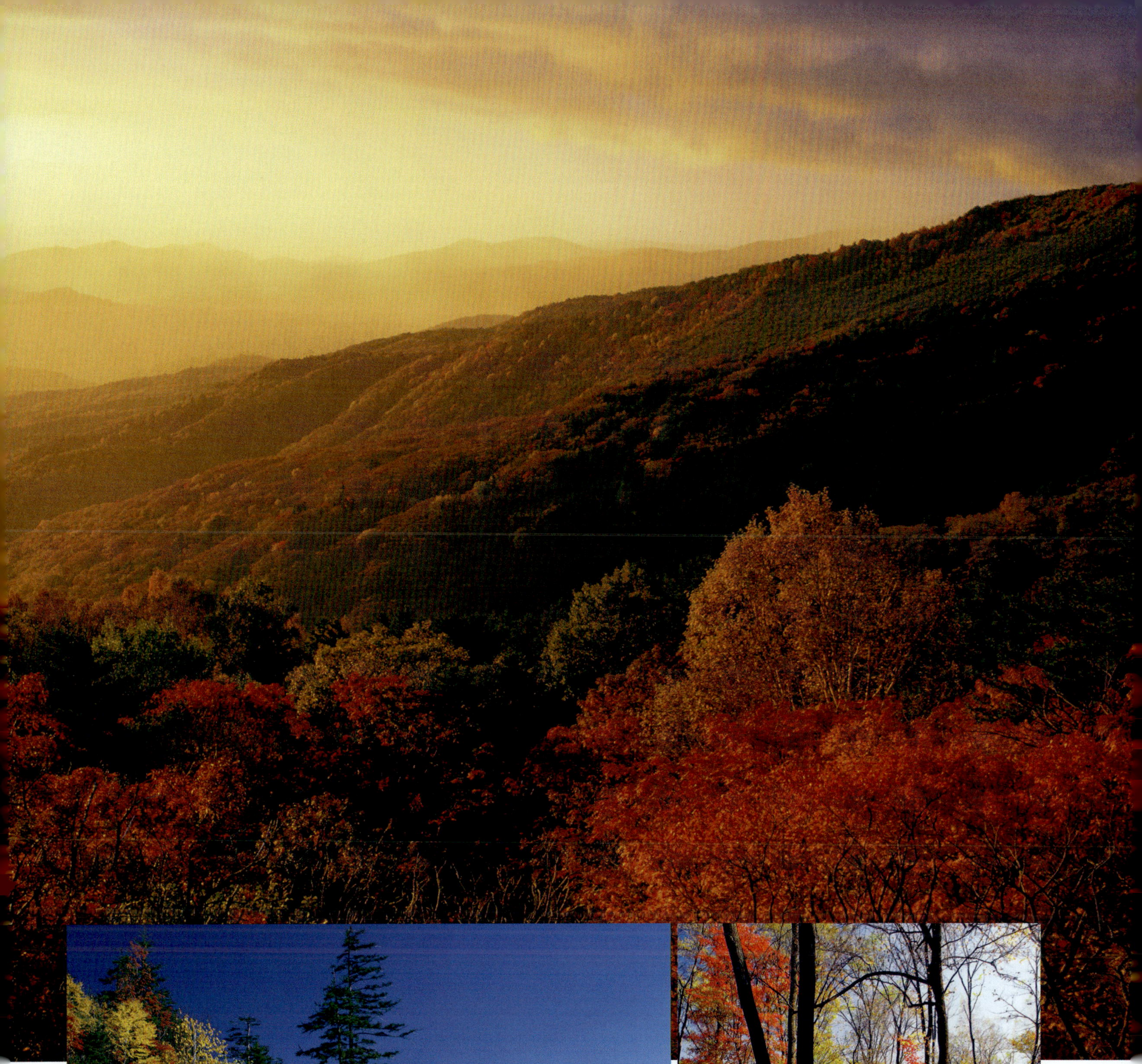
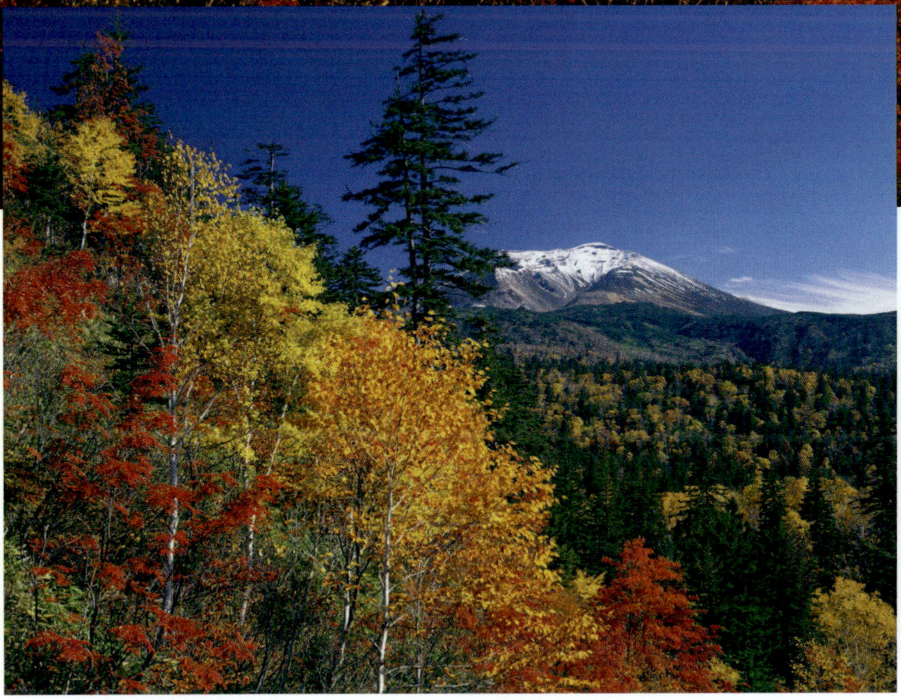
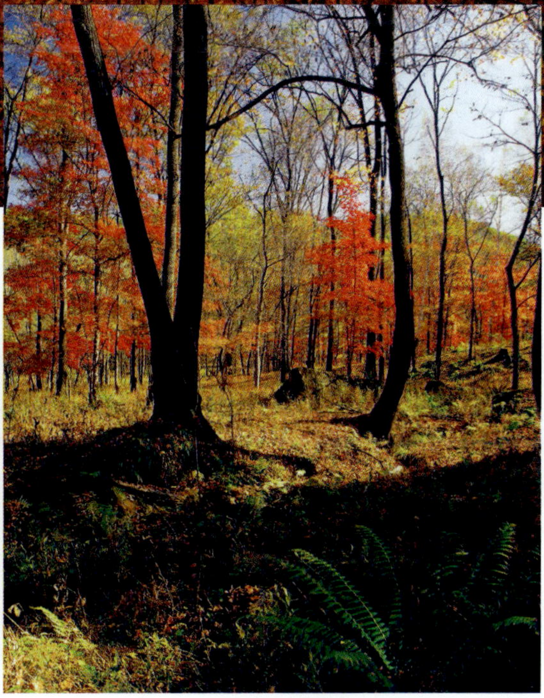

FOUR ELEMENTS

EARTH

# THE WAY TO INNER PEACE

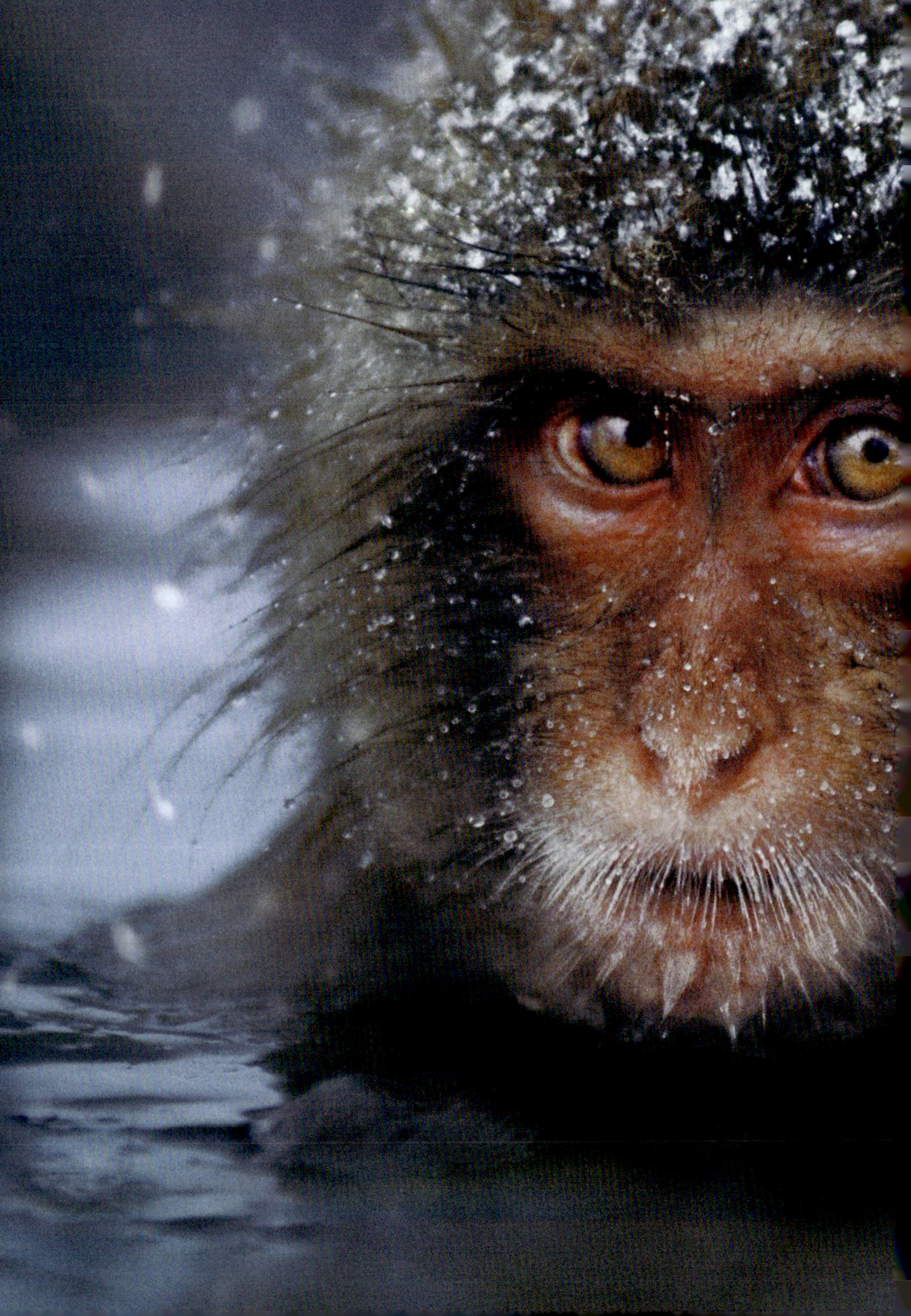

DID YOU KNOW?

### The King of Winter

The species of ape living furthest north in the world is the red-faced macaque, which measures 45 to 75 cm (18 to 30 in). It is also called the Japanese macaque because its habitat extends to Japan, where it is only found on the islands of Honshu, Shikoku and Kyushu. The animals live in various types of forests: in the mountain woodlands in the north and in the subtropical forest areas in the south. The most striking feature of this species of primate is its hairless face which can exhibit a range of shades from pink to deep red. The Japanese macaque is an omnivore, but its diet is mainly plant-based and varies with the seasons. In winter, when the snow can be up to 1.5 m (5 ft) deep, it lives on roots and bark; in summer, it also eats berries, nuts and fruits. The Japanese macaque lives in sizeable groups of 30 to 60 animals with each group having its own territory.

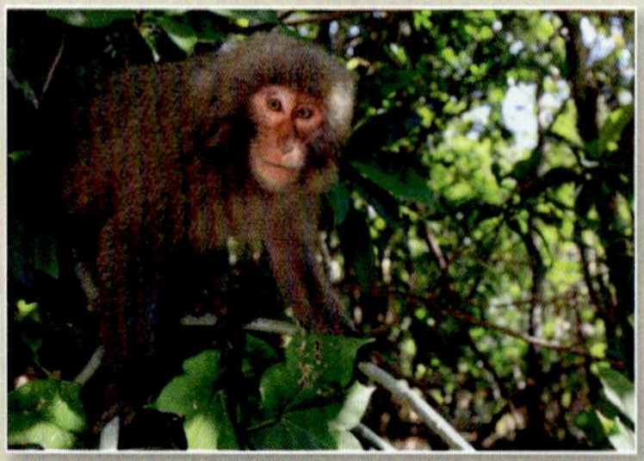

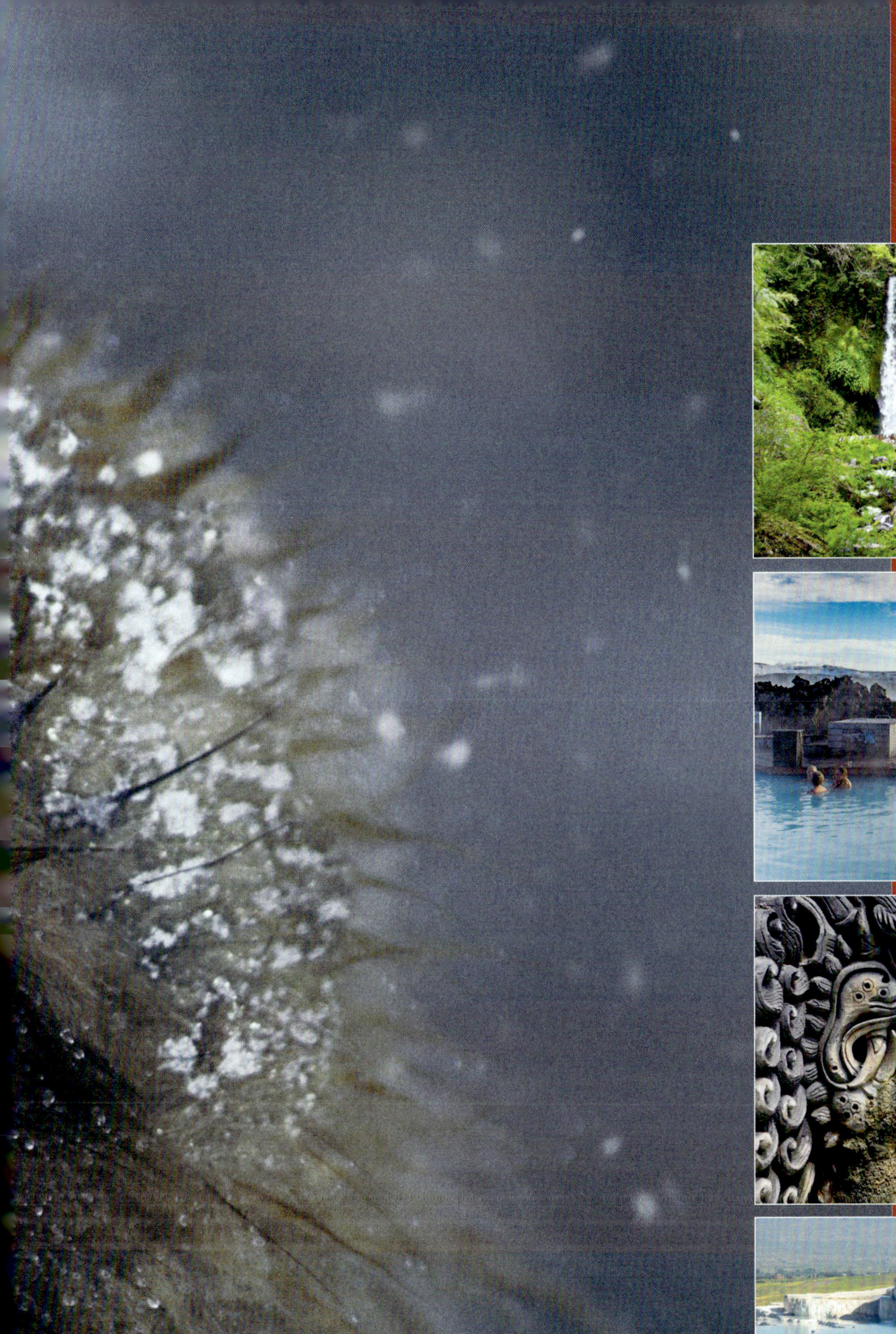

## THE MOST BEAUTIFUL THERMAL SPRINGS FOR HUMANS

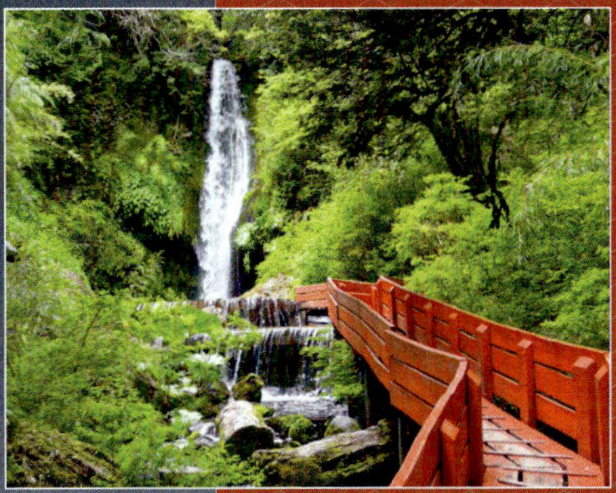

### HOT SPRINGS IN THE JUNGLE

The largest geyser field in the southern hemisphere lies at the foot of the El Tatio volcano. More than 80 geysers are active here, mainly in the morning hours. Close by, there is a thermal spa pool, where you can take a dip in the geothermally heated spring.

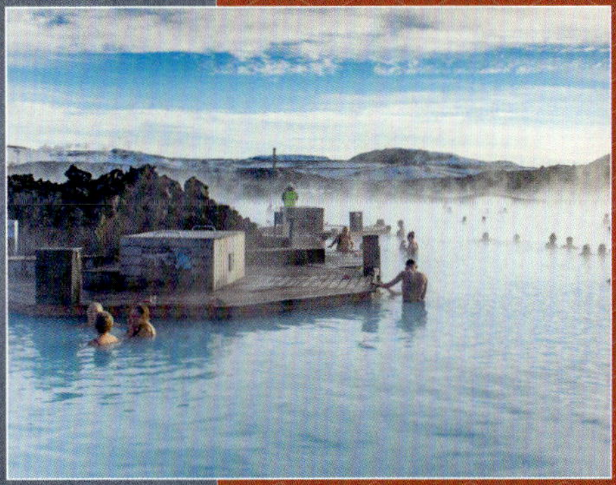

### THE BLUE LAGOON

The Bláa Lónið (Blue Lagoon) in Iceland was created more by accident than by any natural means: the mixture of fresh and sea water, which is used to produce electricity in the nearby geothermal power station, is drained into the lagoon. The water is a pleasant 38 - 39°C (100 - 102°F).

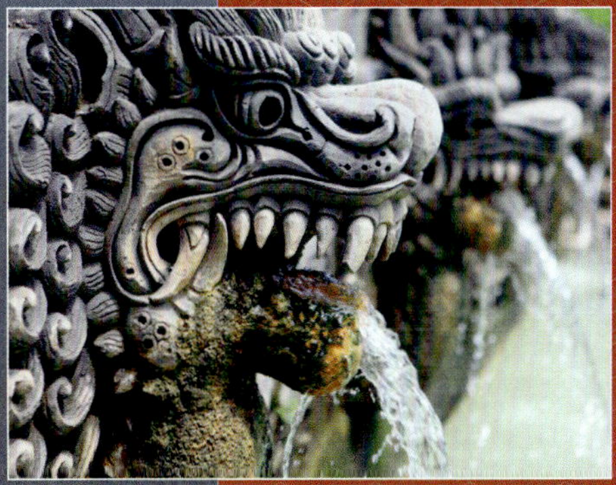

### AIR PANAS BANJAR (BANJAR HOT SPRINGS)

The hot springs in the north of Bali are channelled into a succession of three swimming pools in the middle of the jungle, each pool cooler than the last. The high sulfur content of the water has a relaxing effect and helps promotes health.

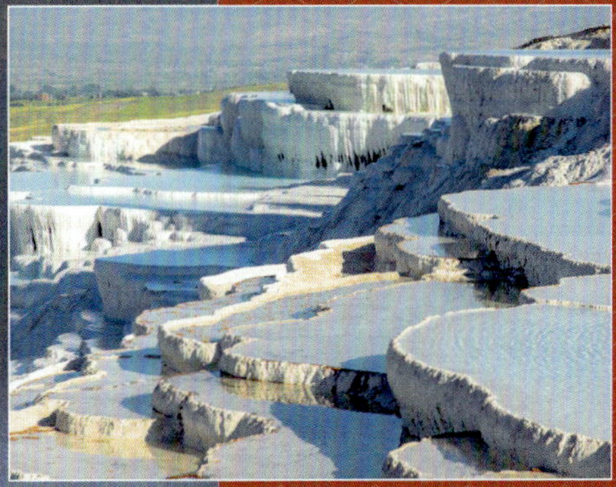

### THE POOLS OF PAMUKKALE

Shallow layers of water on white rock, laid out in terraces – the thermal springs of Pamukkale in Turkey. The warmth of the water means that the calcium carbonate is released and deposited as travertine in the rocks.

### IT'S NICE AND WARM!

Japanese macaques, who live in socially structured groups, are well-known for their love of a sitz bath in the hot thermal springs, which helps regulate their body temperature. A layer of snow on their fur does not spoil their enjoyment!

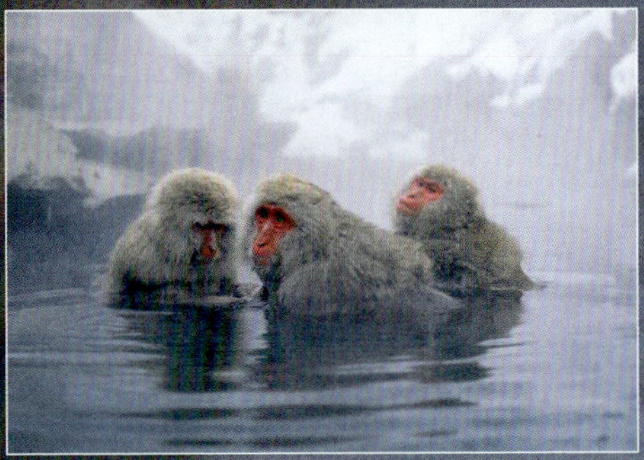

FOUR ELEMENTS

EARTH

# NORTH AMERICA

North America's deciduous woodlands with hickory trees, tulip trees, maples, horse chestnuts, magnolias and more provide habitats for many creatures

## THE APPALACHIANS - THE LARGEST DECIDUOUS AND MIXED WOODLAND REGION IN NORTH AMERICA

The Appalachians, a low, wooded mountain range on the Atlantic coast in the east of North America, run for over 2,400 km (1,491 mi) from the Canadian province of Québec through to the state of Alabama in the USA. The range forms the western part of a chain of mountains that once, before the Atlantic Ocean was formed, continued in northern Scandinavia. The Appalachians are, at 400 million years old, significantly older than, for example, the Himalayas, the Alps or the Rocky Mountains. The principal mountain ranges of the Appalachians are the Green Mountains, the White Mountains, the Catskill Mountains, the Blue Ridge Mountains and the Cumberland Mountains. All these mountain chains and the valleys between the peaks are primarily covered with dense mixed and deciduous woodlands. The Appalachian woodlands are impressive not only on account of their size but also because the huge variety of plants – there are more tree species growing in the southern section than in all of Europe – and animal species here also makes them fascinating.

No less impressive are the dense, extensive deciduous and mixed woodlands covering the ridges and valleys. The flora and fauna here are among the most biodiverse in the world. There are more species of trees in the Great Smoky Mountains National Park, for instance, than in the entire continent of Europe. In addition, there is an enormous wealth of flower species, which transforms the woodland floor into a multicoloured carpet of azaleas, dogwood and mountain laurel in spring. In fall, however, the dense foliage of the woods is bathed in colour; it is the most spectacular time of year in these regions. Ten per cent of all the salamander species in the world make their home here and emerge from their hiding places in the spring when rain is plentiful. Numerous black bears and other large animals, such as the white-tailed deer, live in the Appalachians. The 3,440-km (2,138-mi) long Appalachian Trail, one of the longest hiking trails in the world, runs from Georgia in the southeast up to Maine. It can be traced back to the old routes that the Cherokee used until their forced resettlement in 1838.

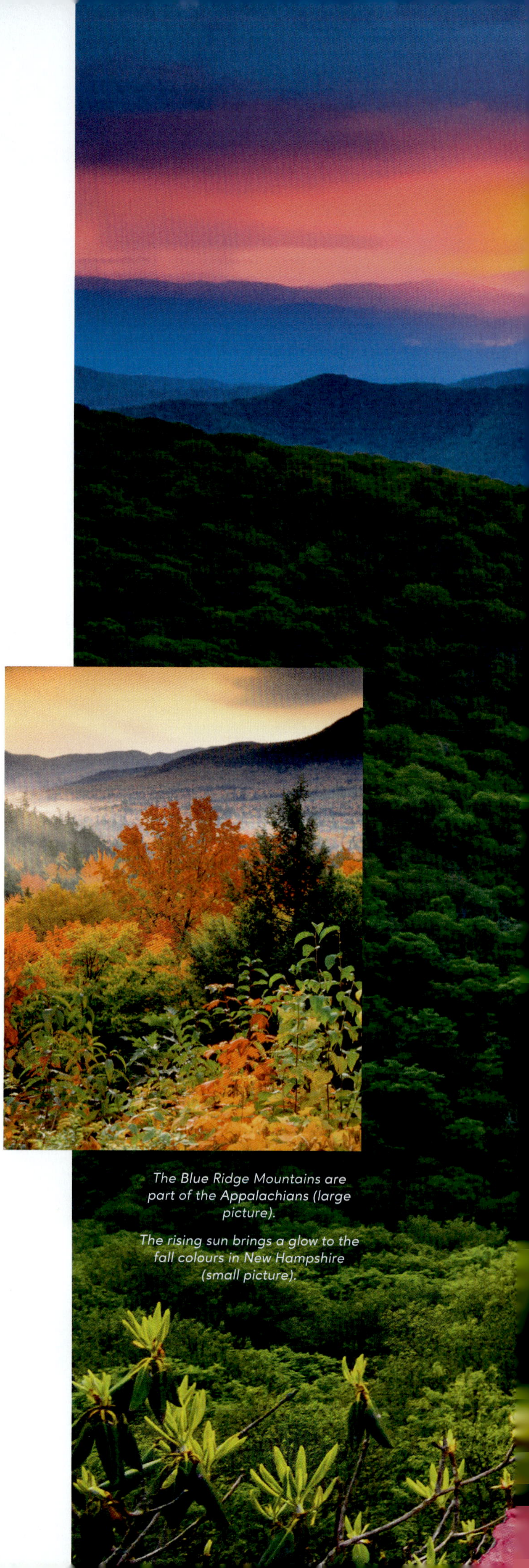

*The Blue Ridge Mountains are part of the Appalachians (large picture).*

*The rising sun brings a glow to the fall colours in New Hampshire (small picture).*

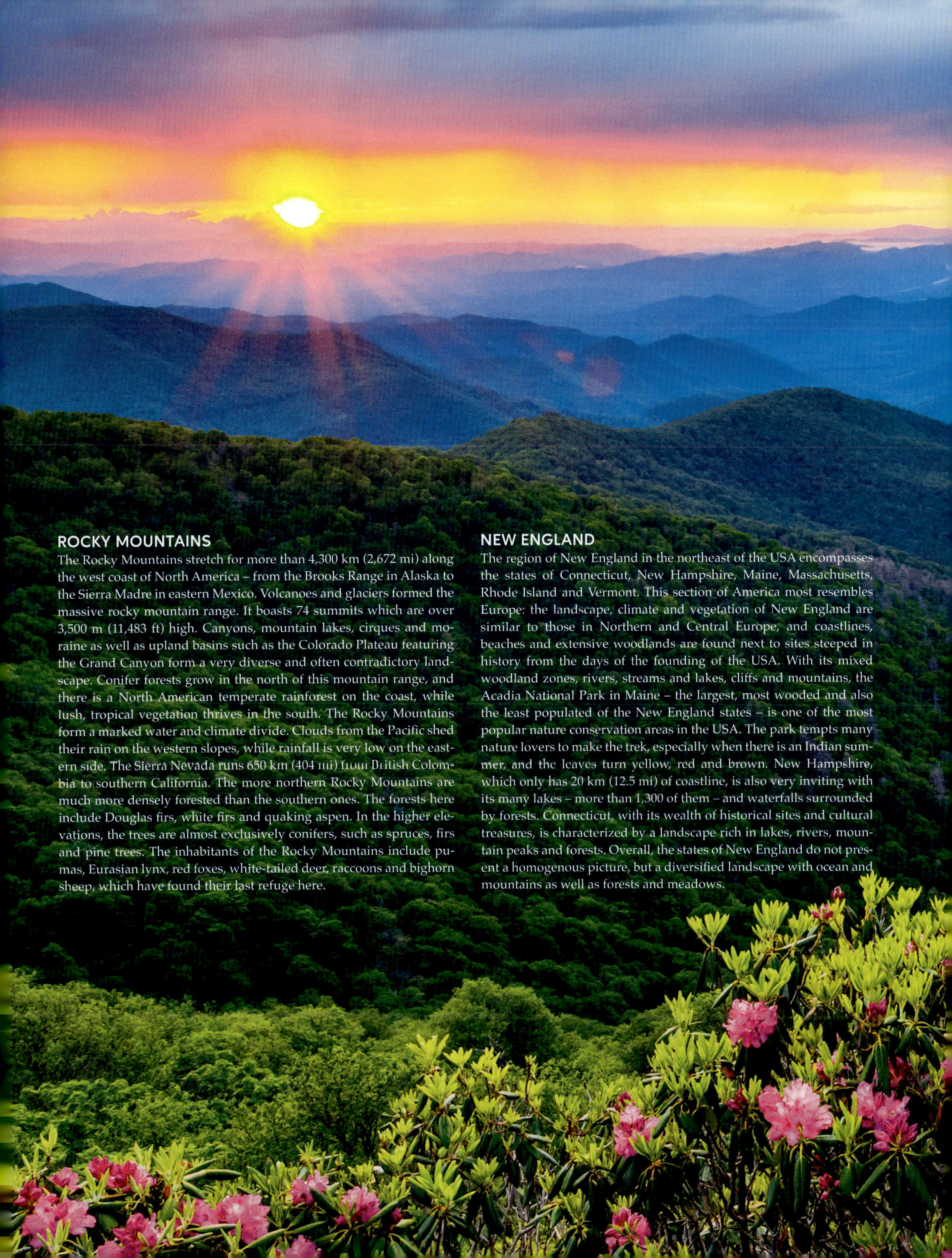

## ROCKY MOUNTAINS

The Rocky Mountains stretch for more than 4,300 km (2,672 mi) along the west coast of North America – from the Brooks Range in Alaska to the Sierra Madre in eastern Mexico. Volcanoes and glaciers formed the massive rocky mountain range. It boasts 74 summits which are over 3,500 m (11,483 ft) high. Canyons, mountain lakes, cirques and moraine as well as upland basins such as the Colorado Plateau featuring the Grand Canyon form a very diverse and often contradictory landscape. Conifer forests grow in the north of this mountain range, and there is a North American temperate rainforest on the coast, while lush, tropical vegetation thrives in the south. The Rocky Mountains form a marked water and climate divide. Clouds from the Pacific shed their rain on the western slopes, while rainfall is very low on the eastern side. The Sierra Nevada runs 650 km (404 mi) from British Colombia to southern California. The more northern Rocky Mountains are much more densely forested than the southern ones. The forests here include Douglas firs, white firs and quaking aspen. In the higher elevations, the trees are almost exclusively conifers, such as spruces, firs and pine trees. The inhabitants of the Rocky Mountains include pumas, Eurasian lynx, red foxes, white-tailed deer, raccoons and bighorn sheep, which have found their last refuge here.

## NEW ENGLAND

The region of New England in the northeast of the USA encompasses the states of Connecticut, New Hampshire, Maine, Massachusetts, Rhode Island and Vermont. This section of America most resembles Europe: the landscape, climate and vegetation of New England are similar to those in Northern and Central Europe, and coastlines, beaches and extensive woodlands are found next to sites steeped in history from the days of the founding of the USA. With its mixed woodland zones, rivers, streams and lakes, cliffs and mountains, the Acadia National Park in Maine – the largest, most wooded and also the least populated of the New England states – is one of the most popular nature conservation areas in the USA. The park tempts many nature lovers to make the trek, especially when there is an Indian summer, and the leaves turn yellow, red and brown. New Hampshire, which only has 20 km (12.5 mi) of coastline, is also very inviting with its many lakes – more than 1,300 of them – and waterfalls surrounded by forests. Connecticut, with its wealth of historical sites and cultural treasures, is characterized by a landscape rich in lakes, rivers, mountain peaks and forests. Overall, the states of New England do not present a homogenous picture, but a diversified landscape with ocean and mountains as well as forests and meadows.

FOUR ELEMENTS

EARTH

# RAINFORESTS OF THE TEMPERATE LATITUDES

Rainforests are not just found along the equator; they can also be found in temperate zones

In temperate zones, the classification as a rainforest – 2,000 mm (79 in) of rainfall – can only be achieved on the slopes of coastal mountains. There are sizeable areas of temperate rainforest in North and South America, Australia and Tasmania, New Zealand, Anatolia and Georgia. Large areas covered in temperate rainforest can also be found on the Pacific coast of North America. They stretch from the coast of Alaska, across Canada and down into northern California. Smaller temperate rainforests can also still be found in Norway, South Africa and East Asia. In many of these regions, their existence is threatened by commercial forestry, as there is a great deal of interest in valuable trees, particularly very distinctive specimens.

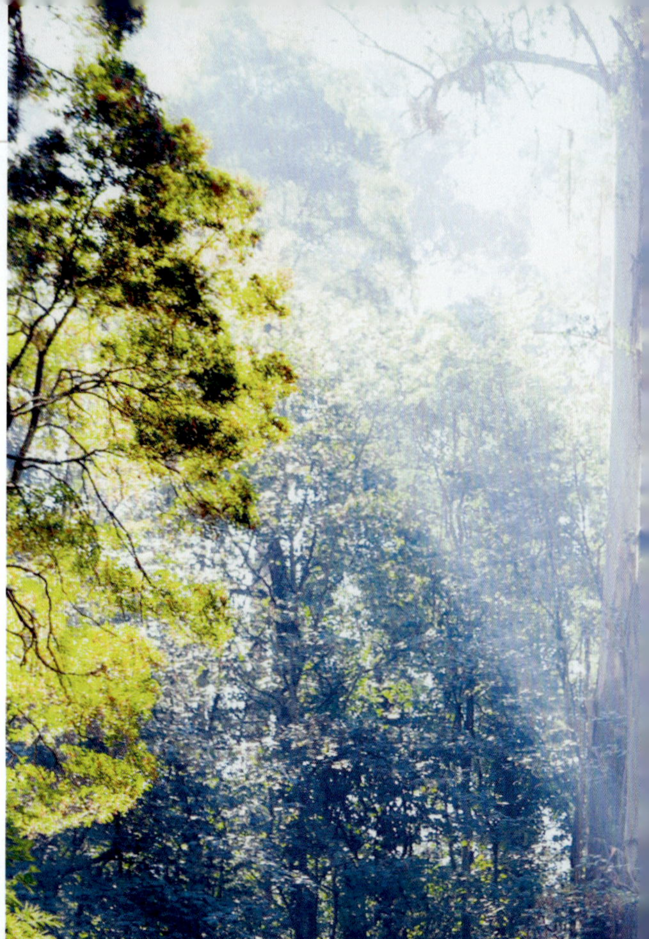

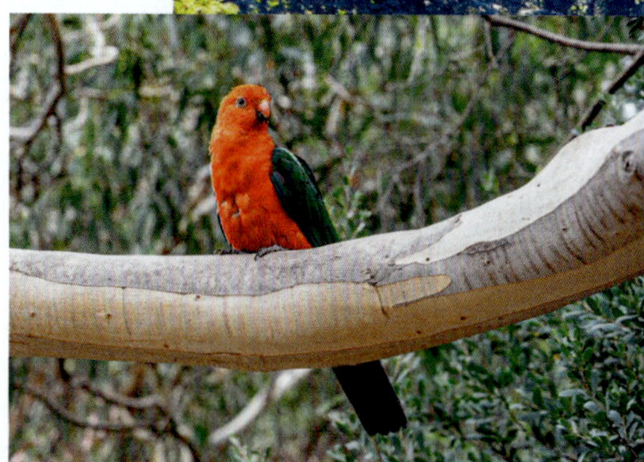

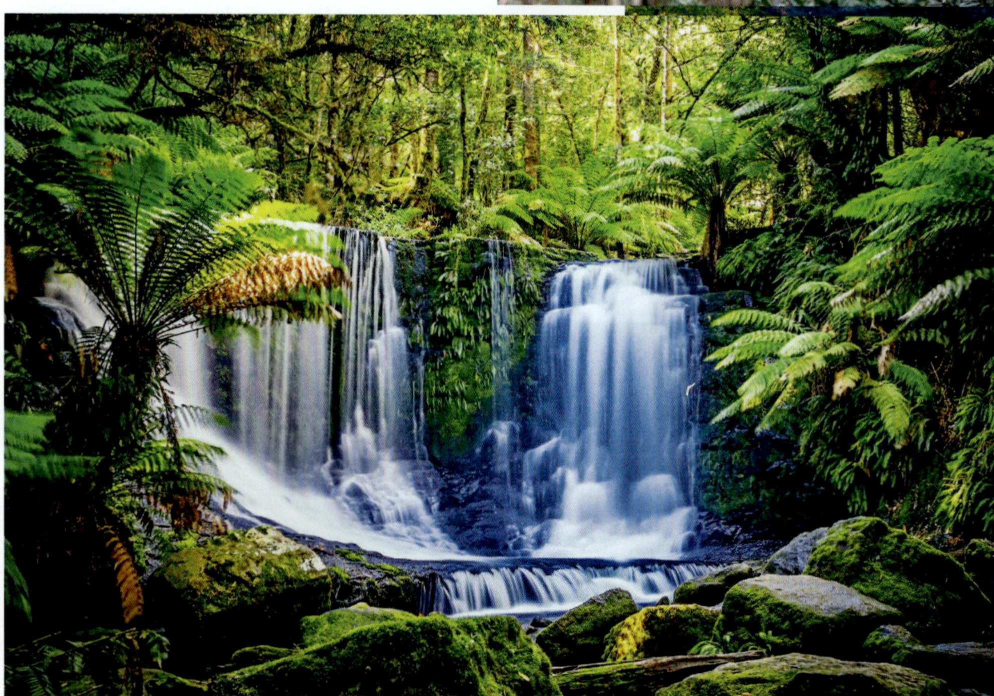

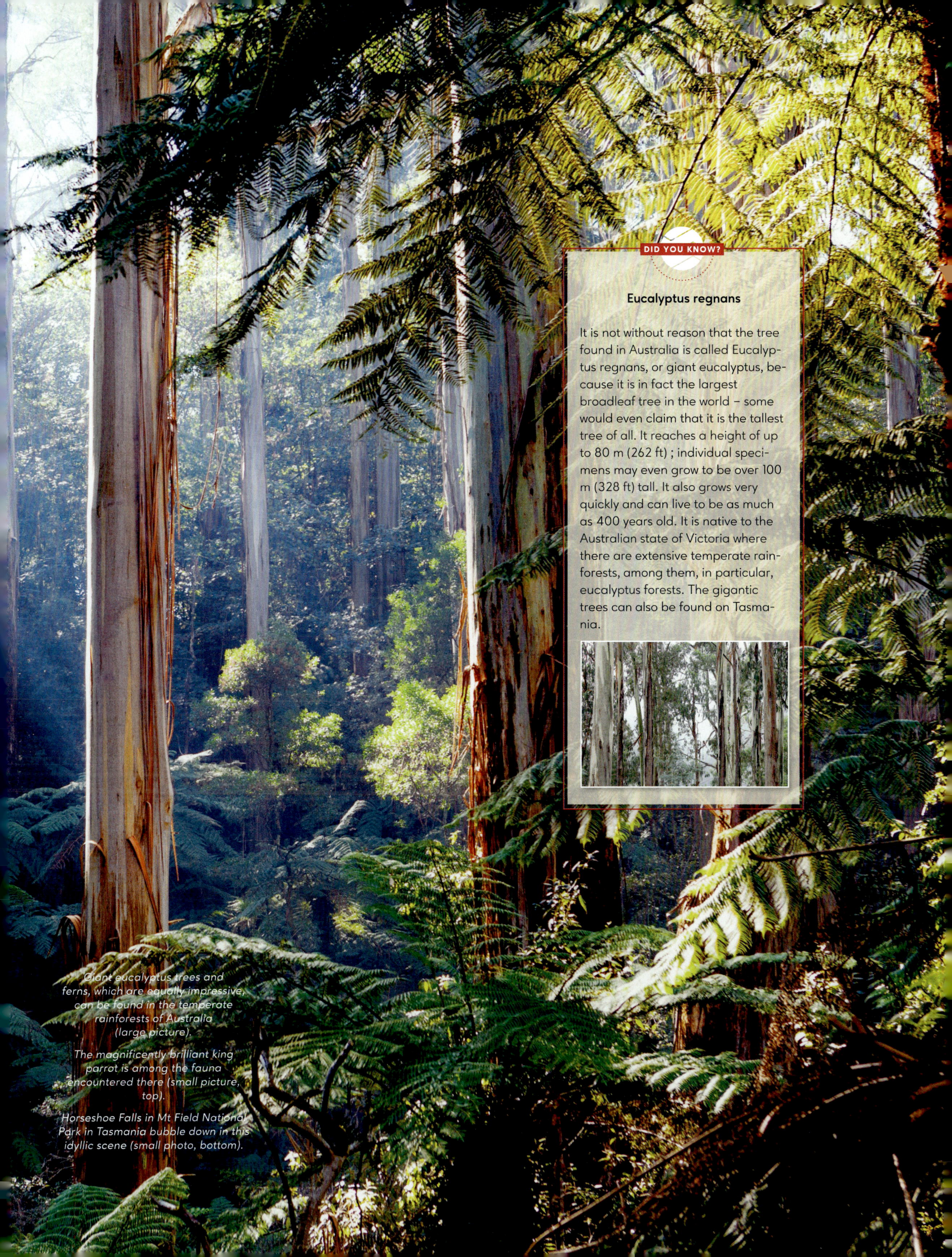

**DID YOU KNOW?**

### Eucalyptus regnans

It is not without reason that the tree found in Australia is called Eucalyptus regnans, or giant eucalyptus, because it is in fact the largest broadleaf tree in the world – some would even claim that it is the tallest tree of all. It reaches a height of up to 80 m (262 ft) ; individual specimens may even grow to be over 100 m (328 ft) tall. It also grows very quickly and can live to be as much as 400 years old. It is native to the Australian state of Victoria where there are extensive temperate rainforests, among them, in particular, eucalyptus forests. The gigantic trees can also be found on Tasmania.

*Giant eucalyptus trees and ferns, which are equally impressive, can be found in the temperate rainforests of Australia (large picture).*

*The magnificently brilliant king parrot is among the fauna encountered there (small picture, top).*

*Horseshoe Falls in Mt Field National Park in Tasmania bubble down in this idyllic scene (small photo, bottom).*

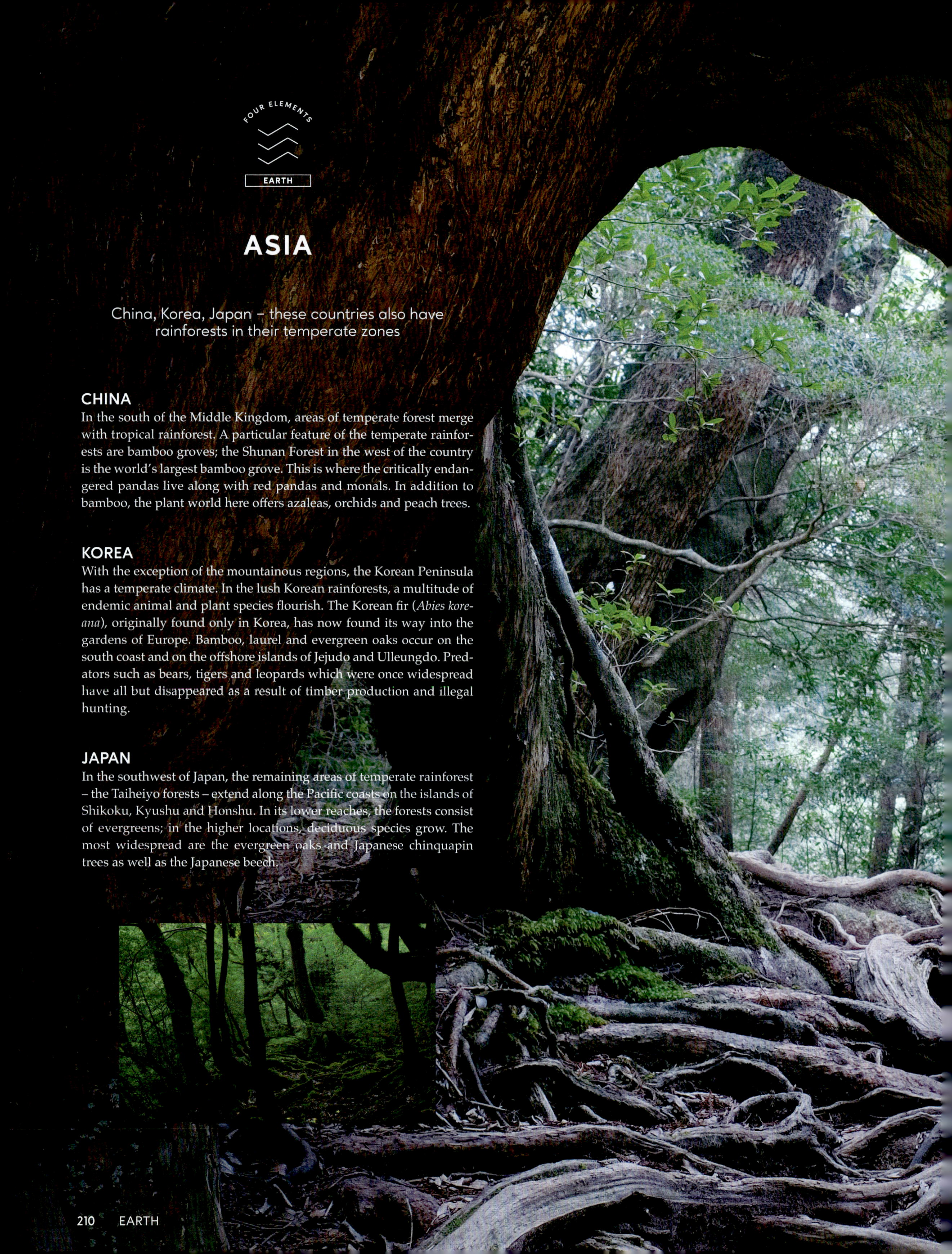



# ASIA

China, Korea, Japan – these countries also have rainforests in their temperate zones

## CHINA

In the south of the Middle Kingdom, areas of temperate forest merge with tropical rainforest. A particular feature of the temperate rainforests are bamboo groves; the Shunan Forest in the west of the country is the world's largest bamboo grove. This is where the critically endangered pandas live along with red pandas and monals. In addition to bamboo, the plant world here offers azaleas, orchids and peach trees.

## KOREA

With the exception of the mountainous regions, the Korean Peninsula has a temperate climate. In the lush Korean rainforests, a multitude of endemic animal and plant species flourish. The Korean fir (*Abies koreana*), originally found only in Korea, has now found its way into the gardens of Europe. Bamboo, laurel and evergreen oaks occur on the south coast and on the offshore islands of Jejudo and Ulleungdo. Predators such as bears, tigers and leopards which were once widespread have all but disappeared as a result of timber production and illegal hunting.

## JAPAN

In the southwest of Japan, the remaining areas of temperate rainforest – the Taiheiyo forests – extend along the Pacific coasts on the islands of Shikoku, Kyushu and Honshu. In its lower reaches, the forests consist of evergreens; in the higher locations, deciduous species grow. The most widespread are the evergreen oaks and Japanese chinquapin trees as well as the Japanese beech.

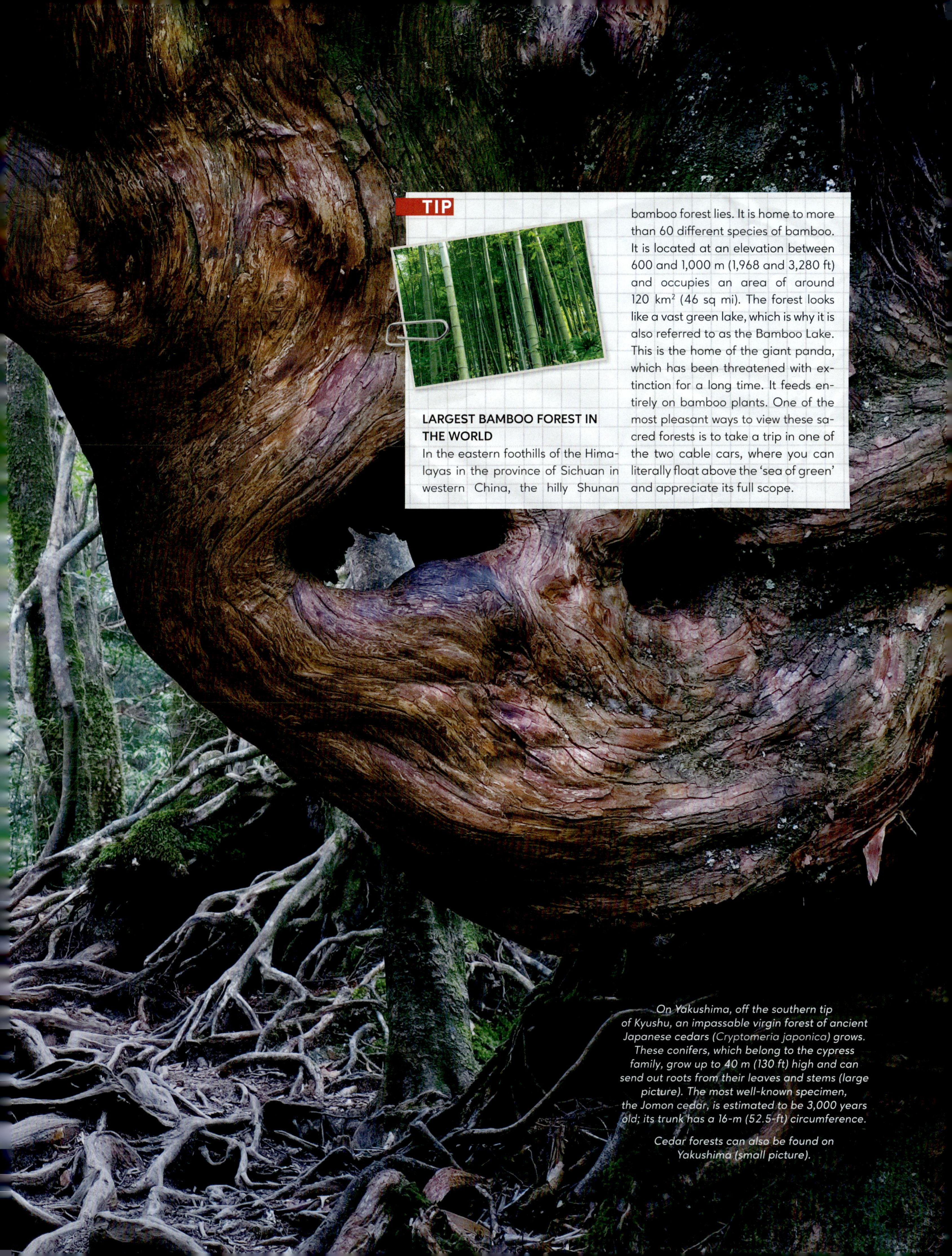

**TIP**

**LARGEST BAMBOO FOREST IN THE WORLD**

In the eastern foothills of the Himalayas in the province of Sichuan in western China, the hilly Shunan bamboo forest lies. It is home to more than 60 different species of bamboo. It is located at an elevation between 600 and 1,000 m (1,968 and 3,280 ft) and occupies an area of around 120 km$^2$ (46 sq mi). The forest looks like a vast green lake, which is why it is also referred to as the Bamboo Lake. This is the home of the giant panda, which has been threatened with extinction for a long time. It feeds entirely on bamboo plants. One of the most pleasant ways to view these sacred forests is to take a trip in one of the two cable cars, where you can literally float above the 'sea of green' and appreciate its full scope.

*On Yakushima, off the southern tip of Kyushu, an impassable virgin forest of ancient Japanese cedars (Cryptomeria japonica) grows. These conifers, which belong to the cypress family, grow up to 40 m (130 ft) high and can send out roots from their leaves and stems (large picture). The most well-known specimen, the Jomon cedar, is estimated to be 3,000 years old; its trunk has a 16-m (52.5-ft) circumference.*

*Cedar forests can also be found on Yakushima (small picture).*

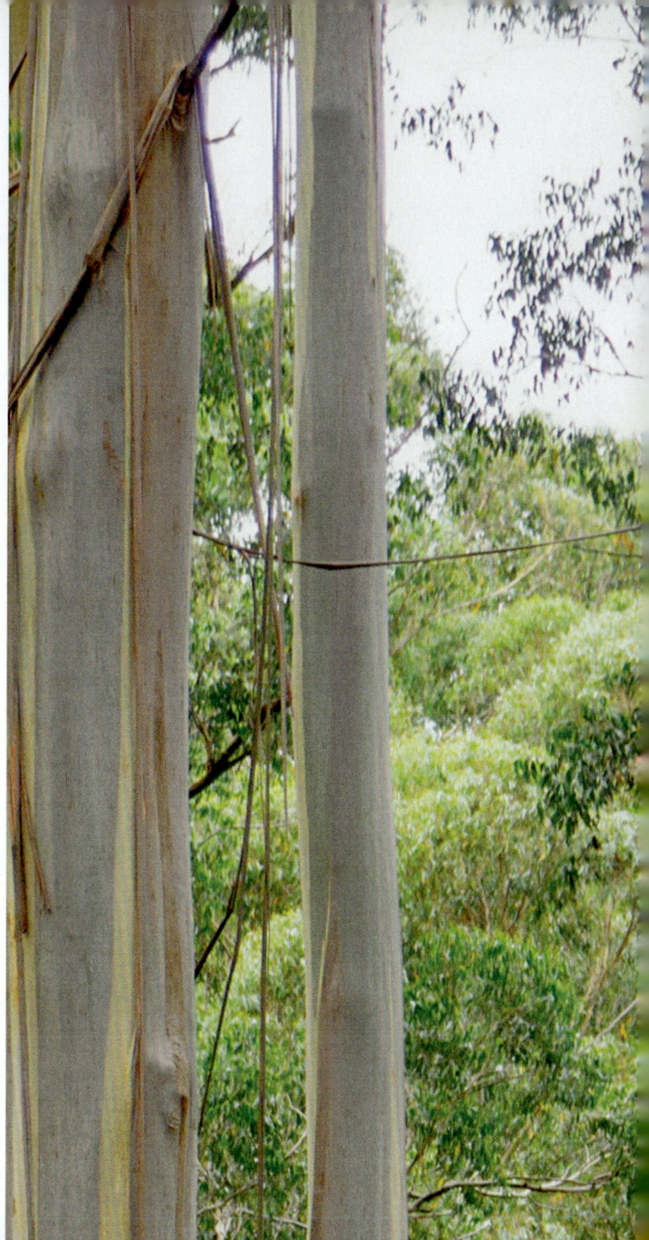

FOUR ELEMENTS

EARTH

# AUSTRALIA/OCEANIA

When people think of Australia, they automatically think of the unique eucalyptus trees, but other trees also grow in this climate.

## TASMANIA

Tasmania has a temperate climate in which rainforests flourish. In the Styx Valley, some of the eucalyptus trees grow over 95 m (312 ft) tall, and some may be over 400 years old. But only a quarter of the original forest still exists – and of that, one-third is threatened by forest clearance. The Tasmanian rainforest is home to, amongst others, pygmy opossums, quolls and the national animal, the Tasmanian devil.

## NEW ZEALAND

More than two per cent of New Zealand's land is taken up by national parks, primeval forests and conservation areas with a diverse range of unspoiled vegetation. The coast of the South Island is covered with dense rainforest hosting many indigenous plant species. In Fiordland National Park alone, there are more than 700 endemic species, including the well-known tree ferns. The Tuatapere Hump Ridge Track in the park passes through rainforest and subalpine vegetation.

## SOUTH AUSTRALIA

In South Australia, a unique range of animals and plants has developed over time, thanks to the temperate climate. Yarra Ranges National Park in the state of Victoria has an ancient tree population that includes tree ferns and mighty giant eucalyptus plants, which are some of the tallest broadleaved trees in the world and can grow to be 400 years old. Over 40 mammal and 120 bird species live here. And in Great Otway National Park, also in Victoria, the koalas feel at home in the light-filled eucalyptus forests.

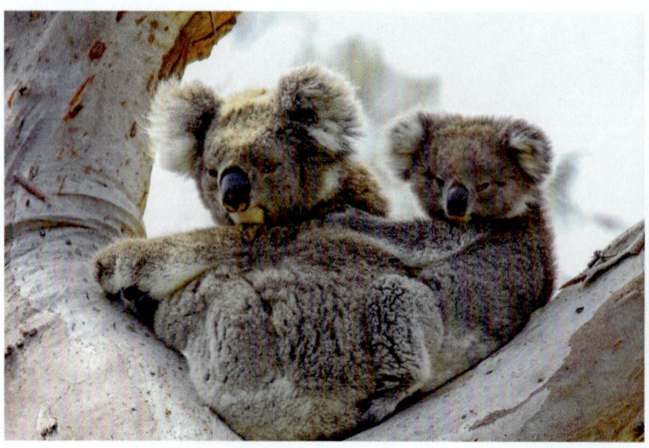

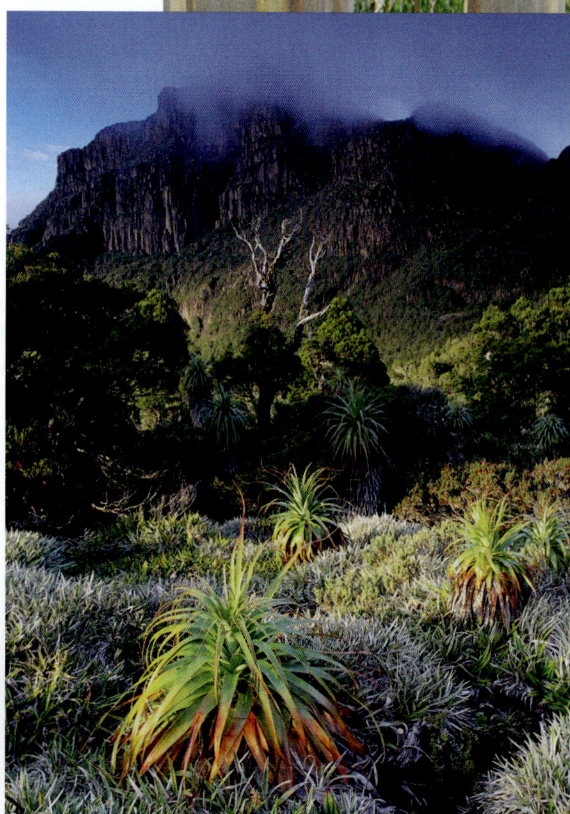

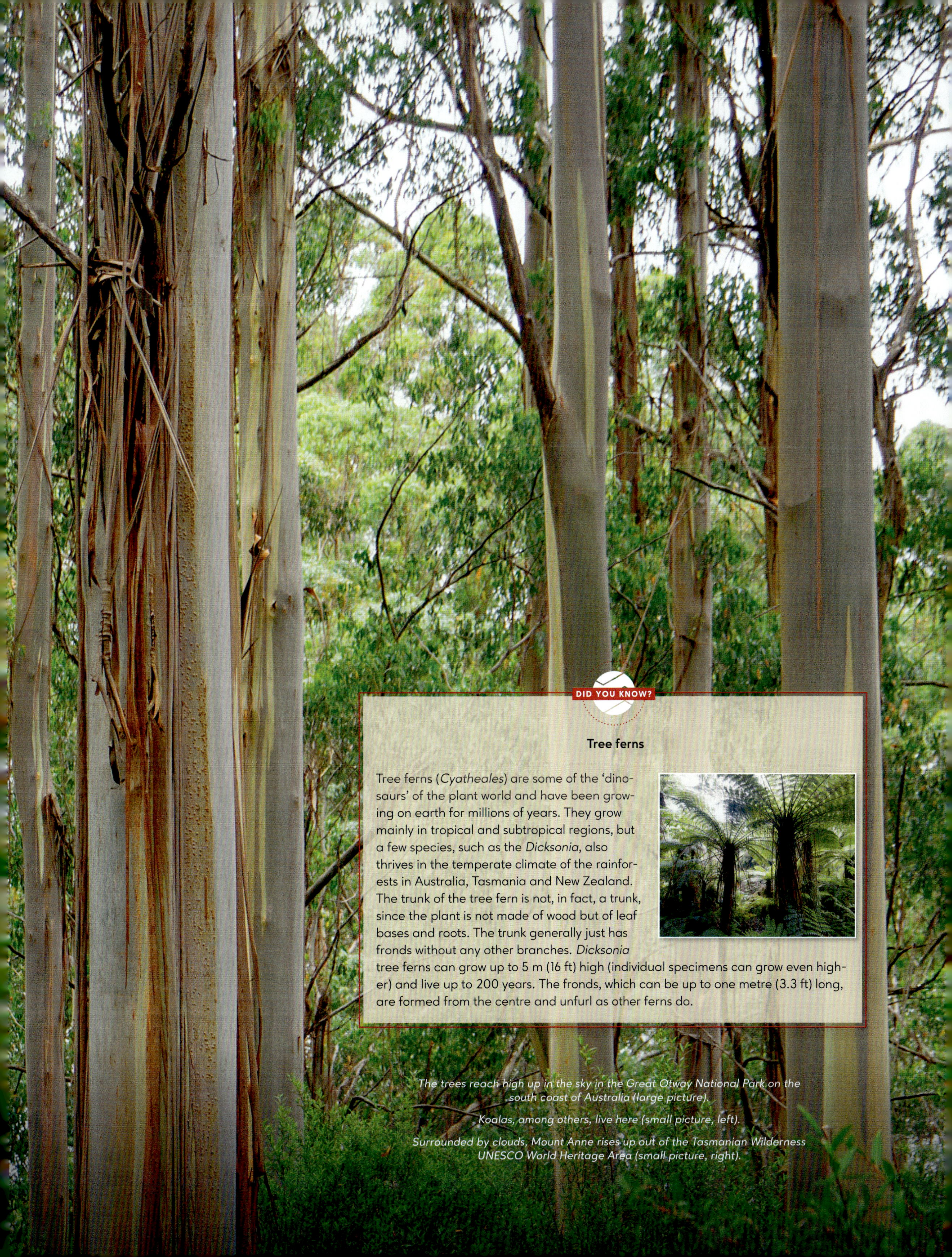

DID YOU KNOW?

**Tree ferns**

Tree ferns (*Cyatheales*) are some of the 'dinosaurs' of the plant world and have been growing on earth for millions of years. They grow mainly in tropical and subtropical regions, but a few species, such as the *Dicksonia*, also thrives in the temperate climate of the rainforests in Australia, Tasmania and New Zealand. The trunk of the tree fern is not, in fact, a trunk, since the plant is not made of wood but of leaf bases and roots. The trunk generally just has fronds without any other branches. *Dicksonia* tree ferns can grow up to 5 m (16 ft) high (individual specimens can grow even higher) and live up to 200 years. The fronds, which can be up to one metre (3.3 ft) long, are formed from the centre and unfurl as other ferns do.

*The trees reach high up in the sky in the Great Otway National Park on the south coast of Australia (large picture).*

*Koalas, among others, live here (small picture, left).*

*Surrounded by clouds, Mount Anne rises up out of the Tasmanian Wilderness UNESCO World Heritage Area (small picture, right).*

DID YOU KNOW?

### Termites as climate activists

Termites build their mound-shaped nests by chewing earth and wood and gluing the mixture together with their saliva. This decomposing of organic material forms a major part of a functioning ecosystem. With their work, the hardworking insects help to keep the forest floor within their biosphere damp.

If a period of drought occurs, even more termites appear in the forest because there is more for them to do then. This helps many more seedlings to survive the drought in the damp soil. In rainforests, this has the effect of reducing the impact of climate change. In the long term, this will help not just the termites but also humankind.

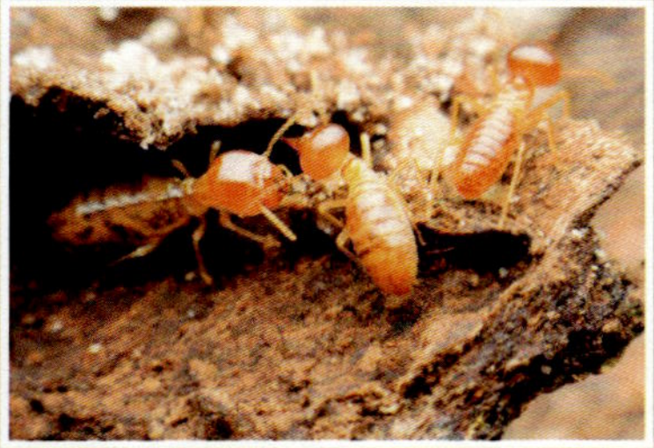

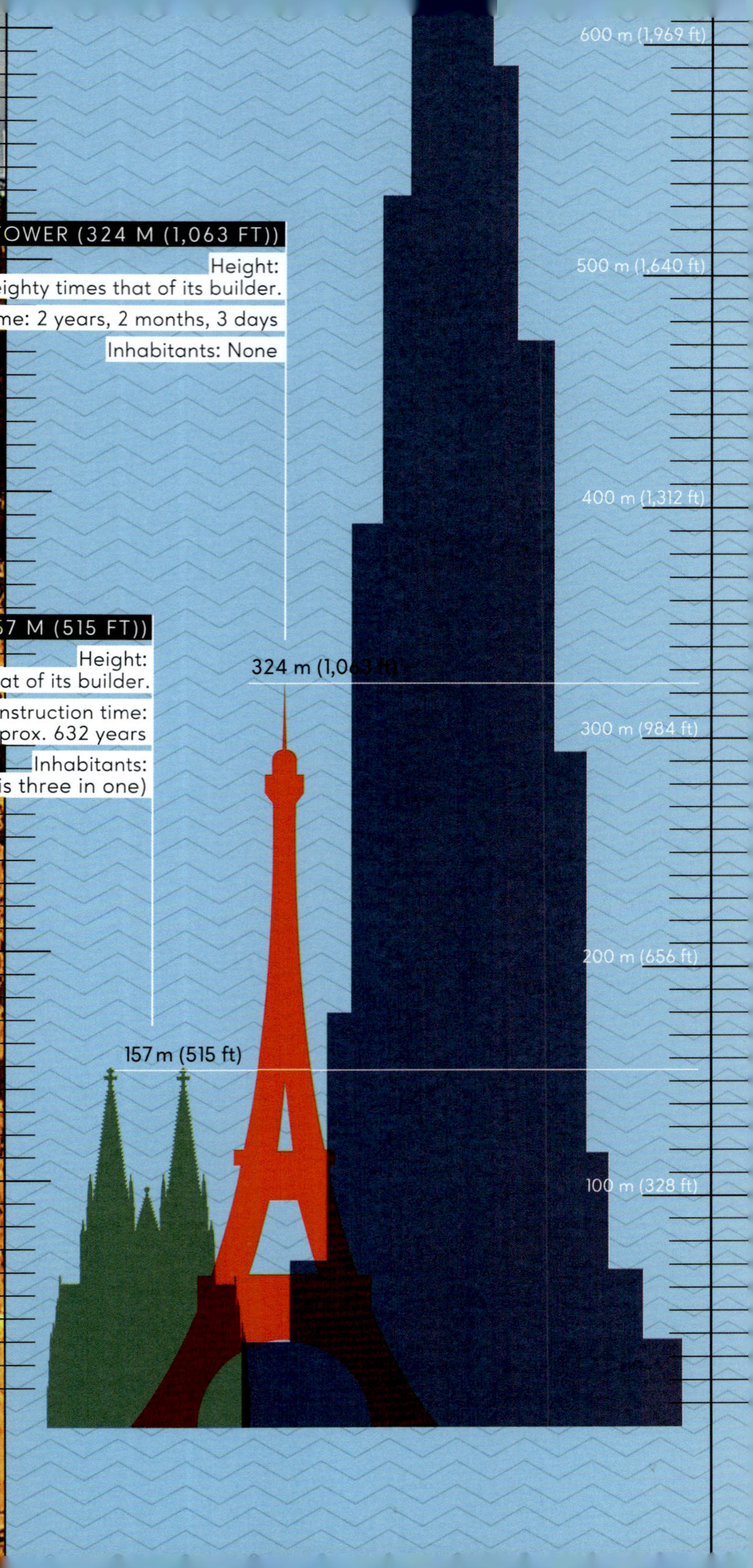

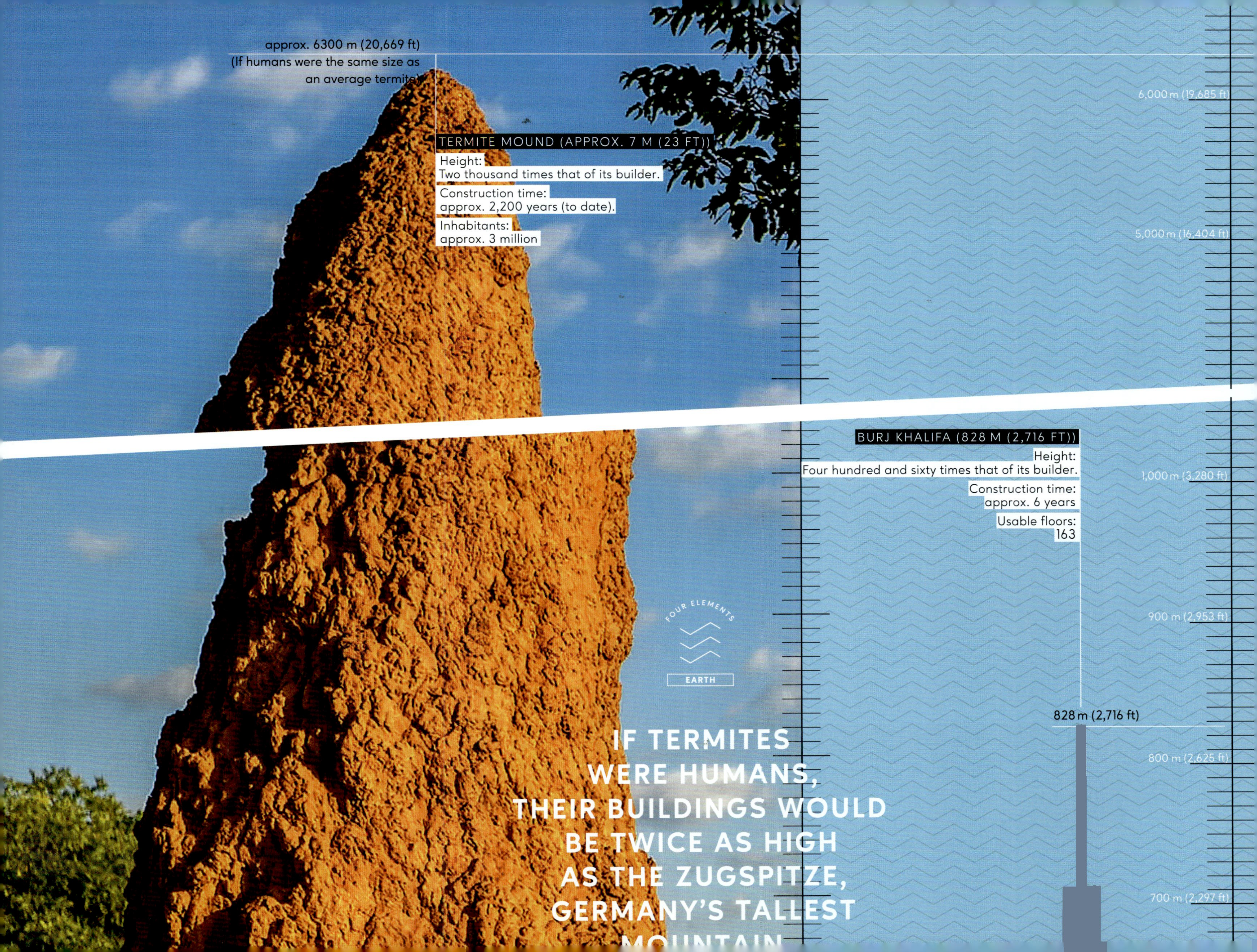
approx. 6300 m (20,669 ft)
(If humans were the same size as an average termite)
TERMITE MOUND (APPROX. 7 M (23 FT))
Height:
Two thousand times that of its builder.
Construction time:
approx. 2,200 years (to date).
Inhabitants:
approx. 3 million
6,000 m (19,685 ft)
5,000 m (16,404 ft)
BURJ KHALIFA (828 M (2,716 FT))
Height:
Four hundred and sixty times that of its builder.
Construction time:
approx. 6 years
Usable floors:
163
1,000 m (3,280 ft)
900 m (2,953 ft)
828 m (2,716 ft)
800 m (2,625 ft)
700 m (2,297 ft)
FOUR ELEMENTS
EARTH
IF TERMITES
WERE HUMANS,
THEIR BUILDINGS WOULD
BE TWICE AS HIGH
AS THE ZUGSPITZE,
GERMANY'S TALLEST

# NORTH AND CENTRAL AMERICA, SOUTH AMERICA

The Americas, which in fact unites three different continents, has a corresponding large number of superlatives in relation to its temperate rainforests

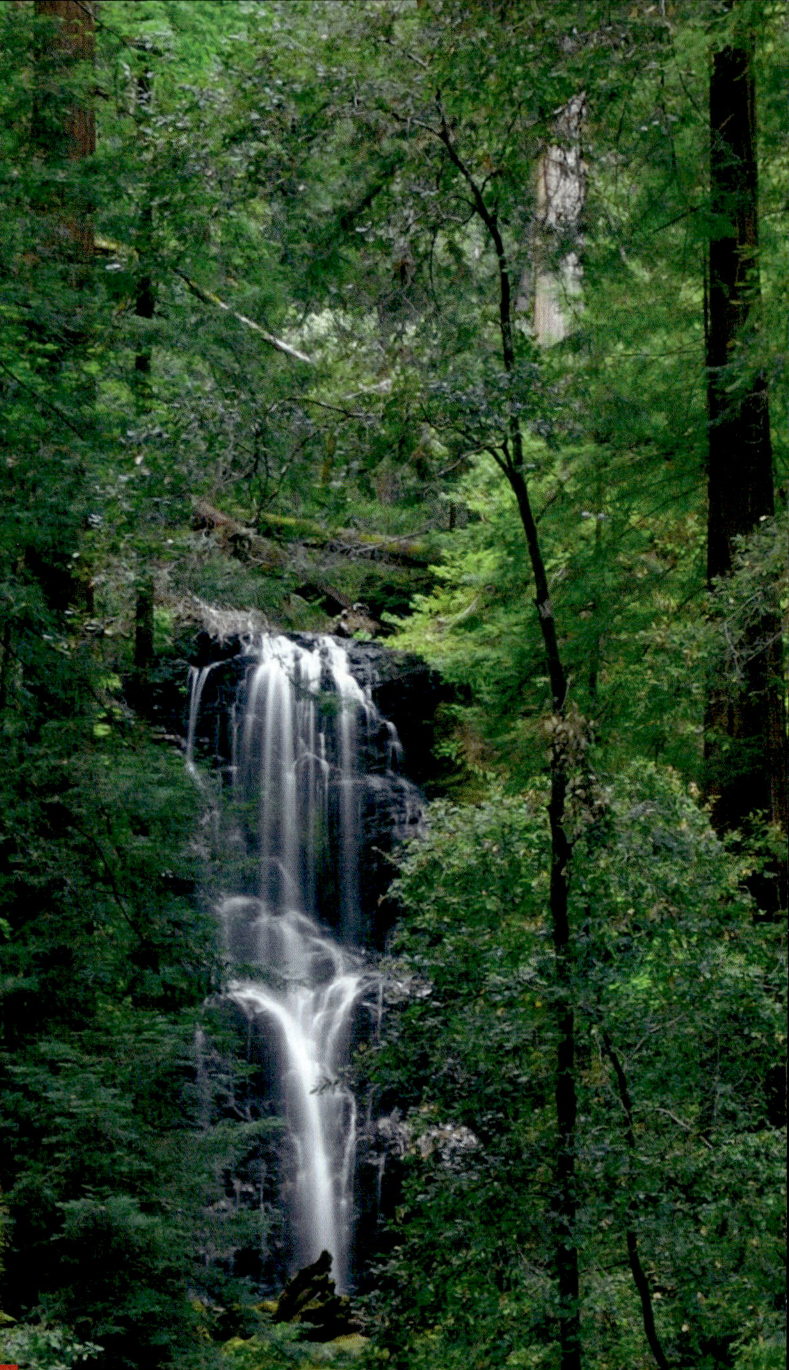

## NORTH AMERICAN PACIFIC COAST

On the North American Pacific coast between California and Alaska, conifers dominate the temperate rainforests. California is also home to the coastal redwood, or sequoia, the tallest tree in the world. The Redwood National Park in California contains many awe-inspiring specimens of this giant tree. Another California park, Big Basin Redwoods State Park in the Santa Cruz Mountains, has the largest number of such coastal redwoods south of San Francisco. Even in Olympic National Park in the US state of Washington, mild temperatures and abundant rainfall provide the conditions for a dense, lush, green rainforest with hemlock conifers and giant cedars as well as ferns, lichens and mosses. Pumas, mule deer, black bears, wildcats and coyotes live in the rainforest as well as approximately 300 different species of birds.

DID YOU KNOW?

### Coastal redwoods – the largest trees in the world

The evergreen sequoias, or giant redwoods, are, in terms of their volume, the mightiest trees in the world. The tree, which towers upwards 90 m (295 ft) or more can reach a trunk diameter of more than 12 m (39 ft) at the base. In older specimens – the oldest of them are over 2,500 years old – the trunks are branchless up to a height of 50 m (164 ft) above the ground. The giant redwoods are endemic to the western slopes of the Californian Sierra Nevada at elevations of 1,350 to 2,500 m (4,430 to 8,200 ft). It is now common across the world, even in Europe. The original populations in the USA have all been cleared, except for about the remaining 10 per cent.

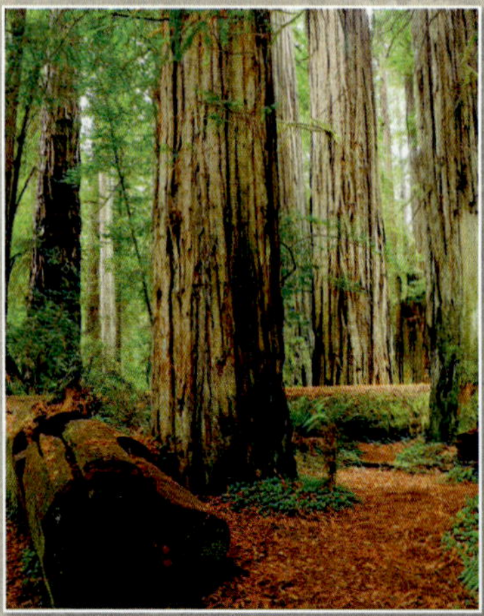

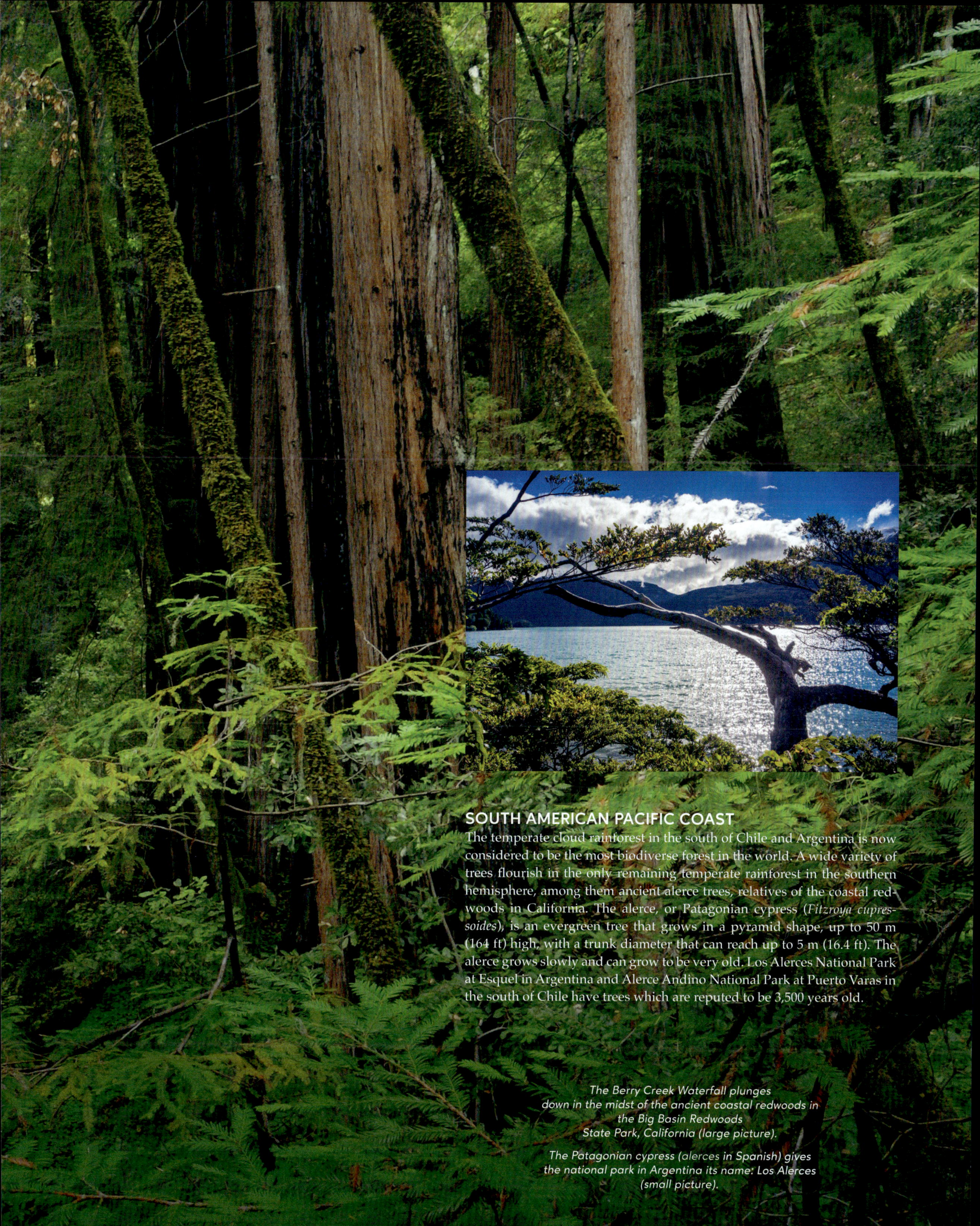

## SOUTH AMERICAN PACIFIC COAST

The temperate cloud rainforest in the south of Chile and Argentina is now considered to be the most biodiverse forest in the world. A wide variety of trees flourish in the only remaining temperate rainforest in the southern hemisphere, among them ancient alerce trees, relatives of the coastal redwoods in California. The alerce, or Patagonian cypress (*Fitzroya cupressoides*), is an evergreen tree that grows in a pyramid shape, up to 50 m (164 ft) high, with a trunk diameter that can reach up to 5 m (16.4 ft). The alerce grows slowly and can grow to be very old. Los Alerces National Park at Esquel in Argentina and Alerce Andino National Park at Puerto Varas in the south of Chile have trees which are reputed to be 3,500 years old.

*The Berry Creek Waterfall plunges down in the midst of the ancient coastal redwoods in the Big Basin Redwoods State Park, California (large picture).*

*The Patagonian cypress (alerces in Spanish) gives the national park in Argentina its name: Los Alerces (small picture).*

# TROPICAL RAINFORESTS AND MONSOON FORESTS

The tropical rainforests
spread out like a green sea
and are home to many different species

The typical types of tropical rainforest vegetation only found in the wet tropical climate zones can be found in South and Central America, Africa, South Asia and Australia. The annual rainfall level is 2,000 to 4,000 mm (79 to 157 in); the temperatures vary only slightly over the course of the year. The maximum daytime temperatures range between 24 and 31°C (75 and 88°F). Characteristic for a tropical rainforest is the 'tiered structure' from the root system up to the understorey or shrub level, which is up to 5 m (16.5 ft) high, and on up to the dense canopy at a height of 40 m (131 ft) and the single giant trees which tower above it. In contrast, monsoon forests do not shoot up so high: the undergrowth is much more pronounced and the number of endemic species is noticeably lower than in tropical rainforests.

Alongside the coral reefs, tropical rainforests have the highest species density in their flora as well as in their fauna. Most of the animal species are arthropods; in addition, mammals such as elephants and tigers, larger reptiles such as crocodiles, small reptiles and amphibians, primates such as orangutans and gorillas as well as countless multi-coloured birds all live here. The rivers of the rainforests are also abundant in fish.

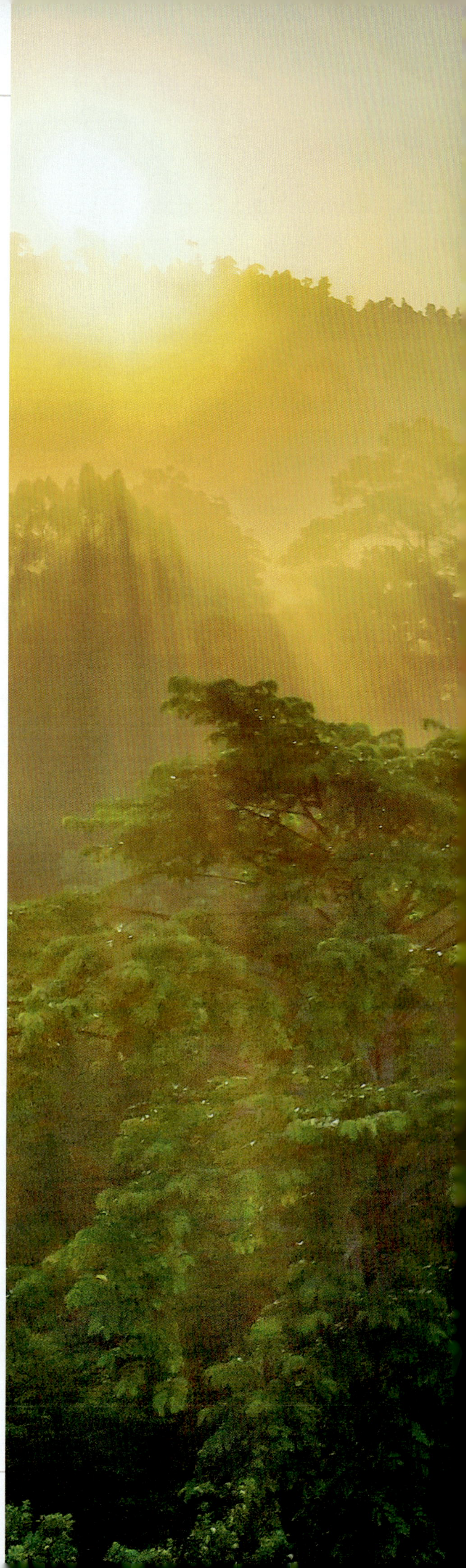

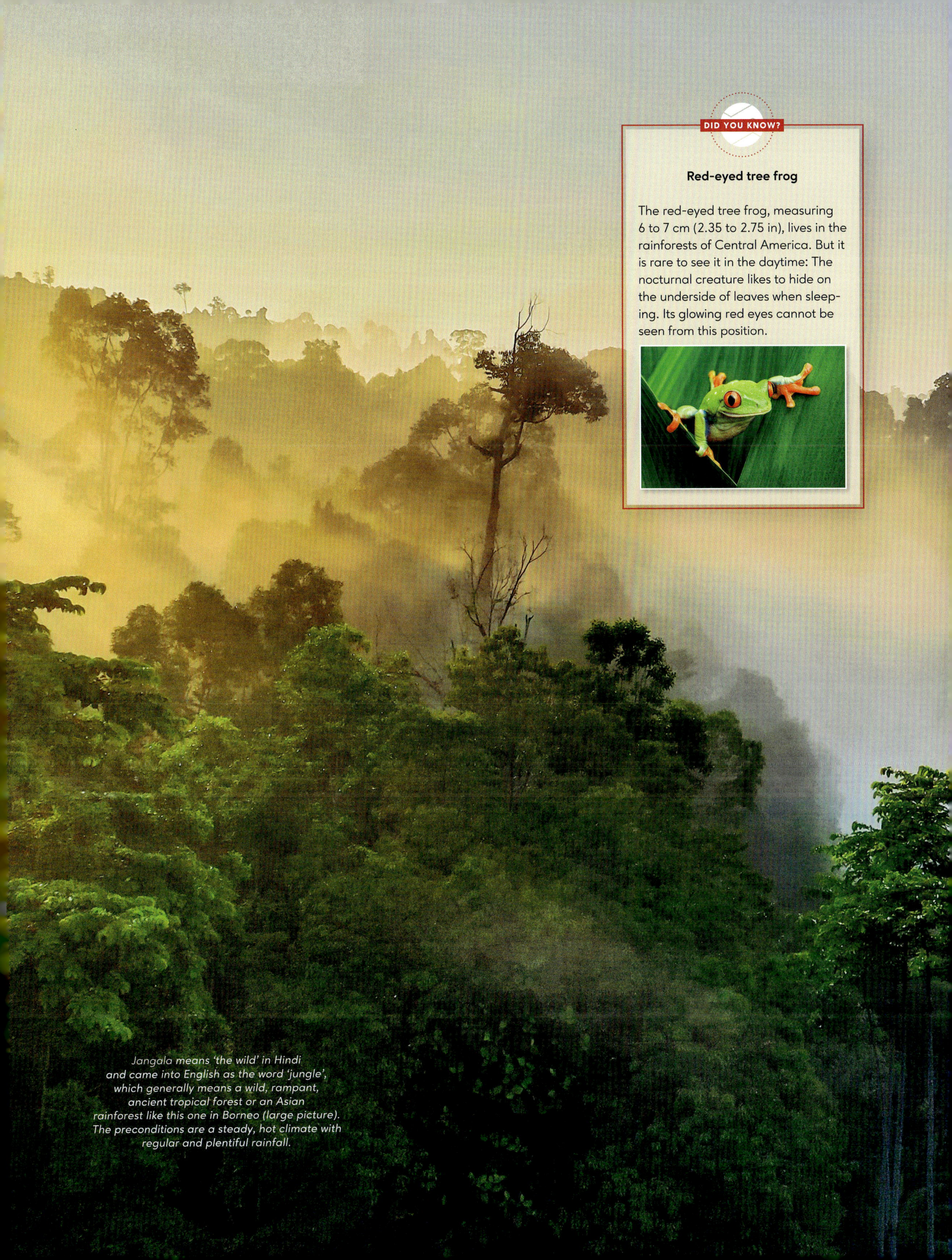

DID YOU KNOW?

**Red-eyed tree frog**

The red-eyed tree frog, measuring 6 to 7 cm (2.35 to 2.75 in), lives in the rainforests of Central America. But it is rare to see it in the daytime: The nocturnal creature likes to hide on the underside of leaves when sleeping. Its glowing red eyes cannot be seen from this position.

*Jangala means 'the wild' in Hindi and came into English as the word 'jungle', which generally means a wild, rampant, ancient tropical forest or an Asian rainforest like this one in Borneo (large picture). The preconditions are a steady, hot climate with regular and plentiful rainfall.*

FOUR ELEMENTS

EARTH

# AMAZON BASIN

The largest rainforest in the world

The Amazon Basin includes regions in Brazil, French-Guiana, Suriname, Venezuela, Guyana, Columbia, Peru, Ecuador and Bolivia. With around 7 million km$^2$ (2,702,715 sq mi), the Amazonian rainforests constitute the largest continuous forest zone in the world.

The basin is defined by the vast river system of the Amazon and its numerous tributaries, of which ten are among the top twenty-five rivers with the highest flow in the world (including the Amazon itself). The tropical rainforests of the region stand out due to their extremely high biodiversity, among other things, but the number of animal species is difficult to gauge because probably only a fraction of these has been discovered so far – it is believed that there are between five and ten million species altogether.

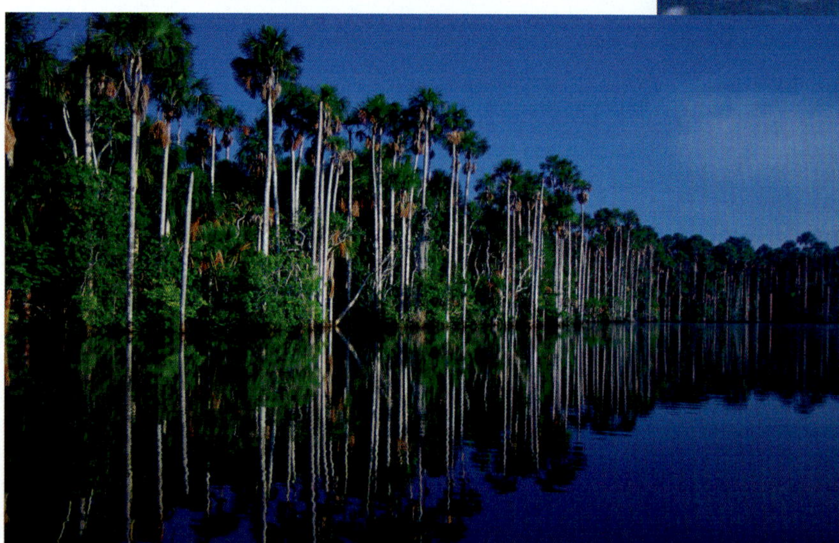

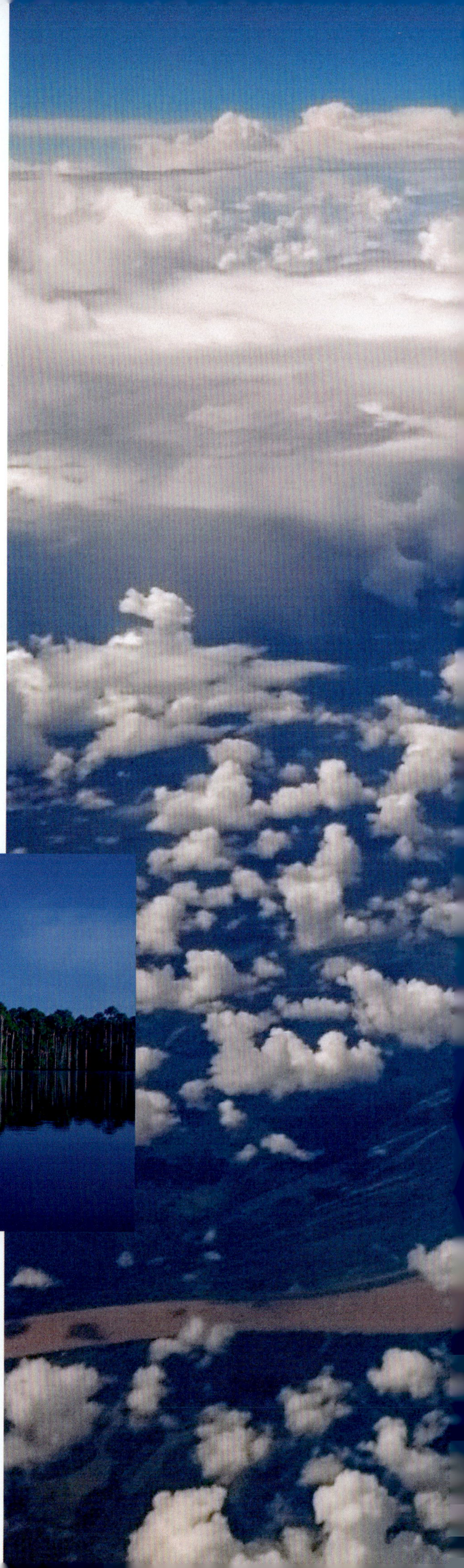

*An aerial view clearly shows how the Amazon meanders through the dense tropical rainforest of Brazil (large picture).*

*At the Lago Sandoval in Peru, the tall trees extend right to the water's edge (small picture).*

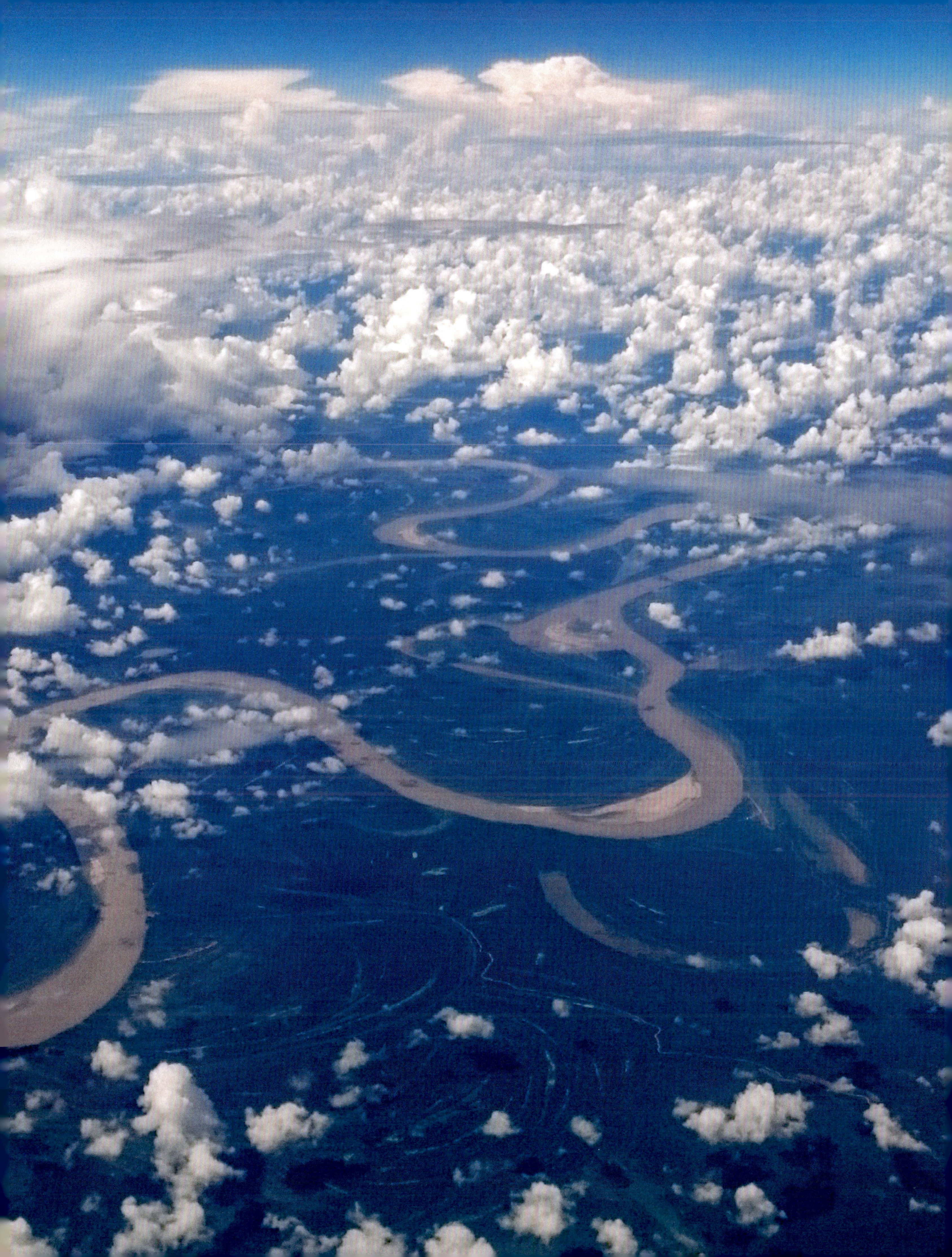

FOUR ELEMENTS

EARTH

# SOUTH AND SOUTHEAST ASIA

In contrast to other regions, there are longer dry seasons in the Asian rainforests.

In South Asia, there is an especially large number of distinct rainforests and monsoon forests. A characteristic feature of the climatic conditions in this region is a steady dry season lasting four to five months.

Due to the permanently wet and semi-humid climate, cloud and tropical mountain rainforests, subtropical rainforests and monsoon forests are all found in Southeast Asia, as are tropical lowland rainforests.

## WESTERN INDIA
**Western Ghats**

The mountain range of the Western Ghats runs 1,600 km (994 mi) along the edge of the Deccan Plateau, separating it from the coastal plains on the Arabian Sea. The mountain range collects the rainfall from the annual monsoon and thus acts as a watershed. The region was originally almost completely covered in rainforest. Now only a small section of it remains in its pristine state. The animal and plant worlds of the Western Ghats are some of the most biodiverse in the world. Many species are only found here. The ecology of the region exhibits great diversity. In general terms, the north is much dryer than the south. Overall, the mountain range is divided into four ecoregions: these are the lower regions of deciduous rainforests in the north and south, respectively, and the cooler humid regions of the northern and southern mountain rainforests. The most diverse region in all of India that of is the mountain rainforests in the south.

## SRI LANKA

The island of Sri Lanka off the southern tip of India is known for having a large number of plant and animal species in a relatively small area. But this abundance is threatened: according to estimates by environmental protection organizations, only 1.5 per cent of the original forests now remain – the rest have been cleared. The main nature conservation area in Sri Lanka and one of the oldest protected areas in South Asia is the Sinharaja Forest Reserve in the southwest of the island, which was declared a national park in 1978 and is now a UNESCO World Natural Heritage Site. The conservation area, which covers 765 km$^2$ (295 sq mi) and has 5,000 mm (197 in) of rainfall per year, is home to 200 different types of wood and a large number of endemic animal species. Of the birds, as many as 95 per cent are endemic to Sri Lanka, and, among the mammals and butterflies, this percentage is well over 50 per cent.

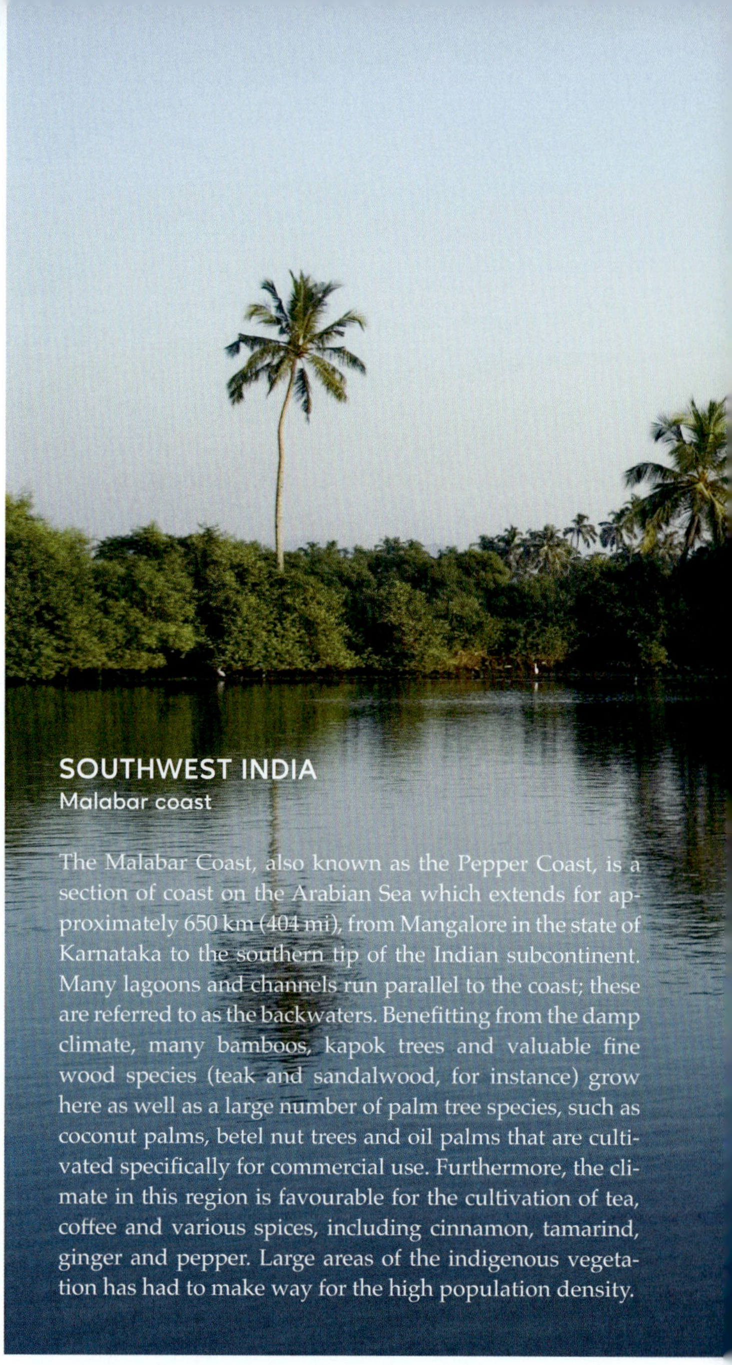

## SOUTHWEST INDIA
**Malabar coast**

The Malabar Coast, also known as the Pepper Coast, is a section of coast on the Arabian Sea which extends for approximately 650 km (404 mi), from Mangalore in the state of Karnataka to the southern tip of the Indian subcontinent. Many lagoons and channels run parallel to the coast; these are referred to as the backwaters. Benefitting from the damp climate, many bamboos, kapok trees and valuable fine wood species (teak and sandalwood, for instance) grow here as well as a large number of palm tree species, such as coconut palms, betel nut trees and oil palms that are cultivated specifically for commercial use. Furthermore, the climate in this region is favourable for the cultivation of tea, coffee and various spices, including cinnamon, tamarind, ginger and pepper. Large areas of the indigenous vegetation has had to make way for the high population density.

## ANDAMAN ARCHIPELAGO

The 204 Andaman islands in the Gulf of Bengal, which belong to India, are still mainly covered in original tropical rainforest. A humid climate prevails on the islands, with temperatures between 18 and 34°C (64 and 93°F) and up to 150 days of rainfall per year. Light winds make the high humidity levels (66 to 85 per cent) bearable. Mangroves flourish on the coasts although these were partly destroyed by the devastating tsunami in December 2004.

## MALAY PENINSULA

The Malay Peninsula, also known as the Golden Peninsula, is a narrow, elongated continuation of the southeast Asian mainland. The territory on the peninsula is divided between the states of Myanmar, Thailand and Malaysia. The landscape and ecosystems are varied and diverse. In addition to mountain regions, there are lowland plains, lakes, beaches and rainforest areas, which are typical of regions near the equator. In Taman Negara, the national park in the north of Peninsular Malaysia, an extensive, unspoiled tropical rainforest spreads out; at 130 million years old, this is the oldest forest on earth.

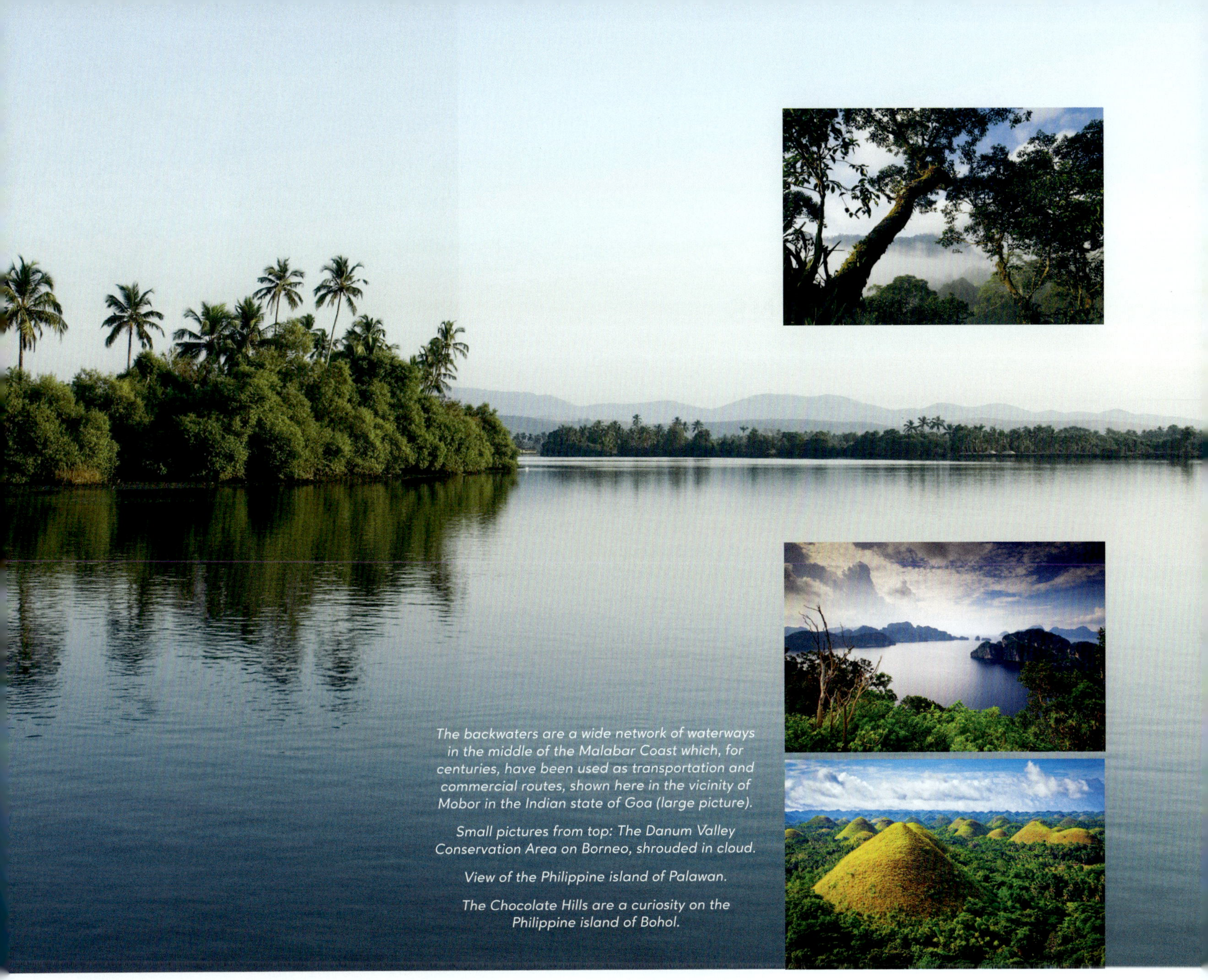

*The backwaters are a wide network of waterways in the middle of the Malabar Coast which, for centuries, have been used as transportation and commercial routes, shown here in the vicinity of Mobor in the Indian state of Goa (large picture).*

*Small pictures from top: The Danum Valley Conservation Area on Borneo, shrouded in cloud.*

*View of the Philippine island of Palawan.*

*The Chocolate Hills are a curiosity on the Philippine island of Bohol.*

## SUMATRA

The tropical rainforests on the Indonesian island of Sumatra are famous the world over, for it is here that some of the rarest and most endangered species live. The Sumatran tiger, Asian tapir, rhinoceros hornbill, great Argus pheasant, Sumatran elephant, white-handed gibbon, long-tailed macaque and Sumatran orangutan all live in the last remaining areas of primeval forest, which have been declared national parks. With 176 mammal, 194 reptile, 62 amphibian and 320 bird species, Sumatra offers more diversity in terms of its fauna than any other Indonesian island.

## BORNEO

Ancient rainforest, mangroves, orangutans, numerous indigenous plants – the Indonesian island of Borneo, the third largest island in the world, is still unspoiled in many places. Among the more than 220 mammal species, there are gibbons, Sumatran rhinoceroses, honey bears, clouded leopards and long-nosed monkeys. The rhinoceros hornbill is one of the 622 bird species living here; in addition, there are 400 reptile and amphibian species, among them the Kapuas mud snake, the only snake that can change its colour.

## LESSER SUNDA ISLANDS

The Lesser Sunda Islands consist of six larger islands – Bali, Lombok, Sumbawa, Flores, Timor and Sumba – and numerous smaller islands lying east of Java in the Indonesian Archipelago. The Wallace Line, the biogeographical demarcation between Asian and Australian flora and fauna, runs between Bali and Lombok. Large areas of the Lesser Sunda Islands were once covered in monsoon forests, but these have been massively pushed back by the cultivation of the land. In western and southern Bali, the forests have remained intact to some extent and have been protected by the Bali Barat National Park since 1984.

## PHILIPPINES

Numbering more than 7,000 islands, the Philippine Islands in the western Pacific Ocean together form one of the most diverse ecosystems in the world. The islands host more than 5,000 animal species as well as 14,000 plant species. A main attraction on the island of Bohol are the Chocolate Hills, an area formed from 1,268 natural conical limestone hills. Due to deforestation in the last century, a strain of very resilient grass vegetation has spread on the hills; in the dry season, this withers and turns the hills a chocolate brown colour.

# SUNDARBANS

The largest mangrove forest in the world

With an area of 6,000 km² (2.317 sq mi), the Sundarbans ('lovely forests') lie on the delta of four major rivers and form the largest mangrove forest in the world. Large parts of the Sundarbans, a UNESCO World Natural Heritage Site, are now part of a national park on the border between India and Bangladesh. Countless birds, fish, deer, crocodiles, pythons and wild boar live in the mangrove forests. The Sundarbans are, moreover, the refuge area for the Bengal tigers, which became the symbol of species extinction in this region. Characteristic of the mangrove swamps are the root systems, which stabilize the plants on hard, fine or loose substrate. Between these, special roots known as pneumatophores, protrude for aeration – without which the roots would die – because the muddy ground is oxygenless, even a few millimetres down.

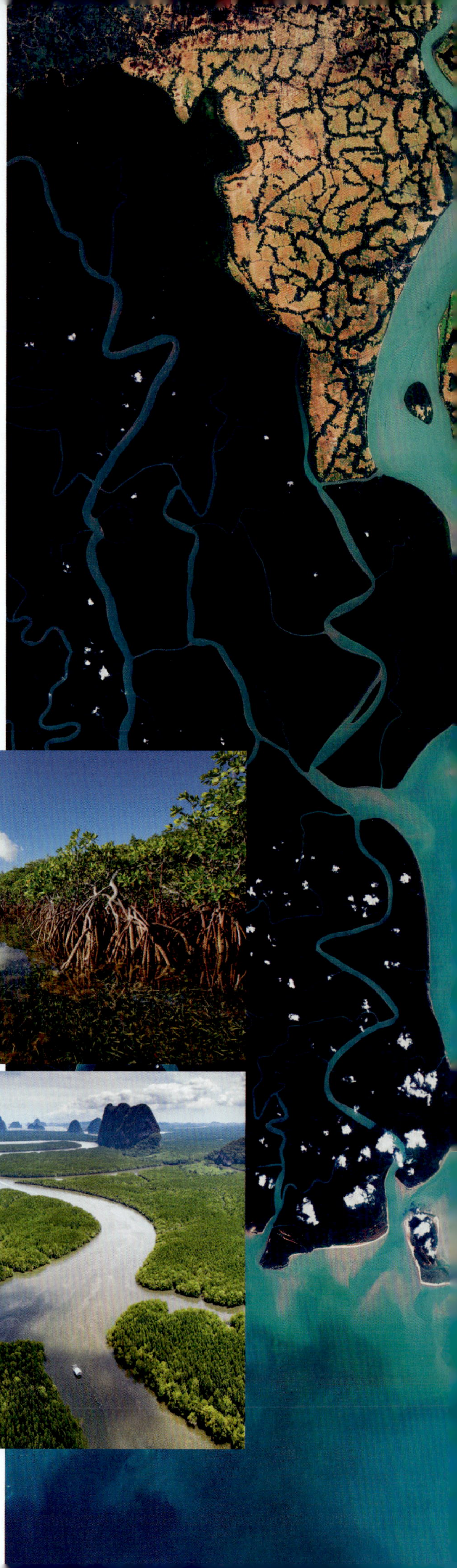

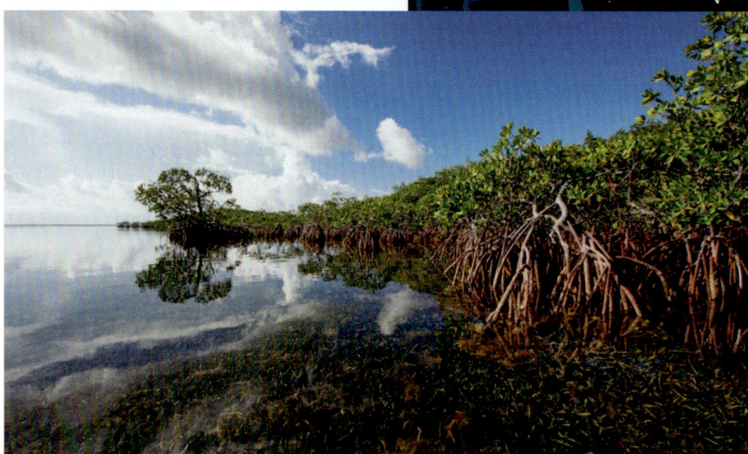

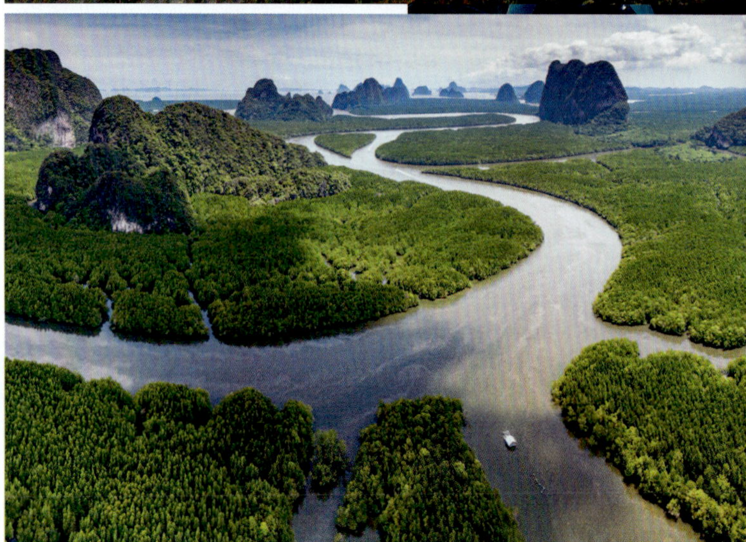

## MANGROVES – AN UNUSUAL TREE SPECIES BETWEEN THE LAND AND SEA

Mangrove forests, which are found in the tidal areas of tropical coasts, consist of salt-tolerant trees and shrubs of various plant families, which have adapted to conditions on the sea coasts and in the brackish estuaries. They store salt in their cells, for example, and some can even store water, as cacti do, in order to be able to dilute the concentration of salt. They excrete the excess salt via glands and then shed the leaves containing high salt concentration. In addition to coral reefs and tropical rainforests, mangroves forests are among the most productive ecosystems in the world. Reptiles, mammals and birds live in the crowns of the mangrove forests. Great numbers of fish, mussels and crabs also live between the underwater roots.

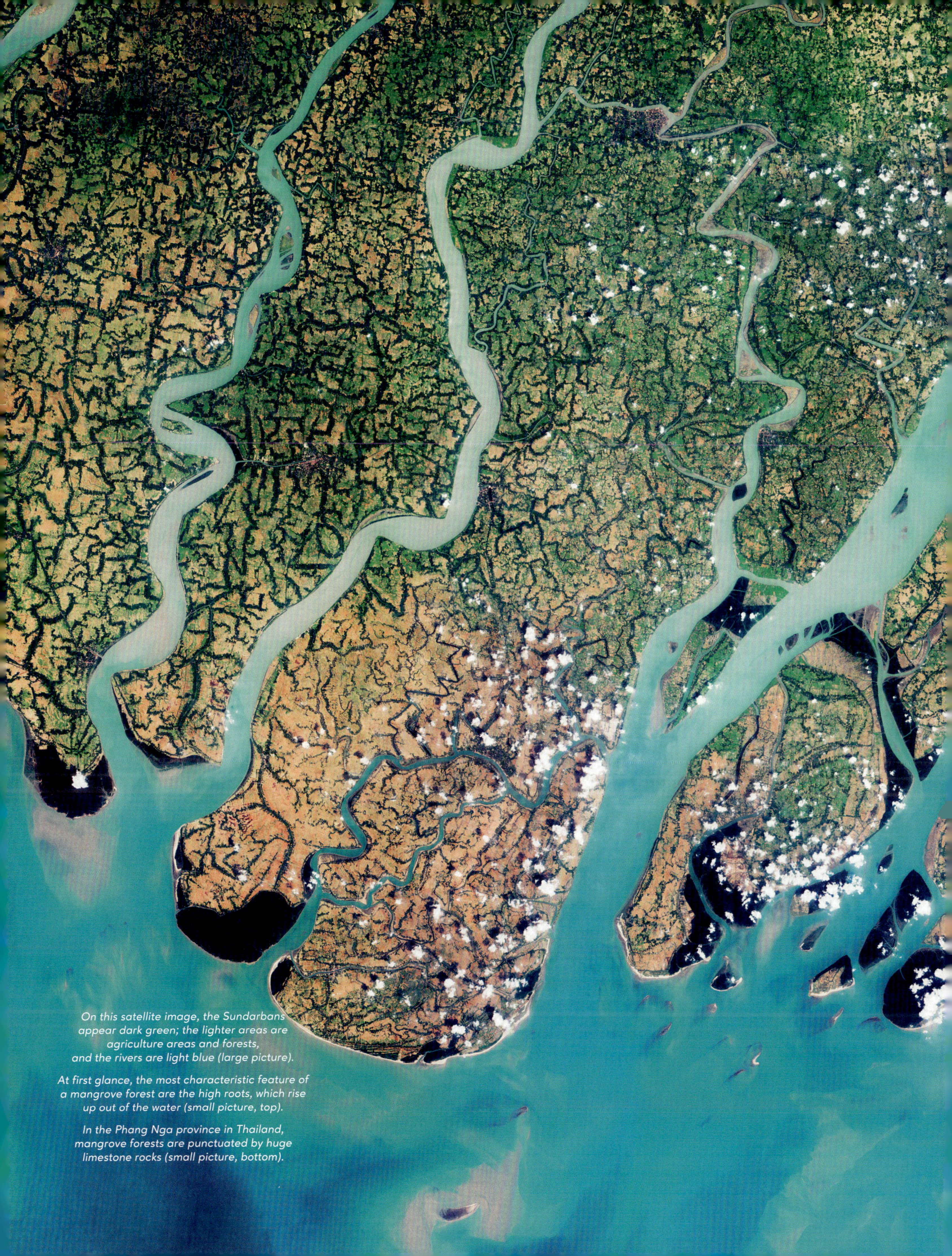

On this satellite image, the Sundarbans appear dark green; the lighter areas are agriculture areas and forests, and the rivers are light blue (large picture).

At first glance, the most characteristic feature of a mangrove forest are the high roots, which rise up out of the water (small picture, top).

In the Phang Nga province in Thailand, mangrove forests are punctuated by huge limestone rocks (small picture, bottom).

# ELEPHANTS, ORANGUTANS, LIZARDS, TIGERS

The largest animals in the Sundarbans

## INDIAN ELEPHANTS – THE LARGEST ANIMALS OF THE ASIAN FORESTS

The Asian or Indian elephant (*Elephas maximus*) is the second largest terrestrial animal after the African elephant. It distinguishes itself from its African cousin by the fact that its ears are smaller and it only has one trunk tip, rather than two.

The cows and their young live in family herds; the bulls live alone or, when they are still young, they form their own associations and temporarily join a herd during the mating season. The habitat of the Asian elephant once included the tropical rainforest as well as open grassland. Today wild Asian elephants, which are classified as highly endangered, can only be found in dense forest; their numbers are estimated to be between 35,000 and 55,000. Of these, approximately forty per cent live on the Indian subcontinent, a further forty per cent on the Southeast Asian mainland and the remainder on Sri Lanka and the southeast Asian islands.

## BENGAL TIGER – THE LARGEST PREDATOR OF THE ASIAN RAINFOREST

The endangered Bengal or Indian tiger, the second largest of the species after the Siberian tiger, could once be found throughout Asia. Now it only occurs in India, Bangladesh, Bhutan, Nepal and Myanmar. The largest Bengal tiger population lives in the Sundarbans mangrove forests.

The majestic Bengal tiger (*Panthera tigris tigris*), with its red, yellow and brown striped coat, powerful frame and lithe gait, is 1.9 to 2.8 m (6 to 9 ft) long and weighs 180 to 260 kg (397 to 573 lb). The predator's canines can measure up to 6 cm (2 in); its claws grow up to 10 cm (4 in) long. Its diet consists of about 9 kg (20 lb) of meat per day.

The tiger population is endangered because its habitat is continually shrinking: the Bengal tiger does, however, have the biggest population – around 4,000 to 5,000 animals.

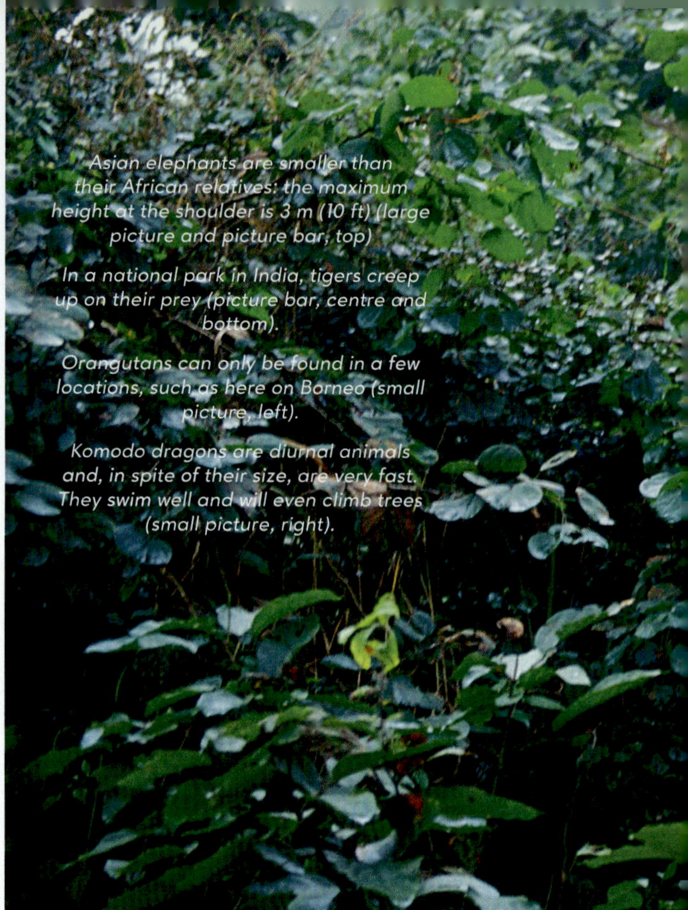

*Asian elephants are smaller than their African relatives: the maximum height at the shoulder is 3 m (10 ft) (large picture and picture bar, top)*

*In a national park in India, tigers creep up on their prey (picture bar, centre and bottom).*

*Orangutans can only be found in a few locations, such as here on Borneo (small picture, left).*

*Komodo dragons are diurnal animals and, in spite of their size, are very fast. They swim well and will even climb trees (small picture, right).*

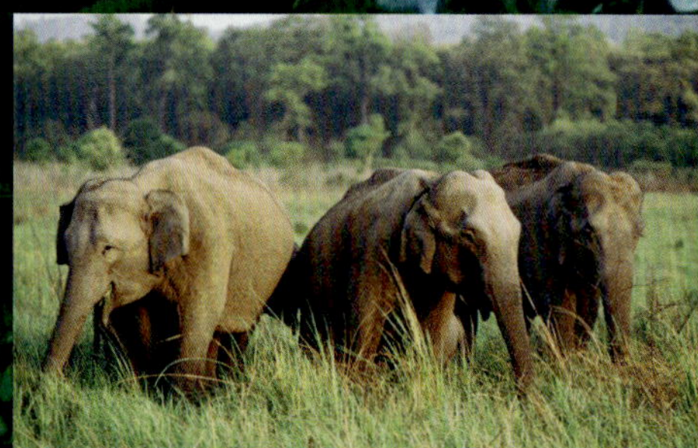

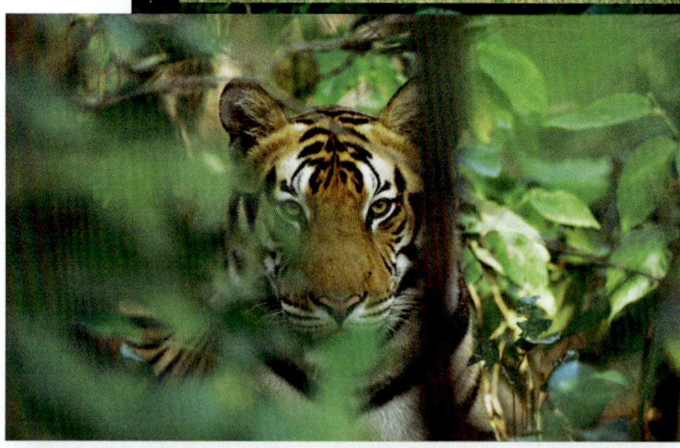

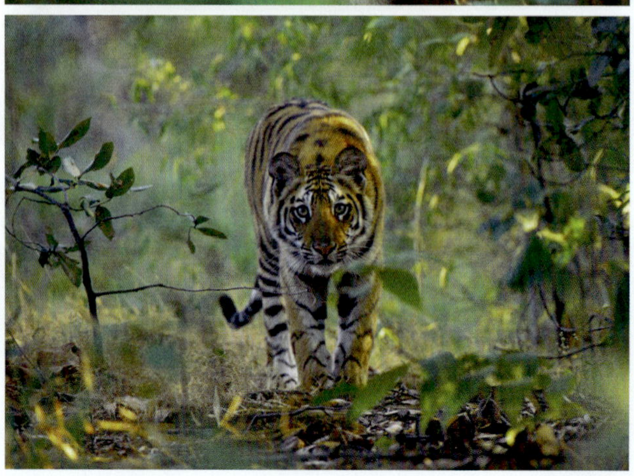

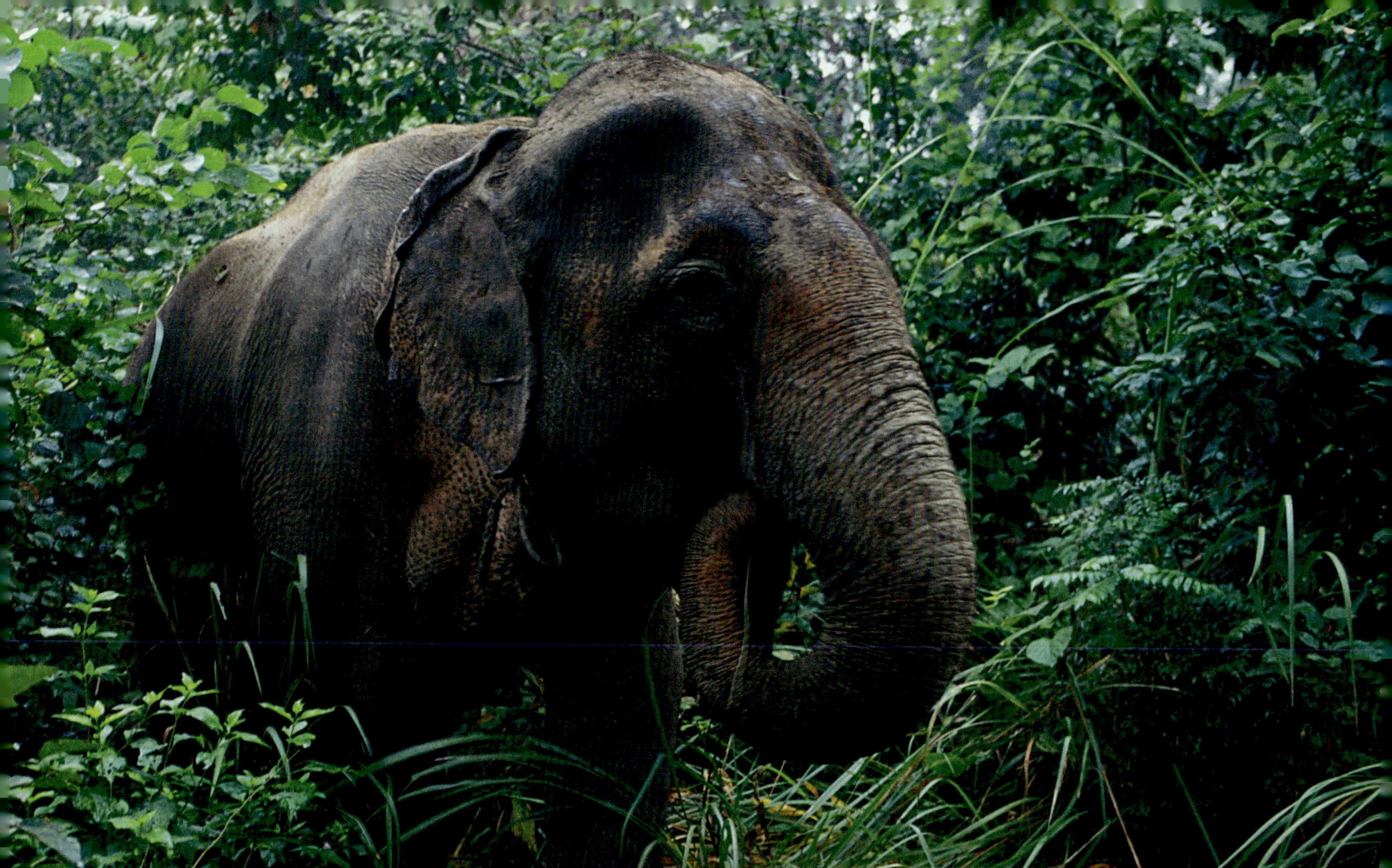

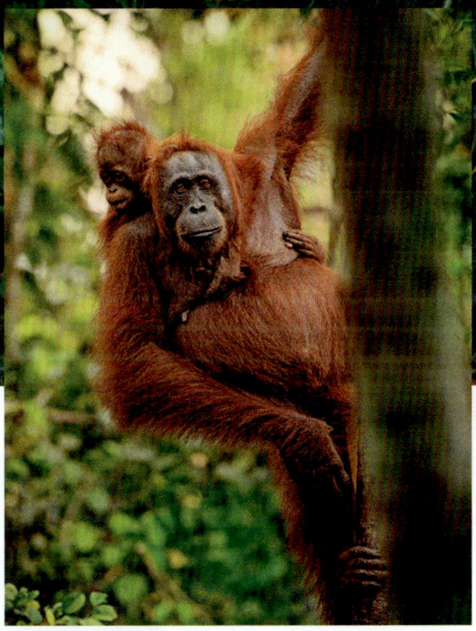

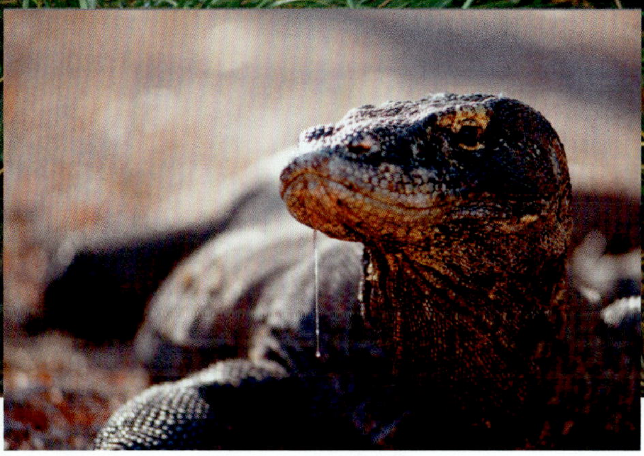

## KOMODO DRAGON – THE LARGEST LIZARDS IN THE WORLD

The Komodo dragon, the largest extant lizard species, can only be found on the Indonesian islands of Komodo, Rinca, Gili Motang and Flores, where it lives in the forest and bush. It can be up to 3 m (9 ft) in length, but most specimens living in the wild are much smaller, reaching, on average, a length of approximately 1.8 m (6 ft). When it has not eaten, the Komodo dragon seldom weighs more than 50 kg (110 lb), but, when it has a full stomach, it can weigh as much as 100 kg (220 lb), as it can consume incredibly large amounts of food within a very short time. The Komodo dragon is also very agile; it possesses excellent sensory organs and is considered one of the most intelligent reptiles. Its diet consists mainly of carrion, but it also enjoys live prey – from small reptiles, birds and mammals to full-grown deer and wild boar. It swallows smaller prey whole.

## ORANGUTANS – THE LARGEST PRIMATES IN ASIA

These reddish-brown primates belong to the ape family. There are now only two extant species in the tropical rainforests on the Southeast Asian islands of Sumatra and Borneo. Fossil remains show that they also previously lived in southern China, Vietnam and on Java. Orangutans are particularly well adapted to living in trees. For instance, when feeding - their main diet is fruit and leaves -, they sit or hang in the trees and bend or even break the thick branches with their powerful arms in order to access the fruit.

FOUR ELEMENTS

EARTH

# THE SMALLEST AND THE LARGEST

DID YOU KNOW?

### The smallest flower in the world

The offshoots of the rootless duckweed are only 0.5 to 1.5 mm (0.02 to 0.06 in) long. Elliptical in shape, they float along on the surface of the water – they do not have roots. In some areas, it is listed as endangered, but the plants, which resemble coarse green granules, can still be found in other places, mainly in standing water which is nutrient-rich and has a large amount of sun exposure. Experiments have already been carried out on the tiny green plants to investigate their use as a food source on space flights. Some of their features make them a good choice: they are rich in protein, comparable to soybeans, and they also grow fast. Under ideal conditions, they can double their biomass in only 48 hours. There is one catch, however, in that, despite their small size, the duckweed plants are very tough and need to be cooked for a long time and then pureed.

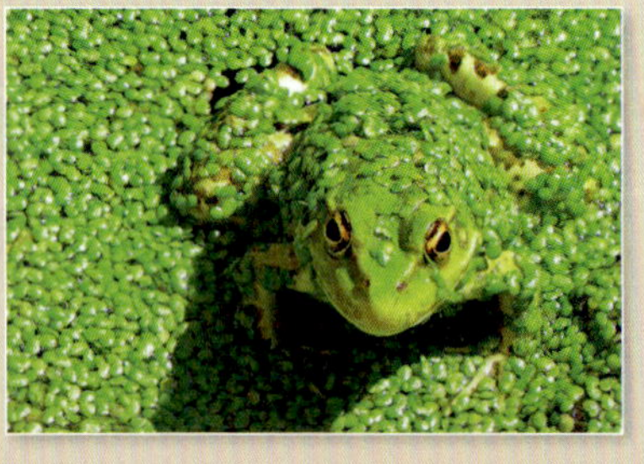

RAFFLESIA

‹‹‹ Diameter: up to 1 m (3.3 ft) ›››

Flower development: up to 1 year

Flowering time: approx. 5 days

**ROOM FOR ALL**

More than 1,250 rootless duckweed plants could fit into one rafflesia flower

ROOTLESS DUCKWEED

Diameter: approx. 0.7 mm (0.03 in)

## RAFFLESIAS – THE LARGEST FLOWERS IN THE WORLD

The *Rafflesia arnoldii*, measuring up to 1 m (3 ft) in diameter and weighing up to 11 kg (24 lb), is the largest single flower in the plant kingdom. Raffliesias are parasitic plants which, with the exception of the flowers, which frequently lie on the ground, live entirely within the host plant. Like the titan arum, they imitate the smell of carrion in order to attract insects for pollination. The *Rafflesia arnoldii* is found in the Malaysian state of Sarawak and in Kalimantan Barat on Borneo in Indonesia as well as on Sumatra.

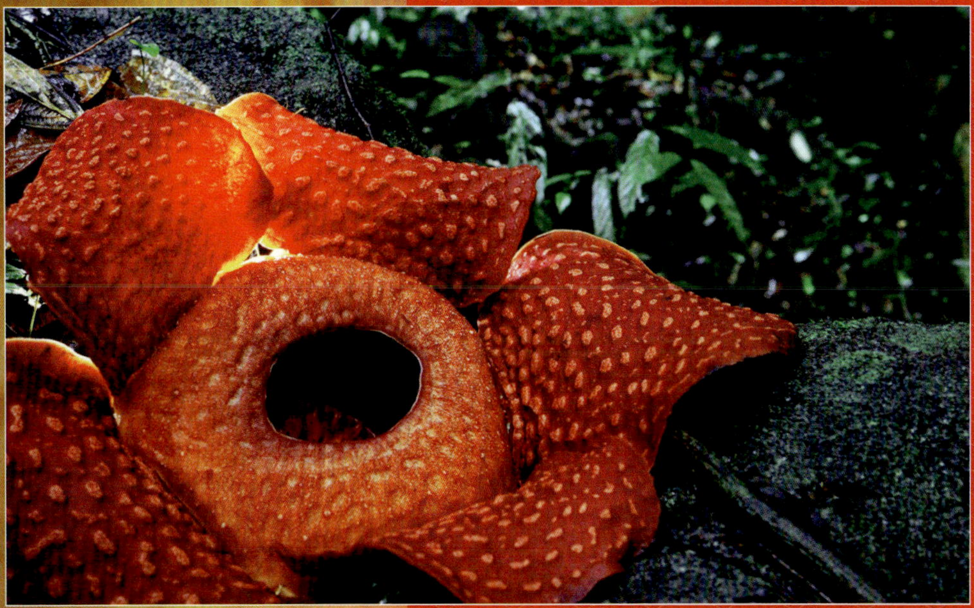

## LIFE AS A PARASITE

It almost seems as if the rafflesia, with its enormous flower, is trying to hide the fact that, apart from its flower, it has nothing else: no stem, no leaves and no roots. In order to survive, it must therefore become a parasite. It anchors itself to another plant and draws nutrients from this host – because without any leaves, the rafflesia cannot carry out photosynthesis.

## IRRESISTIBLE TO FLIES

However impressive the rafflesia flowers might look, their noxious odour of rotting meat is just as intense. Together with the heat they emit and their reddish colour, they attract flies and certain types of beetles that mistake the blooms for carrion. The plant needs these insects for pollination.

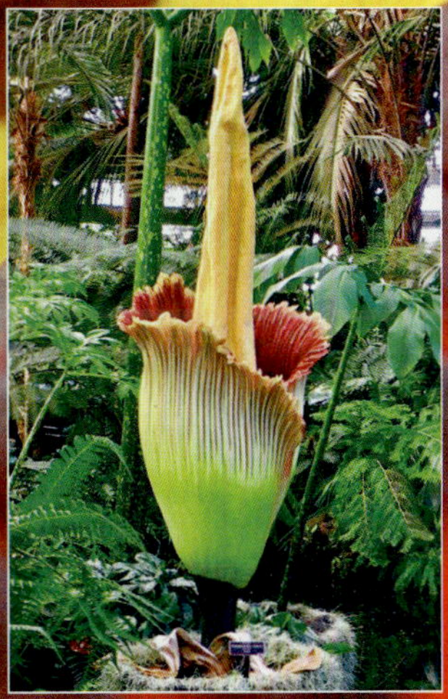

## THE TALLEST FLOWER IN THE WORLD

With the exception of botanical gardens, the titan arum (*Amorphophallus titanum*) is only found on Sumatra. Its inflorescence is, botanically speaking, the largest flower in the world – the record is currently held by a titan arum in the Wilhelma Zoo and Botanical Garden in Stuttgart, Germany, which, in 2005, had an inflorescence measuring 2.94 m (9.6 ft) high with a circumference of 1.5 m (5 ft). However, the titan arum only blooms very seldom and at irregular intervals, roughly every nine to twelve years. Because of its dreadful smell, the plant is called the corpse flower in Indonesia.

FOUR ELEMENTS

EARTH

# AUSTRALIA/OCEANIA

So much for the Red Continent – green forests also await you in Australia and its surrounding islands

Large areas of tropical rain and monsoon forests still exist in Australia, New Zealand and the other island states of Oceania.

## NEW GUINEA

The island of New Guinea, the second largest island in the world after Greenland, is dominated by tropical rainforests with around 20,000 plant species, making it one of the ten most biodiverse regions in the world. Particularly striking is the high number (around 55 per cent) of native species, that is, species that do not appear anywhere else in the world. This wide variety of species is encouraged by the diverse geography of the island and the large enclosed areas of vegetation unspoiled by human activity – approximately seventy-five per cent of the land in New Guinea is only covered in forest. The fauna is also impressive: on New Guinea, 400 to 600 different bird species, over 400 amphibian and butterfly species as well as 180 species of mammals exist. Spiders, reptiles and insects also live here. More than 1,000 different species of fish live in its lagoons, lakes and rivers.

The rainforest in Papua New Guinea in the east of the island is the third largest in the world although its size has been significantly reduced by forest clearance. In the western part of the island, Lorentz National Park covers an area of 23,555 km$^2$ (9,095 sq mi). It is the most extensive national park in all of Southeast Asia. It is the only park that stretches from equatorial glaciers to tropical seas, and it encompasses lowland rainforests, mountain rainforests, subalpine vegetation zones, swamp forests and mangroves.

## THE PACIFIC COAST IN NORTHERN AUSTRALIA

The Atherton Tablelands in Queensland, covering 800 km$^2$ (309 sq mi), form the largest continuous area of rainforest on Australia's northern Pacific coast. Eleven different types of forest can be distinguished here although the main type are the liana forests with many different tree ferns and epiphytes. In Daintree National Park alone, which is also in Queensland and stretches as far as the Great Barrier Reef, thirteen different types of rainforest can be distinguished. In this region, which comprises 0.2 per cent of the landmass of Australia, 30 per cent of the mammals, 65 per cent of all Australian bat and butterfly species and 20 per cent of all bird species can be found. Cape Tribulation, a spit of land where tropical rainforest meets the Great Barrier Reef, is also located in Queensland. The tropical climate and the unspoiled natural environment provide the best conditions for a wide variety of animal and plant worlds.

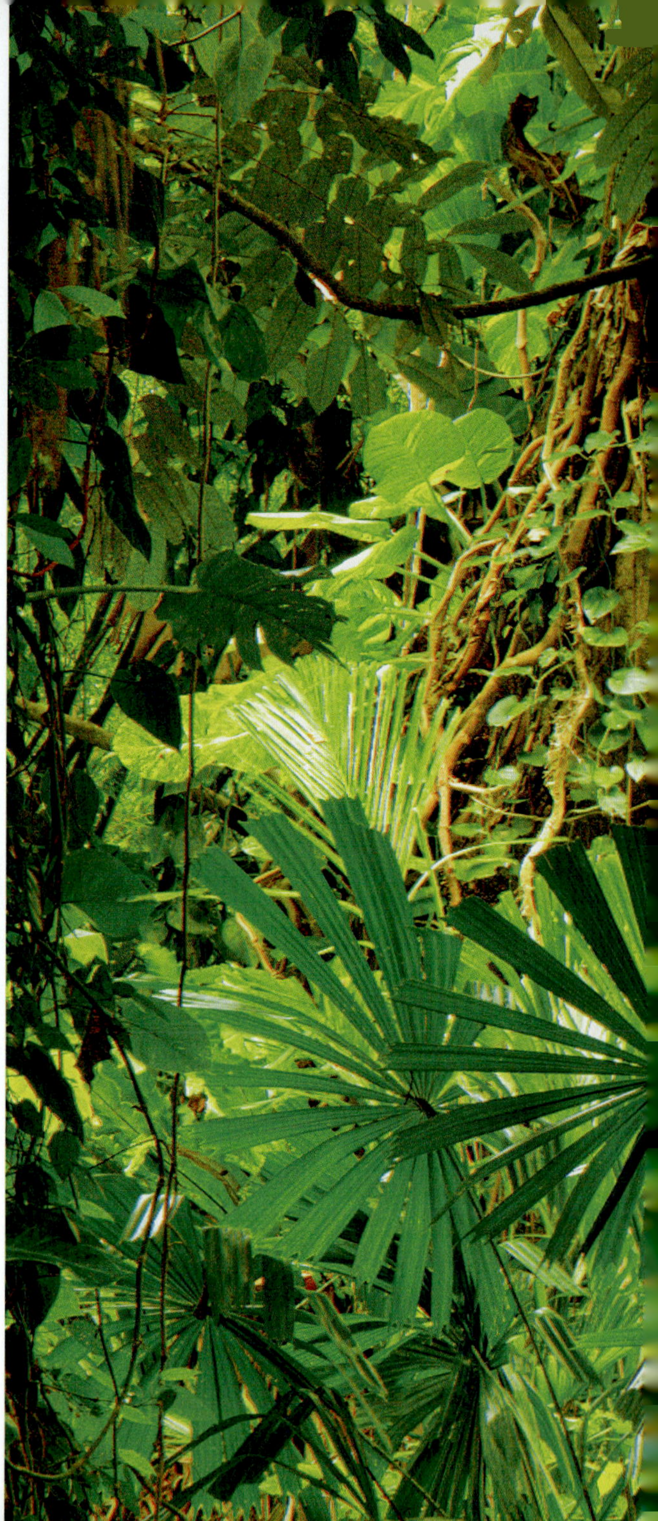

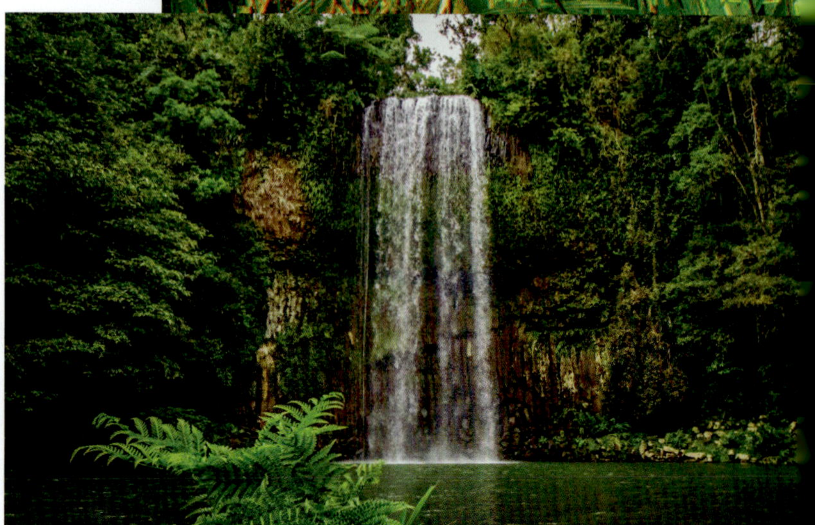

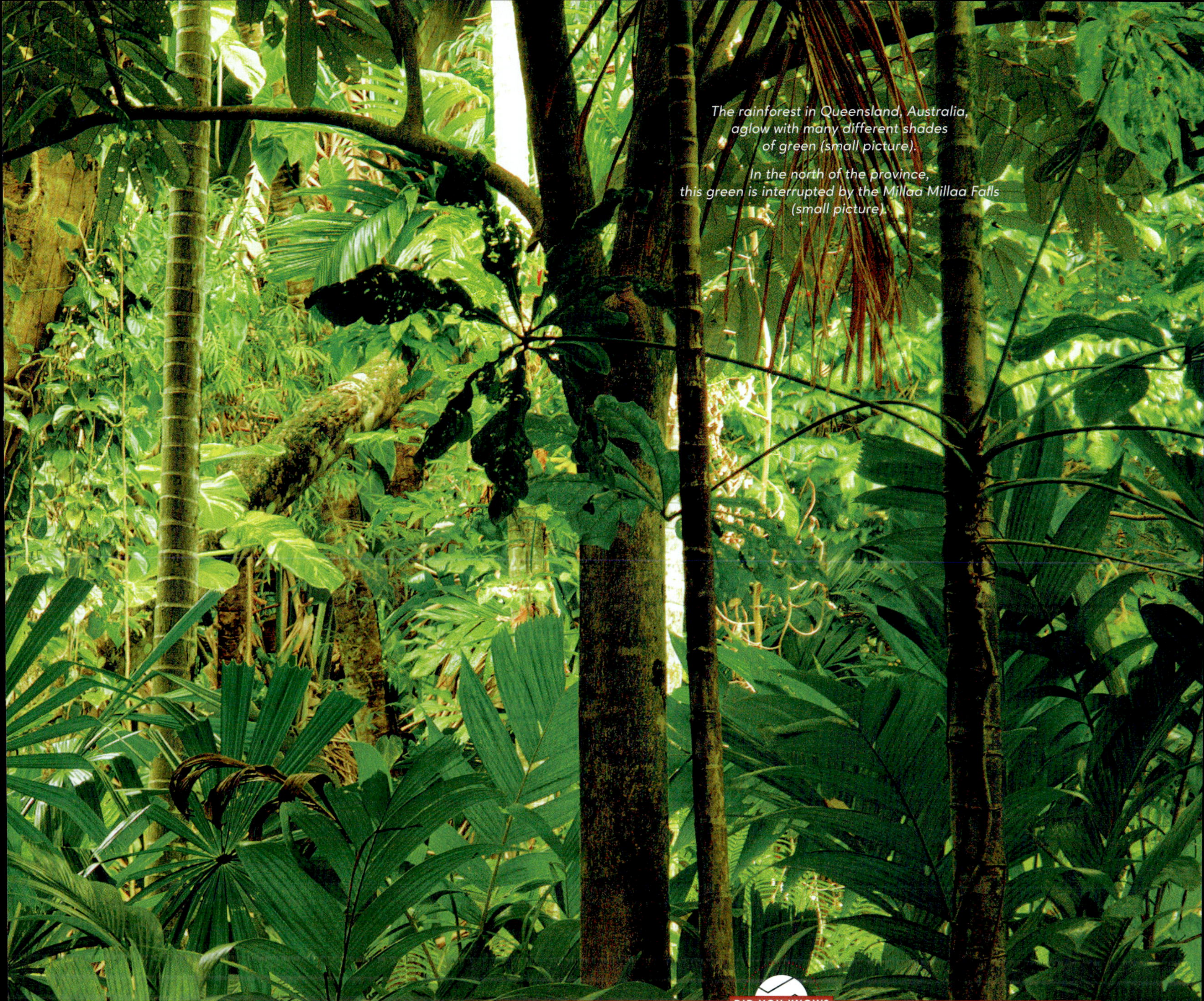

*The rainforest in Queensland, Australia, aglow with many different shades of green (small picture).*

*In the north of the province, this green is interrupted by the Millaa Millaa Falls (small picture).*

DID YOU KNOW?

**The southern cassowary – the largest rainforest bird in the world**

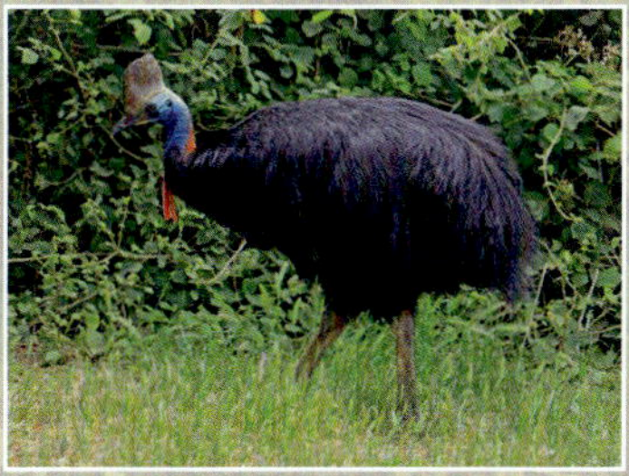

After the African ostrich and the emu, the flightless southern cassowary, or Australian cassowary, is the third largest bird in the world. It is approximately 1.5 m (5 ft) long, has a shoulder height of 90 cm (3 ft) high and weighs up to 70 kg (154 lb). This ratite lives hidden in the forests of New Guinea and Australia; it has a horn-like crest on its head resembling a helmet which protects it in the undergrowth. The southern cassowary is endangered as a result of the loss of its habitat, being hunted for its ornamental feathers and road traffic. Southern cassowaries are solitary creatures and live in an established territory until the breeding season. After the hen has laid between three and eight light green eggs in the nest, she moves on. The male hatches the eggs and looks after the chicks, which are light brown and striped, for around nine months. Its diet consists mainly of fruit, fungi, insects, small mammals as well as birds and their eggs.

FOUR ELEMENTS

EARTH

# AFRICA

The African rainforests are home to many endangered animal species

On both sides of the equator, less than ten per cent of the African continent is now covered in evergreen rainforest – the once closed green belt is now broken up in many places by savanna.

## WEST AFRICA

In the tropical, semi-humid climate in the west of the continent, rainforests continue to flourish. The Ivory Coast, for instance, is defined by forests with baobabs, orchids and epiphytes. The African elephant can now only be found on reservations because hunting and poaching for the coveted tusks have greatly reduced the population. Hippopotamuses, primates, rodents, pangolins, big cats, crocodiles and snakes as well as large numbers of birds live in the forests of the Ivory Coast. In Ghana, too, there are large forest areas with rich plant and animal worlds. However, their populations are being drastically reduced – mainly due to forest clearance for valuable timber, such as mahogany and walnut, destined for the export market.

## CONGO BASIN

The Congo Basin, the second largest continuous rainforest area in the world after the Amazon Basin, extends across the whole of Central Africa and represents the largest area of tropical rainforest in Africa. The rainforests here are still intact and largely undeveloped, in spite of their proximity to the densely populated west coast. The Congo Basin is therefore a 'reserve of unspoiled original nature'. Species which died out a long time ago in other places have survived here. The lowland gorillas and the even rarer mountain gorillas have found a home in this region. There are also chimpanzees, bonobos, forest elephants, forest buffalo, bongo antelopes and okapis. Many of the inhabitants of the rainforest have adapted to the higher levels, where there are proper 'trails' through the branches of the trees. Many of these species have still not been researched nor are they fully known. Some of the ground-dwelling animals have only been discovered recently.

## MADAGASCAR

Approximately 400 km (870 mi) from the East African coast, a very primeval world of animals and plants was able to develop in this isolation, like on a Noah's ark, containing species that have long since become extinct on the African mainland. This unique paradise, however, is now threatened by slash-and-burn land clearance. Around 80 per cent of the flora (of which 86 per cent alone are flowering plants) and numerous animal species are endemic.

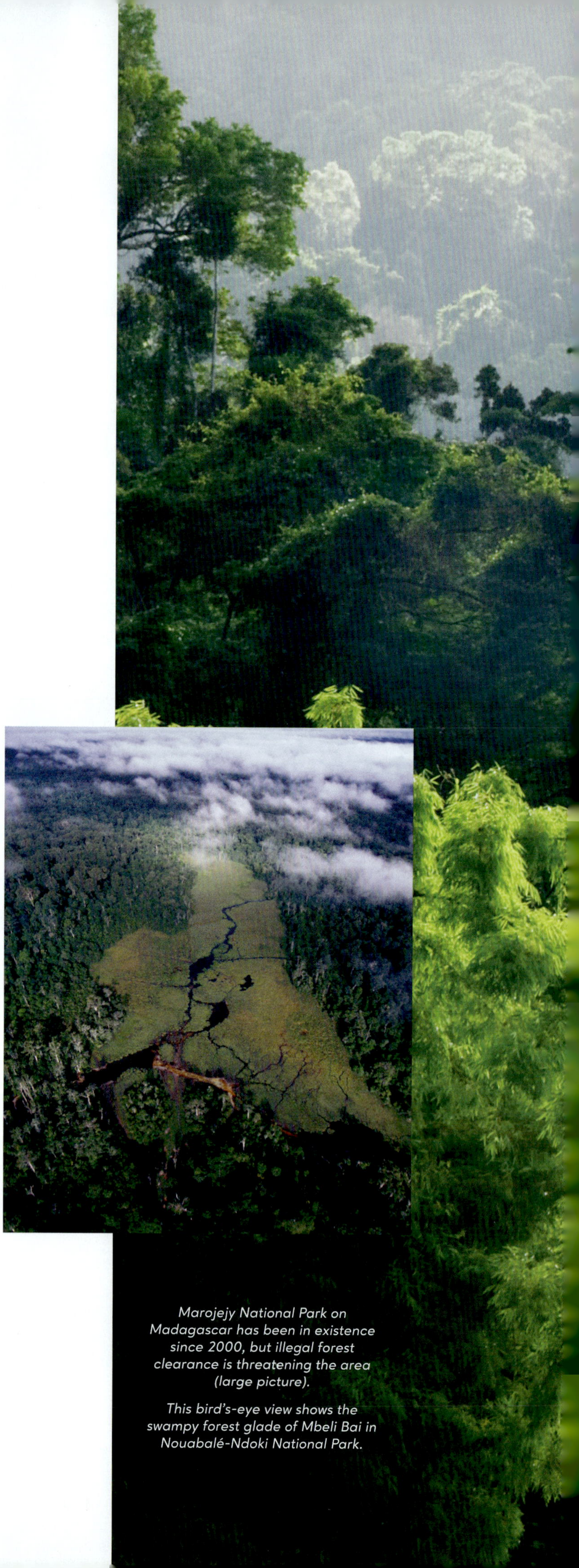

*Marojejy National Park on Madagascar has been in existence since 2000, but illegal forest clearance is threatening the area (large picture).*

*This bird's-eye view shows the swampy forest glade of Mbeli Bai in Nouabalé-Ndoki National Park.*

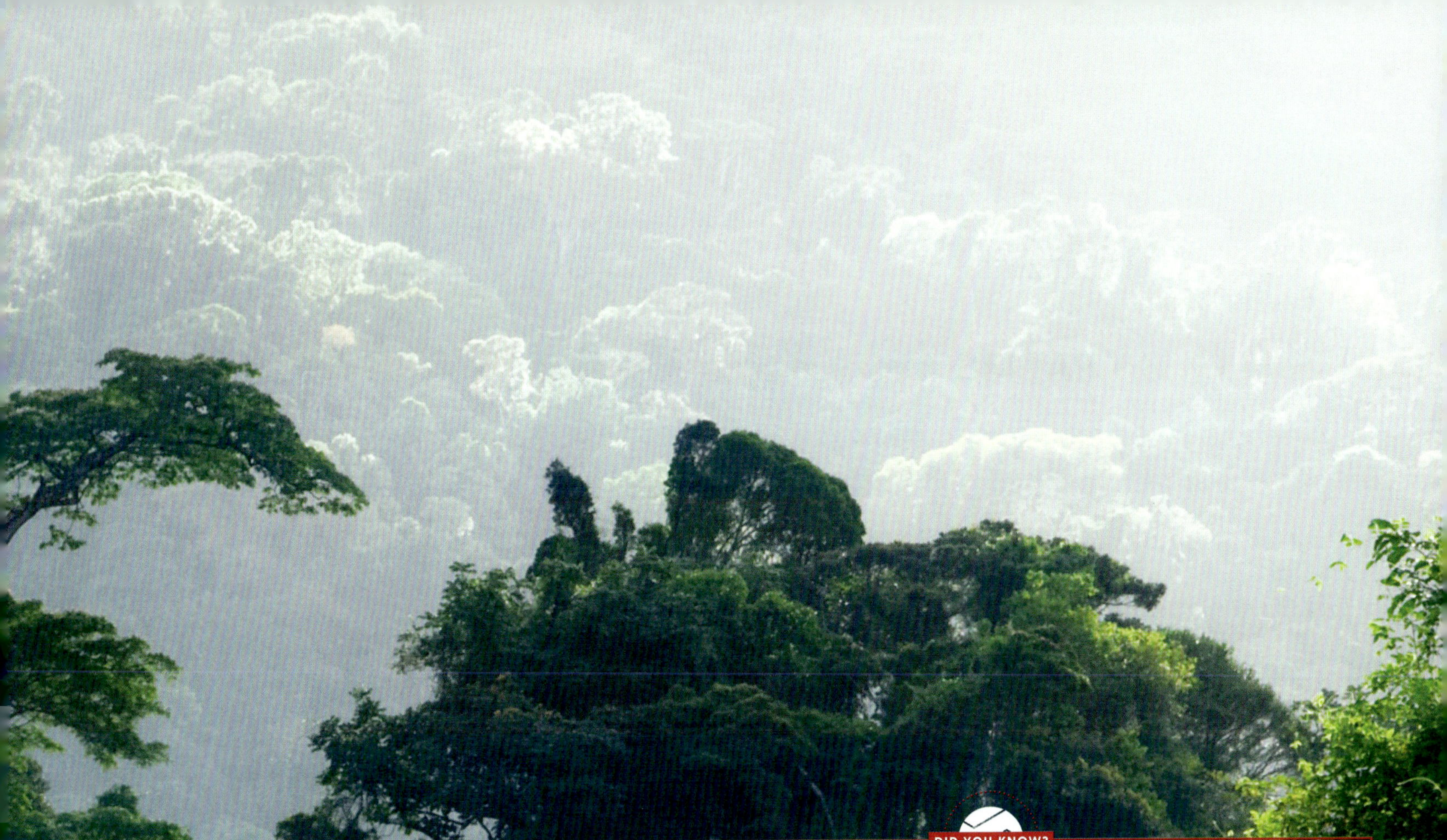

DID YOU KNOW?

### Gorillas - the largest primate on earth

**Mountain Gorillas**

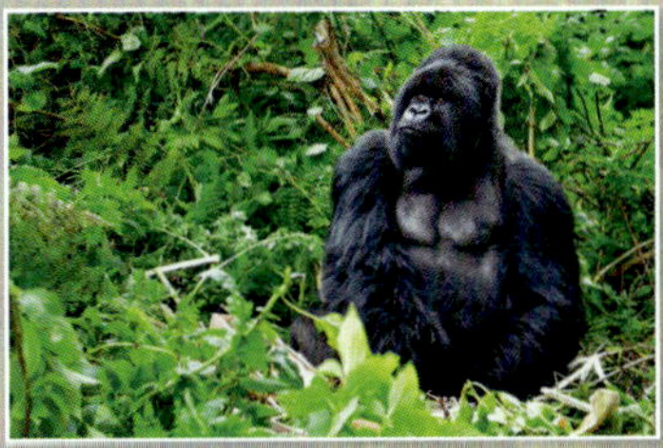

The gorillas' territory once ranged from the coast of Equatorial Guinea through to East Africa. Now they live in two separate populations – the lowland gorillas and the mountain gorillas which are now only found in the Virunga Volcano area in the border region between Uganda, Rwanda and the Democratic Republic of the Congo. It is estimated that there are now only around 400 to 700 of the mountain gorillas. These plant-eating primates live in family-like groups consisting of one adult silverback male, four or so females of reproductive age and their mutual young. Mountain gorillas spend most of their time on the ground and will only occasionally climb trees.

**Lowland Gorillas**

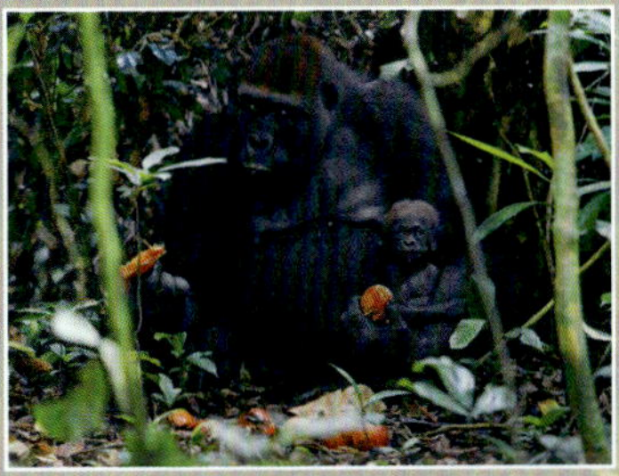

Lowland gorillas are divided into an eastern and western subspecies. The eastern lowland gorillas are the largest primates at 1.75 m (5.5 ft) tall. They live in the lowlands and hill country on the eastern margins of the Congo Basin. The western lowland gorillas, however, are the most widespread species of gorilla. They live in the coastal regions on the Gulf of Guinea, between the Niger and Congo deltas.
Like the mountain gorillas, their relatives in the lowlands are diurnal and build nests of leaves in the trees or on the ground for the night. Due to their size and intelligence, these peaceable nomads have not needed to fear any natural enemies. But now they are threatened by humans. Although their recent habitat has been designated a conservation area, it is being cleared for lumber and firewood as well as for agricultural use.

FOUR ELEMENTS

EARTH

# PALMS

The largest leaves, inflorescences and seeds on earth

The name palm comes from the Latin *palma*, which means 'flat hand', and refers to the similarity between the plant's leaves and a spread-out hand. The trunks, which generally do not have branches, grow to between 0.25 and 60 m (0.8 and 197 ft) high. A cluster of leaves resembling fans or feathers sits on the top with the flowers between them in densely packed inflorescences. The 2,600 species are distributed across tropical and subtropical regions as well as on oceanic islands. The palm family are crop plants: the fruit of around 100 of the species are edible (e.g. the date palm); in others, the seeds are edible (e.g. the coconut); sago can be obtained from the trunk of other palms; and various drinks, sugar, palm wine and palm honey can be extracted from palm juice. The wood from the trunk is used in many places as a building material for houses and furniture.

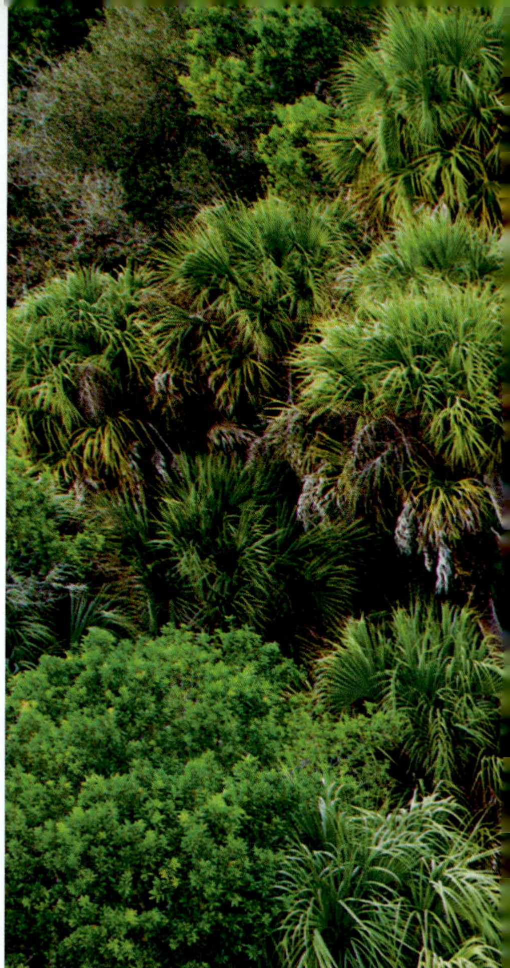

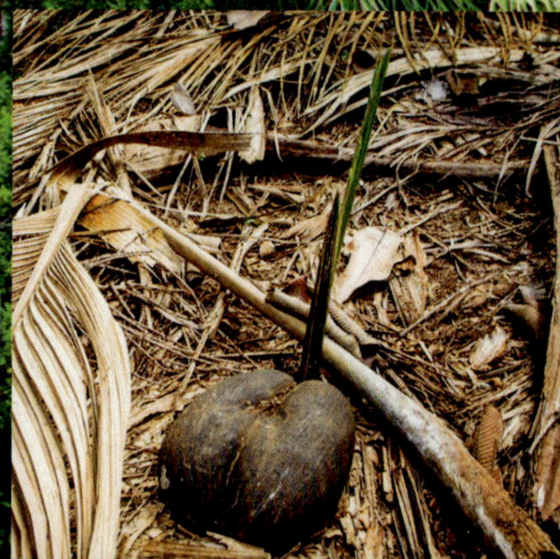

## RAFFIA – THE PALM WITH THE LONGEST LEAVES

This species is mainly found in Africa and on Madagascar; another species occurs in Central and South America (right). The leaves grow up to 25 m (82 ft) long and the trunk reaches a height of 16 m (52 ft).

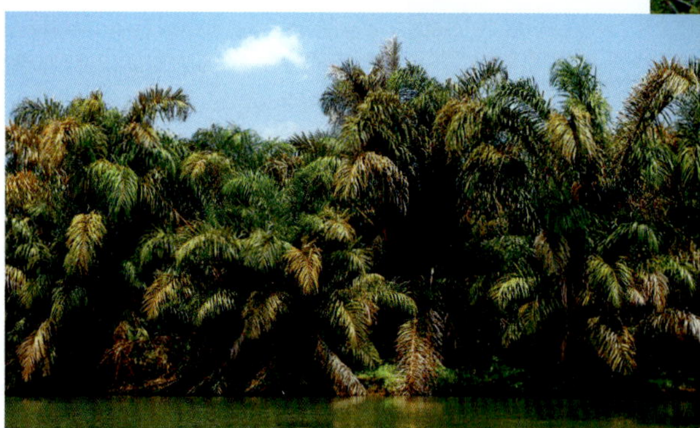

## LODOICEA MALDIVICA – THE PALM TREES WITH THE LARGEST SEEDS

The Seychelle palm (right), also known as the *coco de mer*, develops a green, heart-shaped fruit six to seven years after pollination, which contains one seed weighing up to 25 kg (55 lb).

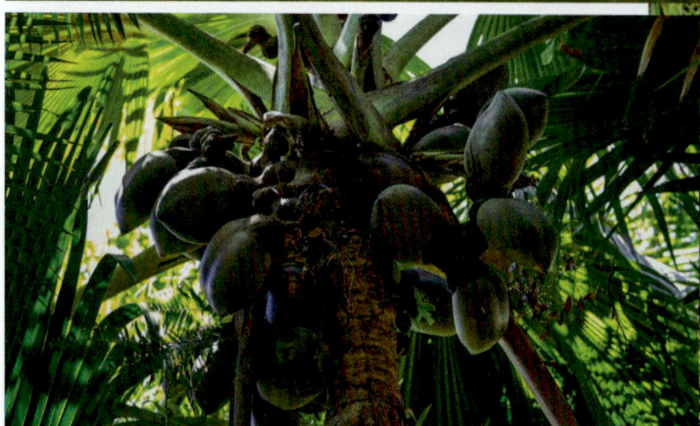

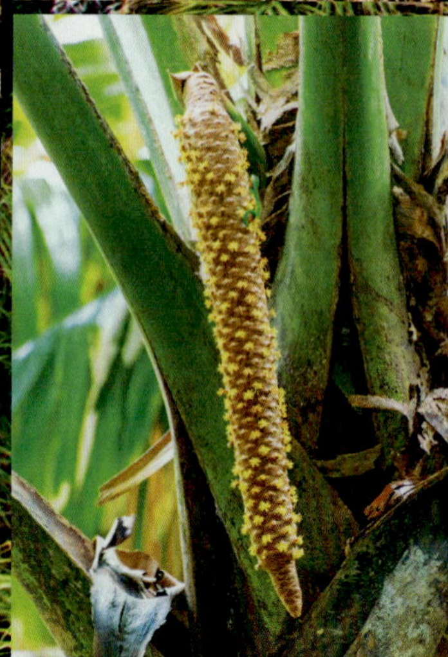

## CORYPHA – THE PALMS WITH THE TALLEST FLOWERS

The tufted or umbrella palm is a 50-m (164-ft) high fan palm found from Southeast Asia to northern Australia. This species develops the largest inflorescence produced by any plant in the world; each inflorescence has approximately ten million flowers. Sago, a thickening agent, can be made from the palm.

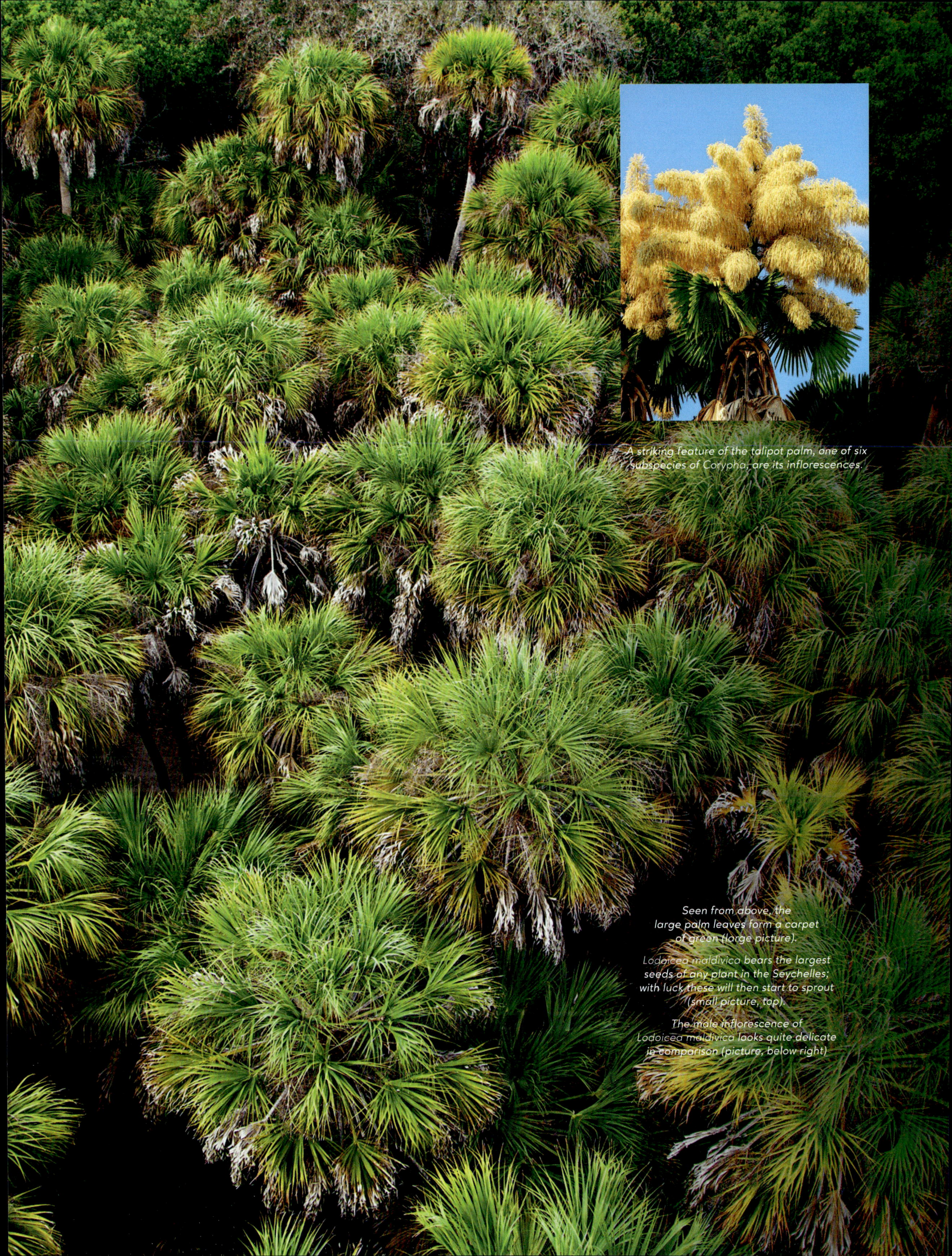

A striking feature of the talipot palm, one of six subspecies of *Corypha*, are its inflorescences.

Seen from above, the large palm leaves form a carpet of green (large picture).

*Lodoicea maldivica* bears the largest seeds of any plant in the Seychelles; with luck these will then start to sprout (small picture, top).

The male inflorescence of *Lodoicea maldivica* looks quite delicate in comparison (picture, below right).

FOUR ELEMENTS

EARTH

# BAOBABS

The largest succulents in the world

One of the most characteristic trees of tropical Africa is the baobab, with its short, but sturdy, trunk and its jutting crown. When it has no leaves, the tree's crown resembles a root – according to legend, the gods were angered at humankind's sins and so pulled up the baobab trees and then replanted them the wrong way up. The baobab, a periodically deciduous tree whose leaves sprout just before the rainy season, occurs south of the Sahara across all of Africa; in Madagascar, some species are endemic. During the rainy season, the tree can store as much as 140,000 litres of water and acts as a valuable water reservoir for humans and animals during the dry months. Its tough trunk can withstand fire with scarcely any damage.

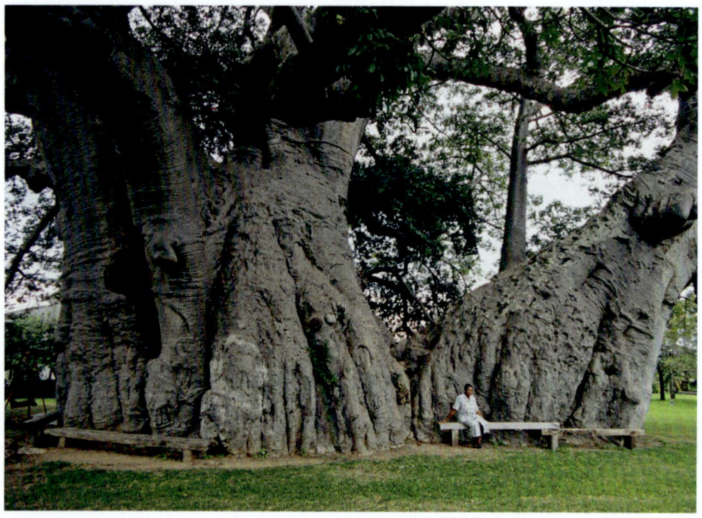

## THE LARGEST BAOBAB IN THE WORLD

The 'Big Baobab' at Tzaneen in South Africa, at 22 m (72 ft) high and with a circumference measuring 47 m (154 ft), is considered the largest baobab in the world. In 1993, the land owner hollowed out the trunk and set up a bar inside. Since years, however, the tree has been split.

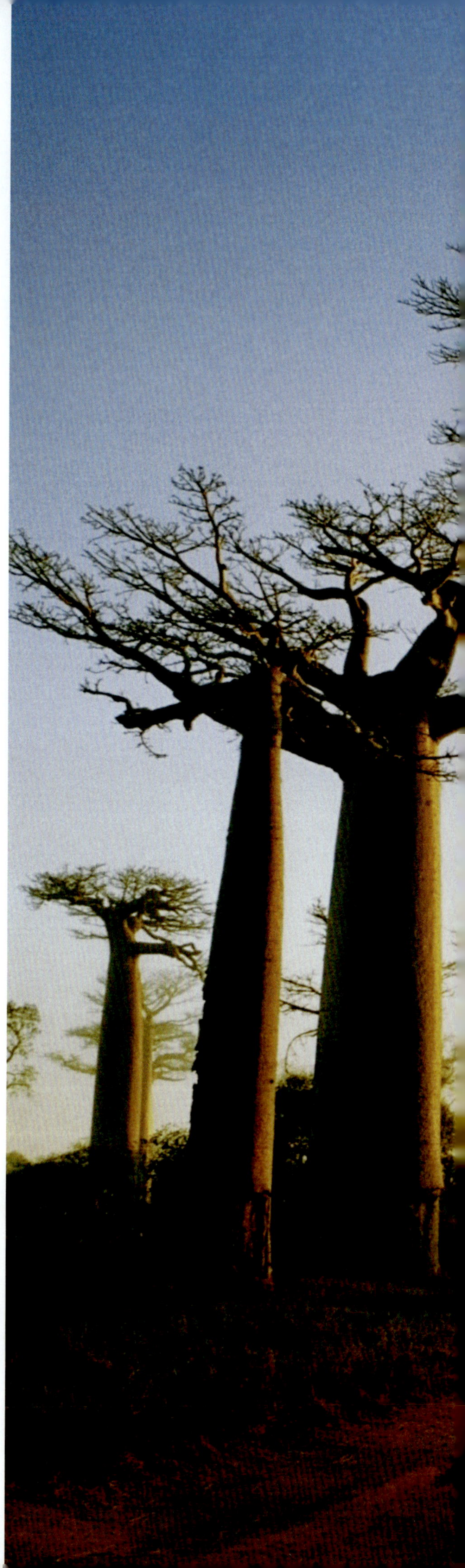

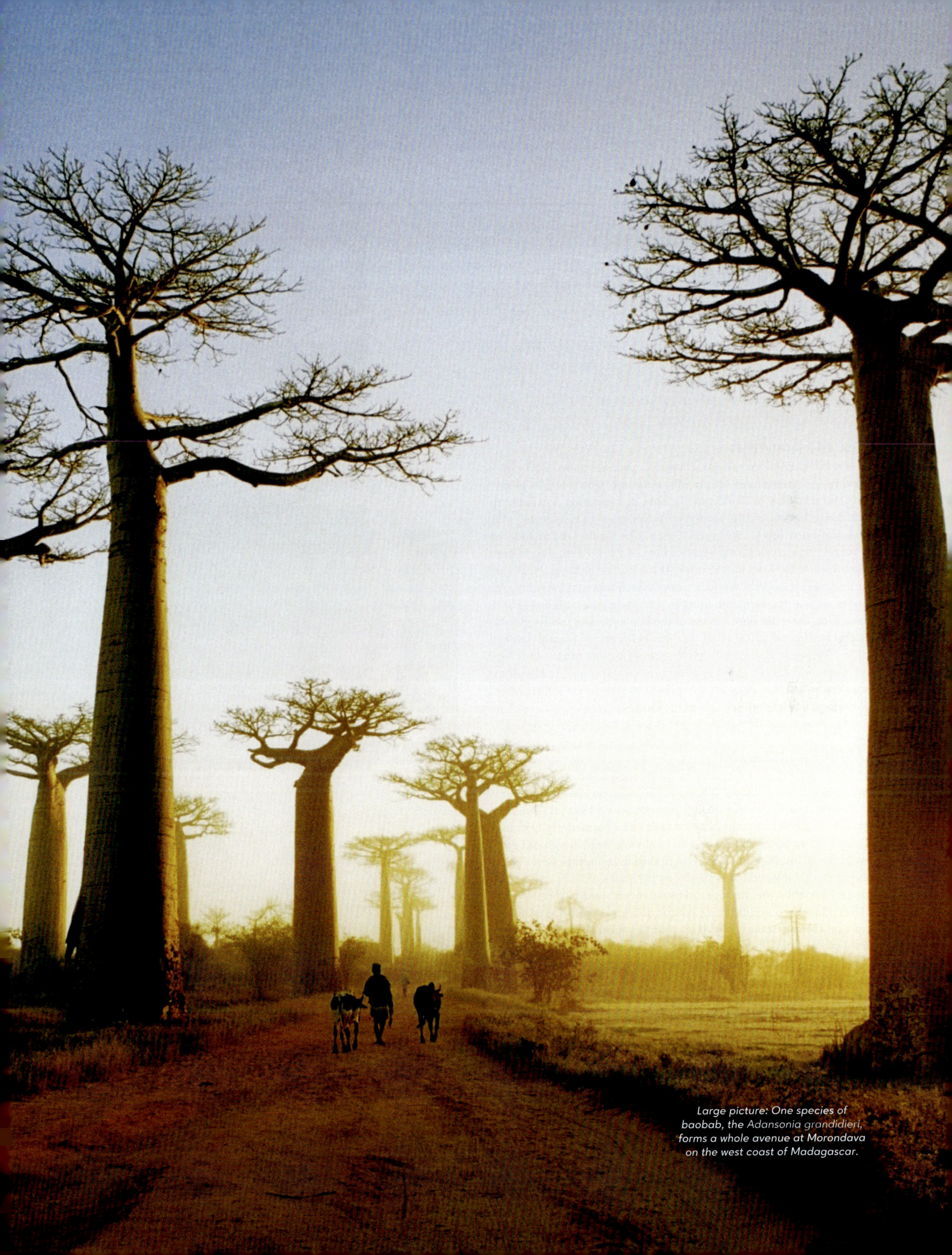

*Large picture: One species of baobab, the Adansonia grandidieri, forms a whole avenue at Morondava on the west coast of Madagascar.*

FOUR ELEMENTS

EARTH

# NORTH AND CENTRAL AMERICA

Tropical rainforests are mainly found in Central America; the climate in North America is better for temperate rainforests

## CENTRAL AMERICA

In several Central American countries, among them Costa Rica, Honduras, Panama and Nicaragua, there are still sizeable rainforests. The province of Darién in Panama, for instance, is a hillscape with dense forests, but its biodiversity continues to be threatened by slash-and-burn forest clearance. Costa Rica, south of Panama, got its name ('rich coast') long ago because of the breathtaking natural treasures that the land had to offer. At the border between North and South America, several animals from both continents, including rare species such as the capuchin monkey and the jaguar, still live here. In Costa Rica, there are now reports of the first successes in restoring the natural forests, which 50 years ago were almost completely deforested. In Nicaragua, there are tropical rainforests with particularly high levels of biodiversity as well as mangrove swamps, and, in Honduras, there are still some areas of unexplored primal forest which offer a haven for countless animal and plant species.

## CARIBBEAN ISLANDS

In Cuba, national parks have been established to protect the ecosystem, among them is Parque Nacional Alejandro de Humboldt with its dry forests, rainforests and mangroves. Jamaica's Cockpit Country is also a lush green forest landscape which has a particularly high level of biodiversity. Some species that have already been driven out of the rest of the island by human activity find their last refuge here. The island of Hispaniola, which is shared by the states of Haiti and the Dominican Republic, still has around fifty per cent rainforest cover. Puerto Rico offers the magnificent El Yunque rainforest, featuring waterfalls and tropical vegetation as well as countless bird species. This is the home of colourful Puerto Rican parrots and the *El Coqui* tree frog. Dominica, with its tropical flora and fauna, is considered the most pristine island of the Lesser Antilles. The island's centre, in particular, features dense rainforest. The island of Guadeloupe has a national park of the same name, which even has mahogany trees growing in its tropical rainforests. Parrots, parakeets, bats, Guadeloupe raccoons, mungos and endangered agoutis all live in this park. On Trinidad and Tobago, rainforests with waterfalls and a wide variety of animals and plants can also be found as well as vast swamplands. The volcanic hill country of St Lucia, traversed by rivers, is covered in rainforests with bougainvillea, hibiscus, wild orchids and roses.

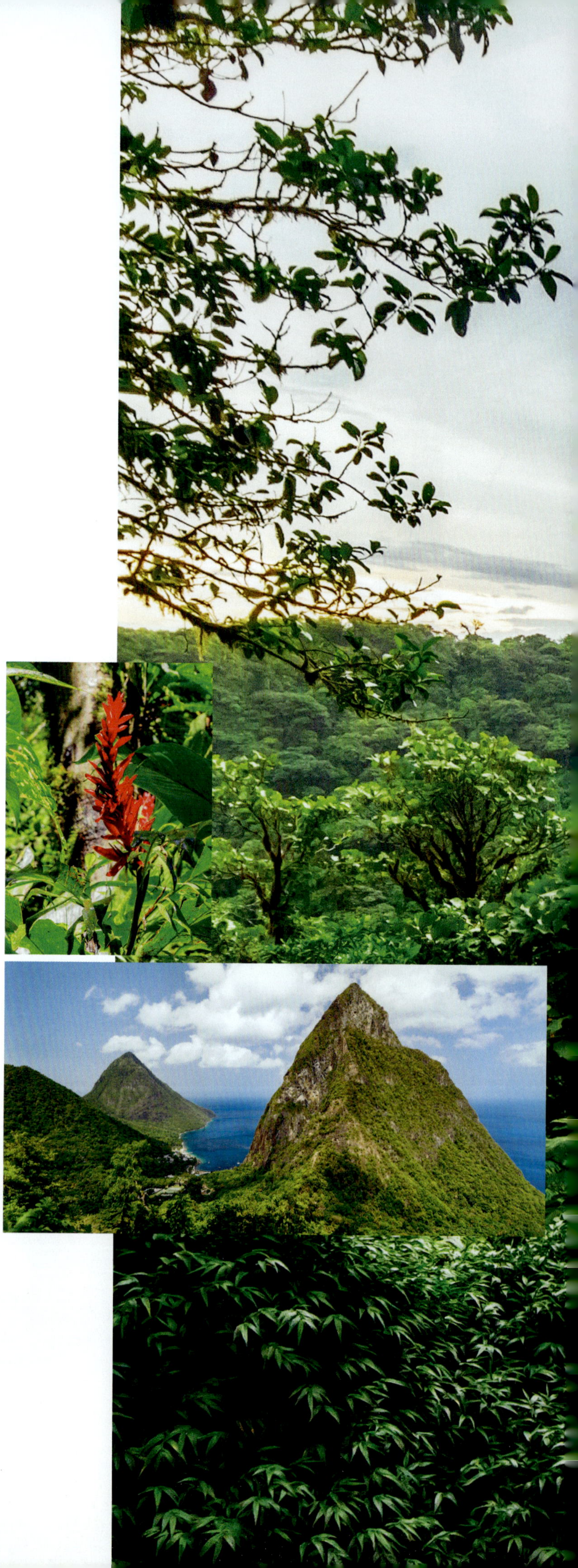

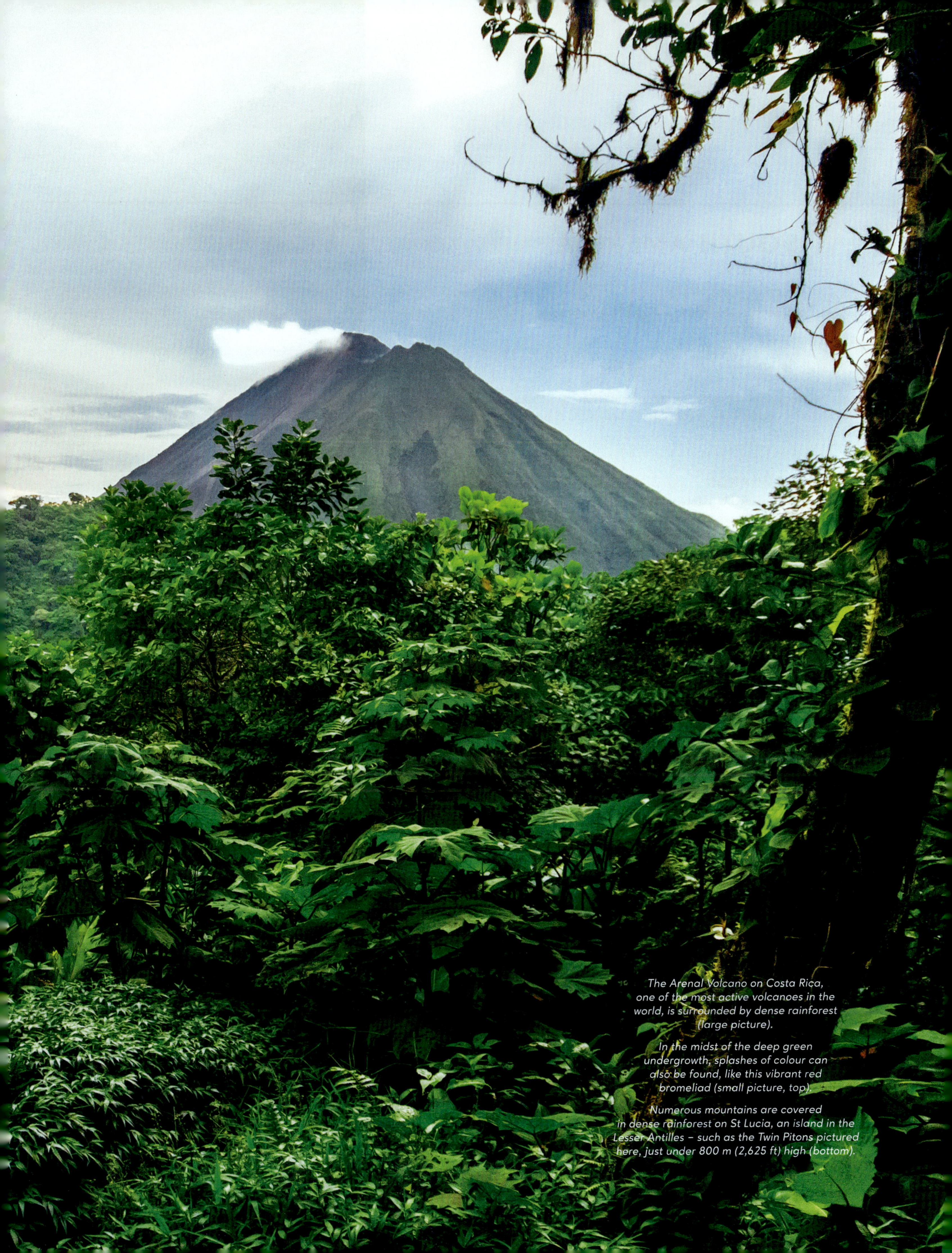

The Arenal Volcano on Costa Rica, one of the most active volcanoes in the world, is surrounded by dense rainforest (large picture).

In the midst of the deep green undergrowth, splashes of colour can also be found, like this vibrant red bromeliad (small picture, top).

Numerous mountains are covered in dense rainforest on St Lucia, an island in the Lesser Antilles – such as the Twin Pitons pictured here, just under 800 m (2,625 ft) high (bottom).

FOUR ELEMENTS

EARTH

# SOUTH AMERICA

The rainforest region in South America is extremely biodiverse

Most tropical rainforests are found in South America. The largest rainforest region in the world extends across the Amazon Basin in South America.

## ORINOCO LOWLANDS

Venezuela can be divided into three main landscape areas – the Andes mountain region, the Guiana uplands and the Orinoco lowlands. The Llanos del Orinoco, a wide plain with an incredibly high number of animal species and lush, prolific vegetation, is one of the most beautiful regions in Venezuela. Numerous rivers and channels meander through various different landscapes, heading for the Atlantic. However, a large part of this region is occupied by tropical rainforests with palm trees, fruit trees, orchids, bromeliads and tree ferns. The rivers are full of mangroves. In the Orinoco lowlands, the animal population includes jaguars, pumas, ocelots, raccoons, giant otters and capybaras as well as howler and capuchin monkeys. In addition, there are many amphibians, reptiles and fish, and, not least, the region is a bird paradise. Among the feathered inhabitants, there are parrots, cormorants, toucans, kingfishers, hummingbirds and birds of paradise.

## GUIANAN HIGHLANDS

The deeply fissured mountain region of the Guianan Highlands stretches over a vast area of 1.5 million km$^2$ (579,153 sq mi) – from eastern Colombia through southern Venezuela, northern Brazil, Guyana and Suriname to French Guiana – and is bordered by the Atlantic Ocean as well as the Amazon and Orinoco lowlands. The highlands, a very old geological formation consisting mainly of granite rock, is characterized primarily by vast table-top mountains referred to as 'tepuis'. The mountains tower over the 1,000-m (3,280-ft) thick, flat sandstone plate called the Gran Sabana; on the high plateaus of the mountains, endemic flora and fauna have developed due to its isolated location away from the rainforest. Mighty waterfalls plunge down into the tropical rainforest from several of these table mountains – among them is Salto Angel, which, with a near 1,000-m (3,280 ft) drop, is considered the highest waterfall in the world. The highest elevation in the region is Pico da Neblina, at 2,994 m (9,823 ft) high. It is also called the Guiana Shield or the Guiana Massif. Tropical rainforests cover the lower part of the mountain region in the east; these have an impressive variety of plant and animal species, among which are many bird species. Further west, savanna and grasslands predominate where crops such as cocoa, rice, sugar canes and cotton are grown.

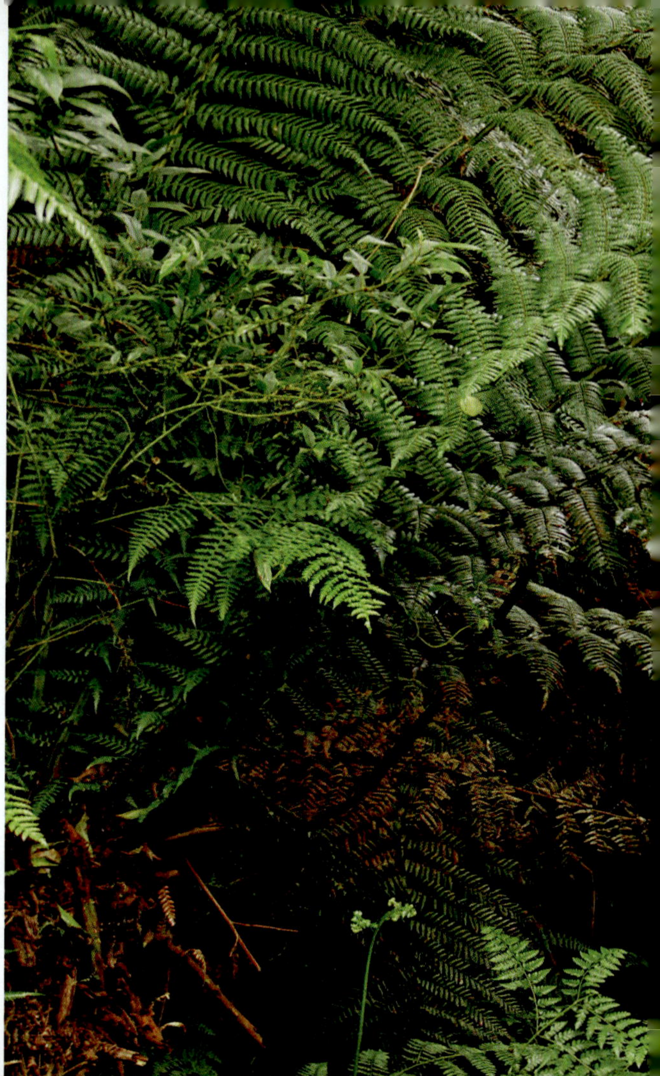

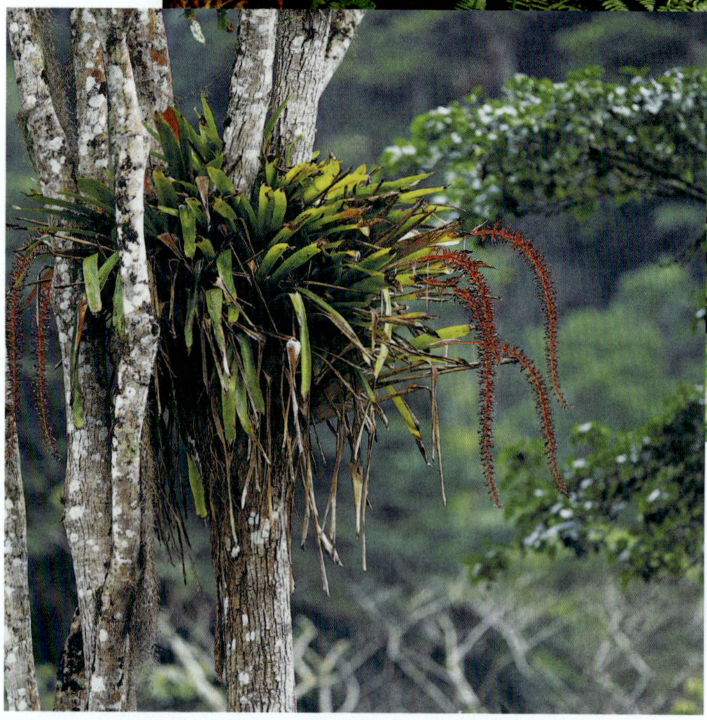

*Large picture: The giant tree ferns (Dicksonia sellowiana) form part of the diverse vegetation in the dense Atlantic Forest in Brazil.*

*In the Venezuelan rainforest, epiphytes, or air plants, can also be found everywhere (small picture, above).*

*The tropical rainforest is reflected in the tranquil lagoon in the Orinoco delta in Venezuela (small picture, right).*

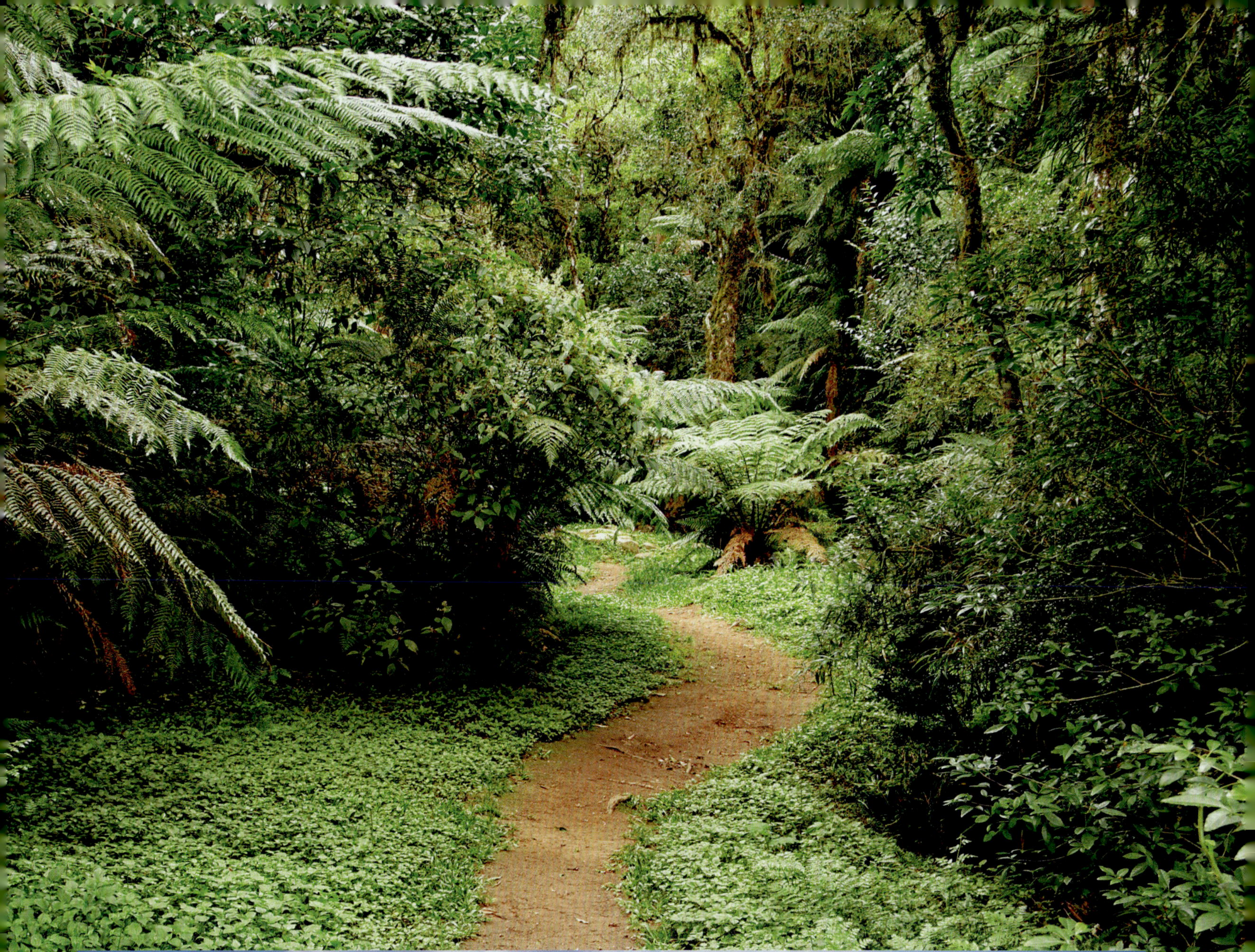

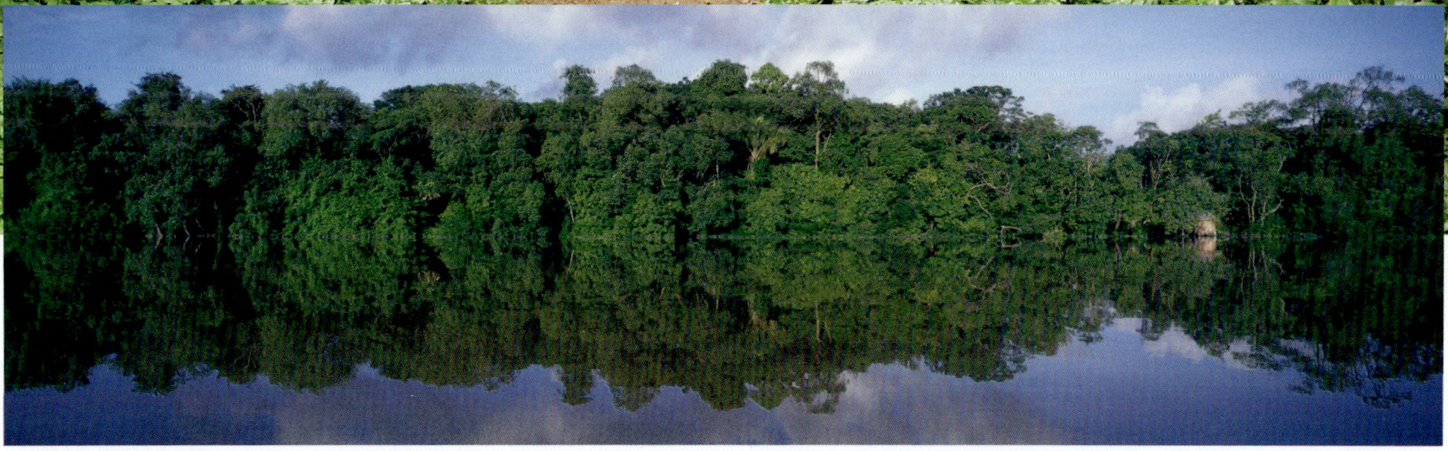

## BRAZIL'S ATLANTIC FORESTS

The Mata Atlântica, the Atlantic Forest, once stretched for more than 3,000 km (1,864 mi) along Brazil's Atlantic coast. Today, only small forest islands remain in the midst of the man-made landscape characterized by human activity, but it is still home to a tremendous number of plants and animals, many of which are indigenous. Because the major cities of Brazil lie in the Atlantic Forest region, these are among the most endangered forest areas in the world. UNESCO has now given them protection, partly as biosphere reserves and partly as World Heritage Sites, but this cannot stop illegal forest clearance. The unique vegetation of Mata Atlântica offers different forms of saltwater lagoons, shifting sand dunes, sandy beaches, mangrove groves, arable land and pasture from the coastal plain at Cananéia to the Serra Paraná Piacaba. Ferns, orchids, begonias and bromeliads grow on the trees of the rainforest, which then serve as habitats for tree frogs. The fauna is further represented by the lowland tapir, the La Plata otter and the harpy eagle, an eagle species.

FOUR ELEMENTS

EARTH

# CHEETAHS

The fastest land animal in the world

As the fastest land mammal in the world, the cheetah can not only reach speeds of 120 km/h (75 mph) over short distances (max. 500 m (310 mi)), it can also accelerate from zero to 100 km/h (62 mph) in five seconds. The incredible speed of the cheetah is due less to its long legs than to its extremely powerful lungs. Even sudden sidestepping gazelles cannot escape the cheetah because it is able to compensate for the necessary zigzag spurt with its long tail. After no more than twenty seconds, it has usually caught its prey, knocked it down with a blow from its paw and killed it with a bite to the neck. It then bolts down its meal very quickly before it is snatched away by lions or hyenas. Its prey are generally smaller hooved animals, such as gazelles, that are easier to overpower than the larger zebras or wildebeest.

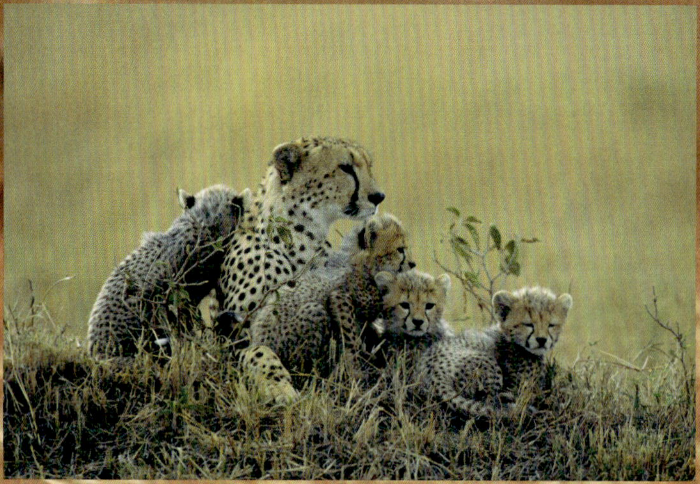

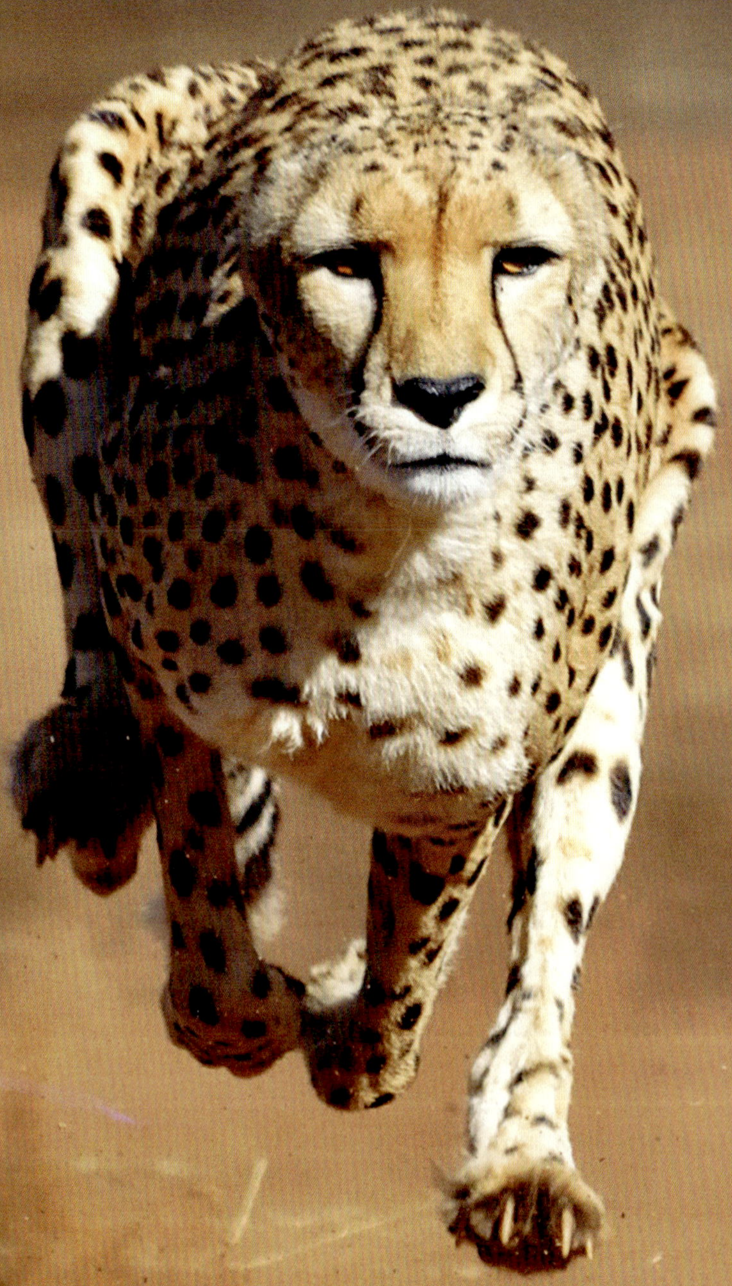

*When cheetahs run, they often have all four paws off the ground at the same time (large picture).*

*Meanwhile, the female cheetah looks after the young in the Kenyan savanna (small picture).*

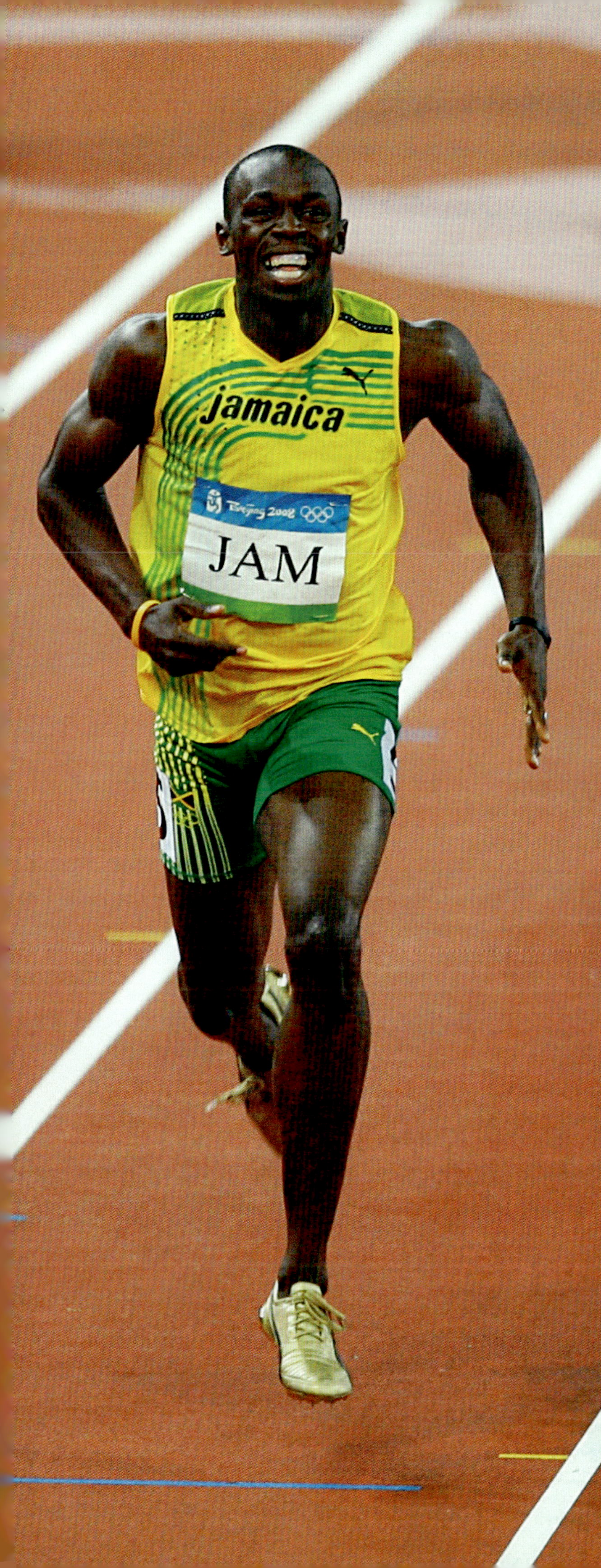


# USAIN BOLT

Fastest sprint by a human:
Usain Bolt

He can run as fast as a galloping horse: Usain Bolt achieves top speeds of up to 44 km p/h (27 mph). But he does not run for an hour at these top speeds; he is well known for his extraordinary sprints. The Jamaican, 1.96 m (6.4 ft) tall, has exceptional talent as a sprinter. He is not only an eight-time Olympic gold medallist and an eleven-time world champion; he is also the only person in the world who has managed to run 100 m (328 ft) in under 9.6 seconds. This makes him the fastest runner of all time. His extremely long legs provide spectacular photos, depicting him behind at the start, but, then with his enormous 2.43 m (8 ft) strides, he very quickly catches up.

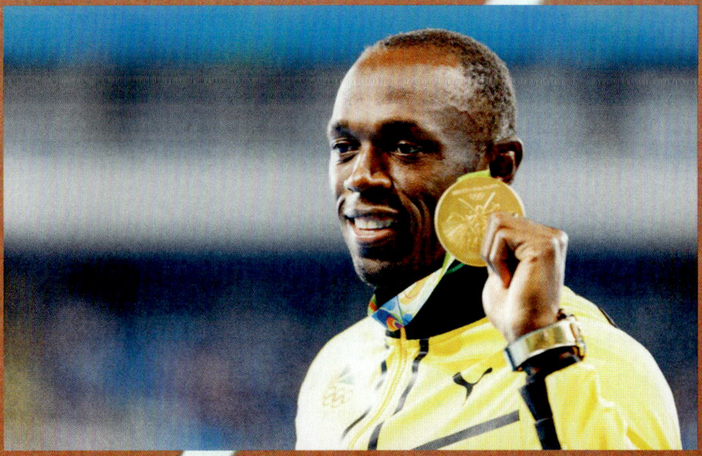

FOUR ELEMENTS

EARTH

# LAND VEHICLES

The fastest cars in the world

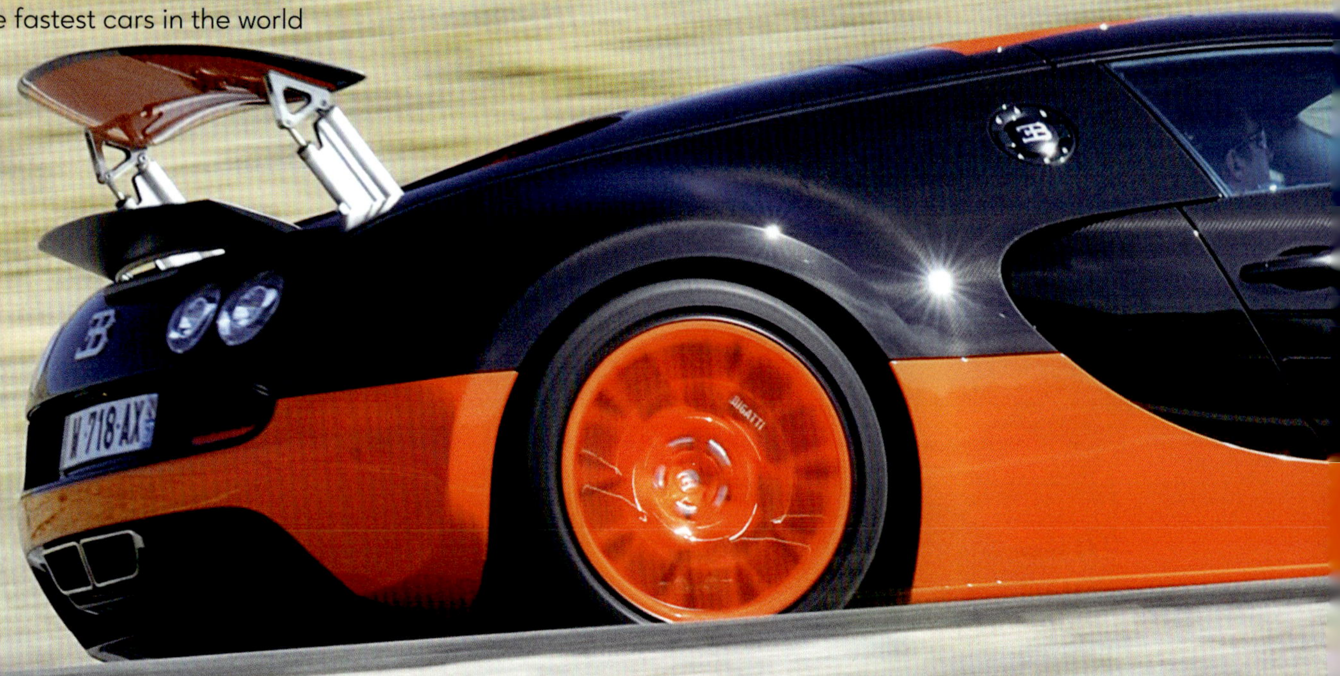

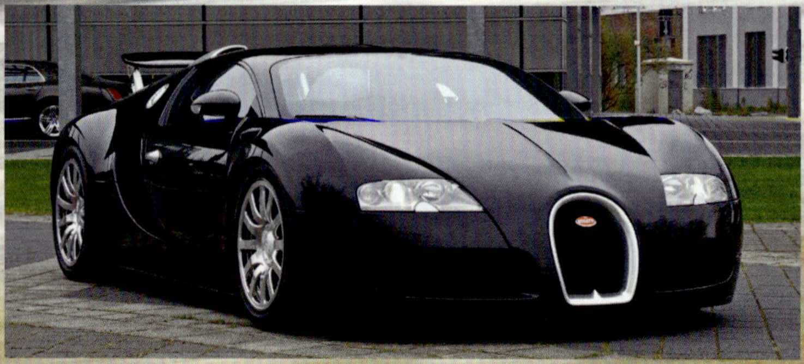

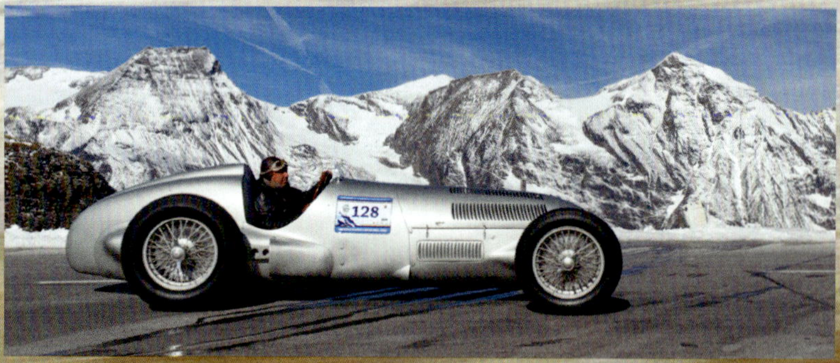


## BUGATTI VEYRON – THE FASTEST ROADWORTHY PRODUCTION VEHICLE IN THE WORLD

In 2.5 seconds, the Bugatti Veyron Super Sport accelerates to 100 km/h (62 mph) and, after 6.7 seconds, to 200 km/h (124 mph). After just under 15 seconds, it has exceeded 300 km/h (186 mph), and, very shortly afterwards, it reaches its top speed of 431 km/h (267 mph). The versions sold, however, can only reach 415 km/h (258 mph) because of the tyres. This makes the Bugatti Veyron Super Sport the fastest production vehicle authorized for the road. The powerful 16-cylinder engine with 7,993 $cm^3$ displacement supplies 882 kilowatt (1200 HP) of power with 6,400 revolutions per minute; the maximum torque of 1,500 newton metres is 3,000 to 5,000 revolutions per minute. With these figures, it also boasts peak value, just as with the consumption data, which, according to the factory specifications, should be an average of 100 km (62 mph) for 23.1 litres (6.1 gal) of premium unleaded gas.

## MERCEDES BENZ W 125

**Rudolf Caracciola, 1938, 432.7 km/h (268.87 mph)**
**Autobahn Frankfurt – Darmstadt**

On 28 January 1938, Rudolf Caracciola achieved the highest speed ever reached on a public highway with 432.7 km/h (268.87 mph). For this record drive, a Grand Prix W125 car was modified – one of the legendary Silver Arrows with which Mercedes-Benz dominated car racing in the 1930s. In place of the normal eight-cylinder engine, the record-breaking car had a V12 compressor engine (5,577 $cm^3$, 736 HP). The record was to stand for almost eighty years – it was first broken by Niklas Lilja in Nevada in 2017. With a Koenigsegg Agera RS, he reached a new speed record of 457.49 km/h (284.27 mph) on a public highway.

As long ago as 29 April 1899, Camille Jenatzy exceeded 100 km/h (62 mph) in *La Jamais Contente*, an electric vehicle. The record of 105.9 km/h (65.8 mph) achieved on the Concours de Vitesse in Paris only held until 13 April 1902 when Léon Serpollet reached 120.8 km/h (75.06 mph) in his steam-powered car, *Œuf de Pâques*. Since then, the race for a new speed record has not ceased. With ever more powerful engines, higher and higher speeds are achieved. Dominated by the combustion engine until the middle of the 20th century, people transitioned to mounting aircraft engines in land vehicles in the 1960s. And because wide, completely flat areas are needed for record-breaking runs, they are now carried out at salt lakes like the Bonneville Salt Flats in Utah.

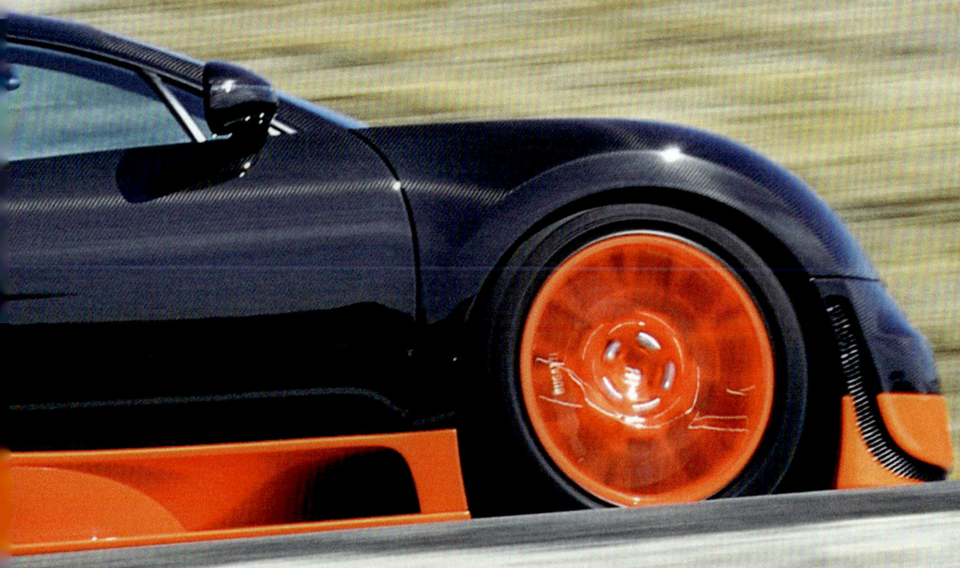

### THE FASTEST LAND VEHICLES IN THE WORLD

1. **THRUST SSC**
   1997, 1,227.985 km/h (763.035 mph)
2. **THRUST2**
   1983, 1,019.47 km/h (633.469 mph)
3. **BLUE FLAME**
   1970, 1,001.67 km/h (622.409 mph)
4. **SPIRIT OF AMERICA SONIC**
   1965, 966.961 km/h (600.842 mph)
5. **GREEN MONSTER**
   1965, 927.872 km/h (576.553 mph)
6. **SPIRIT OF AMERICA**
   1965, 846.861 km/h (526.215 mph)
7. **WINGFOOT EXPRESS**
   1964, 668.027 km/h (415.093 mph)
8. **BLUE BIRD CN7**
   1964, 644.96 km/h (400.760 mph)
9. **RAILTON MOBIL SPECIAL**
   1947, 634.39 km/h (394.192 mph)
10. **RAILTON SPECIAL**
    1939, 595.04 km/h (369.741 mph)

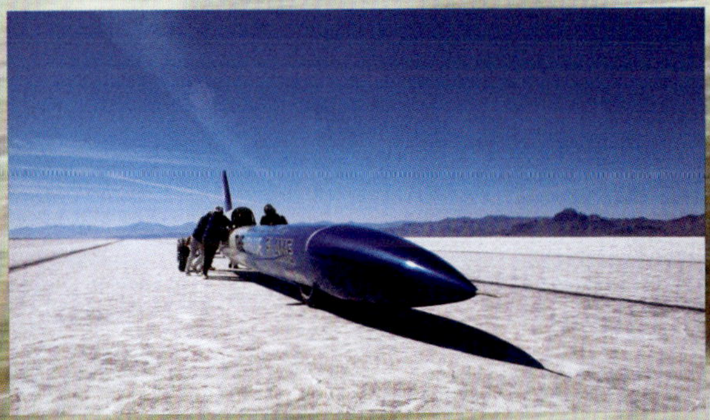

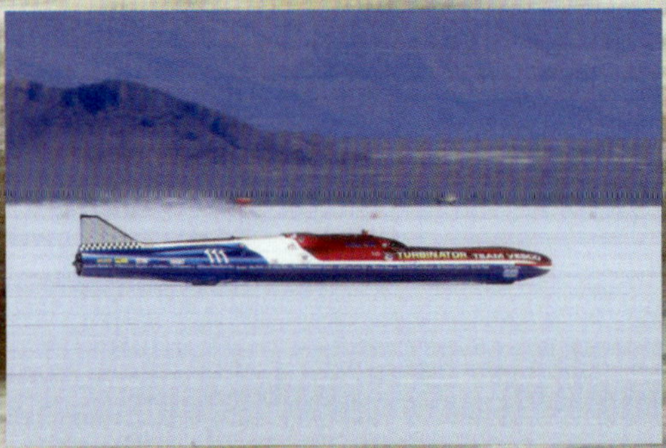

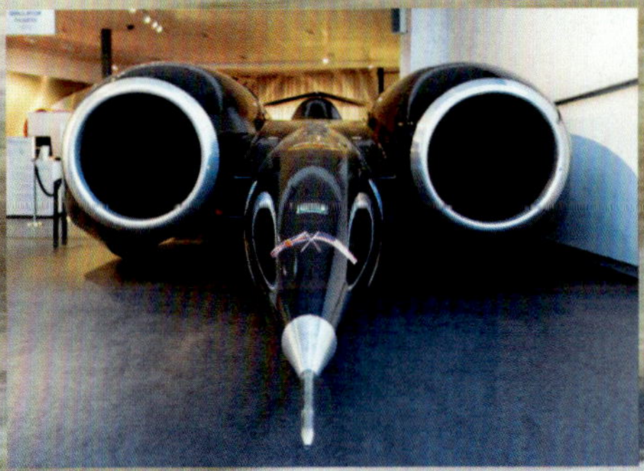

### BLUE FLAME

**Bonneville Salt Flats, Utah**

On 23 October 1970, the *Blue Flame* was the first wheeled vehicle to break the 1,000 km per hour (621 mph) barrier, setting a new speed record. For the engine, a rocket which burned a mixture of liquid natural gas and hydrogen peroxide (which produced the blue flame that gave the car its name) was used. The engine developed a thrust of 7,248 kg (15,980 lb) (= 42,658 kilowatts or 58,000 HP). The record-holding car can be seen in the Museum of Technology in Sinsheim.

### TURBINATOR

**Don Vesco, 2001, 737.8 km/h (478.45 mph)**
**Bonneville Salt Flats, Utah**

Since 18 October 2001, the Turbinator, built by Don and Rick Vesco, has been the fastest wheel-driven car. Its motor is a T-55 gas turbine from a transport helicopter built in 1967; it produces 2,758 kilowatts (3,750 HP) of power. The 1.5-tonne (1.65-ton) Turbinator is 9.5 m (31 ft) long but only 0.91 m (3 ft) wide and 0.81 m (2.7 ft) high. Visually, it is more reminiscent of a plane without wings than a car.

### THRUST SSC

**Andy Green, 1997, 1,227.985 km/h (763.03 mph)**
**Black Rock Desert, Nevada**

The Thrust Supersonic Car was the first land vehicle to break the sound barrier on 15 October 1997. It is 16.5 m (54 ft) long, 3.7 m (12.1 ft) wide, 2.14 m (7 ft) high and weighs 10.5 tonne (11.6 ton). With this, it does not just have the proportions of a fighter jet on wheels – it is also driven by turbofan engines with afterburners, which are also built into fighter jets. The two Rolls Royce Spey 202s develop a thrust of 9,299 kg (20,500 lb) (54,729 kilowatt or 74,411 HP).

FOUR ELEMENTS

EARTH

# MOUNTAINS

We owe our majestic mountains to the shifting of the tectonic plates.

People have always succumbed to the beauty of majestic mountains. Many peoples still view the unreachable peaks as the seat of the gods and the dwelling place of spirits. Even the eight-thousanders among the highest mountain massifs in the world – the Himalayas and the Karakoram – remained unconquered until well into the 20th century. In 1950, Maurice Herzog and Louis Lachenal were the first to reach the summit of Annapurna I; three years later, Edmund Hillary and Tenzing Norgay climbed Mount Everest, the 'throne of the gods' and the highest mountain in the world. These Asian mountains as well as the Alps in Europe, the Rocky Mountains in North America and the Andes in South America were formed over long periods of time through folding caused by the constant movements of the continental plates. If these slide into each other, the landmasses are compressed, folded and piled up to form mountains. Because of the continuous movement of the plates, the earth's summits are continually growing, even now, by a few millimetres every year. Almost every landmass on earth has its superlative mountains.

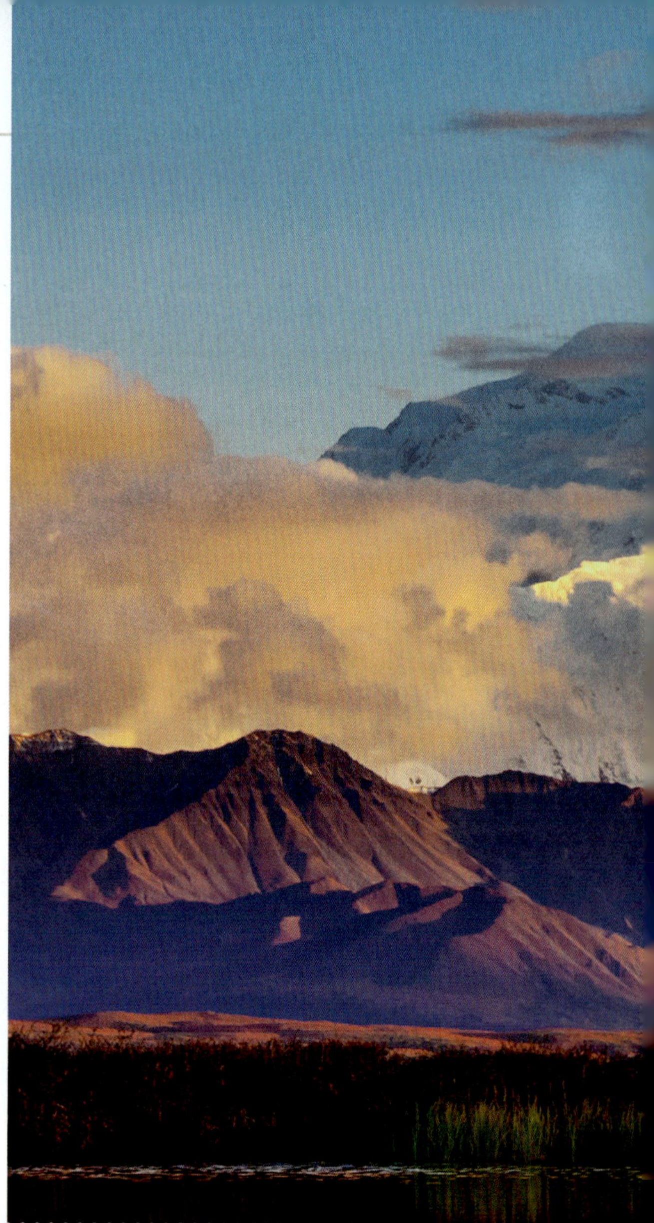

## THE HIGHEST MOUNTAINS ON EARTH

The height above sea level of the highest mountains

1 **HIMALAYAS** Asia, Mount Everest, 8,848 m (29,029 ft)

2 **KARAKORAM** Asia, K2, 8,611 m (28,251 ft)

3 **HINDU KUSH** Asia, Tirich Mir, 7,707 m (25,285 ft)

4 **PAMIR** Asia, Kongur Tagh, 7,649 m (25,095 ft)

5 **HENGDUAN SHAN** Asia, Mount Gongga, 7,556 m (24,790 ft)

6 **TIAN SHAN** Asia, Jengish Chokusu, 7,439 m (24,406 ft)

7 **KUNLUN** Asia, Liushi Shan, 7,167 m (23,514 ft)

8 **TRANSHIMALAYA** Asia, Nyainqêntanglha, 7,162 m (23,497 ft)

9 **ANDES** South America, Aconcagua, 6,963 m (22,844 ft)

10 **HINDU RAJ** Asien, Koyo Zom, 6,873 m (22,549 ft)

11 **ALASKA RANGE** North America, Denali, 6,194 m (20,321 ft)

12 **ST ELLAS MOUNTAINS** North America, Mount Logan, 5,959 m (19,550 ft)

13 **KILIMANJARO** Africa, Kibo, 5,895 m (19,340 ft)

14 **SIERRA NEVADA DE SANTA MARTA** South America, Pico Cristóbal, Colón, 5,775 m (18,946 ft)

15 **CAUCASUS** Europe, Elbrus, 5,642 m (18,510 ft)

16 **SIERRA MADRE ORIENTAL** Central America Citlaltépetl, 5,636 m (18,491 ft)

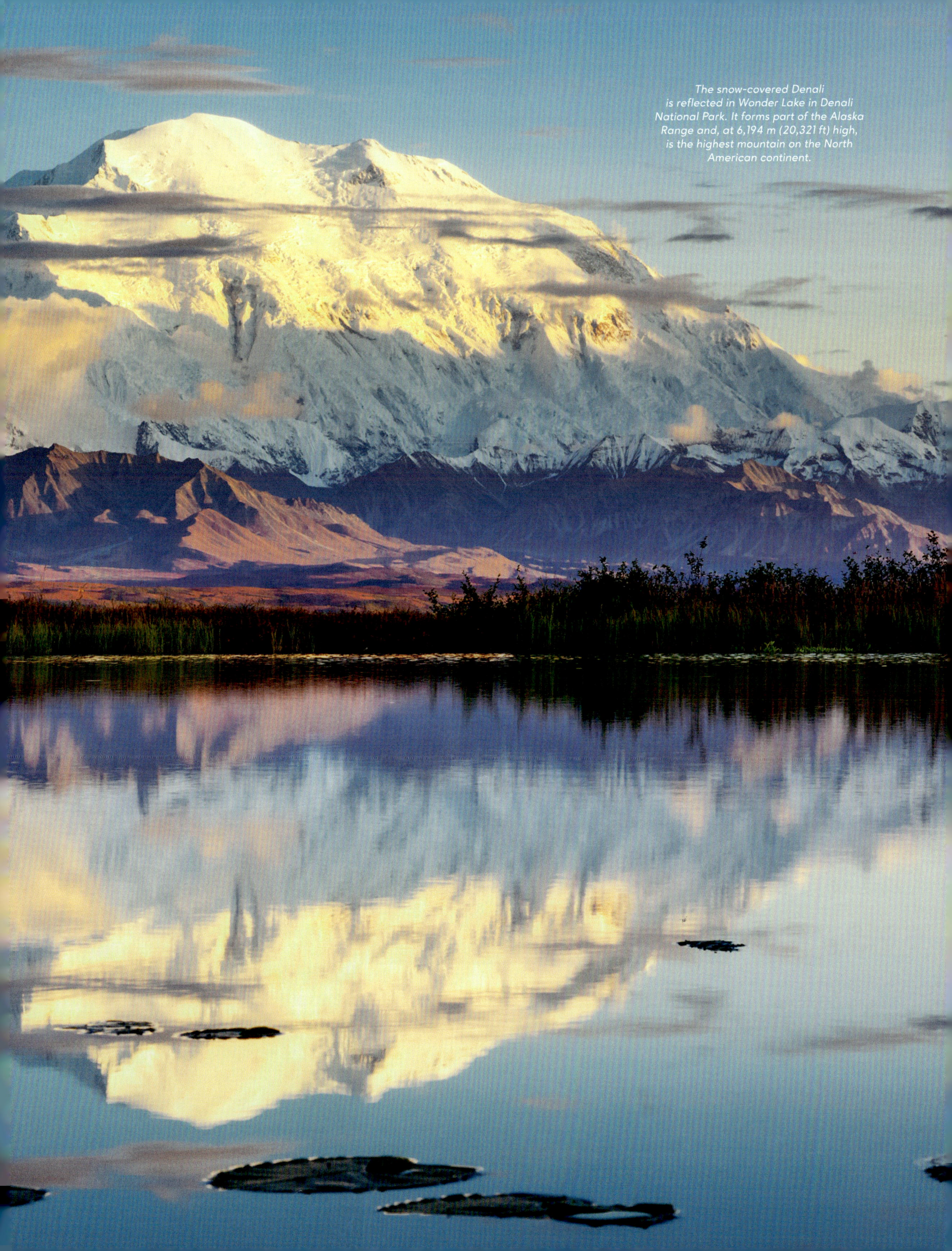

The snow-covered Denali is reflected in Wonder Lake in Denali National Park. It forms part of the Alaska Range and, at 6,194 m (20,321 ft) high, is the highest mountain on the North American continent.

FOUR ELEMENTS

EARTH

# HIMALAYAS

The highest mountains
on earth

The most impressive pictures of the highest mountains in the world are those taken from up high, that is, by satellite. A horseshoe-shaped chain of seven- and eight-thousanders encloses countless, almost snow-free valleys which all lie in Nepalese Sagarmatha National Park. Here, the famous eight-thousanders rise up, among them Cho Oyu (8,201 m (26,906 ft)), Mount Everest (8,850 m (29,035 ft)), Lhotse (8,516 m (27,940 ft)) and Makalu (8,463 m (27,766 ft)). From above, the view clearly shows how the huge ice fronts slide far into the valleys. These glacial rivers discharge into the Dudh Kosi, which flows southwards to the Sunkoshi River and feeds the source regions of the largest rivers in Asia – the Brahmaputra, Ganges and Indus. Some of the 19 rivers which drain the Himalayan system leave the mountain chain to the south through deep breakthrough valleys with steep sides towering up to 5 m (16 ft) high.

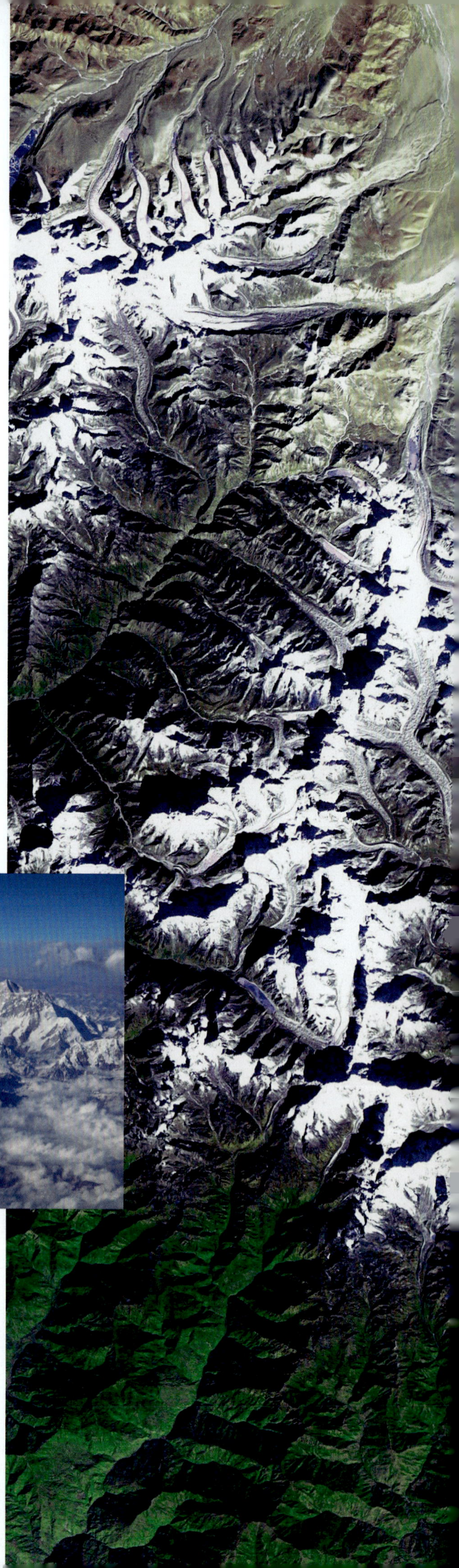

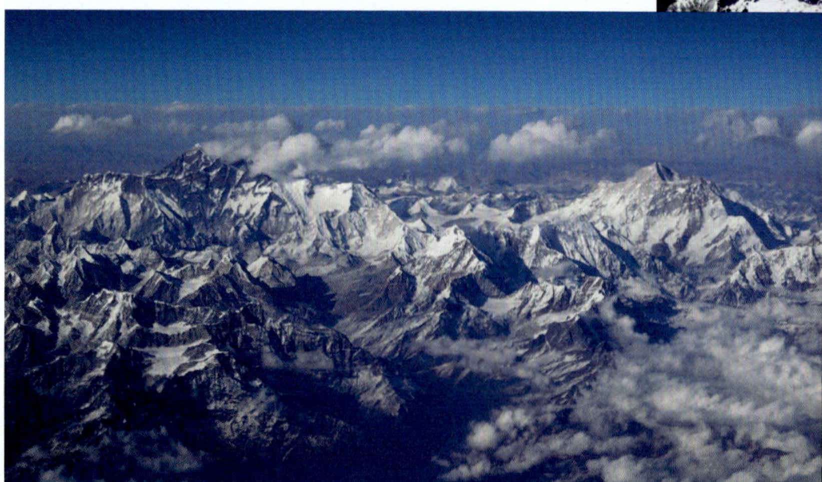

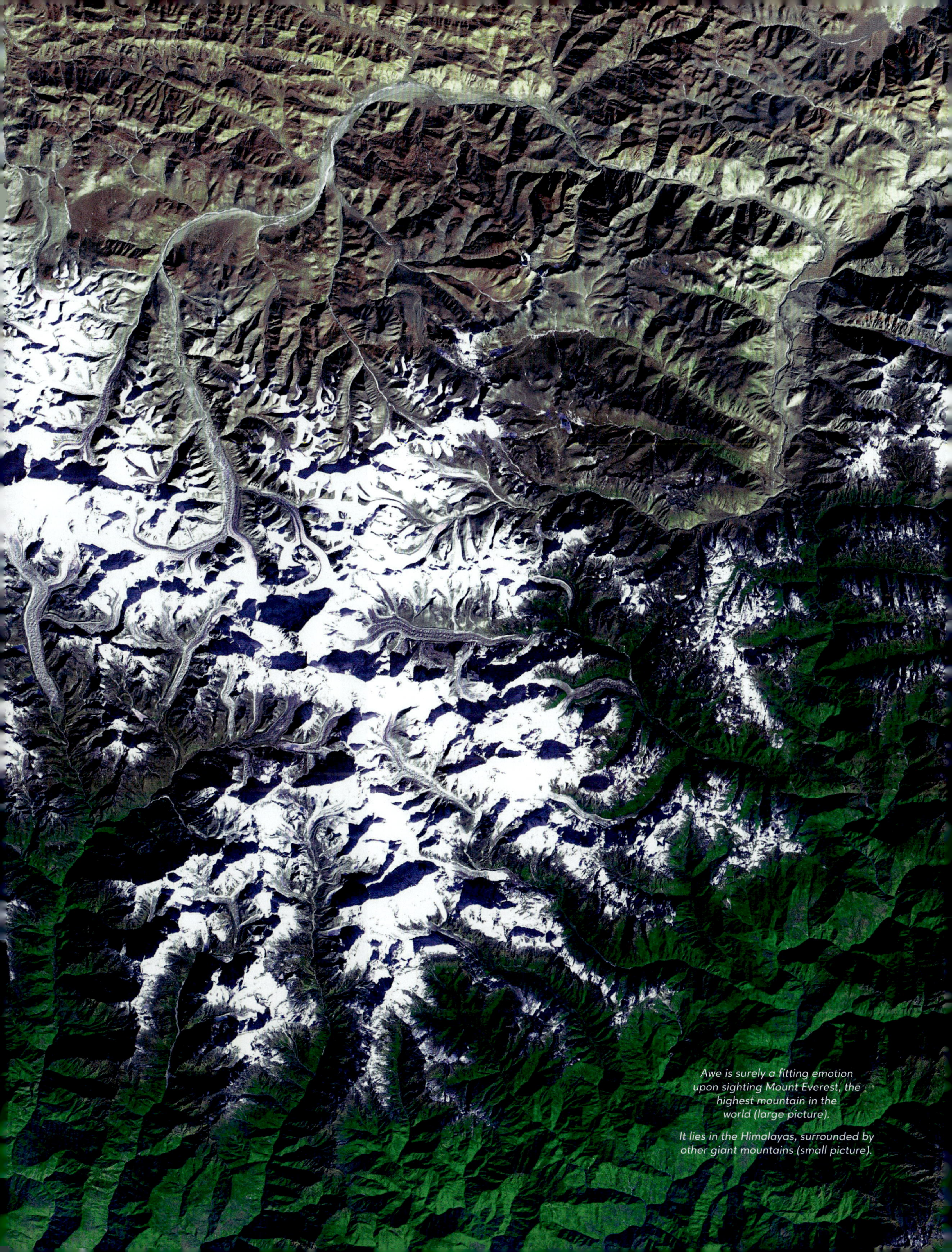

Awe is surely a fitting emotion upon sighting Mount Everest, the highest mountain in the world (large picture).

It lies in the Himalayas, surrounded by other giant mountains (small picture).

FOUR ELEMENTS

EARTH

# MOUNT EVEREST

The highest mountain in the world

Mount Everest, named after the British land surveyor, Sir George Everest, is called *Sagarmatha* ('Goddess of the sky') in Nepalese and *Qomolangma* ('Goddess mother of the world') in Tibetan. And rightly so, for at 8,850 m (29,035 ft) above sea level, it is the highest mountain on earth and is located in the Khumbu Himal region in Nepal on the border with Tibet. On the Nepalese side, it forms part of Sagarmatha National Park; on the northern side, it is part of the Qomolangma National Nature Reserve.

Like the entire Himalayas, Mount Everest emerged through plate tectonics when the Indian plate collided with the Eurasian plate. The smaller Indian plate continues to slide under the Eurasian plate at a rate of around 3 cm (1.2 in) a year. As a consequence, Mount Everest is still growing, albeit only a few millimetres per year.

Its great height means that the air pressure on Mount Everest is barely a third of the normal air pressure at sea level. Temperatures are extremely variable, and strong winds prevail. According to the beliefs of the indigenous population, the steep summits of the Himalayas, constantly covered in snow, are the dwelling place of the gods. Since the beginning of the 20th century, mountaineers have been attempting to climb the eight-thousanders. Following the first ascent by Edmund Hillary and Tenzing Norgay in 1953, Reinhold Messner and Peter Habeler were the first to conquer the summit without supplementary oxygen in 1978. Sherpa Pemba Dorjee achieved the fastest ascent in 2004, climbing from the base camp to the summit in just over eight hours. In 2012, Tamae Watanabe was the oldest person to climb to the summit. Watanabe was 73 years old at the time. Since the 1980s, there has been a wave of Mount Everest tourism, which, as a result of the congestion and large amounts of waste created, has led to environmental problems. Although the technical equipment has improved considerably over the last few decades, the thin air, icy cold and unpredictable weather can still endanger the lives of climbers. Nevertheless, Mount Everest continues to attract many visitors who all want to stand just once on the highest point on earth.

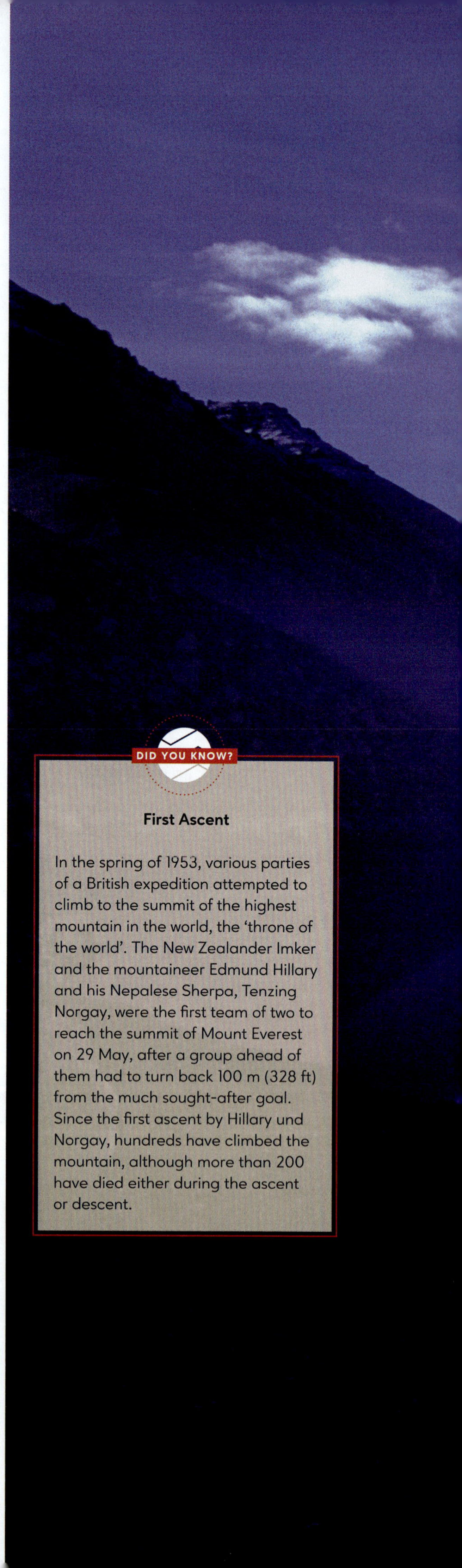

DID YOU KNOW?

**First Ascent**

In the spring of 1953, various parties of a British expedition attempted to climb to the summit of the highest mountain in the world, the 'throne of the world'. The New Zealander Imker and the mountaineer Edmund Hillary and his Nepalese Sherpa, Tenzing Norgay, were the first team of two to reach the summit of Mount Everest on 29 May, after a group ahead of them had to turn back 100 m (328 ft) from the much sought-after goal. Since the first ascent by Hillary und Norgay, hundreds have climbed the mountain, although more than 200 have died either during the ascent or descent.

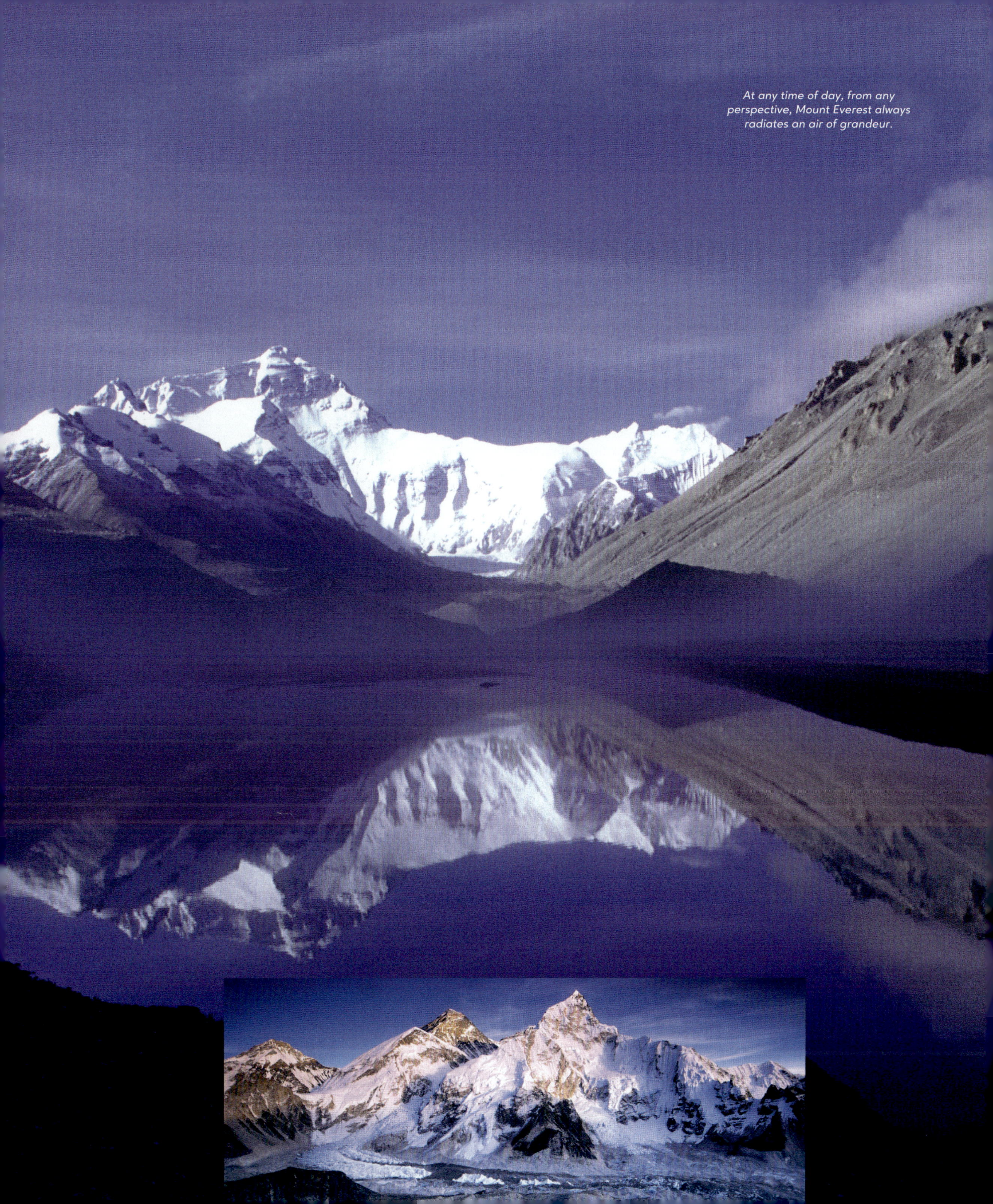

At any time of day, from any perspective, Mount Everest always radiates an air of grandeur.

# EIGHT-THOUSANDERS

The 14 highest mountains in the world

All the mountains in the world over 8,000 m (26,247 ft) are located in Asia – nine in the Himalayas and five in the adjoining Karakoram Mountains. They are scattered across China (including Tibet), India, Nepal and Pakistan. Five of the eight-thousander massifs lie completely within one country; the bases of the remaining nine straddle two different countries. In 1950, Annapurna was the first eight-thousander to be climbed; three years later, the first climbers stood on the summit of Mount Everest, the highest mountain in the world. With more than 4,100 ascents, conquering the summit of Mount Everest is one of the most coveted goals.

## The 14 Eight-Thousanders

(with first ascents)

1 **Mount Everest**
Nepal, China, 8,848 m (29,028 ft)
Hillary, Norgay, 1953

2 **K2**
Pakistan, China, 8,611 m (28,251 ft)
Compagnoni, Lacedelli, 1954

3 **Kangchenjunga**
Nepal, India, 8,586 m (28,169 ft)
Band, Brown, 1955

4 **Lhotse**
Nepal, China, 8,516 m (27,940 ft)
Luchsinger, Reiss, 1956

5 **Makalu**
Nepal, China, 8,485 m (27,838 ft)
Terray, Couzy, 1955

6 **Cho Oyu**
Nepal, China, 8,188 m (26,863 ft)
Tichy, Jöchler, P. Dawa Lama, 1954

7 **Dhaulagiri**
Nepal, 8,167 m (26,794 ft)
Diemberger, Na. Dorje, Forrer, Schelbert, Diener, Ny. Dorje, 1960

8 **Mansalu**
Nepal, 8,163 m (26,781 ft)
Norbu, Imanishi, 1956

9 **Nanga Parbat**
Pakistan, 8,125 m (26,657 ft)
Buhl, 1953

10 **Annapurna I**
Nepal, 8,091 m (26,545 ft)
Lachenal, Herzog, 1950

11 **Gasherbrum I (Hidden Peak),**
Pakistan, China, 8,080 m (26,509 ft)
Schoening, Kauffman, 1958

12 **Broad Peak**
Pakistan, China, 8,051 m (26,414 ft)
Buhl, Diemberger, Schmuck, Wintersteller, 1957

13 **Gasherbrum II**
Pakistan, China, 8,034 m (26,358 ft)
Moravec, Larch, Willenpart, 1956

14 **Shishapangma**
China 8,027 m (26,335 ft)
Chin. Exped. X. Jing, 1964

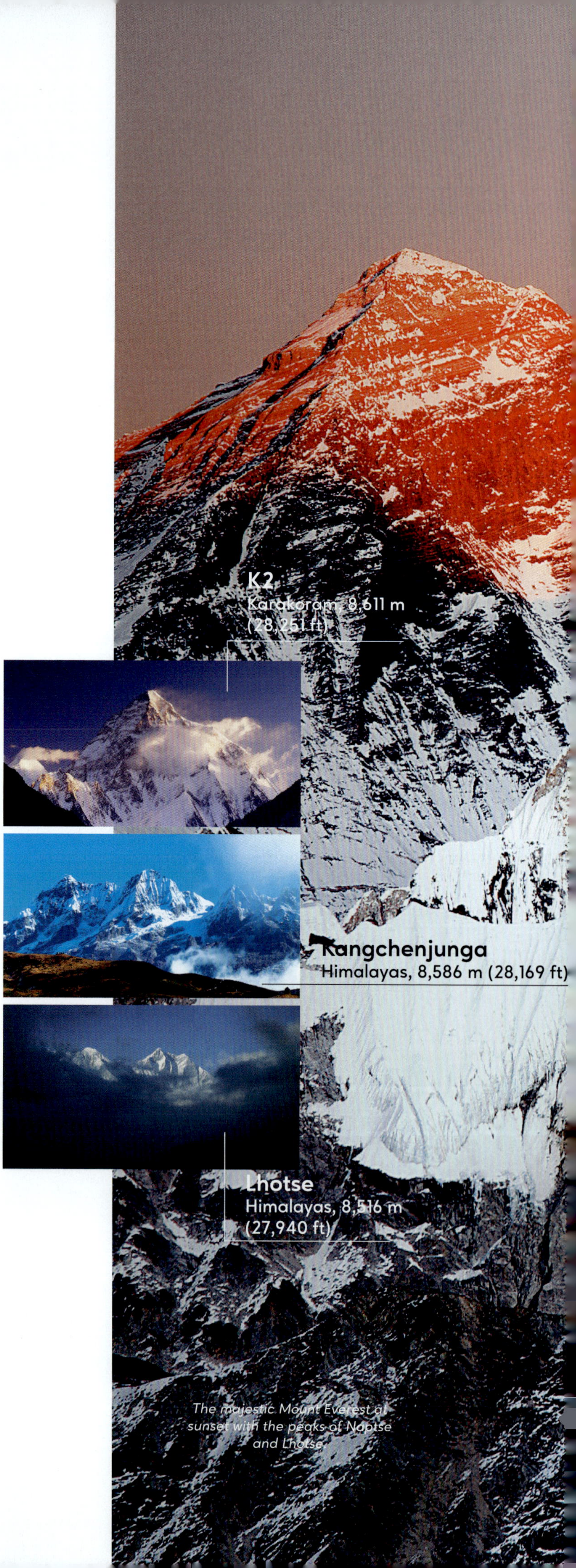


*The majestic Mount Everest at sunset with the peaks of Noptse and Lhotse.*

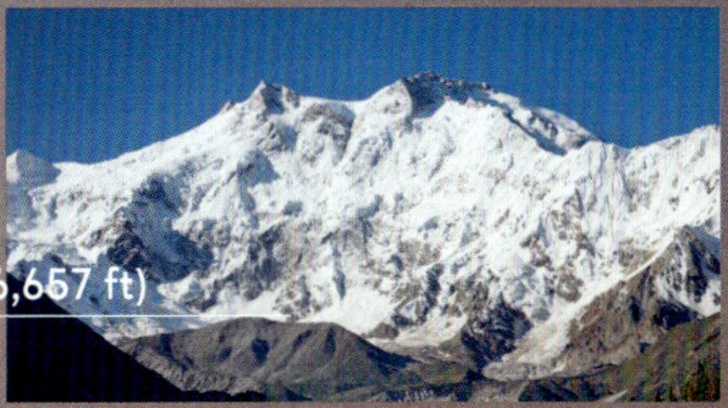

**Nanga Parbat,**
Himalayas, 8,125 m (26,657 ft)

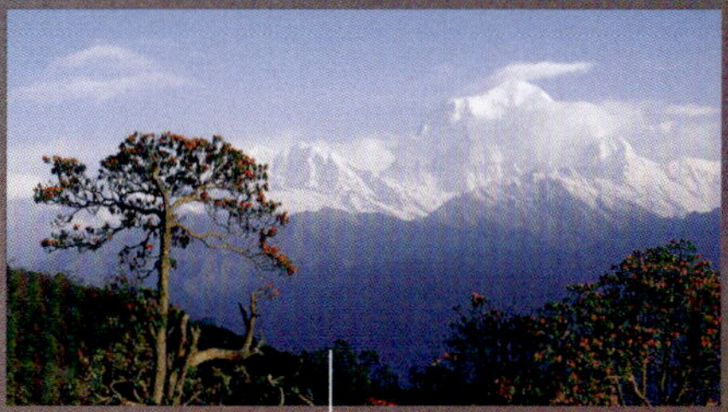

**Annapurna I,**
Himalayas, 8,091 m (26,545 ft)

**Cho Oyu**
Himalayas, 8,188 m
(26,863 ft)

**Makalu,**
Himalayas, 8,485 m (27,838 ft)

**Gasherbrum I (Hidden Peak),**
Karakoram, 8,080 m
(26,509 ft)

**Gasherbrum II,**
Karakoram, 8,034 m
(26,358 ft)

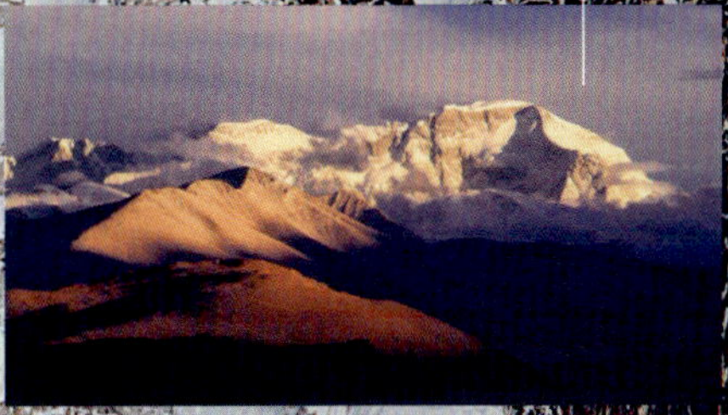

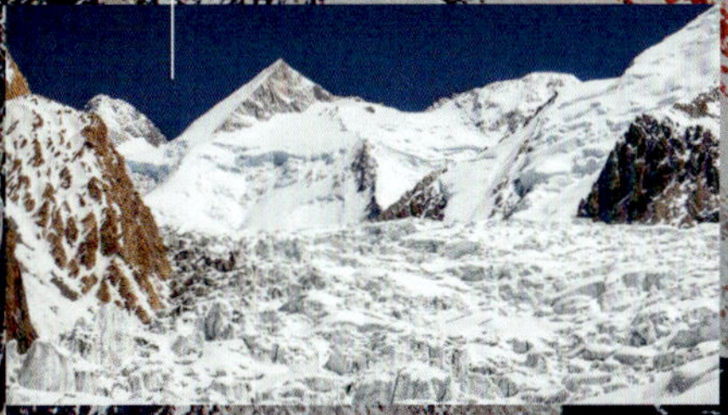

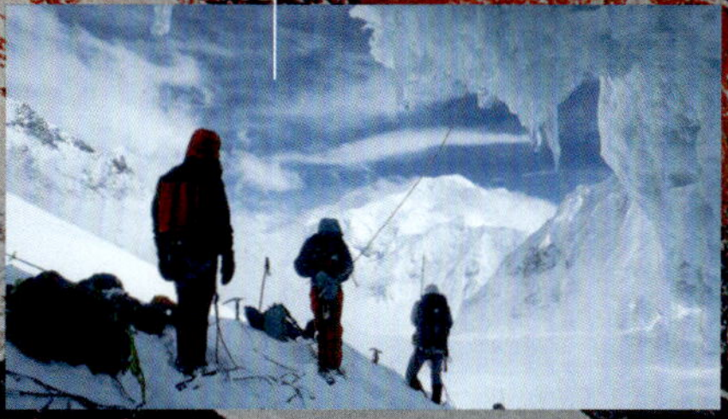

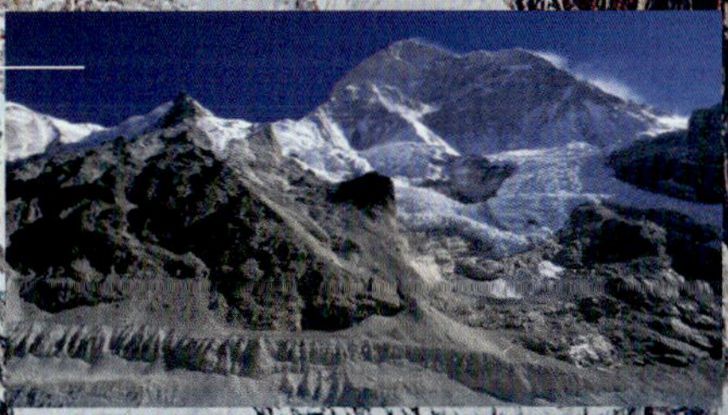

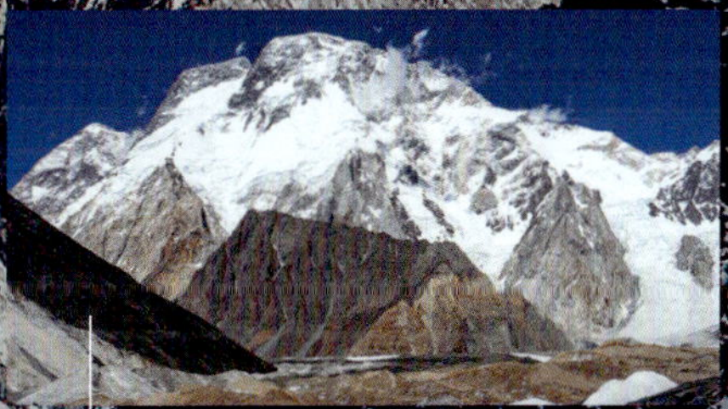

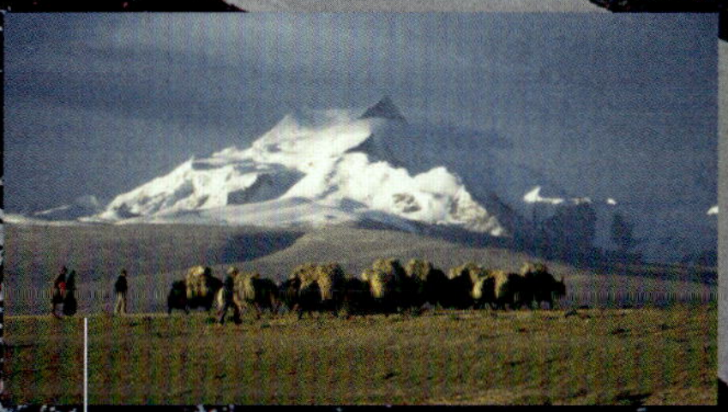

**Broad Peak**
Karakoram, 8,051 m (26,414 ft)

**Shishapangma**
Himalayas, 8,027 m
(26,335 ft)

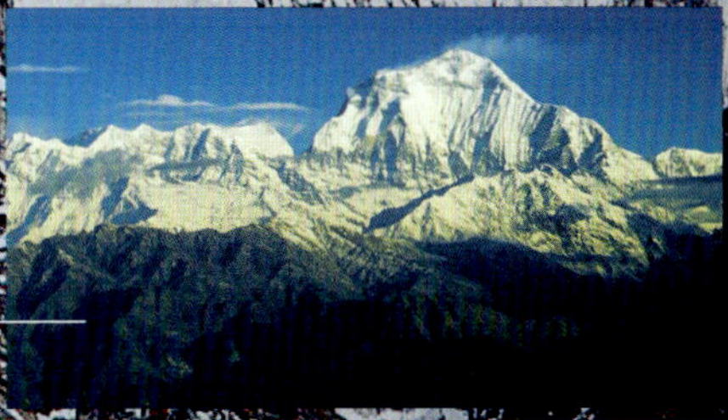

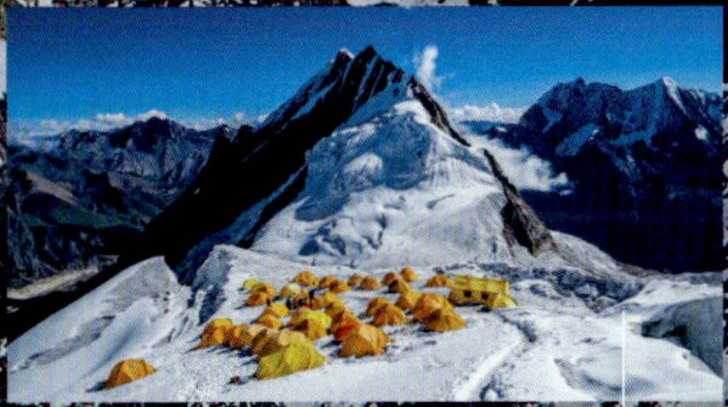

**Dhaulagiri**
Himalayas, 8,167 m
(26,794 ft)

**Manaslu,**
Himalayas, 8,163 m
(26,781 ft)

DID YOU KNOW?

### Climbing All of the Eight-Thousanders

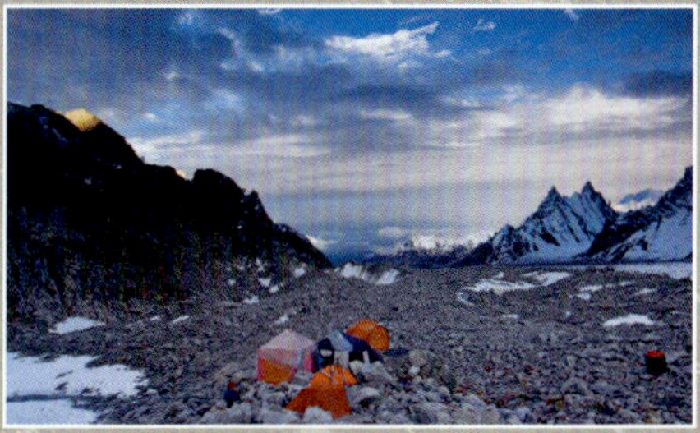

Around 40 climbers have so far succeeded in climbing all fourteen eight-thousanders. Reinhold Messner, perhaps the most famous climber of all, was the first to achieve this. He began in 1970 with his ascent of Nanga Parbat; he finished the series in 1986 with Lhotse. There is some dispute as to which female climber was the first to climb all of the eight-thousanders. However, Edurne Pasaban and Gerline Kaltenbrunner were certainly among the first. A particular challenge is to ascend the eight-thousanders without bottled oxygen. Messner also achieved this for all fourteen peaks of the mountain giants.

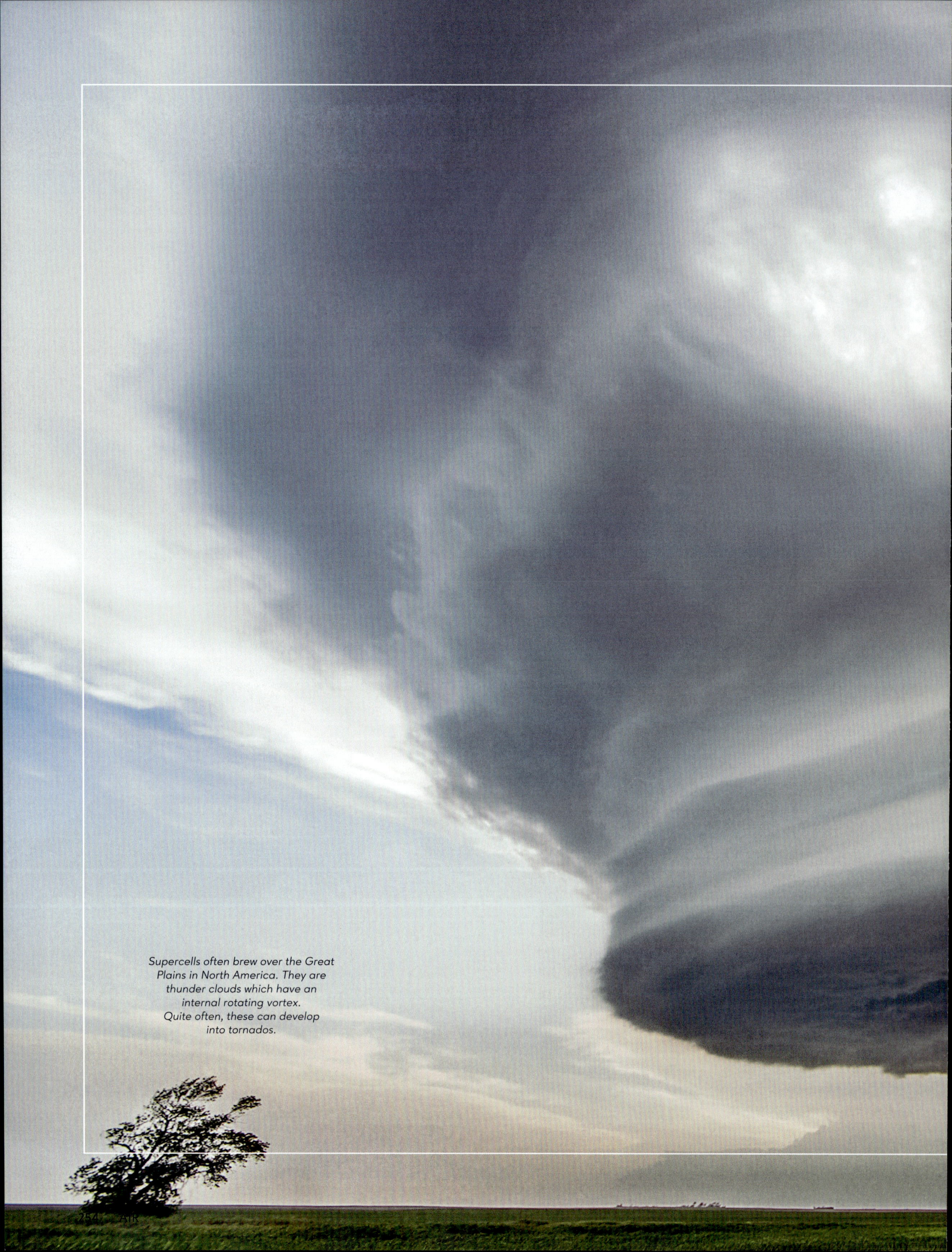

*Supercells often brew over the Great Plains in North America. They are thunder clouds which have an internal rotating vortex. Quite often, these can develop into tornados.*

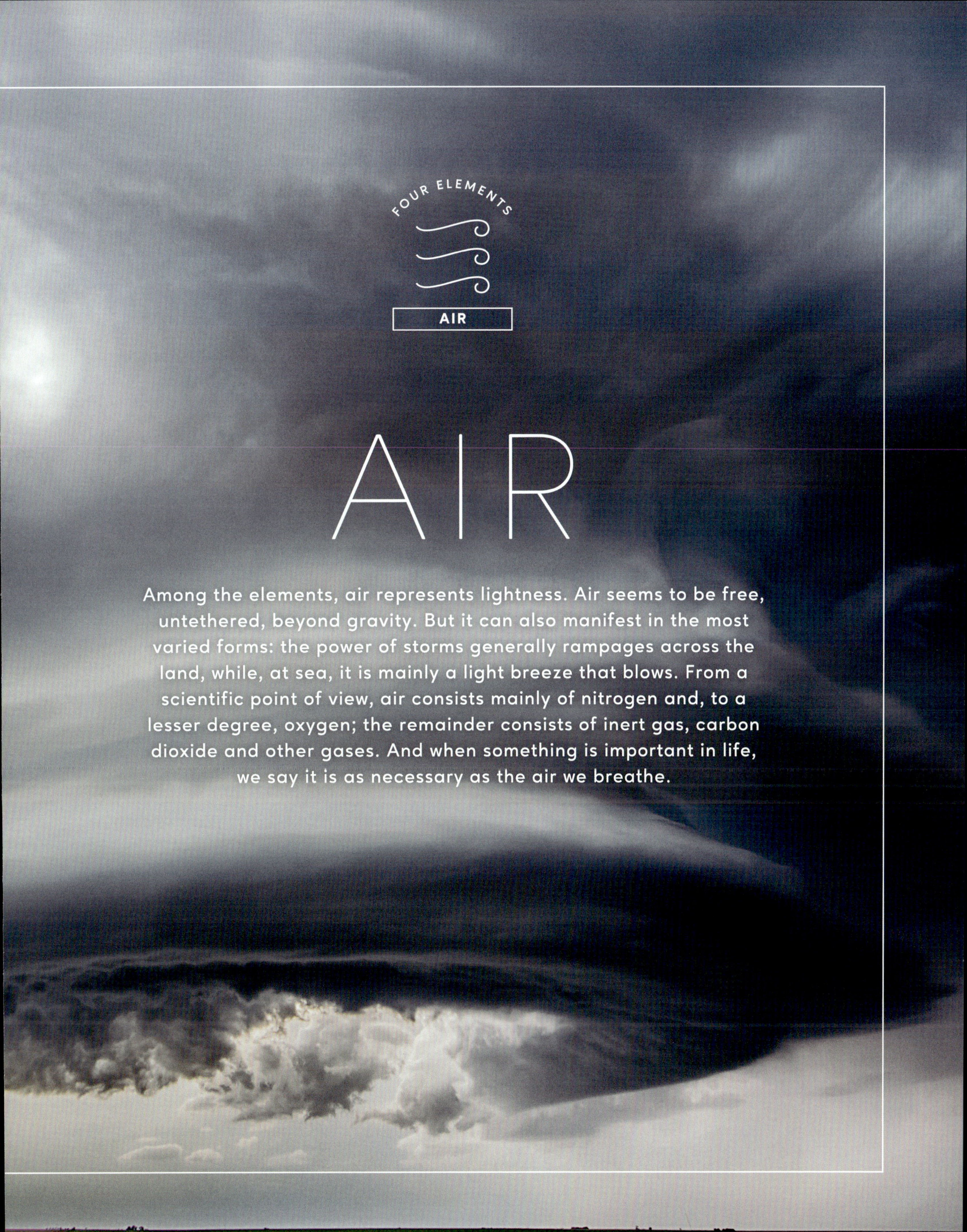

FOUR ELEMENTS

AIR

# AIR

Among the elements, air represents lightness. Air seems to be free, untethered, beyond gravity. But it can also manifest in the most varied forms: the power of storms generally rampages across the land, while, at sea, it is mainly a light breeze that blows. From a scientific point of view, air consists mainly of nitrogen and, to a lesser degree, oxygen; the remainder consists of inert gas, carbon dioxide and other gases. And when something is important in life, we say it is as necessary as the air we breathe.

FOUR ELEMENTS

AIR

# MASTER(PIECE)S OF THE AIR

Humankind invents things with its knowledge and technology that nature has long since mastered

The dream of flight is as old as humankind. Anyone who watches the ease and agility of the birds in the sky knows the longing to take flight and escape earth's gravity. The engineer Otto Lilienthal was probably the first person whose attempts at flight became well-known and famous. The technical capabilities to do this quickly became more professional, and, today, we take it for granted that we can travel on planes and rockets to other countries, continents and even the moon.

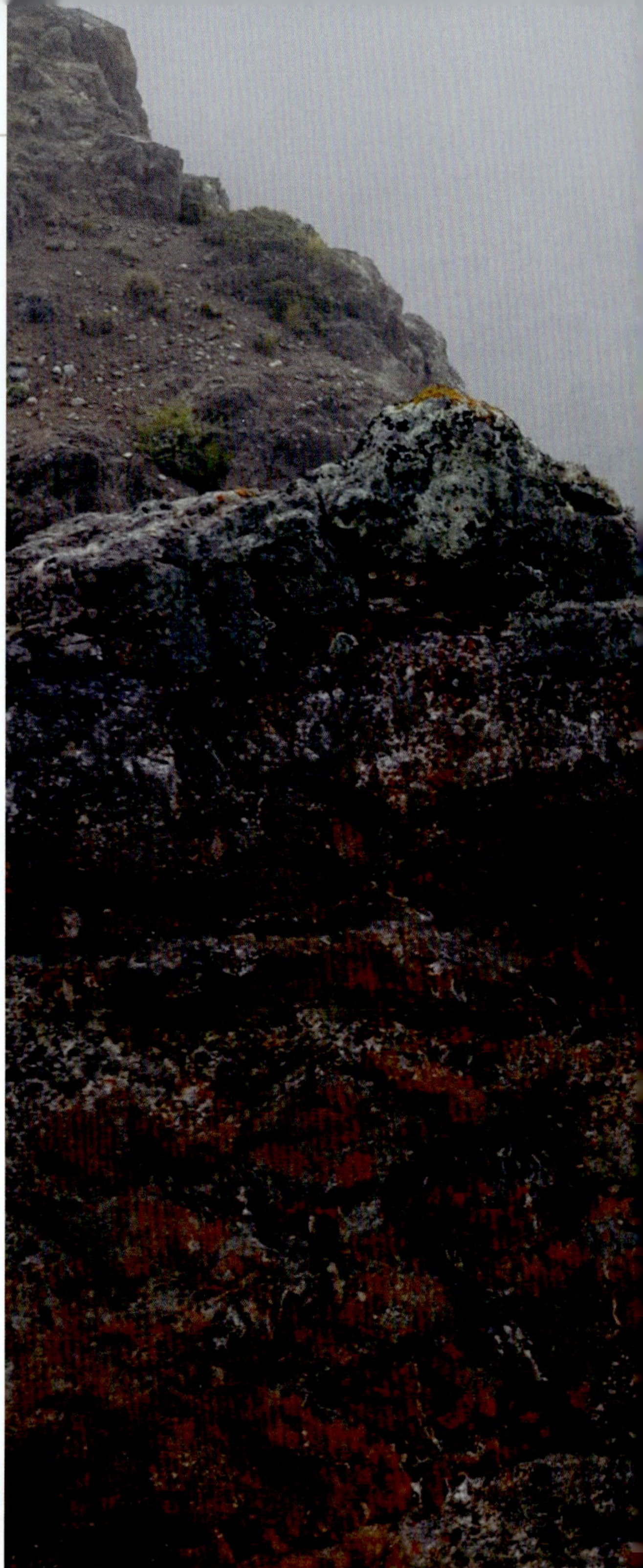

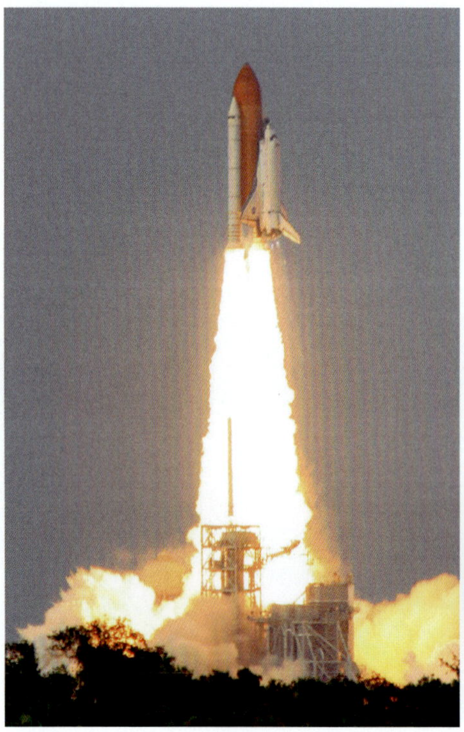

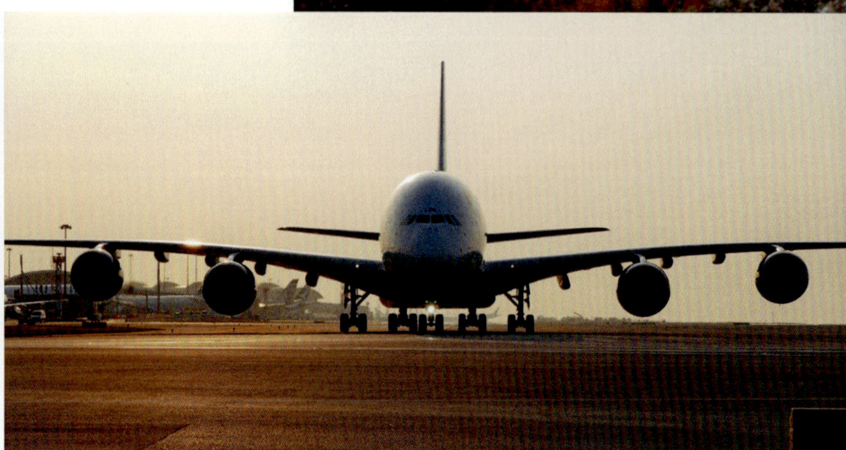

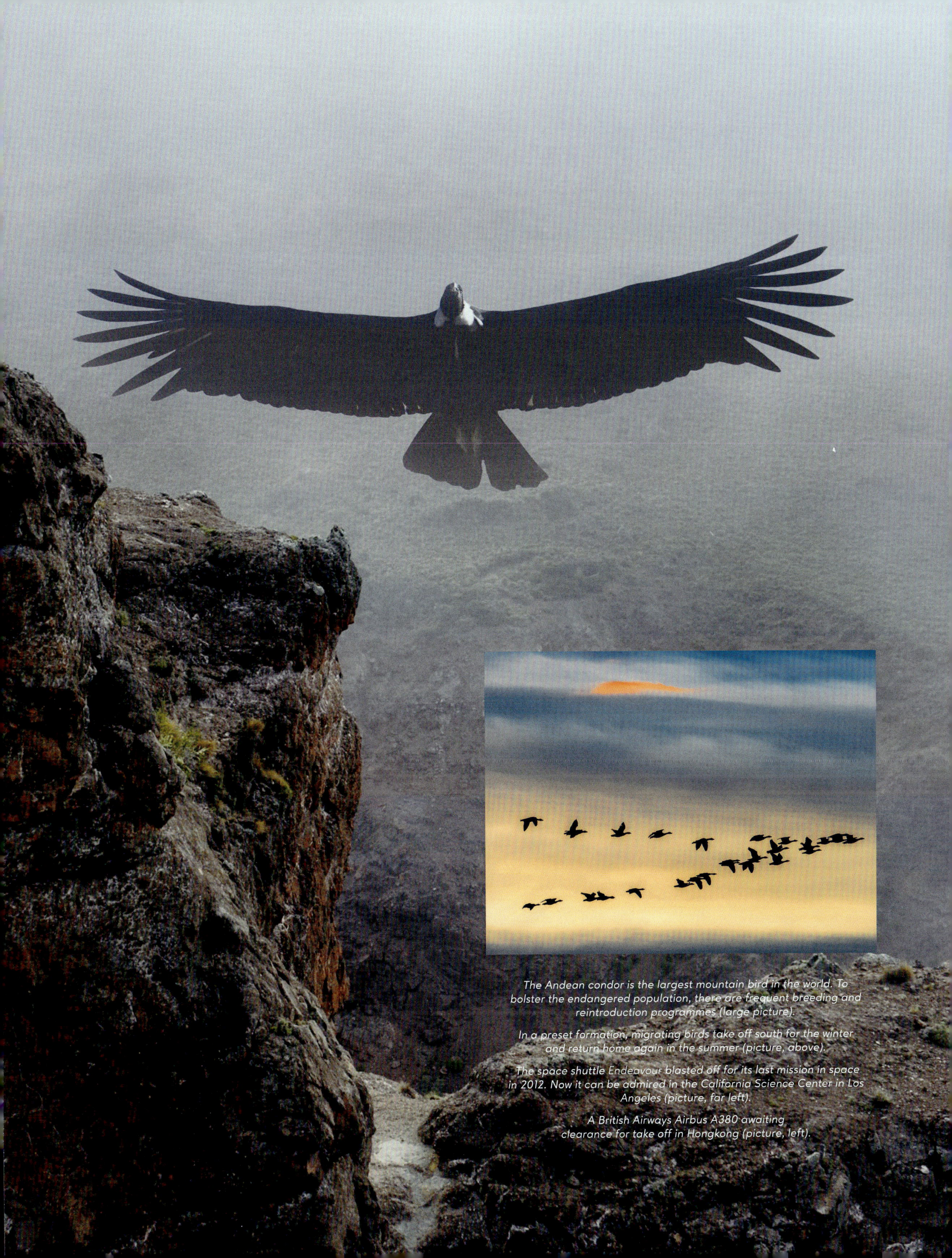

*The Andean condor is the largest mountain bird in the world. To bolster the endangered population, there are frequent breeding and reintroduction programmes (large picture).*

*In a preset formation, migrating birds take off south for the winter and return home again in the summer (picture, above).*

*The space shuttle Endeavour blasted off for its last mission in space in 2012. Now it can be admired in the California Science Center in Los Angeles (picture, far left).*

*A British Airways Airbus A380 awaiting clearance for take off in Hongkong (picture, left).*

# AIRBUS A380

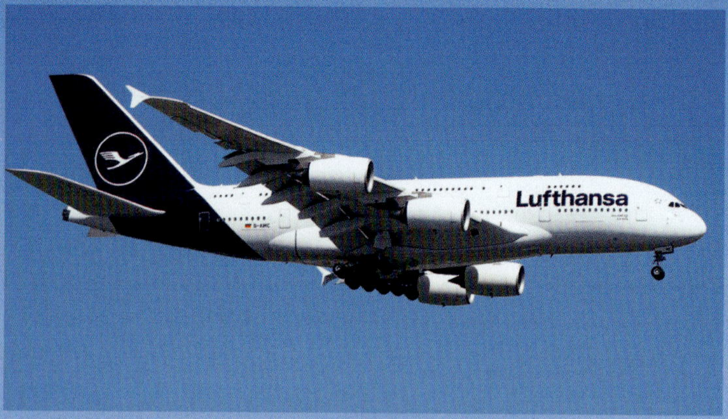


The largest passenger aircraft in the world

With its first flight in 2005, the Airbus A380 took over from the then record holder, the Boeing 747, as the largest passenger aircraft in the world. Although the Airbus is shorter than its competitor, with a fuselage length of 72 m (236 ft), it surpasses it in all other measurements. Due to its 80-m (262-ft) wingspan and a height of 24 m (78 ft), the A380 can only land and take off at a few specially adapted airports. It can hold between 525 and 850 passengers, depending on the version, accommodated in two walk-through passenger decks – recognizable from the outside by the double row of windows on both sides. Thanks to the introduction of new, lighter composites in the structure and components, it has been possible to limit the empty weight to 275 tonne (303 ton); a take-off weight of 560 tonne (617 ton) is the maximum permitted. Four engines with a thrust of 310 to 320 kN each take the A380 to a cruising speed of 920 km/h (572 mph); its range is around 15,000 km (9,320 mi).

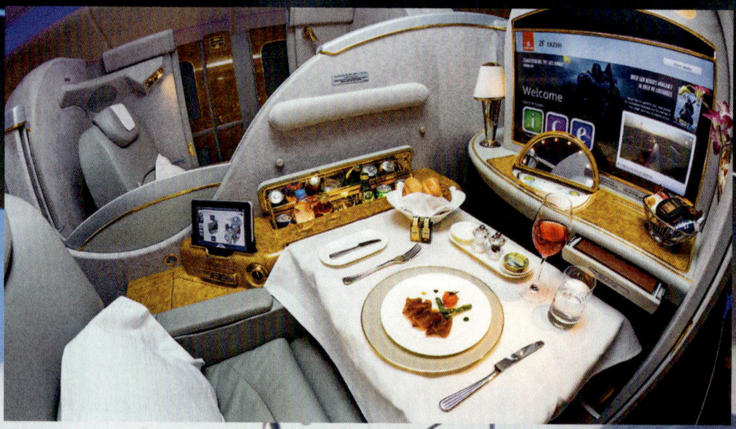

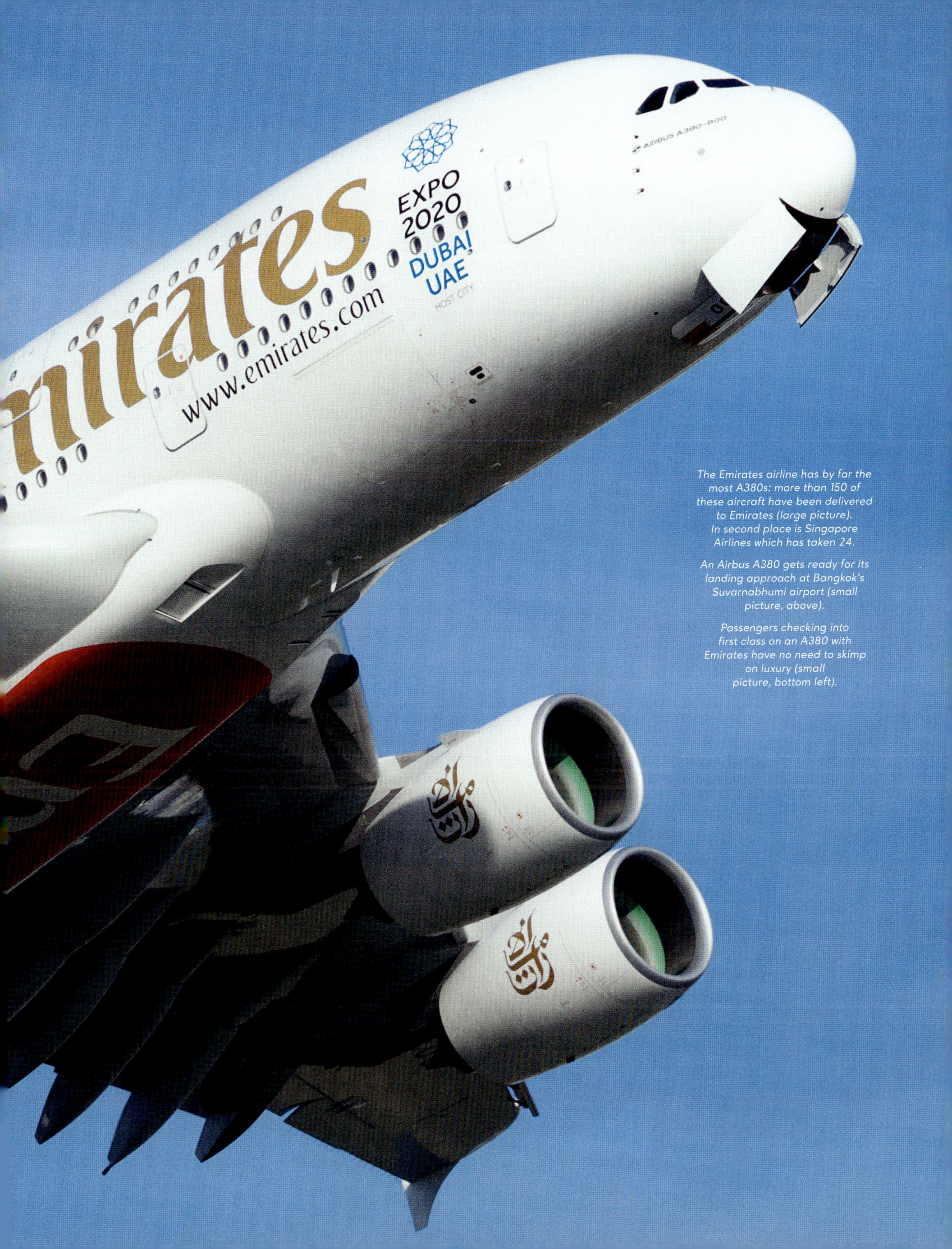


*The Emirates airline has by far the most A380s: more than 150 of these aircraft have been delivered to Emirates (large picture). In second place is Singapore Airlines which has taken 24.*

*An Airbus A380 gets ready for its landing approach at Bangkok's Suvarnabhumi airport (small picture, above).*

*Passengers checking into first class on an A380 with Emirates have no need to skimp on luxury (small picture, bottom left).*

FOUR ELEMENTS

AIR

# ANDEAN CONDOR

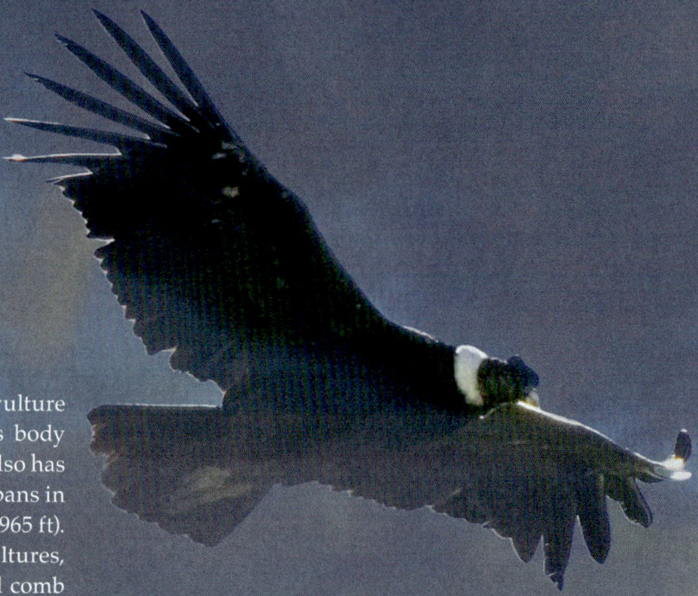

### The largest mountain bird in the world

The Andean condor (*Vultur gryphus*) from the New World vulture family is the largest species of vulture in the world. Its body length is 110 cm (3.5 ft) and it weighs 12 to 15 kg (26 to 33 lb); it also has a wingspan of more than 3 m (9.8 ft) – one of the largest wingspans in the bird world – which enables it to fly at heights of 7,000 m (22,965 ft). The Andean condor has a bald head and neck, typical of vultures, surrounded by a soft white ruff. The male has a fleshy crested comb on its head; the female has no comb and is smaller than the male. Its plumage is black, interspersed with white feathers on its upper body. The Andean condor is an endangered species. In the wild, it can live to be around 40 years old; in captivity, it can reach an astounding 85 years of age. Its habitat stretches from Venezuela along the Andes to Patagonia. The birds generally live in mountainous regions where they often use the variable air currents for gliding. Andean condors frequently cover short distances on foot. However, most of the time they rest and preen each others' plumage, which serves to strengthen the bonds between pairs and within the group.

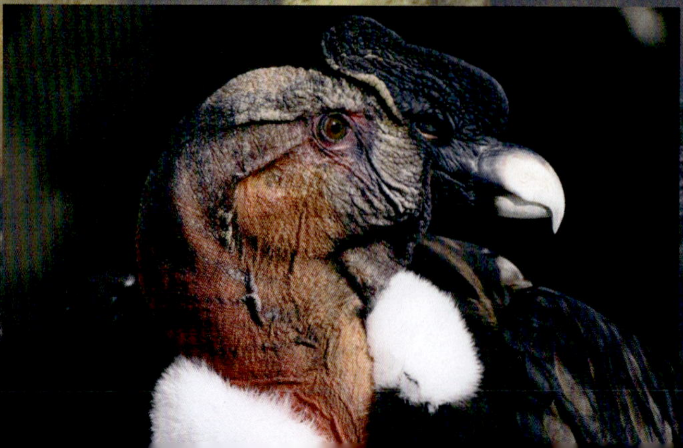

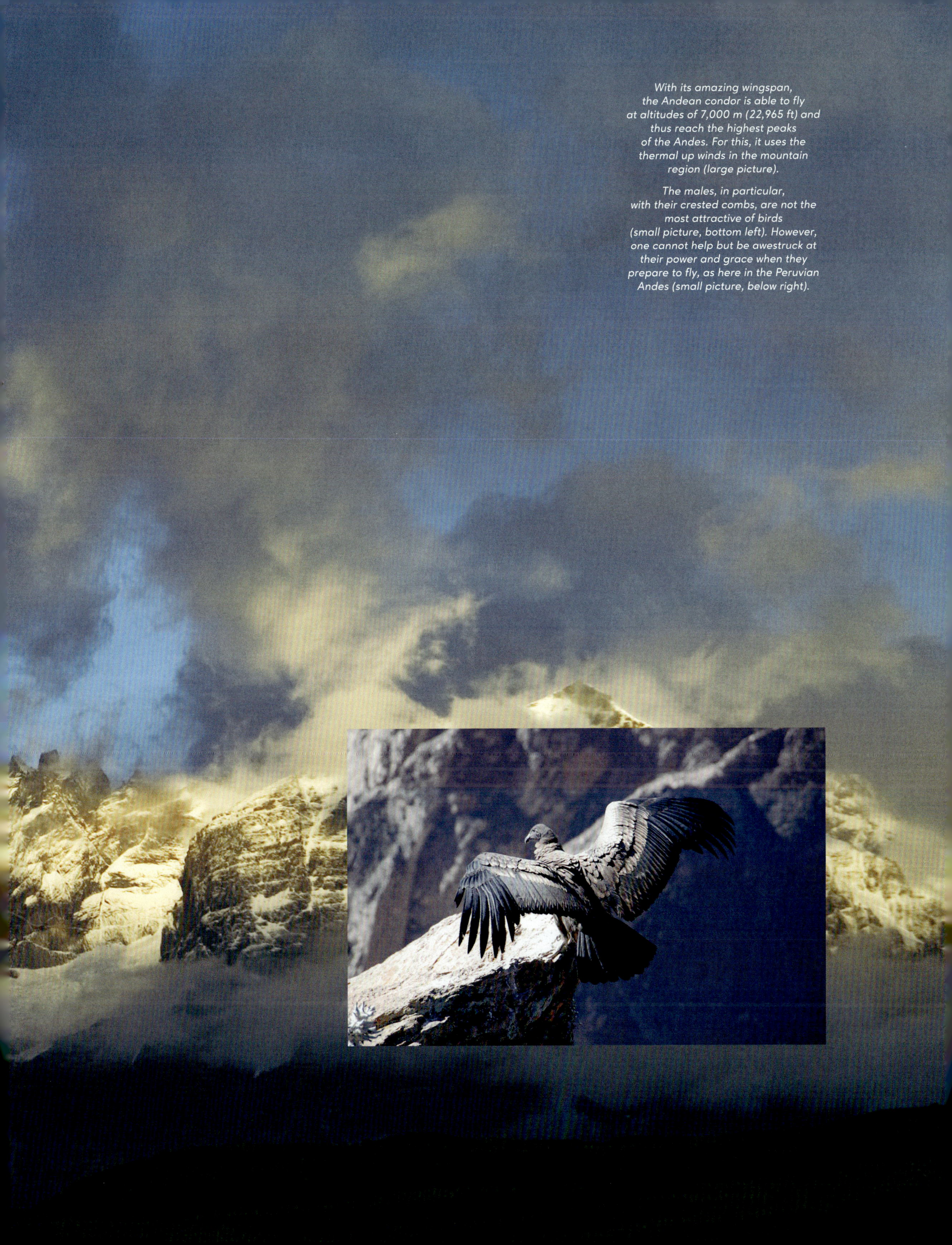

*With its amazing wingspan, the Andean condor is able to fly at altitudes of 7,000 m (22,965 ft) and thus reach the highest peaks of the Andes. For this, it uses the thermal up winds in the mountain region (large picture).*

*The males, in particular, with their crested combs, are not the most attractive of birds (small picture, bottom left). However, one cannot help but be awestruck at their power and grace when they prepare to fly, as here in the Peruvian Andes (small picture, below right).*

# AS FAR AS YOUR WINGS WILL CARRY YOU

Albatrosses are extreme fliers: some of these birds fly around the earth several times over or travel 15,000 km (9,320 mi) without a stopover.

DID YOU KNOW?

**Ready for Takeoff**

Most albatrosses – there are 13 different species – live on or above the polar and subpolar seas in the southern hemisphere. In spite of their considerable weight, the birds are able to fly very large distances over the surface of the ocean. In addition to their long, slender wings, they make use of the technique of dynamic gliding, whereby they choose their flight path so that the energy gain from the wind conditions exceeds the energy loss involved in using their wings. If the wind speed is less than 12 km/h (7.5 mph), however, the albatrosses are unable to take off. They then have to remain on the water or on land and wait for better wind conditions.
An albatross in flight looks very elegant, but they face huge difficulties at takeoff and landing. For takeoff, they need a long start, and, in the glide flight for the equally long landing, they occasionally overshoot. Serious injuries to their wings or legs are not uncommon, particularly on land. So in order to protect the nests in their breeding colonies, the albatrosses arrange a runway and a landing strip where no nests are built.

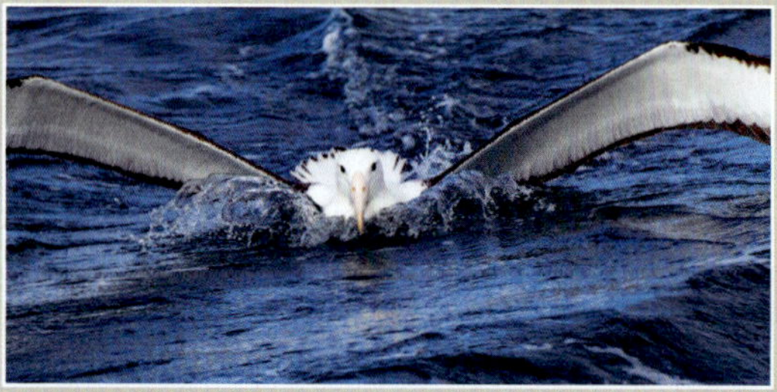

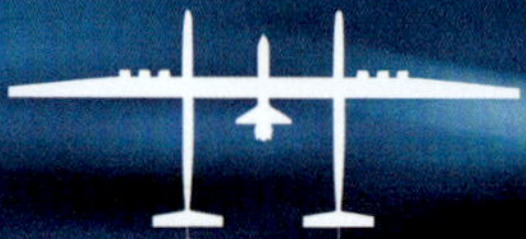

STRATOLAUNCH

Range: 1,852 km (1,150 mi)
Wingspan: 117 m (383 ft)
Weight: 589,670 kg (1,300,000 lb)
Maximum speed: 853 km/h (530 mph)

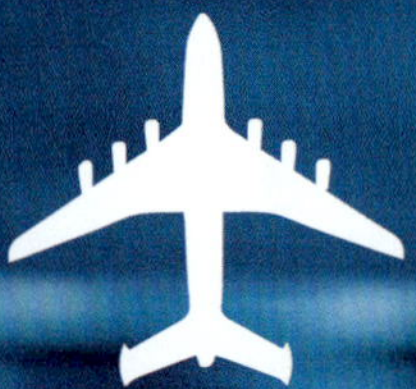

ANTONOV AN-225

Range: 4,500 km (2,796 mi)
Wingspan: 88.40 m (290 ft)
Weight: 285,000 kg (628,317 lb)
Flight speed: 800 km/h (497 mph)

Range in km (mi): 1,000 (621) | 2,000 (1,242) | 3,000 (1,864) | 4,000 (2.485) | 5,000 (3,106) | 6,000 (3,728) | 7,000 (4,350)

## NO AIRCRAFT CURRENTLY HAS A GREATER WINGSPAN

The aircraft with the double fuselage is fitted with six Boeing 747 engines. Comparison: A Boeing 747 has a wingspan of just under 68 m (223 ft). The Stratolaunch, which completed its maiden flight in April 2019, was originally designed to carry rockets with satellites and launch them in space.

## THE LARGEST AIRCRAFT IN THE WORLD

Antonov An-225 is a cargo aircraft with six turbofan engines, built in the Soviet Union in 1988. Until 2012, it completed over 1,000 launches and landings with only around 5,000 flying hours. Until the first flight of the Stratolaunch Roc in 2019, the Antonov was the longest and heaviest aircraft in operation in the world and had the largest wing-

## THE AIRCRAFT WITH THE LARGEST WINGSPAN

The double-fuselage aircraft referred to as Model 351 was developed by Stratolaunch Systems and built by Scaled Composites in Mojave in California. It is designed to shoot carrier rockets into space during flight – and thus launch satellites into orbit around the earth at a much lower cost than previously.

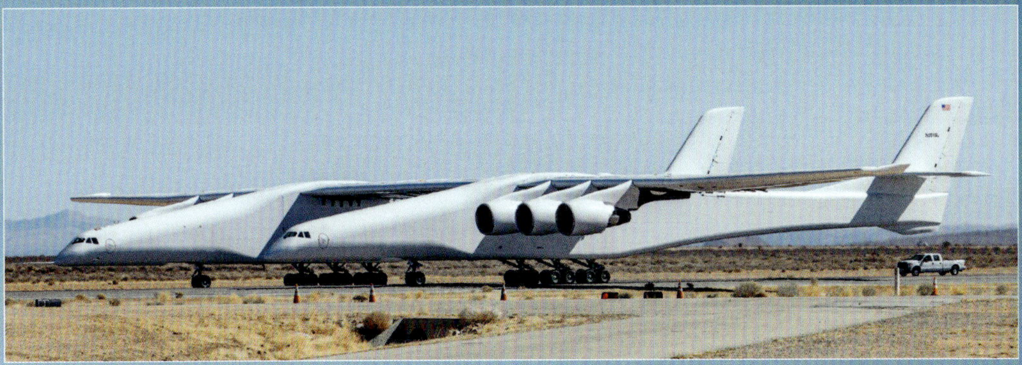

Stratolaunch Systems constructed the mega aircraft from two Boeing 747 jumbo jets and gave it a completely new look. The two fuselages were joined together with a reinforced connector, under which up to three carrier rockets with a weight of up to 227 tonne (250 ton) were fixed. The 73-m (240-ft) long machine is mounted on a chassis with a total of 28 wheels.

Six PW4056 engines provide the required thrust. The cockpit for the three-person crew was built into the right fuselage. At an altitude of around 11,000 m (36,089 ft), the carrier rockets are launched with the satellites. It executes twelve or more launches per year. The maiden flight of the giant aircraft proved successful, but then everything went quiet around the ambitious project.

The aircraft has now been sold – further developments are planned but not yet certain.

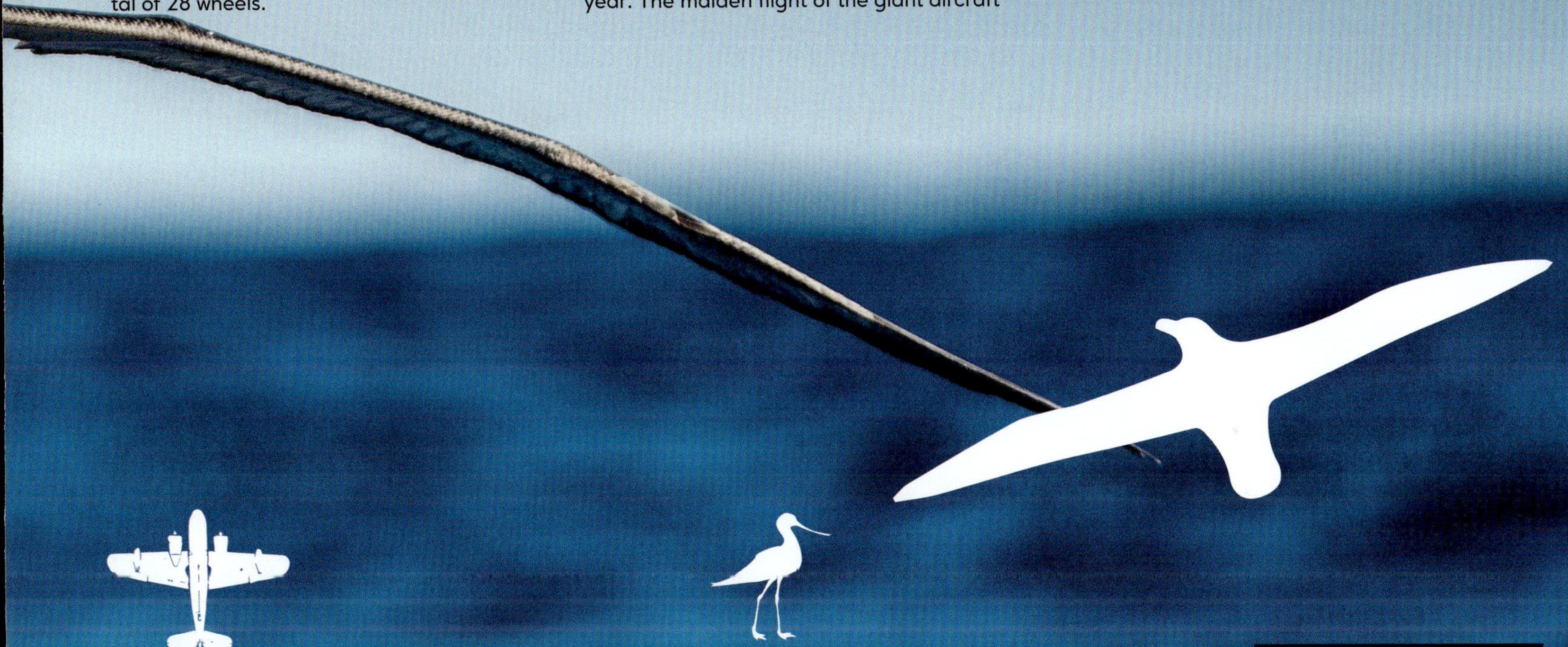

HUGHES H-4 HERCULES

Range: 9,451 km (5,872 mi)
Wingspan: 97.54 m (320 ft)
Weight: 180,000 kg (396,832 lb)
Flight speed: approx. 400 km/h (248 mph)

BAR-TAILED GODWIT

Uninterrupted flight range: over 12,000 km (7,456 mi)
Wingspan: up to 0.4 m (1.3 ft)
Weight: up to 0.5 kg (1.1 lb)
Flight speed: 100 km/h (62 mph)

WANDERING ALBATROSS

Uninterrupted flight range: up to 15,000 km (9,320 mi)
Wingspan: up to 3.5 m (11.5 ft)
Weight: up to 12 kg (26 lb)
Flight speed: 127 km/h (79 mph)

9,000 (5,592) | 10,000 (6,214) | 11,000 (6,835) | 12,000 (7,456) | 13,000 (8,077) | 14,000 (8,699) | 15,000 (9,320) | 16,000 (9,942)

### A SPRUCE GOOSE

The Hughes H-4 Hercules (nicknamed The Spruce Goose) was a flying boat made of wood built between 1944 and 1946 by the Hughes Aircraft Company founded by the legendary millionaire, Howard Hughes. Wood was used because, at the time, only materials not required for the war effort were available for the construction of civilian aircraft. Because the only flight the H-4 made took place at a height of 20 m (65 ft), its flight capacity for higher altitudes has never been demonstrated.

### FIGHTER JET DESIGN

The aerodynamic shape reduces wind resistance, and its long wings help it fly faster. The bar-tailed godwits are classed as long-distance migrants because they fly in large flocks from Alaska to New Zealand to overwinter every year.

### THEY ARE AT HOME IN THE SKY

The wandering albatross – the bird with the largest wingspan in the world – spends most of its life in the air or seeking food on the water. It only seeks out dry land for breeding.

# RUEPPELL'S GRIFFONS, GEESE, SWIFTS

The highest flight altitudes of birds

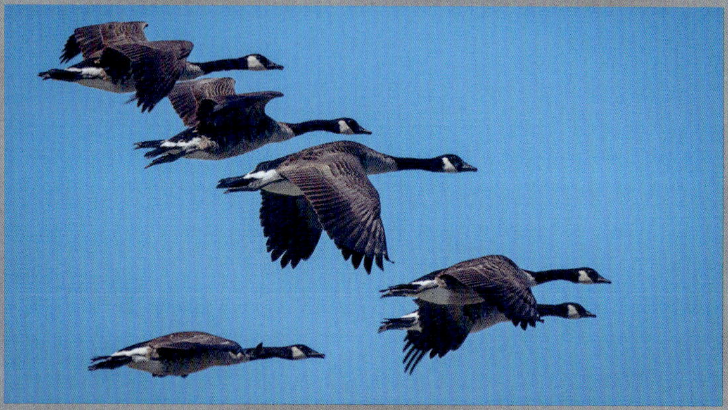

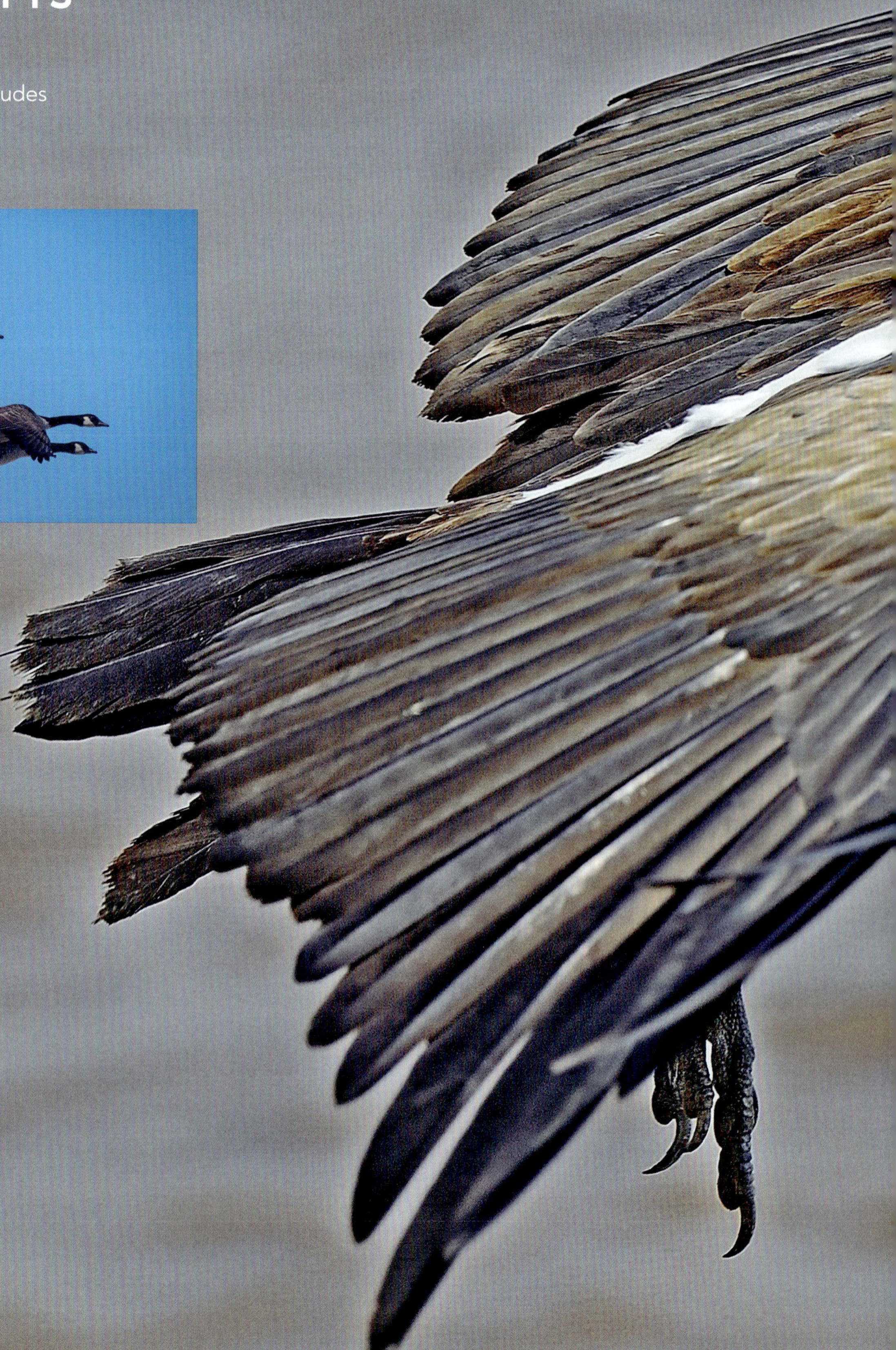

## GEESE

Birds can fly higher than many aircraft – they sometimes reach amazing heights, particularly the bar-headed goose. On its migration route to its winter habitat, it flies over the Himalayas and, in the process, reaches altitudes of up to 6,000 m (19,685 ft). Birds have even been spotted at altitudes of 9,000 m (29,527 ft) – above Mount Everest, for instance.

## RUEPPELL'S GRIFFONS

This, however, is still not everything that birds are capable of in terms of flight altitudes – Rueppell's griffons hold the record for flight altitudes because they reach an altitude of more than 11,000 m (36,089 ft). This was verified because one collided with an aircraft. These ace flyers are threatened with extinction not because of such incidents, but because of habit loss caused by human activity.

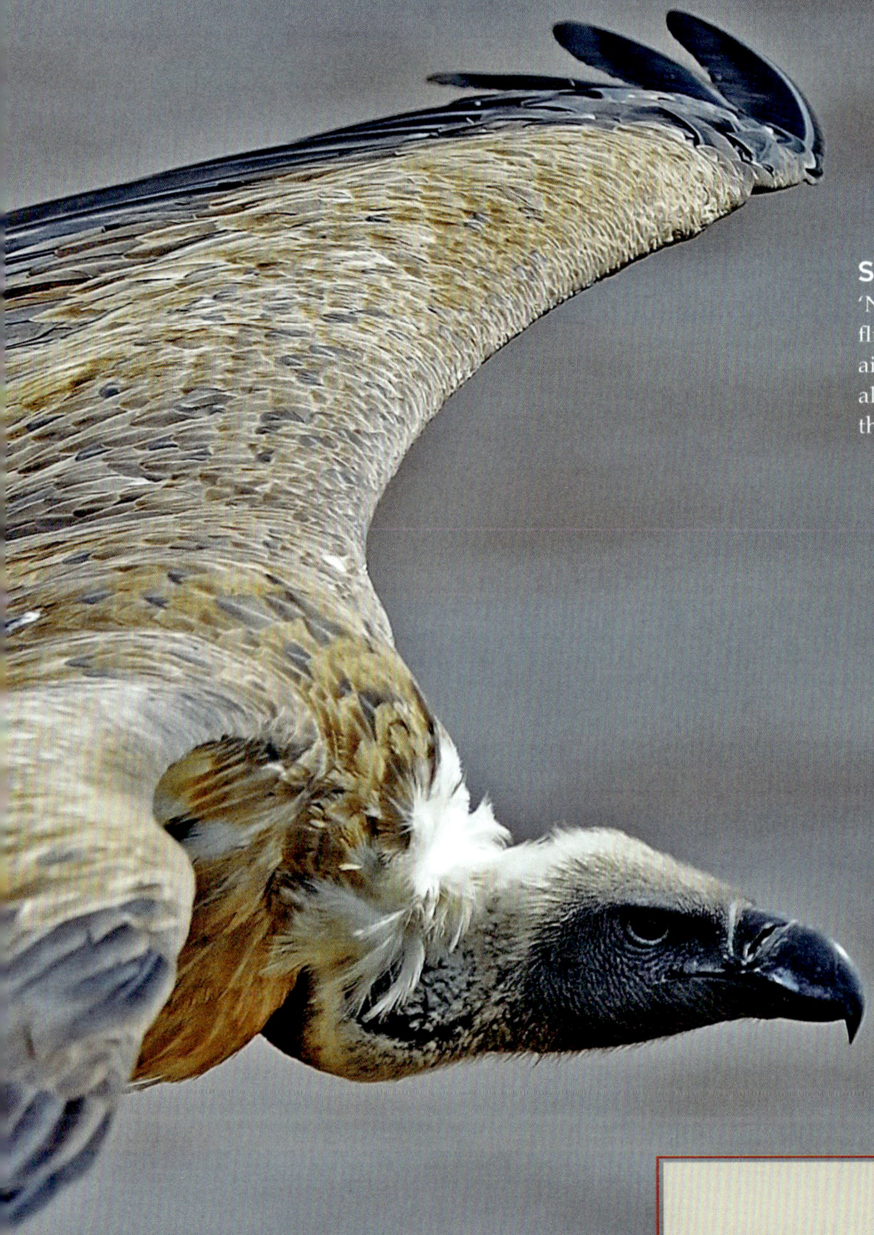

## SWIFTS

'Normal' birds also achieve outstanding feats – like the swift, which flies at altitudes of 2,000 m (6,562 ft). Moreover, the swift stays in the air for longest: it even sleeps in the air and allows itself to be carried along on the thermals. It generally spends over a month at a time in the air; it only comes down to the ground for the breeding season.

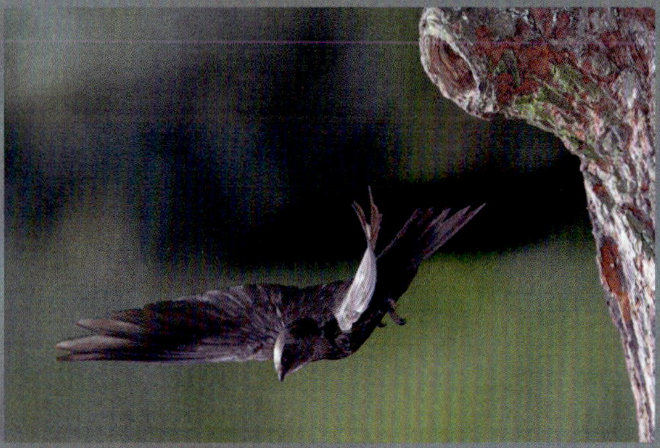

DID YOU KNOW?

### Collisions with aircraft

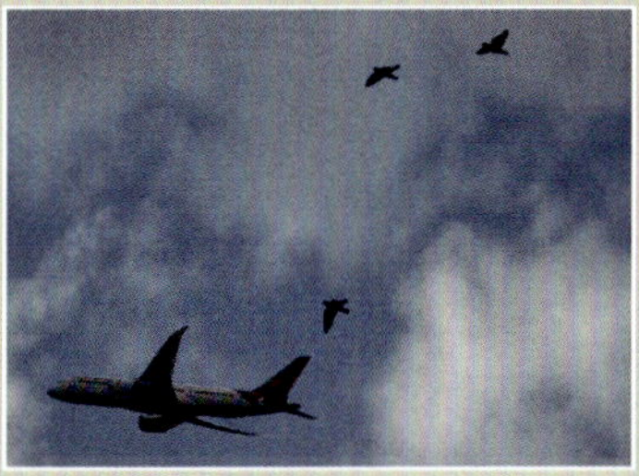

A loud crack and then nothing works. In an aircraft, this scenario is a nightmare which should never happen. But, in the past, this kind of situation used to occur repeatedly. One unforgettable instance is the Airbus that had to make an emergency water landing on the Hudson River in New York. A flock of Canada geese had strayed into the vicinity of the plane and caused the engine to break down. Collisions with flocks of birds are a frequent cause of accidents in flying – primarily when taking off or landing but rarely at higher altitudes. Although the engineering and technology is constantly improving, this problem is something that cannot be prevented. The geese population is growing in North America and Canada, and airports would like to drive them away from the danger areas with deterrents and targeted hunting. This is particularly due to the fact that, at the low altitudes at which most of the collisions occur, there is less time to prepare, making an emergency landing more difficult.

*One Rueppell's griffon that had already been tagged is seen here in the air over Masai Mara National Park (large picture).*

*Geese are amazing not just because of the heights and distances they fly, but also because of the formations they fly in (top left).*

*Swifts generally nest on buildings. But sometimes they make do elsewhere, such as here, where a tree has become their breeding place (top right).*

# FLIGHT ALTITUDES

The aircraft with the highest flight altitudes

## LOCKHEED U-2

When a U-2 was shot down by the Soviet Union in 1960, the whole world learned of its mission. The spy plane can climb to an altitude of 25,900 m (84,974 ft), and, for a long time, the opponents' air force could not reach it.

## MIG-25

Record versions of the MiG-25, the Russian interceptor fighter and scout plane, established a whole series of world records. It still holds the altitude record for a non-rocket-propelled aircraft – 37,650 m (123,523 ft) – although only on a parabolic flight.

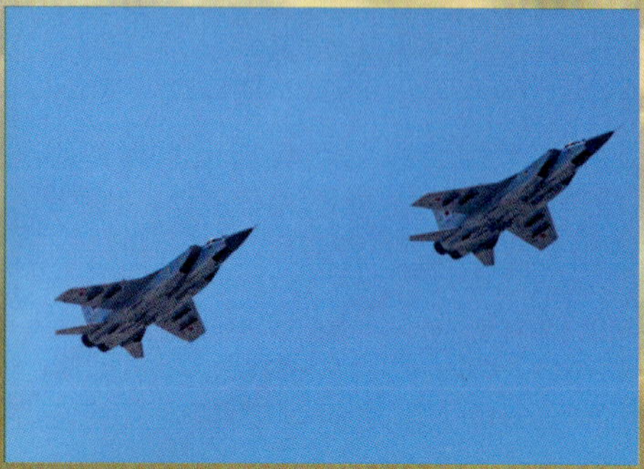

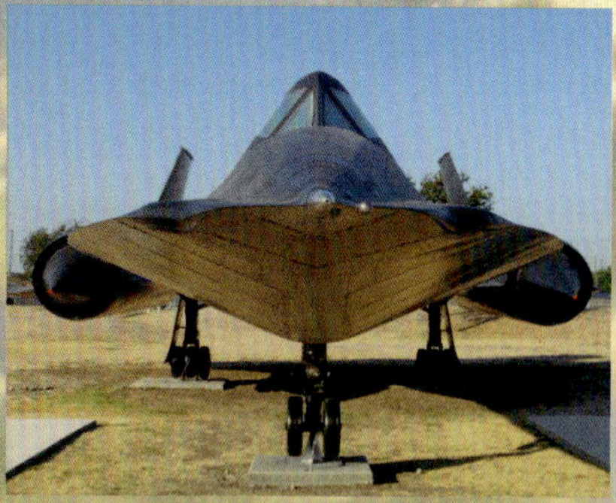

## LOCKHEED SR-71A

Even from its exterior, this strategic reconnaissance aircraft, in service from 1966, can be recognized as a high-end aircraft. Its flowing, undulating forms, absence of right angles, black paint and two vertical tails angled inwards are all 'stealth fighter' features to avoid detection. And its flight performance also packs a punch: the Blackbird is the fastest aircraft without rocket engines in the world, both with and without cargo. Its top speed record is 3,529 km/h (2,193 mph). And all this at almost unachievable altitudes: in level flight, this high flyer reached 26,213 m (86,000 ft), which is also a world record.

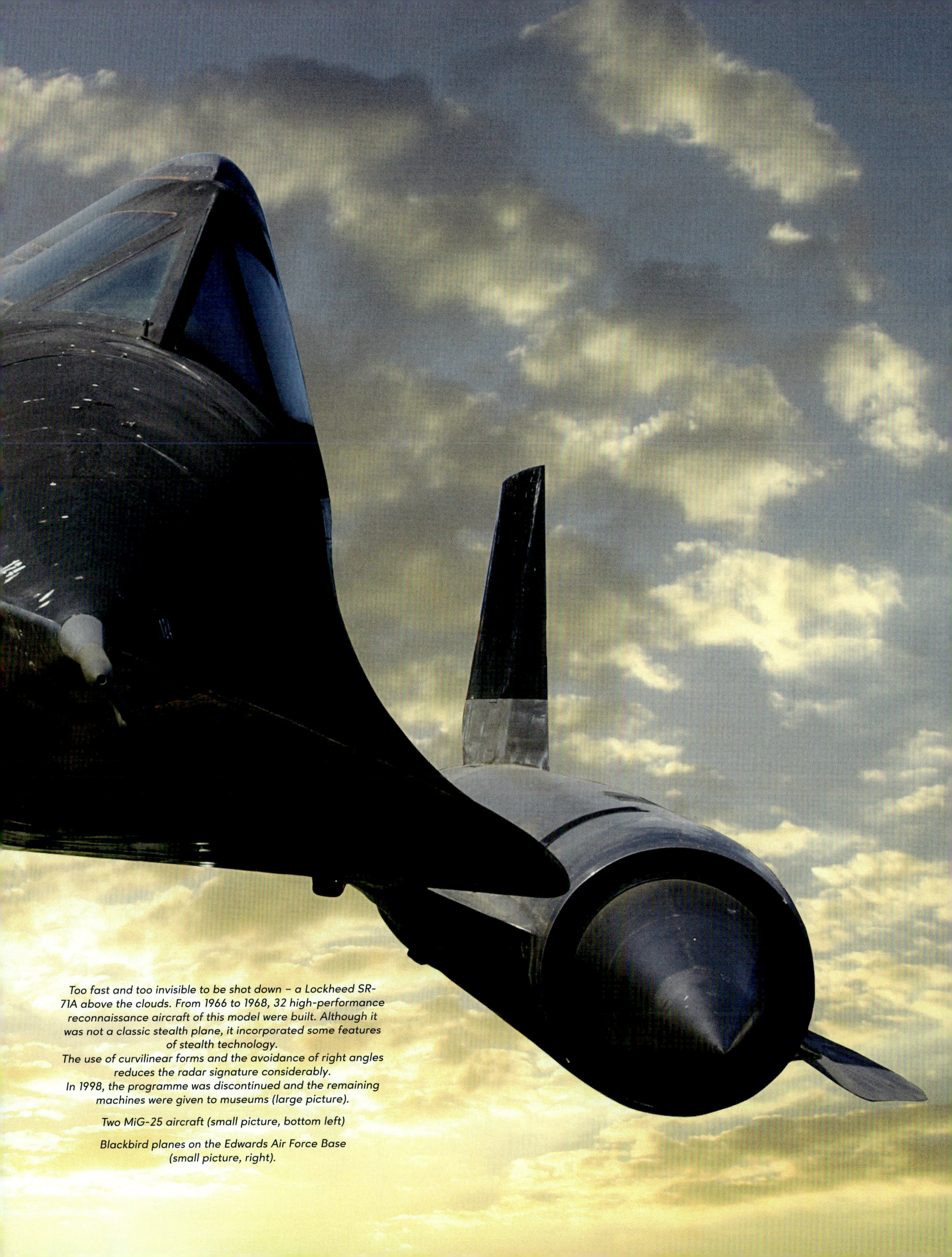

Too fast and too invisible to be shot down – a Lockheed SR-71A above the clouds. From 1966 to 1968, 32 high-performance reconnaissance aircraft of this model were built. Although it was not a classic stealth plane, it incorporated some features of stealth technology.

The use of curvilinear forms and the avoidance of right angles reduces the radar signature considerably.

In 1998, the programme was discontinued and the remaining machines were given to museums (large picture).

Two MiG-25 aircraft (small picture, bottom left)

Blackbird planes on the Edwards Air Force Base (small picture, right).

PEREGRINE FALCON
Maximum speed: 300 km/h (186 mph)
In a dive

SWIFTS
Top speed: 111 km/h (69 mph)

BAT
Top speed: 160 km/h (99 mph)

CONCORDE
Maximum speed: 2,179 km/h (1,354 mph)
First flight: 2 March 1969
Final flight: 24 October 2003

LOCKHEED SR-71 BLACKBIRD
3,529 km/h (2,193 mph)

DID YOU KNOW?

**Mach is a speed designation**

Ernst Mach was an Austrian physicist and philosopher, and he proved the Doppler Effect – and, moreover, his name was immortalized in the Mach number. What information does this provide? In general terms, it gives the speed in relation to the speed of sound for an aircraft, for example. So the Mach number is not an absolute number, such as 10 km per hour, but a ratio value. It is therefore more complicated because the speed of sound fluctuates according to the temperature of the surrounding air. Around freezing point (0°C (32°F)), it is 1,193 km (741.3 mi) per hour, at 100°C (212°F) however, it is 1,394 km (866.2 mi) per hour. According to this, a plane at Mach 1, with an air temperature of around 0°C (32°F) at an altitude of 1,000 m (3,280 ft), travels 1,193 km (741.3 mi) per hour. At a height of 10 km (6.21 mi), with an air temperature of -50°C (-58°F), however, it will only cover 1,076 km (669 mi) per hour. So Mach 1 describes the speed of sound.

**What is the sound or sonic barrier?**

A loud bang announces that a plane has broken the sound barrier. Of course, this is not a proper wall made of bricks that the plane collides with, but rather more of a physical concept. It occurs when a moving source of sound is so fast that it exceeds the speed of sound – and so is faster than its own compression waves, which then intensify at the sound barrier.

**Bell X-1: The first supersonic aircraft in the world**

On 14 October 1947, a special aircraft on a very special mission was launched at an altitude of 7,000 m (22,965 ft) from the missile bay of a B-29 plane. Charles 'Chuck' Yeager, the Bell X-1 pilot, lit the combustion chamber of the rocket engine (powered with methyl alcohol and oxygen) and, for the first time in the history of aviation, broke the sound barrier at an altitude of 13,000 m (42,650 ft) (1,127 km/h (700.28 mph) = Mach 1.06). In 1953, the advanced version, the X-1A, exceeded Mach 2 for the first time.

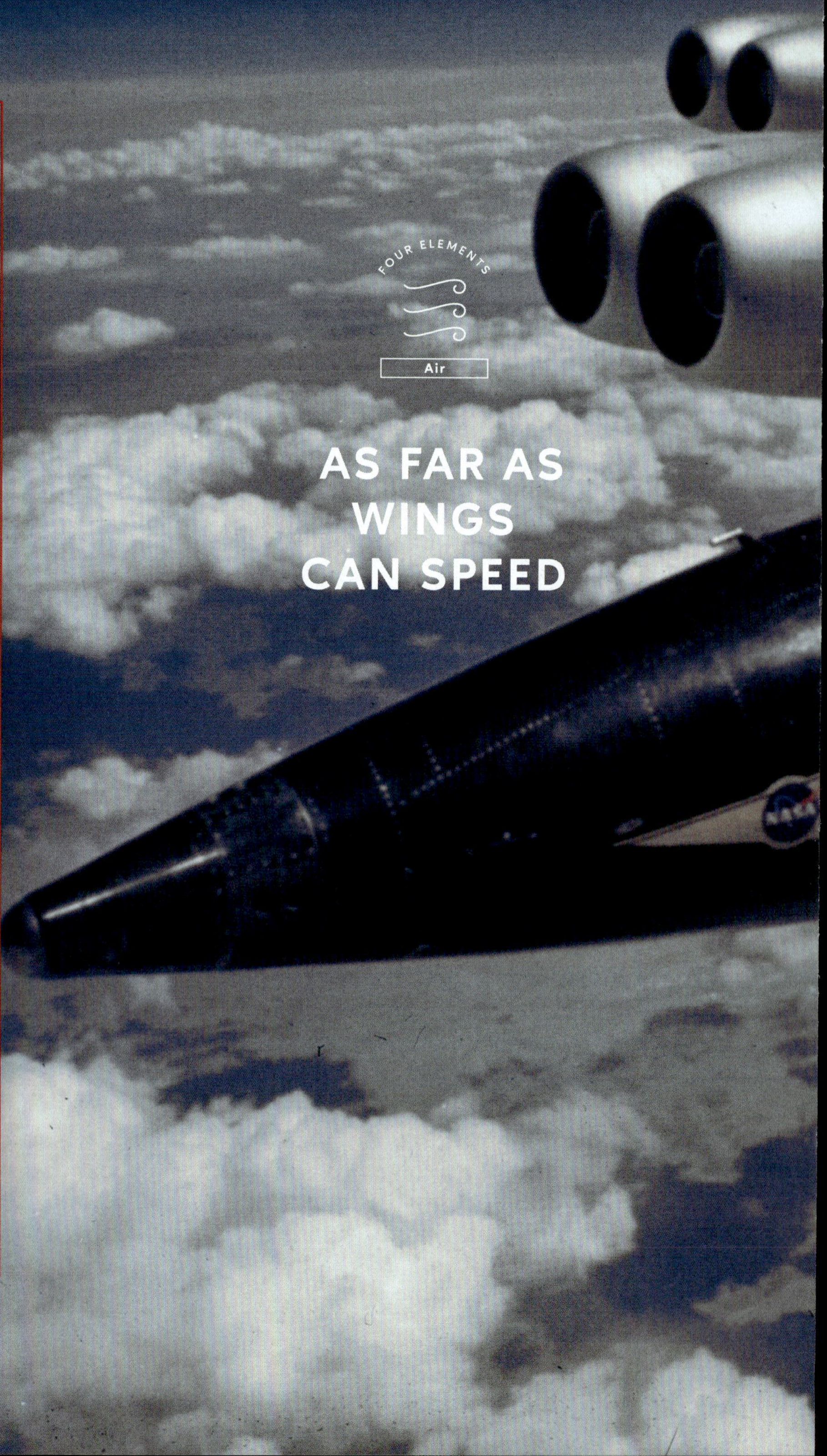

FOUR ELEMENTS
Air

# AS FAR AS WINGS CAN SPEED

5,000 km/h
(3,106 mph)

6,000 km/h
(3,728 mph)

7,000 km/h
(4,350 mph)

X-15

Top speed record: 7,274 km/h ( 4,520 mph)

First test flight: 10 March 1959

Final flight: 24 October 1968

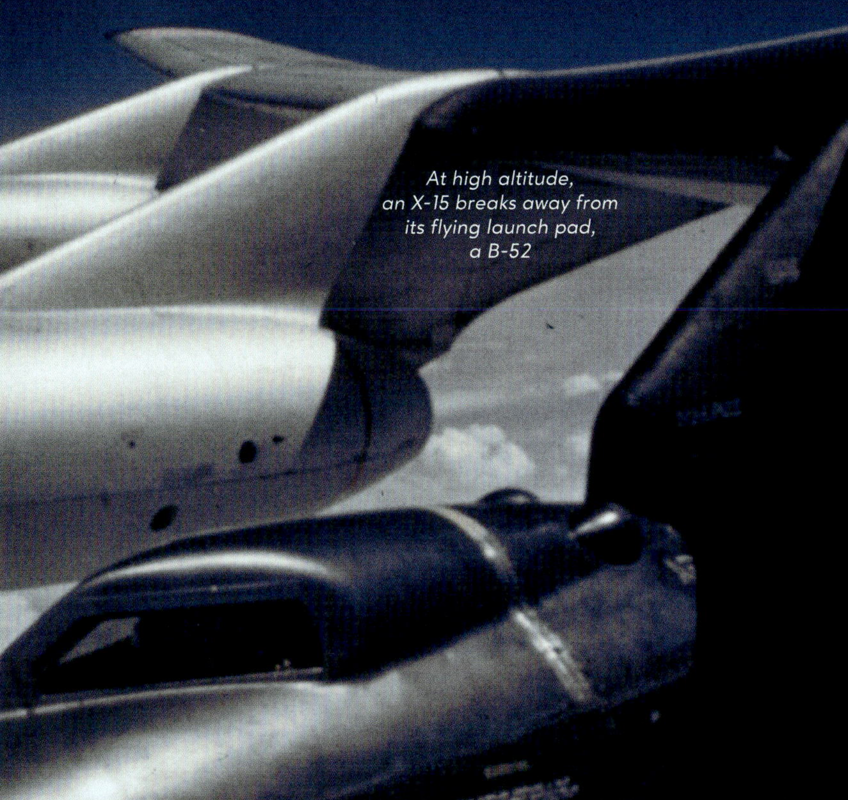

*At high altitude, an X-15 breaks away from its flying launch pad, a B-52*

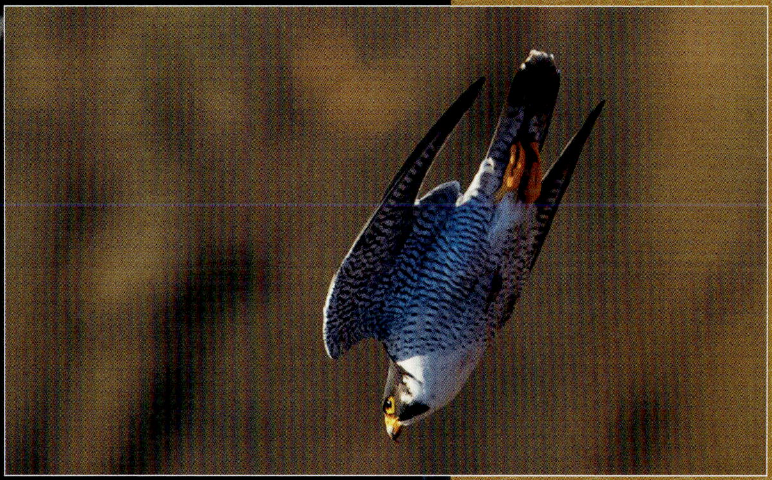

*The peregrine falcon's prey consists almost exclusively of small and medium-sized birds, which it catches in mid-air. This means that it cannot camouflage itself or sneak up on its prey – instead, it uses the circumstances to its advantage by using its speed to gain an element of surprise.*

## THE X-15 – THE FASTED MANNED AIRCRAFT IN THE WORLD

A plane with a pilot that travels at seven times the speed of sound and reaches altitudes of more than 100 km (62 mi)? That has actually been done! In 1959, one of three experimental North American Aviation X-15 aircraft took off on its first flight. The extreme planes were launched at 15,000 m (49,213 ft) from a B-52.
During 199 flights (until 1968), records of 7,274 km/h (4,520 mph) and altitudes of 107,960 m (354,200 ft) were reached.
To withstand the extreme forces and temperatures, advanced materials were used.

## THE PEREGRINE FALCON – THE FASTEST CREATURE ON EARTH

The peregrine falcon can be found on all the continents of the world, except for the Antarctic, making it the most widespread bird in the world. But this is not the only reason for its reputation as a record-breaker. It is the fact that it is the fastest creature on the planet. When it dives, it can reach speeds of over 320 km (199 mi) per hour. It aims for the prey flying below it, which can only escape by wheeling away, otherwise it is caught by the hunter. The force of the impact is generally enough to kill the prey.
Peregrine falcons used to nest mainly on cliffs until about 50 years ago, but, in recent decades, they have been seen nesting more frequently on buildings or in walls.

**DID YOU KNOW?**

**Peregrine falcons or pigeons**

But aren't pigeons faster? Because, if there was a competition, the peregrine falcon would have to fly horizontally to keep up and, in this manner, would soon reach its limits. It can only reach its record speeds when it dives to the ground from a great height. So the short-distance winner would definitely be the pigeon.

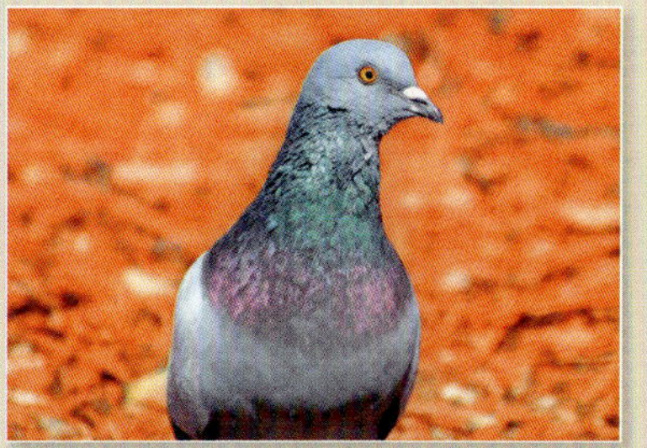

FOUR ELEMENTS

AIR

# BEE HUMMINGBIRDS

The smallest bird
in the world

The bee hummingbird from the Colibri family is the smallest bird in the world. The female bee hummingbird measures 7 cm (2.75 in) from the tip of its beak to the end of its tail; the male measures just 6.3 cm (2.5 in). It weighs approximately 2 g (0.07 oz).

It is only found on the Caribbean island of Cuba in three forest areas widely separated from each other. The female builds its nest from plant fibres; the male does not seem to play much of a role in the breeding process. The young hatch after two weeks; after three weeks, they are fully fledged. The bee hummingbird feeds on nectar and small insects that it finds in the flowers.

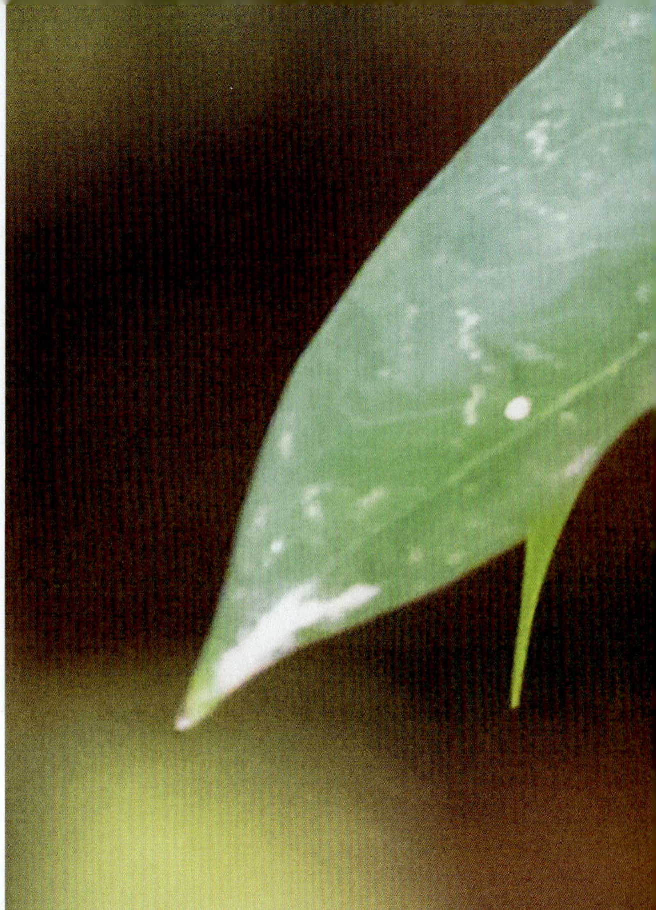

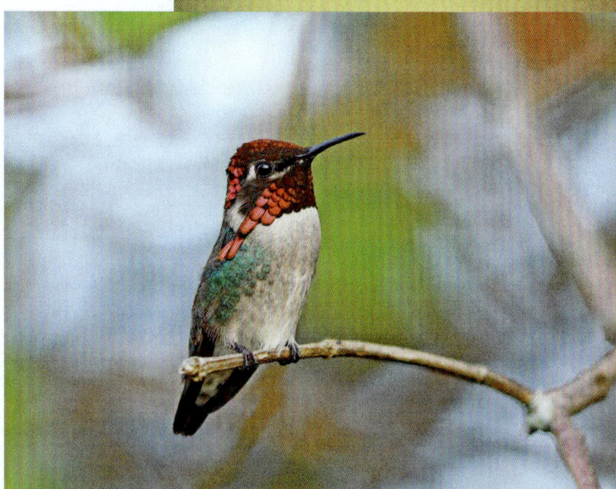

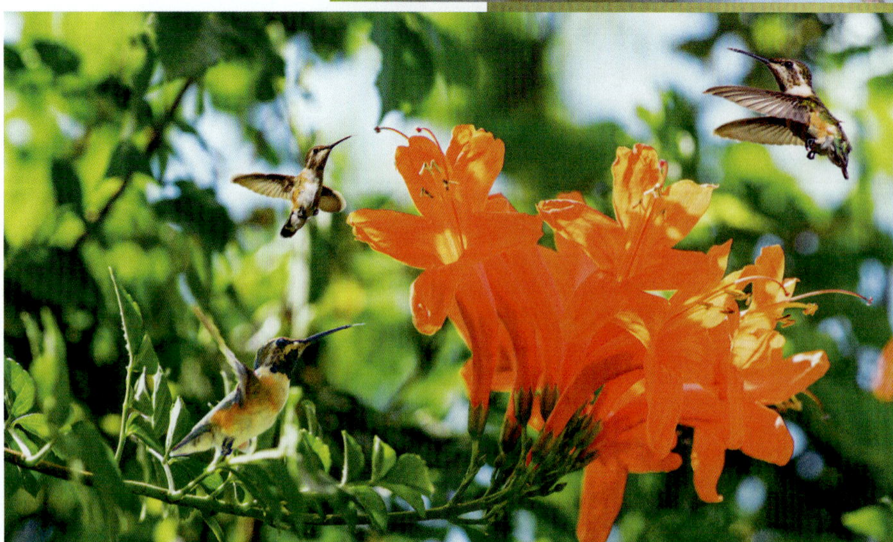

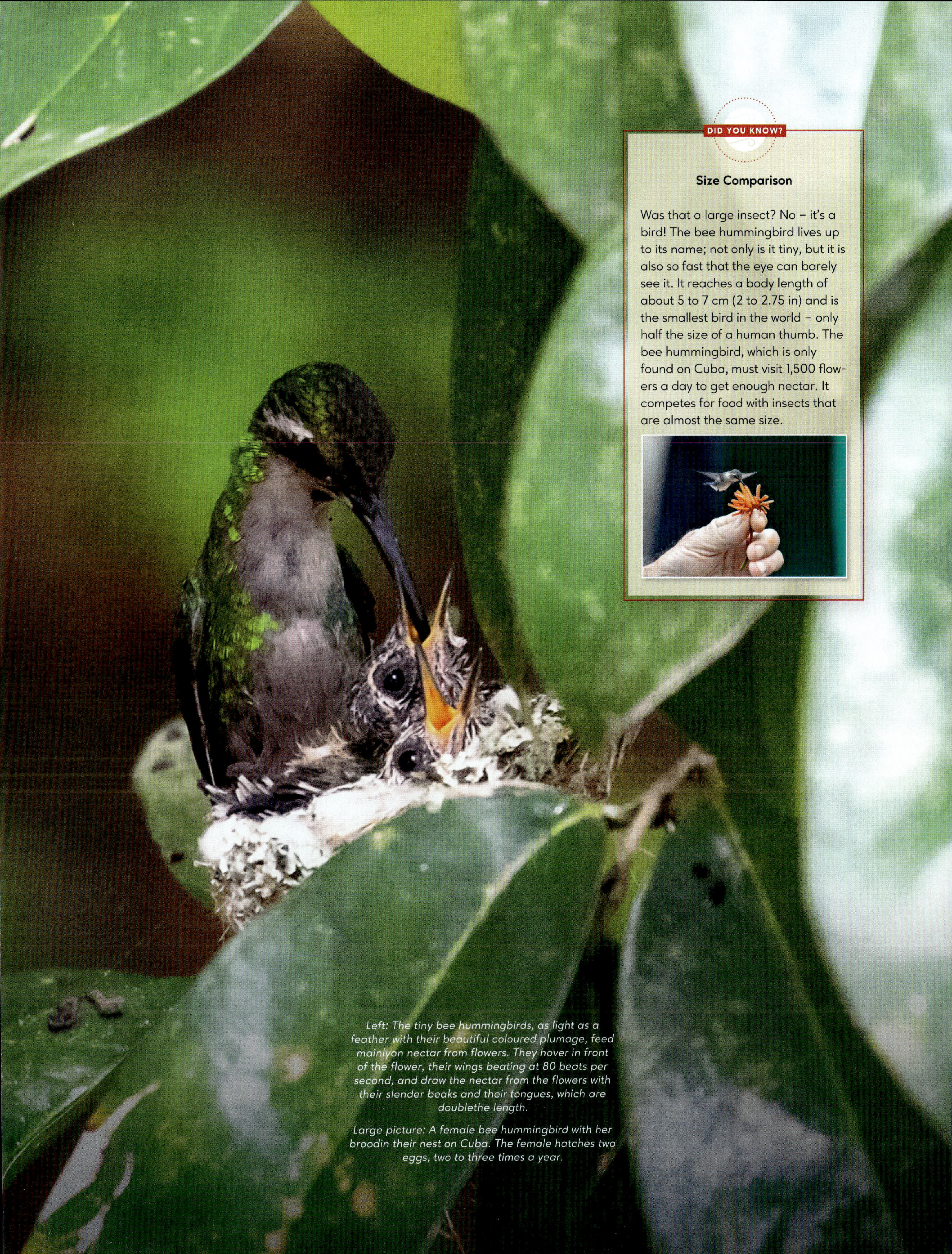

**DID YOU KNOW?**

### Size Comparison

Was that a large insect? No – it's a bird! The bee hummingbird lives up to its name; not only is it tiny, but it is also so fast that the eye can barely see it. It reaches a body length of about 5 to 7 cm (2 to 2.75 in) and is the smallest bird in the world – only half the size of a human thumb. The bee hummingbird, which is only found on Cuba, must visit 1,500 flowers a day to get enough nectar. It competes for food with insects that are almost the same size.

*Left: The tiny bee hummingbirds, as light as a feather with their beautiful coloured plumage, feed mainlyon nectar from flowers. They hover in front of the flower, their wings beating at 80 beats per second, and draw the nectar from the flowers with their slender beaks and their tongues, which are doublethe length.*

*Large picture: A female bee hummingbird with her broodin their nest on Cuba. The female hatches two eggs, two to three times a year.*

# INTERNATIONAL SPACE STATION

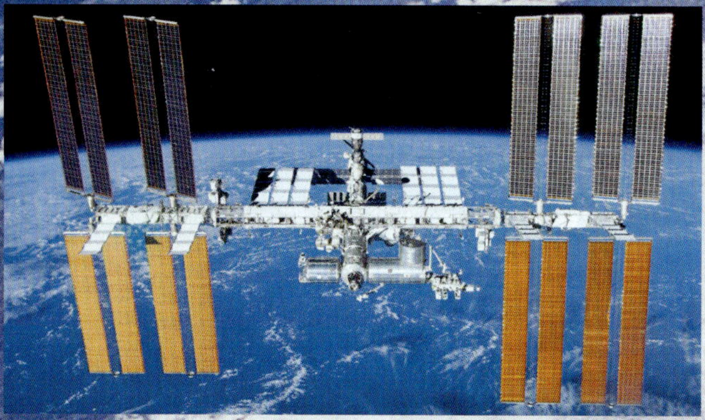

ISS – The largest space station in the world

The ISS is operated jointly by NASA and ESA as well as the space agencies of Russia, Canada and Japan. Starting in 1998, it has been systematically constructed from modules which Russian launcher rockets and space shuttles and, since 2008, unmanned spacecraft from ESA and Japan bring into the orbit. The ISS now weighs around 400 tonne (441 ton) and has a span of 100 m (328 ft). The space station has been manned since 2000. The crew are changed every six months. There were only three crew members initially, but, since 2009, there have been six. Regular flights by cargo rockets from Russia, Japan and Europe ensure they are well-supplied. The Soyuz space capsule stands ready in case of evacuation. Research is actively carried out on the ISS. The nations and organizations involved in the space station have set up laboratories where, among other activities, they observe activity on the sun, carry out tests with atomic clocks and measure cosmic radiation. The ISS will be in operation until at least 2024.

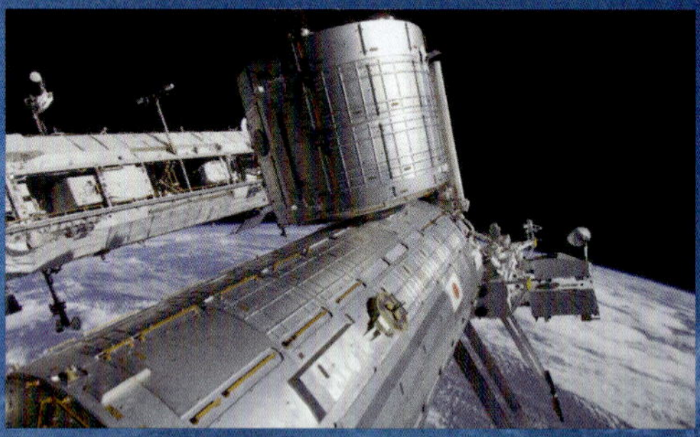

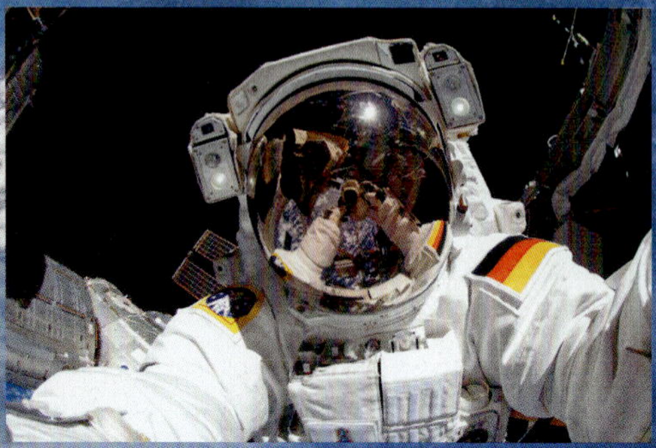

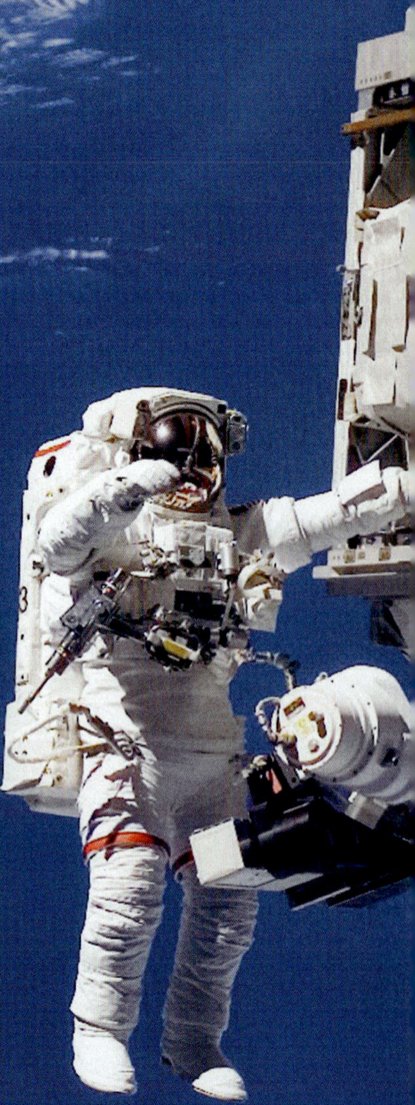

*The Japanese Experiment Module, nicknamed 'Kibo' (small picture, left).*

*One of the most wide-ranging missions in a manned space flight took place with the STS-116 in 2006. Robert Lee Curbeam and Christer Fuglesang took part in the activities outside the space station. Below them, the Pacific and the North and South Islands of New Zealand spread out (large picture).*

*For the crew of the STS-132, uncoupling means bidding farewell to the International Space Station (small picture, top).*

*To date, only eleven German astronauts have been in space. One of these is Alexander Gerst, seen here in his six-hour-long stint outside the space station in October 2014 (small picture left, bottom)*

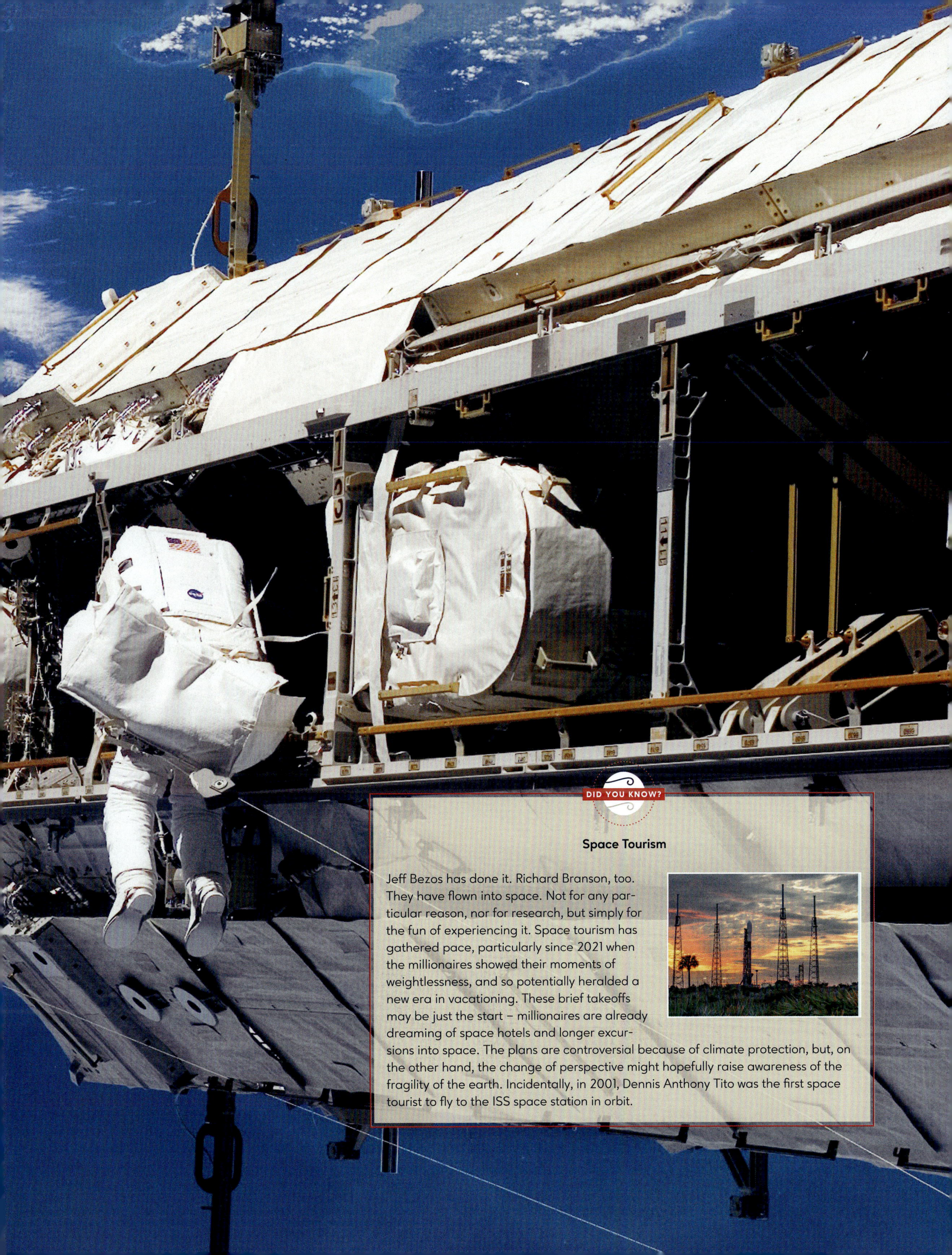

DID YOU KNOW?

### Space Tourism

Jeff Bezos has done it. Richard Branson, too. They have flown into space. Not for any particular reason, nor for research, but simply for the fun of experiencing it. Space tourism has gathered pace, particularly since 2021 when the millionaires showed their moments of weightlessness, and so potentially heralded a new era in vacationing. These brief takeoffs may be just the start – millionaires are already dreaming of space hotels and longer excursions into space. The plans are controversial because of climate protection, but, on the other hand, the change of perspective might hopefully raise awareness of the fragility of the earth. Incidentally, in 2001, Dennis Anthony Tito was the first space tourist to fly to the ISS space station in orbit.

FOUR ELEMENTS

AIR

# SKYSCRAPER

The tallest skyscrapers in the world

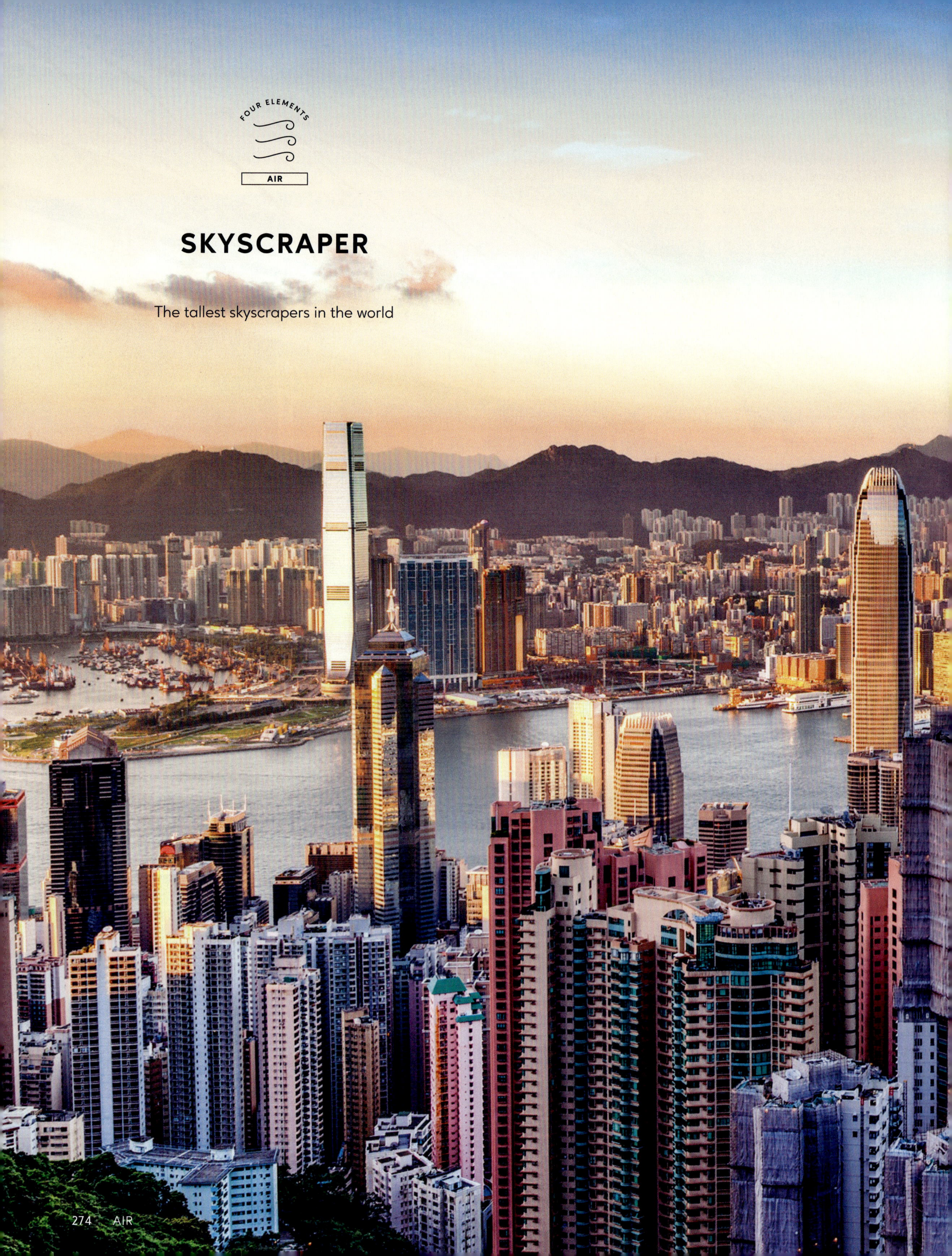

America, with its cities of Chicago and New York, was the centre for high-rise construction throughout the world for well over 100 years. The classic example is still the Empire State Building in New York, which had already reached a height of 381 m (1,250 ft) in 1931 and held the world record for more than 40 years. From 1972, it was outstripped for two years by the World Trade Center (417 m (1,368 ft)), which collapsed in 2001, as this was for 24 years by the Sears/ Willis Tower in Chicago (442 m (1,450 ft)). Since then, the contest to see who has 'the longest' tower has moved to Asia, where in booming metropolises such as Hong Kong, Shanghai and Dubai, whole skyscraper cities are shooting up, and where the current record holder, the Burj Khalifa in Dubai, towers 828 m (2,716 ft) above the ground. The height data refers to the structural height, i.e. spires are counted as part of the structure of the building (even if they are only used to 'artificially' increase the height), but not radio antennas.

*In Hong Kong, the many skyscrapers look out at each other across Victoria Harbour. From Victoria Peak, there is an especially beautiful view of the skyline.*

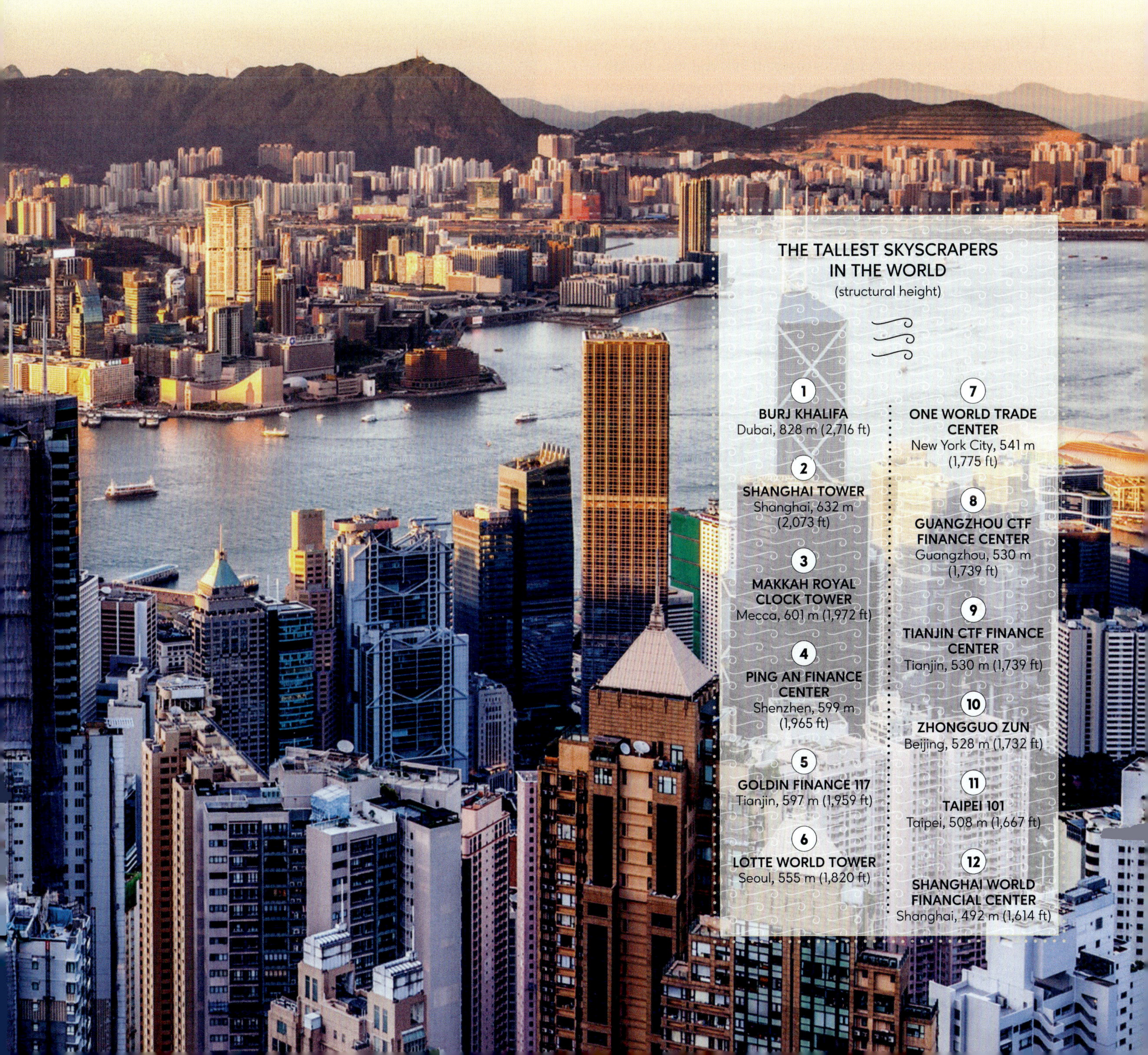

## THE TALLEST SKYSCRAPERS IN THE WORLD

(structural height)

1. **BURJ KHALIFA**
Dubai, 828 m (2,716 ft)
2. **SHANGHAI TOWER**
Shanghai, 632 m (2,073 ft)
3. **MAKKAH ROYAL CLOCK TOWER**
Mecca, 601 m (1,972 ft)
4. **PING AN FINANCE CENTER**
Shenzhen, 599 m (1,965 ft)
5. **GOLDIN FINANCE 117**
Tianjin, 597 m (1,959 ft)
6. **LOTTE WORLD TOWER**
Seoul, 555 m (1,820 ft)
7. **ONE WORLD TRADE CENTER**
New York City, 541 m (1,775 ft)
8. **GUANGZHOU CTF FINANCE CENTER**
Guangzhou, 530 m (1,739 ft)
9. **TIANJIN CTF FINANCE CENTER**
Tianjin, 530 m (1,739 ft)
10. **ZHONGGUO ZUN**
Beijing, 528 m (1,732 ft)
11. **TAIPEI 101**
Taipei, 508 m (1,667 ft)
12. **SHANGHAI WORLD FINANCIAL CENTER**
Shanghai, 492 m (1,614 ft)

FOUR ELEMENTS

AIR

# BURJ KHALIFA

## The tallest building in the world

The Burj Khalifa, which, until its official opening, was called the Burj Dubai, has been the tallest structure in the world since April 2008 – that is, two years before its completion – and also the tallest building to be used for residential purposes. It tops every other tower in the world and also has the most floors of any tower in the world (189). The construction work began in 2004; by January 2009, a structural height of 828 m (2,716 ft) had been reached (including signalling equipment, the overall height was 830 m (2,723 ft)). The official opening ceremony took place on 4 January 2010.

The different floors in the Burj Khalifa are served by a total of 57 elevators and 8 escalators. Two double-deck elevators lead to a viewing platform on the 124th floor that is 452 m (1,483 ft) high. With a speed of 10 m (33 ft) per second (36 km/h (22.37 mph)), they are the fastest double-deck elevators in the world. Furthermore, the building has the world's highest elevator stop (638 m (2,093 ft)). There are 2,909 steps leading from the ground floor to the 160th floor which is 600 m (1,968 ft) high. Up to this height, the building is a reinforced concrete structure; the rest of the central tower (above a height of 200 m (656 ft)) is a steel construction.

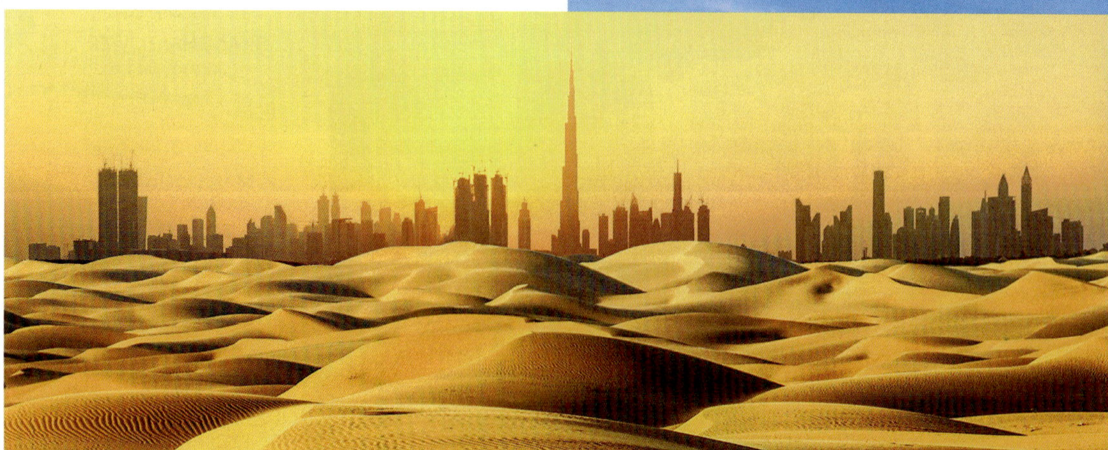

*The US-American, Adrian Smith, designed the architecture of the tallest building in the world. The entire city lies at the foot of the building (large picture).*

*And even from a distance, the unmistakable Burj Khalifa towers in the sky (small picture).*

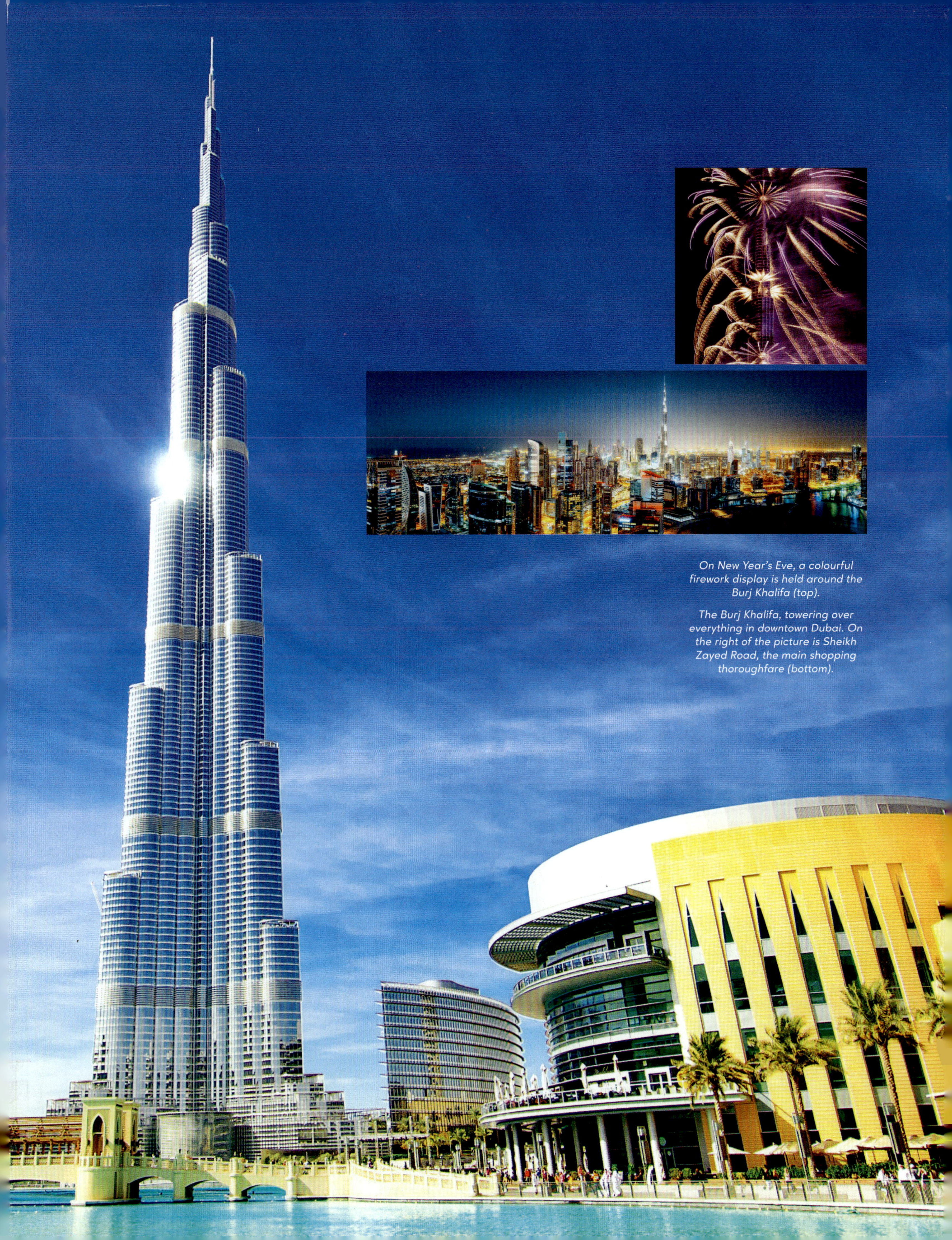

*On New Year's Eve, a colourful firework display is held around the Burj Khalifa (top).*

*The Burj Khalifa, towering over everything in downtown Dubai. On the right of the picture is Sheikh Zayed Road, the main shopping thoroughfare (bottom).*

# INDEX

# PHOTO CREDITS

C = Corbis, G = Getty, M = Mauritius

P. 2 M/Heidi & Hans-Juergen Koch; p. 2 Wead/Shutterstock.com; p. 2 G/Michele Falzone; p. 2 G/John Hay; p. 4 Look/Per-Andre Hoffmann; p. 4 M/Stefan Liebermann; p. 4 kavram/Shutterstock.com; p. 5 A/Jeff Rotman; p. 5 Willyam Bradberry/Shutterstock.com; p. 6 Josemar Franco/Shutterstock.com; p. 6 Nagel Photography/Shutterstock.com; p. 7 HelloRF Zcool/Shutterstock.com; p. 7 Minerva Studio/Shutterstock.com; p. 8–9 M/Stefan Liebermann; p. 10–11 Look/Per-Andre Hoffmann; p. 12 G/Cesar Hugo Storero; p. 12–13 lAguirreV/Shutterstock.com; p. 14–15 C/Macduff Everton; p. 15 Darryl Brooks/Shutterstock.com; p. 15 Wollertz/Shutterstock.com; p. 16 Look/Per-Andre Hoffmann; p. 16 C/Bettmann; p. 16–17 G/Toby Maudsley; p. 17 G/Michael Freeman; p. 17 C/Bettmann; p. 18 C/Michael p. Yamashita; p. 18 C/; p. 18–19 C/Kevin Schafer; p. 19 C/Pablo Corral Vega; p. 19 C/Historical; p. 19 G/Marc Adamus; p. 20 G/Panoramic Images; p. 20– 21 G/Art Wolfe; p. 21 Phillip B. Espinasse/Shutterstock.com; p. 21 G/Severns; p. 22–23 M/Pavel Filatov; p. 23 G/Pubbli Aer Foto; p. 24 Wead/Shutterstock.com; p. 24 G/Simeone Huber; p. 24–25 G/Art Wolfe; p. 25 Traveler17/Shutterstock.com; p. 26 M/Haraldur Stefansson; p. 26 G/Steve Taylor; p. 26–27 Bastien Schneider/Shutterstock.com; p. 27 M/FLPA; p. 27 M/Cavan Images; p. 28 C/Roger Wood; p. 28–29 C/Ren Zhenglai; p. 29 M/Martin Westlake; p. 30 Smelov/Shutterstock.com; p. 30 Vera KL82/Shutterstock.com; p. 30–31 yykkaa/Shutterstock.com; p. 31 Sergey Krasnoshchokov/Shutterstock.com; p. 31 GTW/Shutterstock.com; p. 31 Yuri-D3/Shutterstock.com; p. 32 C/Peter Guttman; p. 32–33 Pav-Pro Photography Ltd/Shutterstock.com; p. 34 C/Richard T. Nowitz; p. 34–35 M/David Keith Jones; p. 35 A/The Africa Image Library; p. 35 A/The Africa Image Library; p. 36 Look/Hauke Dressler; p. 36–37 AndreyFilippov.com/Shutterstock.com; p. 38 C/Fred Hirschmann; p. 38–39 G/Martin Gray; p. 39 Alamy/Jon Arnold; p. 40 G/Bob Stefko; p. 40–41 G/Charles O'Rear; p. 42 M/Tiara Anggamulia; p. 42 M/Harald von Radebrecht; p. 42–43 M/NASA; p. 43 Matyas Rehak/Shutterstock.com; p. 44 Taromon/Shutterstock.com; p. 44 Joaquin Corbalan P/Shutterstock.com; p. 44 Maximillian cabinet/Shutterstock.com; p. 44–45 M/Thomas Ebelt; p. 46 Lijuan Guo/Shutterstock.com; p. 46 Japan Explorers/Shutterstock.com; p. 46–47 norbertv/Shutterstock.com; p. 48 M/Science Source; p. 48–49 Withan Tor/Shutterstock.com; p. 49 xtock/Shutterstock.com; p. 50 Vadim Sadovski/Shutterstock.com; p. 51 M/Massimo Dallaglio; p. 51 Zita/Shutterstock.com; p. 52 marcobrivio.photo/Shutterstock.com; p. 52–53 kavram/Shutterstock.com; p. 54 M/Alamy ; p. 54–55 Patrick Poendl/Shutterstock.com; p. 55 C/Jean du Boisberranger; p. 55 G/Frans Lemmens; p. 55 G/Frans Lemmens; p. 56 aleksander hunta/Shutterstock.com; p. 56 J_K/Shutterstock.com; p. 56 Gonzalo Buzonni/Shutterstock.com; p. 56–57 Kokhanchikov/Shutterstock.com; p. 58 G/Travel Ink; p. 58 G/Travel Ink; p. 58–59 Alamy; p. 60 Dietmar Assmann/Shutterstock.com; p. 61 G/Michael Schwab; p. 62 C/Massimo Borchi; p. 62–63 G/Gordon Wiltsie; p. 63 G/Bob Krist; p. 63 Mohammad.Nouri/Shutterstock.com; p. 64 G/Siegfried Layda; p. 64 G/Jeff Foott; p. 64–65 G/Don Smith; p. 65 G/Jeff Foott; p. 66 M/Clement Philippe; p. 66–67 G/Enrique Aguirre; p. 67 G/Steve Granitz; p. 67 M/Lisa Werner; p. 68 G/Ted Mead; p. 68–69 G/Ted Mead; p. 70–71 Willyam Bradberry/Shutterstock.com; p. 72 Willyam Bradberry/Shutterstock.com; p. 72–73 M/Jeff Rotman; p. 74 A/ClassicStock; p. 74 G/Michael Melford; p. 74 C/Carson Ganci; p. 74–75 BlueOrange Studio/Shutterstock.com; p. 75 nicolasvoisin44/Shutterstock.com; p. 76 Ethan Daniels/Shutterstock.com; p. 76–77 M/Solvin Zankl; p. 77 Sergey Kohl/Shutterstock.com; p. 78 M/Mark Conlin; p. 78 nicolasvoisin44/Shutterstock.com; p. 78 Bangtalay/Shutterstock.com; p. 78 Kristina Vackova/Shutterstock.com; p. 79 Huber/Giovanni; p. 79 Geospace/EDC ; p. 80 C/Gavin Hellier; p. 80–81 Martin M303/Shutterstock.com; p. 81 Mia2you/Shutterstock.com; p. 82–83 elRoce/Shutterstock.com; p. 84 G/WPA Pool ; p. 84 G/Frederic Pacorel; p. 84–85 Altug Galip/Shutterstock.com; p. 86 Mariusz Potocki/Shutterstock.com; p. 86–87 axily/Shutterstock.com; p. 87 G/Sue Flood; p. 88–89 G/Johnny Johnson; p. 89 G/Paul Oomen; p. 90–91 G/Per Breiehagen; p. 91 C/Bettmann; p. 91 C/Bettmann; p. 91 C/Bettmann; p. 91 G/Bettmann; p. 91 C/Bettmann; p. 92 Earth theater/Shutterstock.com; p. 92 Justin Berken/Shutterstock.com; p. 93 Alamy/Stock Connection; p. 93 C/Amos Nachoum; p. 93 M/Franco Banfi; p. 93 M/digitalunderwater.com; p. 94 M/Andrea Izzotti; p. 94–95 G/Jurij Cerednicenko; p. 96 C/Theo Allofs; p. 96 Tatiana Ivkovich/Shutterstock.com; p. 96–97 G/Keith Bradley; p. 98 G/China Tourism Press; p. 98–99 zhao dongfang/Shutterstock.com; p. 100 Ewa Studio/Shutterstock.com; p. 100–101 M/Mediacolor's; p. 101 Lisa-S/Shutterstock.com; p. 102 canon_photographer/Shutterstock.com; p. 102 canon_photographer/Shutterstock.com; p. 103 M/Bailey-Cooper Photography; p. 103 Aldo91/Shutterstock.com; p. 104 C/Skyscan; p. 104–105 M/NASA; p. 105 G/Paul Souders; p. 106–107 Jaka Zvan/Shutterstock.com; p. 107 G/Tim Graham; p. 107 C/Jeffrey L. Rotman; p. 108 M/Fabio Nodari; p. 108 Ivan Kazachenko/Shutterstock.com; p. 108–109 M/Prasit Rodphan; p. 109 Martchan/Shutterstock.com; p. 110–111 M/Manish Lakhani; p. 111 C/Blaine Harrington III; p. 111 C/Alison Wright; p. 111 Prabhjit p. Kalsi/Shutterstock.com; p. 112 M/Wayne Lynch; p. 112 G/Yves Marcoux; p. 112–113 f11photo/Shutterstock.com; p. 113 G/Panoramic Images; p. 114–115 Eshma/Shutterstock.com; p. 115 C/Christophe Boisvieux; p. 115 H. & D. Zielske; p. 115 Laif/Manfred Linke; p. 116 Look/Don Fuchs; p. 116–117 G/John Hay; p. 117 G/Michael Gebicki; p. 118 tonyz20/Shutterstock.com; p. 118 Alchemist777/Shutterstock.com; p. 118–119 David Havel/Shutterstock.com; p. 119 Catchlight Lens/Shutterstock.com; p. 120 Shcherbina Valeriy/Shutterstock.com; p. 120–121 R.M. Nunes/Shutterstock.com; p. 122 G/Ralph Lee Hopkins; p. 122–123 A/Kevin Schafer; p. 123 Vadim Petrakov/Shutterstock.com; p. 124 TRphotos/Shutterstock.com; p. 124 Wikimedia Commons/HansaParkVideos (https://commons.wikimedia.org/wiki/File:Highlander_(Hansa-Park).jpg), https://creativecommons.org/licenses/by-sa/4.0/legalcode; p. 124 Pit Stock/Shutterstock.com; p. 124 M/AC Images; p. 124–125 davide bonaldo/Shutterstock.com; p. 126 A/Peter M. Wilson; p. 126–127 G/Kevin Schafer; p. 128 G/Panoramic Images; p. 128–129 sunsinger/Shutterstock.com; p. 129 Procy/Shutterstock.com; p. 130 Laif/Emmler; p. 130 Laif/Emmler; p. 130–131 Robin Runck/Shutterstock.com; p. 132 G/Christian Hoehn; p. 132–133 GeGiGoggle/Shutterstock.com; p. 133 G/Art Wolfe; p. 134 Johan Knelsen/Shutterstock.com; p. 134 Hue Chee Kong/Shutterstock.com; p. 134–135 TRphotos/Shutterstock.com; p. 135 Linda Harms/Shutterstock.com; p. 135 anderm/Shutterstock.com; p. 136–137 Daniel Bartos/Shutterstock.com; p. 137 Smit/Shutterstock.com; p. 137 A/dfwalls; p. 138 A/View Stock; p. 138 Evgenii Emelianov/Shutterstock.com; p. 138–139 RudiErnst/Shutterstock.com; p. 140 THP Creative/Shutterstock.com; p. 140–141 Piu_Piu/Shutterstock.com; p. 141 A/Radius Images; p. 141 C/Nik Wheeler; p. 142–143 A/Phil Degginger; p. 144 M/Konrad Wothe; p. 144 M/Luis Leamus; p. 144–145 M/Peace portal photo; p. 145 M/Udo Bernhart; p. 146 M/Pilipenko D; p. 146–147 M/B.A.E. Inc.; p. 148 A/ITAR-TASS Photo Agency; p. 148–149 G/Per Breiehagen; p. 149 G/Bettmann; p. 150–151 Josemar Franco/Shutterstock.com; p. 152 Laif/Christian Heeb; p. 152–153 Nagel Photography/Shutterstock.com; p. 154–155 Laif/Berthold Steinhilber; p. 155 ltdedigos/Shutterstock.com; p. 156–157 G/Jon Warburton Lee; p. 157 Lucky-photographer/Shutterstock.com; p. 158–159 G/Ron Watts; p. 159 Simon Dannhauer/Shutterstock.com; p. 160–161 Josemaria Toscano/Shutterstock.com; p. 162 buladeviagens/Shutterstock.com; p. 162–163 cge2010/Shutterstock.com; p. 164–165 Mesa Studios/Shutterstock.com; p. 165 C/H.R. Bowers; p. 165 Sean M Smith/Shutterstock.com; p. 165 A/North Wind Picture Archives; p. 166–167 G/Jochen Schlenker; p. 167 Patryk Kosmider/Shutterstock.com; p. 168–169 NaughtyNut/Shutterstock.com; p. 169 Maurizio De Mattei/Shutterstock.com; p. 170 C/A. L. Murat; p. 170 C/Bettmann; p. 170–171 G/James Balog; p. 171 G/Marc Solomon; p. 171 M/Brand X Pictures; p. 171 C/Liu Jin; p. 171 G/Johansen Krause; p. 172 G/Arctic-Images; p. 172–173 Tomas Pavelka/Shutterstock.com; p. 174 G/Peter Adams; p. 174 G/Tyler Stableford; p. 174–175 M/Steve Allen Travel Photography; p. 176 RUBEN M RAMOS/Shutterstock.com; p. 176–177 Ed Dods/Shutterstock.com; p. 177 GTW/Shutterstock.com; p. 177 Jeff Amantea/Shutterstock.com; p. 177 RUBEN M RAMOS/Shutterstock.com; p. 178 G/John Lawrence; p. 178 G/Sylvain Sonnet; p. 178 Look/; p. 178–179 G/Jake Wyman; p. 179 A/Mikael Utterstrom; p. 179 PhotoVisions/Shutterstock.com; p. 180 G/Peter Adams; p. 180–181 Dennis van de Water/Shutterstock.com; p. 181 G/Ted Mead; p. 182 Stephan Hawks/Shutterstock.com; p. 182–183 C/; p. 183 G/Darryl Torckler; p. 184 M/Danita Delimont/Howie Garber; p. 184 Sophia Granchinho/Shutterstock.com; p. 184–185 Jef Wodniack/Shutterstock.com; p. 186 C/Roger Tidman; p. 186 G/Steve Kaufman; p. 186–187 M/Roland Knauer; p. 188 G/Paul Nicklen; p. 188 G/DC Productions; p. 188–189 M/Thomas Sbampato; p. 190–191 Christopher Meder/Shutterstock.com; p. 191 pticelov/Shutterstock.com; p. 192 G/Macduff Everton; p. 192–193 Vadim Petrakov/Shutterstock.com; p. 194 Nata Naumovec/Shutterstock.com; p. 194 Vladimir Kovalchuk/Shutterstock.com; p. 194 Phuketian.S/Shutterstock.com; p. 194–195 Leonid Ikan/Shutterstock.com; p. 196 Erik Mandre/Shutterstock.com; p. 196 G/Roy Toft; p. 196 Vaclav Sebek/Shutterstock.com; p. 196–197 Vaclav Sebek/Shutterstock.com; p. 198 Rob Christiaans/Shutterstock.com; p. 198–199 Frank Wortmann/Shutterstock.com; p. 199 G/Charles Krebs; p. 200 A/Juniors Bildarchiv; p. 200–201 G/Ronald Wittek; p. 202–203 G/Keisuke Iwamoto; p. 203 A/View Stock; p. 203 G/Hitoshi Osawa; p. 204 G/Nature Expressions; p. 204–205 M/Steve Bloom; p. 205 Mark Green/Shutterstock.com; p. 205 rayints/Shutterstock.com; p. 205 M/Tim Whitby; p. 205 Jakob Fischer/Shutterstock.com; p. 205 Look/Konrad Wothe; p. 206 M/Gary Crabbe; p. 206–207 Dave Allen Photography/Shutterstock.com; p. 208 Andi111/Shutterstock.com; p. 208 MrForever/Shutterstock.com; p. 208–209 Greg Brave/Shutterstock.com; p. 209 Kathie Nichols/Shutterstock.com; p. 210 A/Fabian Gonzales; p. 210–211 G/Karin Slade; p. 211 Yaping/Shutterstock.com; p. 212 Uwe Bergwitz/Shutterstock.com; p. 212 G/Hans Strand; p. 212–213 sma1050/Shutterstock.com; p. 213 A/Christophe Viseux; p. 214 BEJITA/Shutterstock.com; p. 214–215 Artush/Shutterstock.com; p. 216 2009fotofriends/Shutterstock.com; p. 216–217 Alamy/Thomas Hallstein; p. 217 Guaxinim/Shutterstock.com; p. 218–219 BorneoRimbawan/Shutterstock.com; p. 219 Dirk Ercken/Shutterstock.com; p. 220 G/Tim Graham; p. 220–221 huang jenhung/Shutterstock.com; p. 222–223 A/Robert Harding ; p. 223 G/Guido Cozzi; p. 223 G/Mattias Klum; p. 223 Eva Mont/Shutterstock.com; p. 224 Francisco Blanco/Shutterstock.com; p. 224 Richard Whitcombe/Shutterstock.com; p. 224–225 lavizzara/Shutterstock.com; p. 226 G/James Warwick; p. 226 G/Richard Packwood; p. 226 G/P. Kumar; p. 226–227 M/WorldFoto; p. 227 lukaszemanphoto/Shutterstock.com; p. 227 M/Ron Steiner; p. 228 AyhanTuranMenekay/Shutterstock.com; p. 228–229 G/Fadil Aziz; p. 229 Laif/Hoa-Qui; p. 229 AIHA 133/Shutterstock.com; p. 230 Katie Stevens Photography/Shutterstock.com; p. 230–231 AustralianCamera/Shutterstock.com; p. 231 M/Jürgen & Christine Sohns; p. 232 G/Michael Nichols; p. 232–233 M/Kevin Schafer; p. 233 Onyx9/Shutterstock.com; p. 233 G/Ian Nichols; p. 234 M/Jan Csernoch; p. 234 Dmitry Laudin/Shutterstock.com; p. 234 M/Eyal Bartov; p. 234 Filip Fuxa/Shutterstock.com; p. 234–235 AG Technology Solutions/Shutterstock.com; p. 235 Bruno Martins Imagens/Shutterstock.com; p. 236 M/AA World Travel Library; p. 236–237 G/Ray Massey; p. 238 A/Banana Pancake; p. 238 riekephotos/Shutterstock.com; p. 238–239 Kevin Wells Photography/Shutterstock.com; p. 240 G/P. Jaccod; p. 240–241 FMPortella/Shutterstock.com; p. 241 A/Photoshot Holdings Ltd; p. 242 Marcel Brekelmans/Shutterstock.com; p. 242 G/Anup Shah; p. 243 photoyh/Shutterstock.com; p. 243 Petr Toman/Shutterstock.com; p. 244 Wikimedia Commons/PM 93/https://de.wikipedia.org/wiki/Datei:Bugatti_Veyron_16.4_%E2%80%93_Innenraum_(1),_5._April_2012,_D%C3%BCsseldorf.jpg /media/Datei:Bugatti_Veyron_16.4_%E2%80%93_Frontansicht,_5._April_2012,_D%C3%BCsseldorf.jpg/ https://creativecommons.org/licenses/by-sa/3.0/de/deed.en; p. 244 G/Martyn Goddard; p. 244 Wikimedia Commons/Lothar Spurzem/https://de.wikipedia.org/wiki/Datei:LangH-MB-W125-1977.jpg/https://creativecommons.org/licenses/by-sa/2.0/de/deed.de; p. 244–245 Max Earey/Shutterstock.com; p. 245 G/David Taylor; p. 245 patruflo/Shutterstock.com; p. 245 G/Bettmann; p. 246–247 G/Patrick Endres; p. 248 Mathias Putze/Shutterstock.com; p. 248–249 M/NASA Archive; p. 250 G/Bettmann; p. 250–251 A/Robert Preston; p. 251 G/Dan Rafla; p. 252 G/Michael C. Klesium ; p. 252 G/Art Wolfe; p. 252 Beautiful landscape/Shutterstock.com; p. 252–253 Martin Gillespie/Shutterstock.com; p. 253 M/Stock Connection Blue; p. 253 Saulius Damulevicius/Shutterstock.com; p. 253 G/Stockbyte; p. 253 G/Keren Su; p. 253 Piotr Snigorski/Shutterstock.com; p. 253 Punnawit Suwattananun/Shutterstock.com; p. 253 G/Galen Rowell; p. 253 C/Galen Rowell; p. 253 M/CuboImages; p. 253 A/Danita Delimont; p. 253 A/Ian Paterson; p. 254–255 Minerva Studio/Shutterstock.com; p. 256 LPatricK297/Shutterstock.com; p. 256 Jason and Bonnie Grower/Shutterstock.com; p. 256–257 M/G&M Therin-Weise; p. 257 Henk Osinga Photography/Shutterstock.com; p. 258 M101Studio/Shutterstock.com; p. 258 Suparat Chairatprasert/Shutterstock.com; p. 258–259 VanderWolf Images/Shutterstock.com; p. 260 C/Joe McDonald; p. 260–261 C/Chris Hellier; p. 261 M/Gilles Barbier; p. 262 M/M. Guyt; p. 262–263 Sergey Uryadnikov/Shutterstock.com; p. 263 Jakub Vanek PRG/Shutterstock.com; p. 264 James P Mock/Shutterstock.com; p. 264–265 M/Ingo Schulz; p. 265 M/Dieter Mahlke; p. 265 M/Nasir Kachroo; p. 266 Dan Simonsen/Shutterstock.com; p. 266 Meoita/Shutterstock.com; p. 266–267 Jose Gil/Shutterstock.com; p. 268–269 G/Bettmann; p. 269 Ramy Fathalla/Shutterstock.com; p. 269 Harry Collins Photography/Shutterstock.com; p. 270 Al'fred/Shutterstock.com; p. 270 M/M. Woike; p. 270–271 G/Roberto Machado Noa; p. 271 M/Kike Calvo; p. 272 G/Stocktrek Images; p. 272 M/World Book Inc.; p. 272 M/NASA; p. 272–273 M/NASA; p. 273 Guy In Utah/Shutterstock.com; p. 274–275 Ronnie Chua/Shutterstock.com; p. 276 WaitForLight/Shutterstock.com; p. 276–277 Rahhal/Shutterstock.com; p. 277 G/Buena Vista Images; p. 277 Kertu/Shutterstock.com

# IMPRINT

MONACO BOOKS is an imprint of Kunth Verlag – MAIRDUMONT GmbH & Co. KG

For distribution please contact:
Monaco Books
c/o Kunth Verlag – MAIRDUMONT GmbH & Co. KG
Kistlerhofstraße 111 | 81379 Munich, Germany
info@kunth-verlag.de
www.kunth-verlag.de
ISBN 978-3-96965-087-5

Printed in Slovakia

Publishing Director: Grit Müller
Text: Dr. Maria Guntermann, Cornelia Heinrich, Thomas Horsmann, Andrea Lammert (Tipps), Georg Ledig, Dieter Löffler, Heiner Newe, Dr. Reinhard Pietsch, Barbara Rusch, Eckard Schuster, Linde Wiesner
Editor: Isabel Rößler, Annika Voigt, mcp concept gmbh
Layout & Typesetting: Patrick Tümmers, München, Ute Weber, Geretsried
Translation: Travod International Ltd